An Introduction to Chemical Kinetics

An Introduction to Chemical Kinetics

Margaret Robson Wright
Formerly of The University of St Andrews, UK

John Wiley & Sons, Ltd

Telephone (+44) 1243 779777

Email (for orders and customer service enquiries): cs-books@wiley.co.uk
Visit our Home Page on www.wileyeurope.com or www.wiley.com

Reprinted May 2005

Other Wiley Editorial Offices

John Wiley & Sons Inc., 111 River Street, Hoboken, NJ 07030, USA

Jossey-Bass, 989 Market Street, San Francisco, CA 94103-1741, USA

Wiley–VCH Verlag GmbH, Boschstrasse 12, D-69469 Weinheim, Germany

John Wiley & Sons Australia Ltd, 33 Park Road, Milton, Queensland 4064, Australia

John Wiley & Sons (Asia) Pte Ltd, 2 Clementi Loop #02-01, Jin Xing Distripark, Singapore 129809

John Wiley & Sons Canada Ltd, 22 Worcester Road, Etobicoke, Ontario, Canada M9W 1L1

Wiley also publishes its books in a variety of electronic formats. Some content that appears in print may not be available in electronic books.

Library of Congress Cataloging-in-Publication Data

Wright, Margaret Robson.
An introduction to chemical kinetics / Margaret Robson Wright.
p. cm.
Includes bibliographical references and index.
ISBN 0-470-09058-8 (acid-free paper) – ISBN 0-470-09059-6 (pbk. : acid-free paper)
1. Chemical kinetics. I. Title.

QD502.W75 2004
541′.394–dc22 2004006062

British Library Cataloguing in Publication Data

A catalogue record for this book is available from the British Library

ISBN 10: 0 470 09058 8 (HB) ISBN 13: 978 0 470 09058 9 (HB)
ISBN 10: 0 470 09059 6 (PB) ISBN 13: 978 0 470 09059 6 (PB)

Typeset in 10.5/13pt Times by Thomson Press (India) Limited, New Delhi

This book is printed on acid-free paper responsibly manufactured from sustainable forestry in which at least two trees are planted for each one used for paper production.

Dedicated with much love and affection

to

my mother, Anne (in memoriam),

with deep gratitude for all her loving help,

to

her oldest and dearest friends,

Nessie (in memoriam) and Dodo Gilchrist of Cumnock, who,

by their love and faith in me, have always been a source of great encouragement to me,

and last, but not least, to my own immediate family,

my husband, Patrick,

our children Anne, Edward and Andrew and our cats.

Contents

Preface **xiii**

List of Symbols **xvii**

1 Introduction 1

2 Experimental Procedures 5

2.1 Detection, Identification and Estimation of Concentration of Species Present 6
- 2.1.1 Chromatographic techniques: liquid–liquid and gas–liquid chromatography 6
- 2.1.2 Mass spectrometry (MS) 6
- 2.1.3 Spectroscopic techniques 7
- 2.1.4 Lasers 13
- 2.1.5 Fluorescence 14
- 2.1.6 Spin resonance methods: nuclear magnetic resonance (NMR) 15
- 2.1.7 Spin resonance methods: electron spin resonance (ESR) 15
- 2.1.8 Photoelectron spectroscopy and X-ray photoelectron spectroscopy 15

2.2 Measuring the Rate of a Reaction 17
- 2.2.1 Classification of reaction rates 17
- 2.2.2 Factors affecting the rate of reaction 18
- 2.2.3 Common experimental features for all reactions 19
- 2.2.4 Methods of initiation 19

2.3 Conventional Methods of Following a Reaction 20
- 2.3.1 Chemical methods 20
- 2.3.2 Physical methods 21

2.4 Fast Reactions 27
- 2.4.1 Continuous flow 27
- 2.4.2 Stopped flow 28
- 2.4.3 Accelerated flow 29
- 2.4.4 Some features of flow methods 29

2.5 Relaxation Methods 30
- 2.5.1 Large perturbations 31
- 2.5.2 Flash photolysis 31
- 2.5.3 Laser photolysis 32
- 2.5.4 Pulsed radiolysis 32
- 2.5.5 Shock tubes 33
- 2.5.6 Small perturbations: temperature, pressure and electric field jumps 33

2.6 Periodic Relaxation Techniques: Ultrasonics 35

2.7 Line Broadening in NMR and ESR Spectra 38

Further Reading 38

Further Problems 39

3 The Kinetic Analysis of Experimental Data **43**
3.1 The Experimental Data 44
3.2 Dependence of Rate on Concentration 47
3.3 Meaning of the Rate Expression 48
3.4 Units of the Rate Constant, k 49
3.5 The Significance of the Rate Constant as Opposed to the Rate 50
3.6 Determining the Order and Rate Constant from Experimental Data 52
3.7 Systematic Ways of Finding the Order and Rate Constant from Rate/Concentration Data 53
3.7.1 A straightforward graphical method 55
3.7.2 log/log graphical procedures 55
3.7.3 A systematic numerical procedure 56
3.8 Drawbacks of the Rate/Concentration Methods of Analysis 58
3.9 Integrated Rate Expressions 58
3.9.1 Half-lives 59
3.10 First Order Reactions 62
3.10.1 The half-life for a first order reaction 63
3.10.2 An extra point about first order reactions 64
3.11 Second Order Reactions 66
3.11.1 The half-life for a second order reaction 67
3.11.2 An extra point about second order reactions 68
3.12 Zero Order Reaction 68
3.12.1 The half-life for a zero order reaction 70
3.13 Integrated Rate Expressions for Other Orders 71
3.14 Main Features of Integrated Rate Equations 71
3.15 Pseudo-order Reactions 74
3.15.1 Application of pseudo-order techniques to rate/concentration data 75
3.16 Determination of the Product Concentration at Various Times 77
3.17 Expressing the Rate in Terms of Reactants or Products for Non-simple Stoichiometry 79
3.18 The Kinetic Analysis for Complex Reactions 79
3.18.1 Relatively simple reactions which are mathematically complex 81
3.18.2 Analysis of the simple scheme $A \xrightarrow{k_1} I \xrightarrow{k_2} P$ 81
3.18.3 Two conceivable situations 82
3.19 The Steady State Assumption 84
3.19.1 Using this assumption 84
3.20 General Treatment for Solving Steady States 86
3.21 Reversible Reactions 89
3.21.1 Extension to other equilibria 90
3.22 Pre-equilibria 92
3.23 Dependence of Rate on Temperature 92
Further Reading 95
Further Problems 95

4 Theories of Chemical Reactions **99**
4.1 Collision Theory 100
4.1.1 Definition of a collision in simple collision theory 100
4.1.2 Formulation of the total collision rate 102
4.1.3 The p factor 108
4.1.4 Reaction between like molecules 110
4.2 Modified Collision Theory 110

4.2.1 A new definition of a collision 110
4.2.2 Reactive collisions 111
4.2.3 Contour diagrams for scattering of products of a reaction 112
4.2.4 Forward scattering: the stripping or grazing mechanism 117
4.2.5 Backward scattering: the rebound mechanism 118
4.2.6 Scattering diagrams for long-lived complexes 119
4.3 Transition State Theory 122
4.3.1 Transition state theory, configuration and potential energy 122
4.3.2 Properties of the potential energy surface relevant to transition state theory 123
4.3.3 An outline of arguments involved in the derivation of the rate equation 131
4.3.4 Use of the statistical mechanical form of transition state theory 135
4.3.5 Comparisons with collision theory and experimental data 136
4.4 Thermodynamic Formulations of Transition State Theory 140
4.4.1 Determination of thermodynamic functions for activation 142
4.4.2 Comparison of collision theory, the partition function form and the thermodynamic form of transition state theory 142
4.4.3 Typical approximate values of contributions entering the sign and magnitude of $\Delta S^{\neq *}$ 144
4.5 Unimolecular Theory 145
4.5.1 Manipulation of experimental results 148
4.5.2 Physical significance of the constancy or otherwise of k_1, k_{-1} and k_2 151
4.5.3 Physical significance of the critical energy in unimolecular reactions 152
4.5.4 Physical significance of the rate constants k_1, k_{-1} and k_2 153
4.5.5 The simple model: that of Lindemann 153
4.5.6 Quantifying the simple model 154
4.5.7 A more complex model: that of Hinshelwood 155
4.5.8 Quantifying Hinshelwood's theory 155
4.5.9 Critique of Hinshelwood's theory 157
4.5.10 An even more complex model: that of Kassel 158
4.5.11 Critique of the Kassel theory 159
4.5.12 Energy transfer in the activation step 159
4.6 The Slater Theory 160
Further Reading 162
Further Problems 162

5 Potential Energy Surfaces **165**
5.1 The Symmetrical Potential Energy Barrier 165
5.2 The Early Barrier 167
5.3 The Late Barrier 167
5.4 Types of Elementary Reaction Studied 168
5.5 General Features of Early Potential Energy Barriers for Exothermic Reactions 170
5.6 General Features of Late Potential Energy Surfaces for Exothermic Reactions 172
5.6.1 General features of late potential energy surfaces where the attacking atom is light 173
5.6.2 General features of late potential energy surfaces for exothermic reactions where the attacking atom is heavy 175
5.7 Endothermic Reactions 177
5.8 Reactions with a Collision Complex and a Potential Energy Well 178
Further Reading 180
Further Problems 180

6 Complex Reactions in the Gas Phase **183**
6.1 Elementary and Complex Reactions 184
6.2 Intermediates in Complex Reactions 186
6.3 Experimental Data 188
6.4 Mechanistic Analysis of Complex Non-chain Reactions 189
6.5 Kinetic Analysis of a Postulated Mechanism: Use of the Steady State Treatment 192
6.5.1 A further example where disentangling of the kinetic data is necessary 195
6.6 Kinetically Equivalent Mechanisms 198
6.7 A Comparison of Steady State Procedures and Equilibrium Conditions in the Reversible Reaction 202
6.8 The Use of Photochemistry in Disentangling Complex Mechanisms 204
6.8.1 Kinetic features of photochemistry 204
6.8.2 The reaction of H_2 with I_2 206
6.9 Chain Reactions 208
6.9.1 Characteristic experimental features of chain reactions 209
6.9.2 Identification of a chain reaction 210
6.9.3 Deduction of a mechanism from experimental data 211
6.9.4 The final stage: the steady state analysis 213
6.10 Inorganic Chain Mechanisms 213
6.10.1 The H_2/Br_2 reaction 213
6.10.2 The steady state treatment for the H_2/Br_2 reaction 214
6.10.3 Reaction without inhibition 216
6.10.4 Determination of the individual rate constants 217
6.11 Steady State Treatments and Possibility of Determination of All the Rate Constants 218
6.11.1 Important points to note 221
6.12 Stylized Mechanisms: A Typical Rice–Herzfeld Mechanism 221
6.12.1 Dominant termination steps 223
6.12.2 Relative rate constants for termination steps 224
6.12.3 Relative rates of the termination steps 224
6.12.4 Necessity for third bodies in termination 227
6.12.5 The steady state treatment for chain reactions, illustrating the use of the long chains approximation 229
6.12.6 Further problems on steady states and the Rice–Herzfeld mechanism 233
6.13 Special Features of the Termination Reactions: Termination at the Surface 240
6.13.1 A general mechanism based on the Rice–Herzfeld mechanism used previously 240
6.14 Explosions 243
6.14.1 Autocatalysis and autocatalytic explosions 244
6.14.2 Thermal explosions 244
6.14.3 Branched chain explosions 244
6.14.4 A highly schematic and simplified mechanism for a branched chain reaction 246
6.14.5 Kinetic criteria for non-explosive and explosive reaction 247
6.14.6 A typical branched chain reaction showing explosion limits 249
6.14.7 The dependence of rate on pressure and temperature 250
6.15 Degenerate Branching or Cool Flames 252
6.15.1 A schematic mechanism for hydrocarbon combustion 254
6.15.2 Chemical interpretation of 'cool' flame behaviour 257
Further Reading 259
Further Problems 260

7 Reactions in Solution **263**
7.1 The Solvent and its Effect on Reactions in Solution 263
7.2 Collision Theory for Reactions in Solution 265
7.2.1 The concepts of ideality and non-ideality 268
7.3 Transition State Theory for Reactions in Solution 269
7.3.1 Effect of non-ideality: the primary salt effect 269
7.3.2 Dependence of $\Delta S^{\neq *}$ and $\Delta H^{\neq *}$ on ionic strength 279
7.3.3 The effect of the solvent 280
7.3.4 Extension to include the effect of non-ideality 284
7.3.5 Deviations from predicted behaviour 284
7.4 $\Delta S^{\neq *}$ and Pre-exponential A Factors 289
7.4.1 A typical problem in graphical analysis 292
7.4.2 Effect of the molecularity of the step for which $\Delta S^{\neq *}$ is found 292
7.4.3 Effect of complexity of structure 292
7.4.4 Effect of charges on reactions in solution 293
7.4.5 Effect of charge and solvent on $\Delta S^{\neq *}$ for ion–ion reactions 293
7.4.6 Effect of charge and solvent on $\Delta S^{\neq *}$ for ion–molecule reactions 295
7.4.7 Effect of charge and solvent on $\Delta S^{\neq *}$ for molecule–molecule reactions 296
7.4.8 Effects of changes in solvent on $\Delta S^{\neq *}$ 296
7.4.9 Changes in solvation pattern on activation, and the effect on A factors for reactions involving charges and charge-separated species in solution 296
7.4.10 Reactions between ions in solution 297
7.4.11 Reaction between an ion and a molecule 298
7.4.12 Reactions between uncharged polar molecules 299
7.5 $\Delta H^{\neq *}$ Values 301
7.5.1 Effect of the molecularity of the step for which the $\Delta H^{\neq *}$ value is found 301
7.5.2 Effect of complexity of structure 302
7.5.3 Effect of charge and solvent on $\Delta H^{\neq *}$ for ion–ion and ion–molecule reactions 302
7.5.4 Effect of the solvent on $\Delta H^{\neq *}$ for ion–ion and ion–molecule reactions 303
7.5.5 Changes in *solvation pattern* on activation and the effect on $\Delta H^{\neq *}$ 303
7.6 Change in Volume on Activation, $\Delta V^{\neq *}$ 304
7.6.1 Effect of the molecularity of the step for which $\Delta V^{\neq *}$ is found 305
7.6.2 Effect of complexity of structure 306
7.6.3 Effect of charge on $\Delta V^{\neq *}$ for reactions between ions 306
7.6.4 Reactions between an ion and an uncharged molecule 308
7.6.5 Effect of solvent on $\Delta V^{\neq *}$ 308
7.6.6 Effect of change of *solvation pattern* on activation and its effect on $\Delta V^{\neq *}$ 308
7.7 Terms Contributing to Activation Parameters 310
7.7.1 $\Delta S^{\neq *}$ 310
7.7.2 $\Delta V^{\neq *}$ 310
7.7.3 $\Delta H^{\neq *}$ 311
Further Reading 314
Further Problems 314

8 Examples of Reactions in Solution **317**
8.1 Reactions Where More than One Reaction Contributes to the Rate of Removal of Reactant 317
8.1.1 A simple case 318

8.1.2 A slightly more complex reaction where reaction occurs by two concurrent routes, and where both reactants are in equilibrium with each other 321
8.1.3 Further disentangling of equilibria and rates, and the possibility of kinetically equivalent mechanisms 328
8.1.4 Distinction between acid and base hydrolyses of esters 331
8.2 More Complex Kinetic Situations Involving Reactants in Equilibrium with Each Other and Undergoing Reaction 336
8.2.1 A further look at the base hydrolysis of glycine ethyl ester as an illustration of possible problems 336
8.2.2 Decarboxylations of β-keto-monocarboxylic acids 339
8.2.3 The decarboxylation of β-keto-dicarboxylic acids 341
8.3 Metal Ion Catalysis 344
8.4 Other Common Mechanisms 346
8.4.1 The simplest mechanism 346
8.4.2 Kinetic analysis of the simplest mechanism 347
8.4.3 A slightly more complex scheme 351
8.4.4 Standard procedure for determining the expression for k_{obs} for the given mechanism (Section 8.4.3) 352
8.5 Steady States in Solution Reactions 359
8.5.1 Types of reaction for which a steady state treatment could be relevant 359
8.5.2 A more detailed analysis of Worked Problem 6.5 360
8.6 Enzyme Kinetics 365
Further Reading 368
Further Problems 369

Answers to Problems **373**

List of Specific Reactions **427**

Index **431**

Preface

This book leads on from elementary basic kinetics, and covers the main topics which are needed for a good working knowledge and understanding of the fundamental aspects of kinetics. It emphasizes how experimental data is collected and manipulated to give standard kinetic quantities such as rates, rate constants, enthalpies, entropies and volumes of activation. It also emphasizes how these quantities are used in interpretations of the mechanism of a reaction. The relevance of kinetic studies to aspects of physical, inorganic, organic and biochemical chemistry is illustrated through explicit reference and examples. Kinetics provides a unifying tool for all branches of chemistry, and this is something which is to be encouraged in teaching and which is emphasized here.

Gas studies are well covered with extensive explanation and interpretation of experimental data, such as steady state calculations, all illustrated by frequent use of worked examples. Solution kinetics are similarly explained, and plenty of practice is given in dealing with the effects of the solvent and non-ideality. Students are given plenty of practice, via worked problems, in handling various types of mechanism found in solution, and in interpreting ionic strength dependences and enthalpies, entropies and volumes of activation.

As the text is aimed at undergraduates studying core physical chemistry, only the basics of theoretical kinetics are given, but the *fundamental concepts* are clearly explained. More advanced reading is given in my book *Fundamental Chemical Kinetics* (see reading lists).

Many students veer rapidly away from topics which are quantitative and involve mathematical equations. This book attempts to allay these fears by guiding the student through these topics in a step-by-step development which explains the logic, reasoning and actual manipulation. For this reason a large fraction of the text is devoted to worked examples, and each chapter ends with a collection of further problems to which detailed and explanatory answers are given. If through the written word I can help students to understand and to feel confident in their ability to learn, and to teach them, in a manner which gives them the feeling of a direct contact with the teacher, then this book will not have been written in vain. It is the teacher's duty to show students how to achieve understanding, and to think scientifically. The philosophy behind this book is that this is best done by detailed explanation and guidance. It is understanding, being able to see for oneself and confidence which help to stimulate and sustain interest. This book attempts to do precisely that.

This book is the result of the accumulated experience of 40 very stimulating years of teaching students at all levels. During this period I regularly lectured to students, but more importantly I was deeply involved in devising tutorial programmes at all levels where consolidation of lecture material was given through many problem-solving exercises. I also learned that providing detailed explanatory answers to these exercises proved very popular and successful with students of all abilities. During these years I learned that being happy to help and being prepared to give extra explanation and to spend extra time on a topic could soon clear up problems and difficulties which many students thought they would never understand. Too often teachers forget that there were times when they themselves could not understand, and when a similar explanation and preparedness to give time were welcome. To all the many students who have provided the stimulus and enjoyment of teaching I give my grateful thanks.

I am very grateful to John Wiley & Sons for giving me the opportunity to publish this book, and to indulge my love of teaching. In particular, I would like to thank Andy Slade, Rachael Ballard and Robert Hambrook of John Wiley & Sons who have cheerfully, and with great patience, guided me through the problems of preparing the manuscript for publication. Invariably, they all have been extremely helpful.

I also extend my very grateful thanks to Martyn Berry who read the whole manuscript and sent very encouraging, very helpful and constructive comments on this book. His belief in the method of approach and his enthusiasm has been an invaluable support to me.

Likewise, I would like to thank Professor Derrick Woollins of St Andrews University for his continued very welcome support and encouragement throughout the writing of this book.

To my mother, Mrs Anne Robson, I have a very deep sense of gratitude for all the help she gave me in her lifetime in furthering my academic career. I owe her an enormous debt for her invaluable, excellent and irreplaceable help with my children when they were young and I was working part-time during the teaching terms of the academic year. Without her help and her loving care of my children I would never have gained the continued experience in teaching, and I could never have written this book. My deep and most grateful thanks are due to her.

My husband, Patrick, has, throughout my teaching career and throughout the thinking about and writing of this book, been a source of constant support and help and encouragement. His very high intellectual calibre and wide-ranging knowledge and understanding have provided many fruitful and interesting discussions. He has read in detail the whole manuscript and his clarity, insight and considerable knowledge of the subject matter have been of invaluable help. I owe him many apologies for the large number of times when I have interrupted his own activities to pursue a discussion of aspects of the material presented here. It is to his very great credit that I have never been made to feel guilty about doing so. My debt to him is enormous, and my most grateful thanks are due to him.

Finally, my thanks are due to my three children who have always encouraged me in my teaching, and have encouraged me in the writing of my books. In particular, Anne and Edward have been around during the writing of this book and have given me every encouragement to keep going.

Margaret Robson Wright
Formerly Universities of Dundee and St Andrews
October, 2003

List of Symbols

A	absorbance $\{=\log_{10}(I_0/I)\}$
A	A factor in Arrhenius equation
A	Debye-Huckel constant
a	activity
B	constant in extended Debye-Huckel equation
B'	kinetic quantity related to B
b	number of molecules reacting with a molecule
b	impact parameter
c	concentration
$c^{\neq}$	concentration of activated complexes
d	length of path in spectroscopy
d	distance along a flow tube
E_A	activation energy
E_0	activation energy at absolute zero
E_1, E_{-1}	activation energy of reaction 1 and of its reverse
e	electronic charge
f	constant of proportionality in expression relating to fluorescence
G	Gibbs free energy
H	enthalpy
h	Planck's constant
I	intensity of radiation
I_0	initial intensity of radiation
I_{abs}	intensity of radiation absorbed
I	moment of inertia
I	ionic strength
K	equilibrium constant
$K^{\neq}$	equilibrium constant for formation of the activated complex from reactants
$K^{\neq *}$	equilibrium constant for formation of the activated complex from reactants, with one term missing
K_M	Michaelis constant
k	rate constant

k_1, k_{-1}	rate constant for reaction 1 and its reverse
k	Boltzmann's constant
m	mass
n	order of a reaction
$N^{\neq}$	number of activated complexes
p	pressure
p_i	partial pressure of species i
p	p factor in collision theory
Q	molecular partition function per unit volume
$Q^{\neq}$	molecular partition function per unit volume for the activated complex
$Q^{\neq *}$	molecular partition function per unit volume for the activated complex, but with one term missing
R	the gas constant
r	internuclear distance
$r_{\neq}$	distance between the centres of ions in the activated complex
S	entropy
s	half of the number of squared terms, in theories of Hinshelwood and Kassel
T	absolute temperature
t	time
U	energy
V	volume
V	velocity of sound
V_s	term in Michaelis-Menten equation
v	velocity
v	relative velocity
v	vibrational quantum number
Z	collision number
$\mathcal{Z}$	collision rate
z	number of charges
α	order with respect to one reactant
α	branching coefficient
α	polarisability
β	order with respect to one reactant
γ	activity coefficient
Δ	change in
ΔG^{θ}	standard change in free energy
ΔH^{θ}	standard change in enthalpy
ΔS^{θ}	standard change in entropy
ΔV^{θ}	standard change in volume

$\Delta G^{\neq *}$	free energy of activation with one term missing
$\Delta H^{\neq *}$	enthalpy of activation with one term missing
$\Delta S^{\neq *}$	entropy of activation with one term missing
$\Delta V^{\neq *}$	volume of activation with one term missing
δ	distance along the reaction coordinate specifying the transition state
ε	molar absorption coefficient (Beer's Law)
ε	energy of a molecule
ε_0	energy of a molecule in its ground state
ε_0	universal constant involved in expressions for electrostatic interactions
ε_r	relative permittivity
η	viscosity
θ	angle of approach of an ion to a dipole
κ	transmission coefficient
λ	wavelength
λ	collision efficiency
μ	chemical potential
μ	dipole moment
μ	absorption coefficient for ultrasonic waves
ν	frequency
σ	cross section
$\sigma_{(\mathrm{R})}$	cross section for chemical reaction
τ	relaxation time
τ	time taken to pass through the transition state
τ	lifetime

1 Introduction

Chemical kinetics is conventionally regarded as a topic in physical chemistry. In this guise it covers the measurement of rates of reaction, and the analysis of the experimental data to give a systematic collection of information which summarises all the quantitative kinetic information about any given reaction. This, in turn, enables comparisons of reactions to be made and can afford a kinetic classification of reactions. The sort of information used here is summarized in terms of

- the factors influencing rates of reaction,
- the dependence of the rate of the reaction on concentration, called the order of the reaction,
- the rate expression, which is an equation which summarizes the dependence of the rate on the concentrations of substances which affect the rate of reaction,
- this expression involves the rate constant which is a constant of proportionality linking the rate with the various concentration terms,
- this rate constant collects in one quantity all the information needed to calculate the rate under specific conditions,
- the effect of temperature on the rate of reaction. Increase in temperature generally increases the rate of reaction. Knowledge of just exactly how temperature affects the rate constant can give information leading to a deeper understanding of how reactions occur.

All of these factors are explained in Chapters 2 and 3, and problems are given to aid understanding of the techniques used in quantifying and systematizing experimental data.

However, the science of kinetics does not end here. The next task is to look at the chemical steps involved in a chemical reaction, and to develop a mechanism which summarizes this information. Chapters 6 and 8 do this for gas phase and solution phase reactions respectively.

An Introduction to Chemical Kinetics. Margaret Robson Wright
 ISBNs: 0-470-09058-8 (hbk) 0-470-09059-6 (pbk)

The final task is to develop theories as to why and how reactions occur, and to examine the physical and chemical requirements for reaction. This is a very important aspect of modern kinetics. Descriptions of the fundamental concepts involved in the theories which have been put forward, along with an outline of the theoretical development, are given in Chapter 4 for gas phase reactions, and in Chapter 7 for solution reactions.

However, kinetics is not just an aspect of physical chemistry. It is a unifying topic covering the whole of chemistry, and many aspects of biochemistry and biology. It is also of supreme importance in both the chemical and pharmaceutical industries. Since the mechanism of a reaction is intimately bound up with kinetics, and since mechanism is a major topic of inorganic, organic and biological chemistry, the subject of kinetics provides a unifying framework for these conventional branches of chemistry. Surface chemistry, catalysis and solid state chemistry all rest heavily on a knowledge of kinetic techniques, analysis and interpretation. Improvements in computers and computing techniques have resulted in dramatic advances in quantum mechanical calculations of the potential energy surfaces of Chapters 4 and 5, and in theoretical descriptions of rates of reaction. Kinetics also makes substantial contributions to the burgeoning subject of atmospheric chemistry and environmental studies.

Arrhenius, in the 1880s, laid the foundations of the subject as a rigorous science when he postulated that not all molecules can react: only those which have a certain critical minimum energy, called the activation energy, can react. There are two ways in which molecules can acquire energy or lose energy. The first one is by absorption of energy when radiation is shone on to the substance and by emission of energy. Such processes are important in photochemical reactions. The second mechanism is by energy transfer during a collision, where energy can be acquired on collision, activation, or lost on collision, deactivation. Such processes are of fundamental importance in theoretical kinetics where the 'how' and 'why' of reaction is investigated. Early theoretical work using the Maxwell–Boltzmann distribution led to collision theory. This gave an expression for the rate of reaction in terms of the rate of collision of the reacting molecules. This collision rate is then modified to account for the fact that only a certain fraction of the reacting molecules will react, that fraction being the number of molecules which have energy above the critical minimum value. As is shown in Chapter 4, collision theory affords a physical explanation of the exponential relationship between the rate constant and the absolute temperature.

Collision theory encouraged more experimental work and met with considerable success for a growing number of reactions.

However, the theory appeared not to be able to account for the behaviour of unimolecular reactions, which showed first order behaviour at high pressures, moving to second order behaviour at low pressures. If one of the determining features of reaction rate is the rate at which molecules collide, unimolecular reactions might be expected always to give second order kinetics, which is not what is observed.

This problem was resolved in 1922 when Lindemann and Christiansen proposed their hypothesis of time lags, and this mechanistic framework has been used in all the more sophisticated unimolecular theories. It is also common to the theoretical framework of bimolecular and termolecular reactions. The crucial argument is that molecules which are activated and have acquired the necessary critical minimum energy do not *have to* react immediately they receive this energy by collision. There is sufficient time after the final activating collision for the molecule to lose its critical energy by being deactivated in another collision, or to react in a unimolecular step.

It is the existence of this time lag between activation by collision and reaction which is basic and crucial to the theory of unimolecular reactions, and this assumption leads inevitably to first order kinetics at high pressures, and second order kinetics at low pressures.

Other elementary reactions can be handled in the same fundamental way: molecules can become activated by collision and then last long enough for there to be the same two fates open to them. The only difference lies in the molecularity of the actual reaction step:

- in a unimolecular reaction, only one molecule is involved at the actual moment of chemical transformation;
- in a bimolecular reaction, two molecules are involved in this step;
- in a termolecular reaction, three molecules are now involved.

$A + A \xrightarrow{k_1} A^* + A$	bimolecular activation by collision
$A^* + A \xrightarrow{k_{-1}} A + A$	bimolecular deactivation by collision
$A^* + bA \xrightarrow{k_2}$ products	reaction step

where A^* is an activated molecule with enough energy to react.

$b = 0$ defines spontaneous breakdown of A^*,

$b = 1$ defines bimolecular reaction involving the coming together of A^* with A and

$b = 2$ defines termolecular reaction involving the coming together of A^* with two As.

This is a mechanism common to all chemical reactions since it describes *each individual* reaction step in a complex reaction where there are many steps.

Meanwhile, in the 1930s, the idea of reaction being defined in terms of the spatial arrangements of all the atoms in the reacting system crystallized into transition state theory. This theory has proved to be of fundamental importance. Reaction is now defined as the acquisition of a certain critical geometrical configuration of all the atoms involved in the reaction, and this critical configuration was shown to have a critical maximum in potential energy with respect to reactants and products.

The lowest potential energy pathway between the reactant and product configurations represents the changes which take place during reaction, and is called the reaction coordinate or minimum energy path. The critical configuration lies on this pathway at the configuration with the highest potential energy. It is called the transition state or activated complex, and it must be attained before reaction can take place. The rate of reaction is the rate at which the reactants pass through this critical configuration. Transition state theory thus deals with the third step in the master mechanism above. It does not discuss the energy transfers of the first two steps of activation and deactivation.

Transition state theory, especially with its recent developments, has proved a very powerful tool, vastly superior to collision theory. It has only recently been challenged by modern advances in molecular beams and molecular dynamics which look at the microscopic details of a collision, and which can be regarded as a modified collision theory. These developments along with computer techniques, and modern experimental advances in spectroscopy and lasers along with fast reaction techniques, are now revolutionizing the science of reaction rates.

2 Experimental Procedures

The five main components of any kinetic investigation are

1. product and intermediate detection,
2. concentration determination of all species present,
3. deciding on a method of following the rate,
4. the kinetic analysis and
5. determination of the mechanism.

Aims

This chapter will examine points 1–3 listed above. By the end of this chapter you should be able to

- decide on a method for detecting and estimating species present in a reaction mixture,
- distinguish between fast reactions and the rest,
- explain the basis of conventional methods of following reactions,
- convert experimental observations into values of [reactant] remaining at given times,
- describe methods for following fast reactions and
- list the essential features of each method used for fast reactions.

An Introduction to Chemical Kinetics. Margaret Robson Wright
 ISBNs: 0-470-09058-8 (hbk) 0-470-09059-6 (pbk)

2.1 Detection, Identification and Estimation of Concentration of Species Present

Modern work generally uses three major techniques, *chromatography*, *mass spectrometry* and *spectroscopy*, although there is a wide range of other techniques available.

2.1.1 Chromatographic techniques: liquid–liquid and gas–liquid chromatography (GLC)

When introduced, chromatographic techniques completely revolutionized analysis of reaction mixtures and have proved particularly important for kinetic studies of complex gaseous reactions. A complete revision of most gas phase reactions proved necessary, because it was soon discovered that many intermediates and minor products had not been detected previously, and a complete re-evaluation of gas phase mechanisms was essential.

Chromatography refers to the separation of the components in a sample by distribution of these components between two phases, one of which is stationary and one of which moves. This takes place in a column, and once the components have come off the column, identification then takes place in a detector.

The main virtues of chromatographic techniques are versatility, accuracy, speed of analysis and the ability to handle complex mixtures and separate the components accurately. Only very small samples are required, and the technique can detect and measure very small amounts e.g. 10^{-10} mol or less. Analysis times are of the order of a few seconds for liquid samples, and even shorter for gases. However, a lower limit around 10^{-3} s makes the technique unsuitable for species of shorter lifetime than this.

Chromatography is often linked to a spectroscopic technique for liquid mixtures and to a mass spectrometer for gaseous mixtures. Chromatography separates the components; the other technique identifies them and determines concentrations.

2.1.2 Mass spectrometry (MS)

In mass spectrometry the sample is vaporized, and bombarded with electrons so that the molecules are ionized. The detector measures the mass/charge ratio, from which the molecular weight is determined and the molecule identified. Radicals often give the same fragment ions as the parent molecules, but they can be distinguished because lower energies are needed for the radical.

Most substances can be detected provided they can be vaporized. Only very small samples are required, with as little as 10^{-12} mol being detected. Samples are leaked directly from the reaction mixture; the time of analysis is short, around 10^{-5} s, and so fairly reactive species can be studied. For highly complex mixtures, MS is linked to chromatography, which first separates the components.

Many reaction types can be studied in the mass spectrometer: e.g. flash photolysis, shock tube, combustion, explosions, electric discharge and complex gas reactions. Mass spectrometry is ideal for ion–ion and ion–molecule reactions, isotopic analysis and kinetic isotope effect studies.

2.1.3 Spectroscopic techniques

There are three important features of spectra which are of prime interest to the kineticist.

1. The *frequency* and the *fine structure* of the lines give the identity of the molecule; this is particularly important in detecting intermediates and minor products.

2. The *intensity* of the lines gives the concentration; this is useful for monitoring the concentrations of reactants and intermediates with time.

3. The *line width* enables kinetic features of the transition and the excited state to be determined.

In spectroscopic experiments, radiation is absorbed (*absorption spectra*) or emitted (*emission spectra*). The frequency of absorption, or emission, is a manifestation of transitions occurring within the molecule, and the frequency of a line in the spectrum is related to the energy change as the molecule moves from one energy state to another.

$$h\nu = \varepsilon' - \varepsilon'' = \Delta\varepsilon \tag{2.1}$$

where ε' and ε'' are the energies of the upper and lower levels involved in the transition. These states are unique to any given molecule, radical or ion, and so the lines in a spectrum are a unique fingerprint of the molecule in question. Regions of the electromagnetic spectrum are characteristic of types of transition occurring within the molecule. These can be changes in rotational, vibrational, electronic and nuclear and electronic spin states.

Identification of species present during a reaction

Microwave, infrared, Raman, visible and UV spectra are all used extensively for identification. In the gas phase these show sharp lines so that identification is easy. In solution, the complexity of the spectra gives them sufficient features to make them recognizably specific to the molecule in question.

Concentration determination

Using Beer's law, concentrations can be found from the *change in intensity* of the radiation passed through the sample. The absorbance, A, of the sample at a given wavelength is defined as

$$\mathrm{A} = \log_{10} I_0/I \tag{2.2}$$

where I_0 is the incident intensity and I the final intensity. Beer's law relates this absorbance to the concentration of the species being monitored:

$$A_\lambda = \varepsilon_\lambda cd \tag{2.3}$$

where ε_λ is the absorption coefficient for the molecule in question.

ε_λ depends on the wavelength and the identity of the molecule.
λ is the wavelength, d is the path-length and c is the concentration of the absorbing species.

Provided that ε_λ and d are known, the concentration of absorbing species can be found. A calibration graph of A versus c should be linear with slope $\varepsilon_\lambda d$ and zero intercept. Microwave, infra-red, Raman, visible and UV spectra are all used.

Special features of absorbance measurements

If the reaction being monitored is first order, i.e. has rate $\propto$ [reactant]^1, or is being studied under pseudo-first order conditions, Section 3.15, the absorbance can be used directly, eliminating the need to know the value of ε_λ. Chapter 3 will show that such reactions can be quantified by plotting $\log_e$[reactant] against time. Since absorbance $\propto$ [reactant], then $\log_e$ absorbance can be plotted directly against time without the need to convert absorbance to concentration using Beer's law.

$$A = \varepsilon cd \quad \text{and so} \quad c = \frac{A}{\varepsilon d} \tag{2.4}$$

$$\log_e c = \log_e \frac{A}{\varepsilon d} = \log_e A - \log_e \varepsilon d \tag{2.5}$$

Since εd is a constant in any given experiment, then $\log_e \varepsilon d$ is also a constant, and

$$\log_e A = \log_e c + \text{constant} \tag{2.6}$$

A plot of $\log_e A$ versus time differs from a plot of $\log_e c$ versus time only in so far as it is displaced up the y-axis by an amount equal to $\log_e \varepsilon d$ (see Figure 2.1). The *slope* of

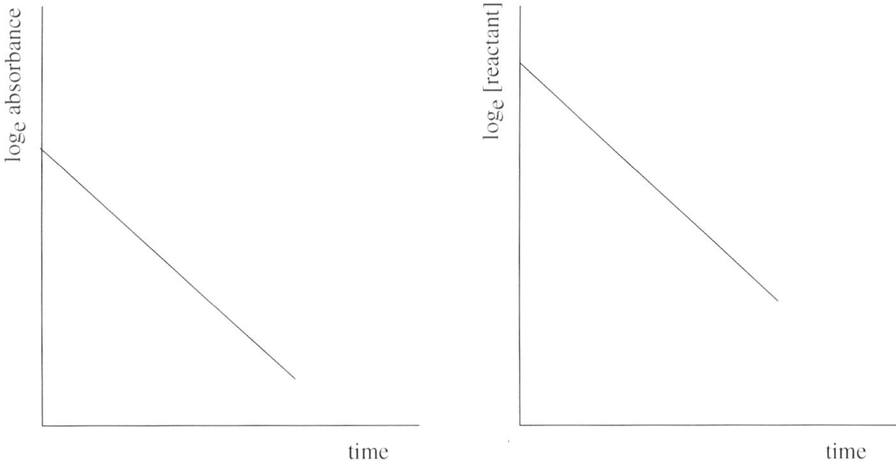

Figure 2.1 Graphs of $\log_e$ absorbance vs time, $\log_e$ [reactant] vs time

the graph of $\log_e$ absorbance versus time is the same as the *slope* of the graph of $\log_e$[reactant] versus time (Figure 2.1). The slope of this latter graph gives the quantity which characterizes the reaction.

Special features of fluorescence intensity

Fluorescence is a special type of emission of radiation (Section 2.1.5). In emission, a molecule is moving from an excited state to a lower state, and the frequency of emission is a manifestation of the energy change between the two states. When the excited molecules have been created by absorption of radiation shone on the molecules, the resultant return to lower states is termed *fluorescence*. The initial exciting radiation can either be a conventional source or a laser (Section 2.1.4).

The fluorescence intensity is generally proportional to [reactant], so that

$$\text{intensity} = f\ [\text{reactant}] \tag{2.7}$$

where f is a constant of proportionality.

$$\log_e \text{intensity} = \log_e f + \log_e [\text{reactant}] \tag{2.8}$$

and so again there is no need to convert to concentrations (see Figure 2.2). This can be sometimes be particularly useful, e.g. when laser-induced fluorescence is being used for monitoring concentrations.

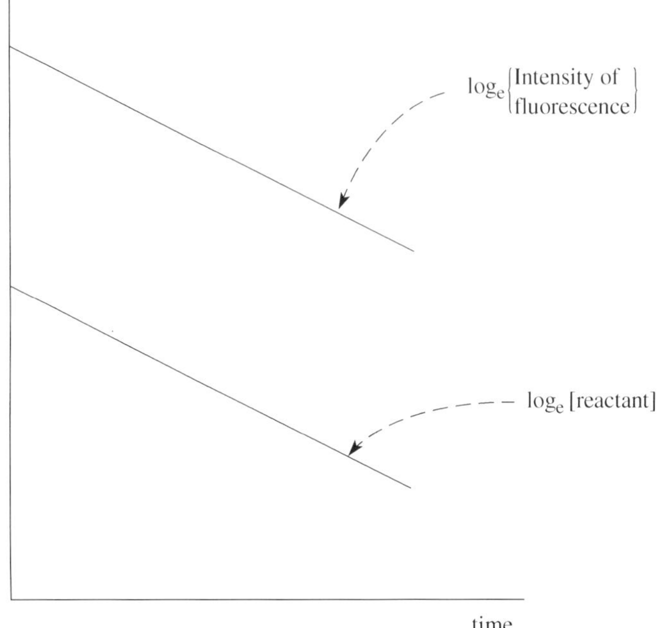

Figure 2.2 Graphs of $\log_e$ intensity of fluorescence versus and, $\log_e$ [reactant] versus time

Worked Problem 2.1

Question. In a 1.000 cm spectrophotometric cell, solutions of $C_6H_5CH{=}CHCCl_2$ in ethanol of known concentration give the following values of the absorbance, *A*.

$\dfrac{10^5 \times \text{conc.}}{\text{mol dm}^{-3}}$	0.446	0.812	1.335	1.711	2.105
A	0.080	0.145	0.240	0.305	0.378

Note.

$$\frac{10^5\, c}{\text{mol dm}^{-3}} = 0.446$$

Divide both sides by 10^5, and multiply both sides by mol dm^{-3}. This gives

$$c = \frac{0.446\,\text{mol dm}^{-3}}{10^5} = 0.446 \times 10^{-5}\ \text{mol dm}^{-3} = 4.46 \times 10^{-6}\,\text{mol dm}^{-3}.$$

This is a standard way of presenting tabulated data, and it is necessary to be completely at ease in performing the above manipulation.

It is also necessary to be careful when drawing graphs of tabulated data presented in this manner, so as to get the powers of ten correct.

Using Beer's law, draw a calibration graph and determine the value of ε_λ.

The reaction of this substance with $C_2H_5O^-$ in ethanolic solution is followed spectrophotometrically. The following results are found. Plot a graph of reactant concentration against time.

$\frac{\text{time}}{\text{min}}$	0	40	80	120	140
A	0.560	0.283	0.217	0.168	0.149

Answer. A graph of A versus c is a straight line through the origin. From the slope $\varepsilon = 1.79 \times 10^4\ \text{dm}^3\ \text{mol}^{-1}\ \text{cm}^{-1}$ (Figure 2.3).

This can be used to calculate the [reactant] corresponding to each absorbance, since $A = \varepsilon cd$. Since $d = 1.000$ cm the following table can be drawn up:

$\frac{\text{time}}{\text{min}}$	0	40	80	120	140
$\frac{10^5 \times \text{conc.}}{\text{mol dm}^{-3}}$	3.13	1.58	1.21	0.939	0.832

Alternatively, values of the concentration corresponding to the measured absorbances can be read off from the calibration curve.

Figure 2.4 shows the smooth curve of [reactant] versus time.

Techniques for concentration determination when two species absorb at the same wavelength

When more than one of the reaction species absorbs at around the same wavelength, conventional measurements must be carried out at two or more wavelengths, with consequent additional calculations.

Let the two species be A, concentration c_A with molar absorption coefficient $\varepsilon_{A\lambda(1)}$ at wavelength $\lambda(1)$, and B, concentration c_B with molar absorption coefficient $\varepsilon_{B\lambda(1)}$ at wavelength $\lambda(1)$.

$$A_{\lambda(1)} = \varepsilon_{A\lambda(1)} c_A d + \varepsilon_{B\lambda(1)} c_B d \tag{2.9}$$

There are two unknowns in this equation, c_A and c_B, and so it cannot be solved. A second independent equation must be set up. This utilizes measurements on the same solution, but at another wavelength, $\lambda(2)$, where both species absorb.

$$A_{\lambda(2)} = \varepsilon_{A\lambda(2)} c_A d + \varepsilon_{B\lambda(2)} c_B d \tag{2.10}$$

These two equations can now be solved to give c_A and c_B.

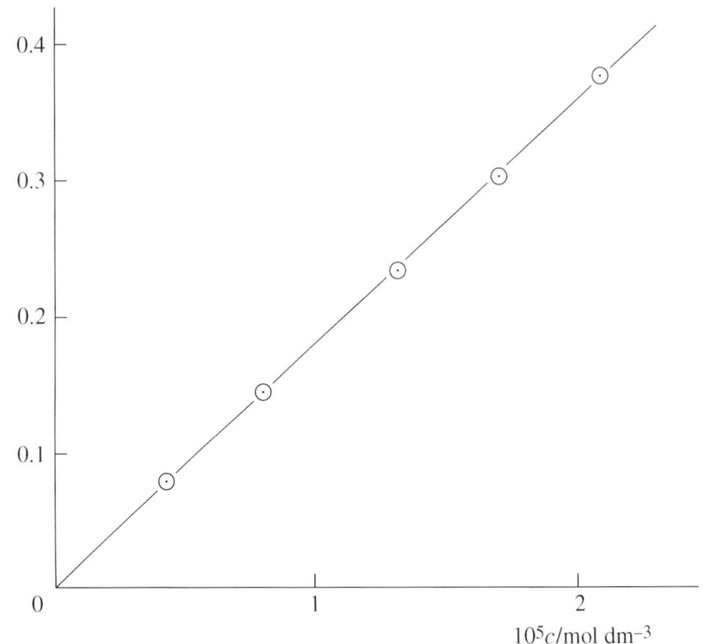

Figure 2.3 Beer's law plot of absorbance versus concentration

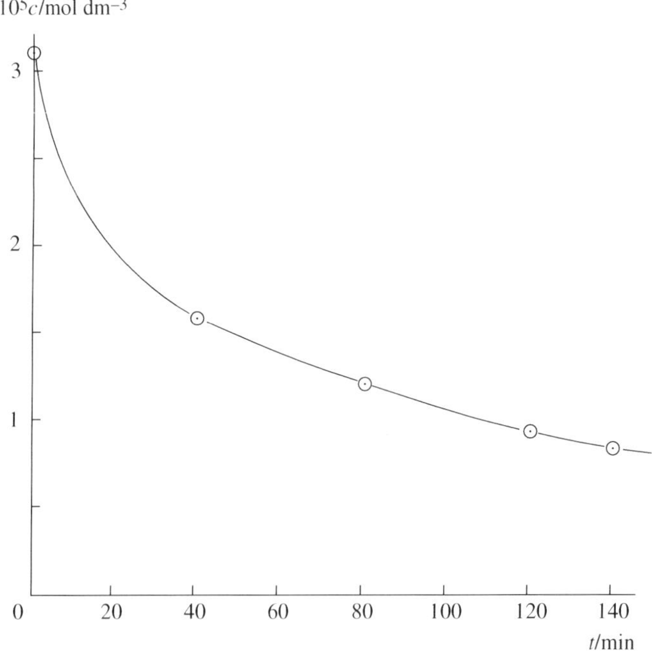

Figure 2.4 Graph of concentration versus time

The necessity of making measurements at two wavelengths can be overcome for gases by using lasers, which have highly defined frequencies compared with the conventional radiation used in standard absorption analyses. Different species will absorb at different frequencies. A laser has such a precise frequency that it will only excite the species which absorbs at precisely that frequency, and no other. Hence close lying absorptions can be easily separated. This gives a much greater capacity for singling and separating out absorptions occurring at very closely spaced frequencies; see Section 2.1.4 below.

Highly sensitive detectors, coupled with the facility to store each absorption signal digitally for each separate analysis time in a microcomputer, have enabled absorbance changes as small as 0.001 to be accurately measured.

Spectroscopic techniques are often linked to chromatograph columns for separation of components, or to flow systems, flash photolysis systems, shock tubes, molecular beams and other techniques for following reaction.

2.1.4 Lasers

Lasers cover the range from microwave through infrared and visible to the UV. The following lists the properties of lasers which are of importance in kinetics.

1. Lasers are highly coherent beams allowing reflection through the reaction cell very many times. This increases the path length and hence the sensitivity.

2. Conventional monochromatic radiation has a span of frequencies, and will thus excite simultaneously all chemical species which absorb within that narrow range. Lasers have a precisely defined frequency. This allows species with absorptions close to each other to be identified and monitored by separate lasers, or by a tunable laser, in contrast to the indiscriminate absorption which would occur with conventional sources of radiation.

3. *In any spectrum* the intensity of absorption is proportional to the concentration of the molecule in the energy level which is being excited. This is generally the ground state. If the concentration is very low, as is the case for many gas phase intermediates, then the intensity of absorption may not be measurable. The same problem arises in fluorescence where the intensity is proportional to the concentration of the molecule in the level to which it has been previously excited. Lasers, *on the other hand*, have a very high intensity, allowing accurate concentration determination of intermediates present in very, very low concentration, and enabling short lived species and processes occurring within 10^{-15} second to be picked up. Low intensity conventional sources of radiation cannot do this.

4. Lasers can follow reactions of the free radical and short-lived intermediates found in complex reactions. Many intermediates in complex gas phase reactions are

highly reactive and are removed from the reaction almost as soon as they are formed. If the rate at which they are produced is almost totally balanced by the total rate of their removal, then the intermediates are present in very, very low and almost constant concentrations. If these conditions are met, the species are said to be *in steady state concentrations*. Since rates of reactions are followed by studying how concentration varies with time, and since the concentrations of these steady state intermediates do not change with time, then the rates of their formation and removal cannot be studied. In conventional kinetic experiments it is impossible to study these species directly because steady state concentrations remain virtually constant throughout a kinetic experiment. If these species are produced in isolation in high concentrations their decay or build-up with time can be studied. With flash and laser photolysis (Sections 2.5.2 and 2.5.3), high intensity photolytic flashes produce radicals in high concentrations well above low steady state concentrations. Subsequent reactions of specific radicals can then be studied under non-steady state conditions. In flash photolysis, the higher the intensity of the flash the longer is its duration, limiting the highest intensities to those giving flashes of around 10^{-6} s. This limits the accessible time span of reactions to those with lifetimes greater than this. This limitation does not occur with lasers, and high intensity flashes of 10^{-15} s duration are now possible.

5. Early lasers gave single pulses of short duration, e.g. 10^{-9} s. Pulsed lasers give a train of equally spaced, high intensity, short duration pulses with intervals ranging from 10^{-9} to 10^{-12} to 10^{-15} s. These can be used as probes to monitor the changes in concentration during very fast reactions which are over in times as short as 10^{-9} to 10^{-12} s. Isolation of one of the pulses is now possible, giving a single very short duration initial photolysing flash.

2.1.5 Fluorescence

The frequencies and intensities of fluorescence enable identification and concentration determinations to be made. The technique is often around 10^4 times as sensitive as infra-red, visible or UV absorption spectrophotometry.

In *absorption* it is a *difference* in intensity which is being measured (Section 2.1.3). In absorption the intensity depends on the ground state concentration, and if this is low then the absorption of radiation is low and accurate measurement is difficult. The amount absorbed $= I_0 - I_f$. If the change in intensity is small then $I_0 - I_f \approx 0$, and no absorption will be found.

In *fluorescence* it is the *actual* intensity of the emitted radiation which is being measured against a *zero background*, and this is much easier to measure.

Fluorescence intensity depends on the intensity of the exciting radiation, and also depends on the concentration of the ground state prior to excitation. Calibration is necessary unless the reaction is first order (Section 2.1.3).

In fluorescence experiments only a small proportion of the incident radiation results in fluorescence, and so a very intense source of incident radiation is required. Modern lasers do this admirably, and fluorescence techniques are now routine for determining very low concentrations.

By using lasers of suitable frequency, fluorescence can be extended into the infrared and microwave. Offshoots of laser technology include resonance fluorescence for detecting atoms, and laser magnetic resonance for radicals.

2.1.6 Spin resonance methods: nuclear magnetic resonance (NMR)

In optical spectroscopy the concentration of an absorbing species can be followed by monitoring the change in intensity of the signal with time. The same can be done in NMR spectroscopy, but with limitations as indicated.

For the kineticist, the major use of NMR spectroscopy lies in identifying products and intermediates in reaction mixtures. Concentrations greater than 10^{-5} mol dm^{-3} are necessary for adequate absorption, so low concentration intermediates cannot be studied.

NMR spectra show considerable complexity, which makes the spectra unambiguous, highly characteristic and immediately seen to be unique. This dramatically increases the value of NMR spectra for identification.

2.1.7 Spin resonance methods: electron spin resonance (ESR)

ESR is an excellent way to study free radicals and molecules with unpaired electrons present in complex gas reactions, and is used extensively by gas phase kineticists since it is the one technique which can be applied so directly to free radicals. It is also used in kinetic studies of paramagnetic ions such as those of transition metals. Again the change in intensity of the signal from the species can be monitored with time.

Chromatographic techniques are often used to separate free radicals formed in complex gas reactions, and ESR is used to identify them. The extreme complexity of the spectra results in a unique fingerprint for the substance being analysed. With ESR it is possible to detect radicals and other absorbing species in very, very low amounts such as 10^{-11} to 10^{-12} mol, making it an ideal tool for detecting radical and triplet intermediates present in low concentrations in chemical reactions.

2.1.8 Photoelectron spectroscopy and X-ray photoelectron spectroscopy

Two more modern spectroscopic techniques for detection, identification and concentration determination are photoelectron spectroscopy and X-ray photoelectron spectroscopy. Essentially these techniques measure how much energy is required to remove an electron from some orbital in a molecule, and a photoelectron spectrum shows a series of bands, each one corresponding to a particular ionization energy. This spectrum is highly typical of the molecule giving the spectrum, and identification can be made. At present the method is in its infancy, and a databank of spectra specific to given molecules must be produced before the technique can rival e.g. infrared methods.

When X-rays are used rather than vacuum UV radiation, electrons are emitted from inner orbitals, and the spectrum obtained reflects this. These spectra also give much scope as an analytical technique.

Worked Problem 2.2

Question. Suggest ways of detecting and measuring the concentration of the following.

1. The species in the gas phase reaction

$$Br^{\bullet} + HCl \rightarrow HBr + Cl^{\bullet}$$

2. The many radical species formed in the pyrolysis of an organic hydrocarbon.
3. The blood-red ion pair $FeSCN^{2+}$ formed by the reaction

$$Fe^{3+}(aq) + SCN^{-}(aq) \rightleftarrows FeSCN^{2+}(aq)$$

4. The catalytic species involved in the acid-catalysed hydrolysis of an ester in aqueous solution.
5. The charged species in the gas phase reaction

$$N_2^{+\bullet} + H_2 \rightarrow N_2H^{+} + H^{\bullet}$$

6. A species whose concentration is 5×10^{-10} mol dm^{-3}.
7. The species involved in the reaction

$$H_2O(g) + D_2(g) \rightarrow HDO(g) + HD(g)$$

Answer.

1. $Br^{\bullet}$ and $Cl^{\bullet}$ are radicals: detection and concentration determination by ESR. HCl and HBr are gaseous: spectroscopic (IR) detection and estimation.
2. Separate by chromatography, detect and analyse spectroscopically or by mass spectrometry.
3. The ion pair is coloured: spectrophotometry for both detection and estimation.
4. The catalyst is $H_3O^{+}(aq)$: estimated by titration with $OH^{-}(aq)$.

5. $N_2^{+\bullet}$ and N_2H^+ are ions in the gas phase: mass spectrometry for detection and estimation.
6. This is a very low concentration species: detection and estimation by laser induced fluorescence.
7. Three of the species involve deuterium: mass spectrometry is ideal, and can also be used for H_2O.

2.2 Measuring the Rate of a Reaction

Rates of reaction vary from those which seem to be instantaneous, e.g. reaction of $H_3O^+(aq)$ with $OH^-(aq)$, to those which are so slow that they appear not to occur, e.g. conversion of diamond to graphite. Intermediate situations range from the slow oxidation of iron (rusting) to a typical laboratory experiment such as the bromination of an alkene. But *in all cases* the reactant concentration shows a smooth decrease with time, and the reaction rate describes how rapidly this decrease occurs.

The *reactant* concentration remaining at various times is the fundamental quantity which requires measurement in any kinetic study.

2.2.1 Classification of reaction rates

Reactions are roughly classified as fast reactions – and the rest. The borderline is indistinct, but the general consensus is that a 'fast' reaction is one which is over in one second or less. Reactions slower than this lie in the conventional range of rates, and any of the techniques described previously can be adapted to give rate measurements. Fast reactions require special techniques.

A very rough general classification of rates can also be given in terms of the time taken for reaction to appear to be virtually complete, or in terms of half-lives.

Type of reaction	Time span for apparent completion	Half-life
very fast rate	microseconds or less	10^{-12} to 10^{-6} second
fast rate	seconds	10^{-6} to 1 second
moderate rate	minutes or hours	1 to 10^3 second
slow rate	weeks	10^3 to 10^6 second
very slow rate	weeks or years	$>10^6$ second

The *half-life* is the time taken for the concentration to drop to one-half of its value. If the concentration is 6×10^{-2} mol dm^{-3}, then the first half-life is the time taken for the concentration to fall to 3×10^{-2} mol dm^{-3}. The second half-life is the time taken for the concentration to fall from 3×10^{-2} mol dm^{-3} to 1.5×10^{-2} mol dm^{-3}, and so on.

The dependence of the half-life on concentration reflects the way in which the rate of reaction depends on concentration.

Care must be taken when using the half-life classification. First order reactions are the only ones where the half-life is *independent* of concentration (Sections 3.10.1, 3.11.1 and 3.12.1).

Worked Problem 2.3

Question. If reaction rate $\propto$ [reactant]^2, the half-life, $t_{1/2} \propto 1/\text{conc}$. Such a reaction is called *second order* (Sections 3.11 and 3.11.1). For a *first order* reaction, reaction rate $\propto$ [reactant] and the half-life is independent of concentration (Sections 3.10 and 3.10.1).

(a) Show how it is possible to bring a fast second order reaction into the conventional rate region.

(b) Show that this is not possible for first order reactions.

Answer.

(a) If the initial concentration of reactant is decreased by e.g. a factor of 10^3, then the half-life is increased by a factor of 10^3, and if the decrease is by a factor of 10^6 the half-life increases by a factor of 10^6. This latter decrease could bring a reaction with a half-life of 10^{-4} s into the moderate rate category, with half-life now 10^2 s.

(b) Since the half-life for a first order reaction is independent of concentration, there is no scope for increasing the half-life by altering the initial concentration.

2.2.2 Factors affecting the rate of reaction

- The standard variables are concentration of reactants, temperature and catalyst, inhibitor or any other substance which affects the rate.
- Chemical reactions are generally very sensitive to temperature and must be studied at constant temperature.

- Rates of reactions in solution and unimolecular reactions in the gas phase are dependent on pressure.
- Some gas phase chain reactions have rates which are affected by the surface of the reaction vessel. Heterogeneous catalysis occurs when a surface increases the rate of the reaction.
- Photochemical reactions occur under the influence of radiation. Conventional sources of radiation, and modern flash and laser photolysis techniques, are both extensively used.
- Change of solvent, permittivity, viscosity and ionic strength can all affect the rates of reactions in solution.

2.2.3 Common experimental features for all reactions

- Chemical reactions must be studied at constant temperature, with control accurate to $\pm$ 0.01 °C or preferably better. The reactants must be very rapidly brought to the experimental temperature at zero time so that reaction does not occur during this time.
- Mixing of the reactants must occur very much faster than reaction occurs.
- The start of the reaction must be pinpointed exactly and accurately. A stop-watch is adequate for timing conventional rates; for faster reactions electronic devices are used. If spectroscopic methods of analysis are used it is simple to have flashes at very short intervals, e.g. 10^{-6} s, while with lasers intervals of 10^{-12} s are common. Recent advances give intervals of 10^{-15} s.
- The method of analysis must be very much faster than the reaction itself, so that virtually no reaction will occur during the period of concentration determination.

2.2.4 Methods of initiation

Normally, thermal initiation is used and the critical energy is acquired by collisions. In photochemical initiation the critical energy is accumulated by absorption of radiation. This can only be used if the reactant molecule has a sufficiently strong absorption in an experimentally accessible region, though modern laser techniques for photochemical initiation increase the scope considerably.

Absorption of radiation excites the reactant to excited states, from which the molecule can be disrupted into various radical fragments. Conventional sources produce steady state concentrations. Flash and laser sources produce much higher concentrations, enabling more accurate concentration determination, and allowing monitoring of production and removal by reaction of these radicals; this is something which is not possible with either thermal initiation or conventional photochemical initiation.

Lasers have such a high intensity that they can give significant absorption of radiation, even though ε for the absorbing species and its concentration are both very low. With conventional sources, absorption will be very low, if only one or other of ε and the concentration is low. The crucial point is that the absorbance also depends on the intensity of the exciting source, and so with lasers this can outweigh a low concentration and/or a low ε.

Other useful features of photochemical methods include the following. They

- enable reaction to occur at temperatures at which the thermal reaction does not occur,
- allow the rate of initiation to be varied at constant temperature, impossible with thermal initiation,
- allow the rate of initiation to be held constant while the temperature is varied, so that the temperature effects on the subsequent reactions can be studied independently of the rate of initiation (again this is impossible in the thermal reaction) and
- give selective initiation. Frequencies can be chosen at which known excited states are produced, and the rates of reaction of these excited states can then be studied. Thermal initiation is totally unselective.

Radiochemical and electric discharge initiation are also used, though these are much less common. These are much higher energy sources, and they have a much more disruptive effect on the reactant molecules, producing electrons, atoms, ions and highly excited molecular and radical species.

2.3 Conventional Methods of Following a Reaction

These determine directly changes in concentrations of reactants and products with time, but they may have the disadvantages of sampling and speed of analysis.

When reaction is sufficiently fast to result in significant reaction occurring during the time of sampling and analysis, the rate of reaction is slowed down by reducing the temperature of the sample drastically, called 'quenching'; reaction rate generally decreases dramatically with decreasing temperature. Alternatively, reaction can be stopped by adding a reagent which will react with the remaining reactant. The amount of this added reagent can be found analytically, and this gives a measure of the amount of reactant remaining at the time of addition.

2.3.1 Chemical methods

These are mainly titration methods and they can be highly accurate. They are generally reserved for simple reactions in solution where either only the reactant or the product concentrations are being monitored. Here sampling errors and speed of analysis are crucial. Chemical methods have been largely superseded by modern black box techniques, though, in certain types of solution reaction, they can still be very useful.

Worked Problem 2.4

Question. The catalysed decomposition of hydrogen peroxide, H_2O_2, is easily followed by titrating 10.0 cm^3 samples with 0.0100 mol dm^{-3} $KMnO_4$ at various times.

$$2H_2O_2(aq) \rightarrow 2H_2O(l) + O_2(g)$$

time/min	5	10	20	30	50
volume of 0.0100 mol dm^{-3} $KMnO_4/cm^3$	37.1	29.8	19.6	12.3	5.0

$$5/2H_2O_2(aq) + MnO_4^-(aq) + 3H^+(aq) \rightarrow Mn^{2+}(aq) + 5/2O_2(g) + 4H_2O(l)$$

1 mol MnO_4^- reacts with 5/2 mol H_2O_2

Calculate the $[H_2O_2]$ at the various times, and show that these values lie on a smooth curve when plotted against time.

Answer.

$10^4 \times$ number of mol MnO_4^- used	3.71	2.98	1.96	1.23	0.50
$10^4 \times$ number of mol H_2O_2 present in sample	9.28	7.45	4.90	3.08	1.25
$10^2 \times [H_2O_2]/\text{mol dm}^{-3}$	9.28	7.45	4.90	3.08	1.25

A graph of $[H_2O_2]$ versus time is a smooth curve showing the progressive decrease in reactant concentration with time (Figure 2.5).

2.3.2 Physical methods

These use a physical property dependent on concentration and must be calibrated, but are still much more convenient than chemical methods. Measurement can often be made *in situ*, and analysis is often very rapid. Automatic recording gives a continuous trace. It is vital to make measurements *faster* than reaction is occurring.

The following problems illustrate typical physical methods used in the past.

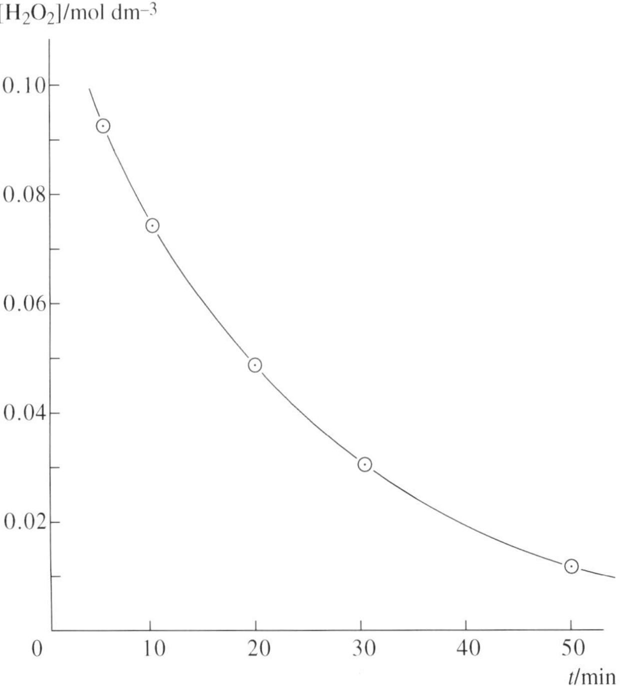

Figure 2.5 Graph of $[H_2O_2]$ versus time

Pressure changes in gas phase reactions

Worked Problem 2.5

Question. Which of the following reactions can be followed in this way and why?

1. $CH_3CHO(g) \rightarrow CH_4(g) + CO(g)$
2. $Br_2(g) + CH_2{=}CH_2(g) \rightarrow C_2H_4Br_2(g)$
3. $Cl_3CCOOH(aq) \rightarrow CHCl_3(aq) + CO_2(g)$
4. $2HI(g) \rightarrow H_2(g) + I_2(g)$
5. $(CH_3CHO)_3(g) \rightarrow 3CH_3CHO(g)$

Which reaction would give the largest change in total pressure and why? Suggest an alternative method for reaction 3.

Answer. This method can be used for reactions which occur with an *overall change* in the number of molecules in the *gas phase* and which consequently show a change in total pressure with time.

$$pV = nRT \quad \text{and} \quad p \propto n \qquad \text{if } V \text{ and } T \text{ are constant} \tag{2.11}$$

1. Gives an increase in the number of gaseous molecules: pressure increases.
2. Pressure decreases with time.
3. Pressure increases with time.
4. Pressure remains constant, so this method cannot be used.
5. Pressure increases.

Reaction 5 gives the greatest change in the number of molecules in the gas phase, and hence the largest change in pressure.

In reaction 3, CO_2 is the only gaseous species present and the reaction could be followed by measuring the volume of gas evolved at constant pressure with time.

An actual problem makes the stoichiometric arguments clearer.

Worked Problem 2.6

Question. The gas phase decomposition of ethylamine produces ethene and ammonia,

$$C_2H_5NH_2(g) \rightarrow C_2H_4(g) + NH_3(g)$$

If p_0 is the initial pressure of reactant and p_{total} is the total pressure at time t, show how the partial pressure of reactant remaining at time t can be found. Using this and the following data, find the partial pressure of $C_2H_5NH_2$ remaining at the various times, and plot an appropriate graph showing the progress of reaction.

What would the final pressure be?

The following total pressures were found for a reaction at 500 °C with an initial pressure of pure ethylamine equal to 55 mmHg.

$\frac{p_{\text{total}}}{\text{mm Hg}}$	55	64	72	89	93
$\frac{\text{time}}{\text{min}}$	0	2	4	10	12

Answer. If p_0 is the initial pressure of ethylamine, and an amount of ethylamine decomposes so that the decrease in its partial pressure is y, then at time t there will be $p_0 - y$ of ethylamine left, y of C_2H_4 formed and y of NH_3 formed.

$\therefore$ total pressure at time t,

$$p_{total} = p_0 - y + y + y$$
$$= p_0 + y \quad (2.12)$$

$$\therefore y = p_{total} - p_0 \quad (2.13)$$

$$\therefore p(C_2H_5NH_2)_{remaining} = p_0 - y$$
$$= 2p_0 - p_{total} \quad (2.14)$$

Using this gives

$\frac{p_{total}}{mmHg}$	55	64	72	89	93
$p(C_2H_5NH_2) = \frac{2p_0 - p_{total}}{mmHg}$	55	46	38	21	17
$\frac{time}{min}$	0	2	4	10	12

When reaction is over, the total pressure $= p(C_2H_4) + p(NH_3) = 2p_0 = 110$ mmHg. A graph of $p(C_2H_5NH_2)_{remaining}$ versus time is a smooth curve (Figure 2.6).

Conductance methods

These are useful when studying reactions involving ions. Again this can be illustrated by a problem.

Worked Problem 2.7

Question. Which of the following reactions can be studied in this way, and why?

1. $(CH_3)_3CCl(aq) + H_2O(l) \rightarrow (CH_3)_3COH(aq) + H^+(aq) + Cl^-(aq)$
2. $NH_4^+(aq) + OCN^-(aq) \rightarrow CO(NH_2)_2(aq)$
3. $H_3O^+(aq) + OH^-(aq) \rightarrow 2H_2O(l)$

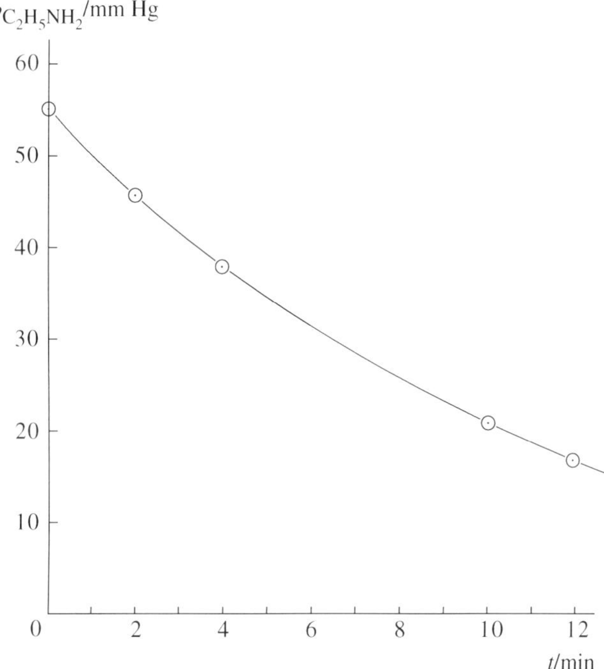

Figure 2.6 Graph of $p(C_2H_5NH_2)$ versus time

4. $CH_3COOCH_3(aq) + OH^-(aq) \rightarrow CH_3COO^-(aq) + CH_3OH(aq)$
5. $CH_3COOH(aq) + H_2O(l) \rightarrow CH_3COO^-(aq) + H_3O^+(aq)$

Answer. Conductance changes are ideal for reactions involving ions.

1. Ions are produced, and the conductance rises with time.
2. Ions are removed, and the conductance decreases with time.
3. Ions are removed as reaction occurs, but this is a very rapid reaction and special fast reaction techniques are needed. Ion–ion reactions are often very fast.
4. Here a highly conducting ion is replaced by a less conducting ion, and the conductance will fall with time.
5. Ions are produced, but this is an ionization of an acid, and these are generally very fast reactions and cannot be studied in this way.

pH and EMF methods using a glass electrode sensitive to H_3O^+

Worked Problem 2.8

Question. What types of reaction could be studied in this way?

Answer. In these methods a glass electrode sensitive to H_3O^+ enables reactions which occur with change in $[H_3O^+]$ or a change in $[OH^-]$ to be followed with ease. A pH meter measures pH directly, and a millivoltmeter measures EMFs directly, and these are related to $[H_3O^+]$, e.g.

- $Br_2(aq) + CH_3CO$-
  $COCH_3(aq) + H_2O(l) \rightarrow CH_3COCH_2Br(aq) + H_3O^+(aq) + Br^-(aq)$
  Here $H_3O^+(aq)$ is produced and the pH will decrease with time, with a corresponding change in EMF,
- $CH_3COOCH_3(aq) + OH^-(aq) \rightarrow CH_3COO^-(aq) + CH_3OH(aq)$
  Since $OH^-(aq)$ is removed, $[H_3O^+]$ will increase and the pH will decrease with time, with a corresponding change in EMF and
- $CH_3COOCH_2CH_3(aq) + H_3O^+(aq) \rightarrow CH_3COOH(aq) + CH_3CH_2OH(aq)$
  In the acid hydrolyses of esters $H_3O^+(aq)$ is removed, and the pH increases with time, with a corresponding change in EMF.

Other EMF methods

'Ion-selective' electrodes sensitive to ions other than H_3O^+ can be used to follow reactions involving these ions. Silver halide electrodes are used to determine Cl^-, Br^-, I^- and CN^-; lanthanum fluoride to determine F^-; Ag_2S electrodes to determine S^{2-}, Ag^+ and Hg^{2+}; and electrodes made from a mixture of divalent metal sulphides and Ag_2S are used to determine Pb^{2+}, Cu^{2+} and Cd^{2+}. Other types determine K^+, Ca^{2+}, organic cations, and anions such as NO_3^-, ClO_4^- and organic anions. Ion-selective electrodes which monitor the concentration of ions other than H_3O^+ or OH^- have proved particularly suitable for biological systems.

A variant on pH methods: the pH–stat and Br_2–stat

Monitoring of reactions carried out at constant pH is achieved by a pH-stat device which adds H_3O^+ or OH^- automatically to the reaction mixture, depending on whether H_3O^+ or OH^- is being used up, and in amounts necessary to maintain constant pH throughout reaction. The volume of solution added is continuously recorded, and is converted to give the amount of reactant remaining at various times throughout the reaction. A similar Br_2-stat device has been used to follow bromination reactions. Beside being very useful methods for carrying out routine experiments, pH-stat, Br_2-stat or any other kind of stat are very useful for studying reactions under pseudo-first order conditions where the other reactant is held in constant concentration.

Chromatographic, mass spectrometric and spectroscopic methods

These can all be used for following the reaction by measuring concentrations at various times. Often they will be linked up to an automatic recording device and can be used for measurements *in situ*. These methods have been discussed earlier.

Dilatometric methods

These measure the change in volume of a solution with time. If this change is sufficiently large and exaggerated by the use of a dilatometer, it can be used to follow the rate of reaction. Polymerization reactions are often particularly suited to this technique.

2.4 Fast Reactions

With fast reactions it is very important to ensure that there is rapid mixing, fast initiation and fast analysis. Special timing devices are needed. Consequently techniques special to fast reactions had to be developed.

2.4.1 Continuous flow

Flow systems replace the *clock* as a timer by *a distance along a tube*. Two gases or two solutions are rapidly mixed, and flow rapidly along a tube along which there are observation chambers at accurately known distances at which the concentration is determined, either *in situ* or by leaking samples.

The time for mixing limits the rate of reaction which can be followed, e.g. if mixing takes 10^{-3} second, then reactions over in around 10^{-3} seconds or less cannot be studied. The moment of mixing gives the zero time. The observation points are at accurately known distances along the tube and larger distances correspond to longer reaction times (Figure 2.7).

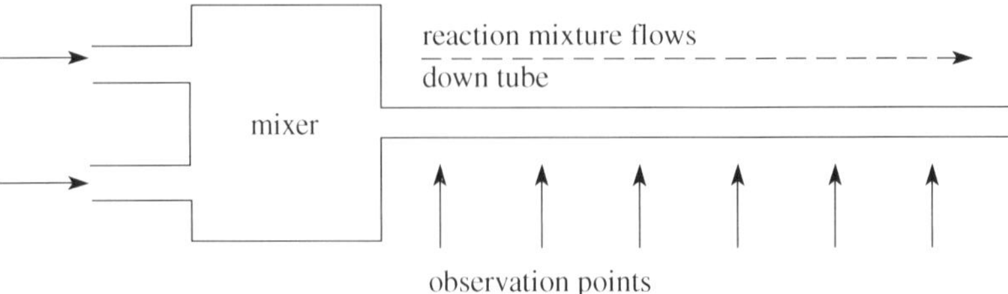

Figure 2.7 Schematic diagram of a flow apparatus

Worked Problem 2.9

Question.

(a) Calculate the reaction times when analysis is made at 1, 5 and 10 cm distances along a tube when the flow rate is 20 m s^{-1}.

(b) Altering the flow rate allows different times to be observed. Calculate the times which corresponds to a distance of 1 cm along the tube when the flow rate is 10 m s^{-1} and 50 m s^{-1}.

Be careful with units. The distances along the tube are given in cm, but the velocity of flow is given in m s^{-1}. When using these in the equation $t = d/v$, both d and v must have the same units. In the answer given, the flow rate is converted to cm s^{-1}. 1 m = 100 cm, and so 20 m s^{-1} = 20 × 100 cm s^{-1}.

Answer.

(a) If an observation point is at a distance d along the tube and the flow rate along the tube is v, then the time taken to reach the observation point is d/v,

1 cm corresponds to a time = d/v = 1 cm/20 × 100 cm s^{-1} = 5×10^{-4} s.

5 cm corresponds to a time = d/v = 5 cm/20 × 100 cm s^{-1} = 2.5×10^{-3} s.

10 cm corresponds to a time = d/v = 10 cm/20 × 100 cm s^{-1} = 5×10^{-3} s.

(b) If v = 10 m s^{-1} (easily attained), then 1 cm corresponds to 10^{-3} s.

If v = 50 m s^{-1} then 1 cm corresponds to 2×10^{-4} s, about the smallest time interval attainable on a flow apparatus.

2.4.2 Stopped flow

Developed for solutions, this has the advantage that only small samples are required. Two solutions are rapidly forced through a mixing chamber from which there is an exit into a tube. The flow is rapidly halted and a detector at the position of halting of the flow gives the concentration. This apparatus in effect acts as an efficient and very rapid mixing device. The smaller the distance along the tube at which the flow is halted, the shorter is the time from the start of the reaction at the time of mixing. The reaction is then followed with time at this point.

The time of analysis must be very rapid. Stopped flow methods always require an inbuilt timing device, and time intervals at which analysis is made are dictated by the speed at which successive analyses can be carried out. Spectroscopic methods using pulsed radiation are very useful here because they can be both analytical and timing devices.

2.4.3 Accelerated flow

Hypodermic plungers are pushed into syringes increasingly rapidly, and so accelerate the flow as the mixed reactants pass out of the mixing chamber. Consequently, the velocity with which the mixed reactants emerge from the mixing chamber increases with time. When the mixture passes a fixed point on the tube, it will have been reacting for progressively shorter times. Only small samples are required, but fast detection is vital.

Worked Problem 2.10

Question. A reaction is carried out in an accelerated flow tube and observations are made spectrophotometrically.

If the initial flow velocity is 400 cm s^{-1}, calculate the time to reach an observation point 1 cm along the reaction tube. If the acceleration is such that the velocity increases progressively to 800, 1200 and 1600 cm s^{-1}, calculate the reaction times corresponding to these flow rates. If the absorbances corresponding to these times are 0.124, 0.437, 0.655 and 0.812, draw a graph showing the progress of reaction.

Answer. For observation at 1 cm along the tube

$\frac{v}{\text{cm s}^{-1}}$	400	800	1200	1600
$\frac{\text{time}}{\text{ms}}$	2.50	1.25	0.83	0.625
A	0.124	0.437	0.655	0.812

For a graph of A versus time, see Figure 2.8.

2.4.4 Some features of flow methods

Analysis is generally spectroscopic and done *in situ*, though mass spectrometry, chromatography and conductivity methods have been used. Gas and liquid phase reactions

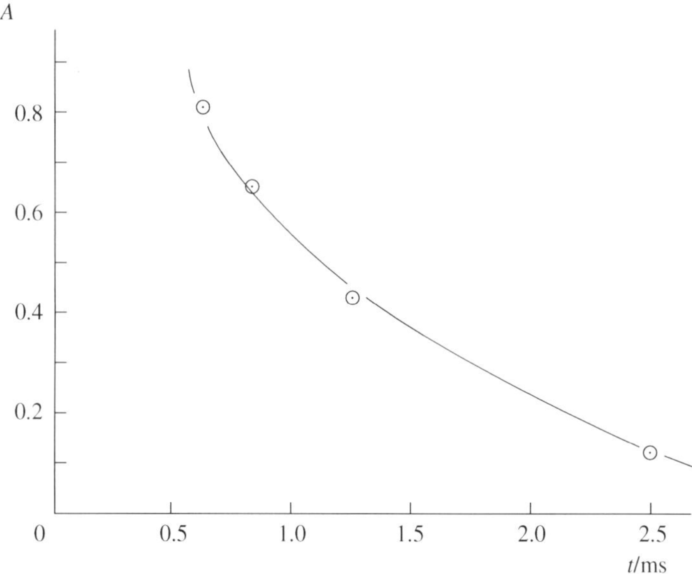

Figure 2.8 Graph of absorbance versus time

can be studied, and in photochemical reactions the mixture passes through an illuminated region. Reaction times are relatively limited, and are of the order of 1 to 10^{-3} s. Flow methods are used extensively in kinetics, especially for solution reactions.

2.5 Relaxation Methods

These are specific to reactions which *come to equilibrium*.

The reaction is *initially at equilibrium*, and so the start of the reaction can be pinpointed very accurately: something which is essential for fast reactions. Also there will be no mixing problems, and the temperature can be established before reaction occurs.

The conditions are then altered very rapidly and the old equilibrium position is now a *non-equilibrium position*. The system has then to adjust to this new equilibrium position by reaction. The consequent concentration changes are rapid and are followed with time. Very rapid detection is needed.

In these studies, chemical analysis and methods such as spectroscopy, mass spectrometry and chromatography are superior to physical methods such as conductivity, because the species and chemical equilibria can be identified; otherwise the species present have to be inferred.

The perturbations can be a *single impulse*, with either large or very small displacements from equilibrium; or they can be the *periodic* displacements of an ultrasonic wave. Because the reactants are premixed and reaction is initiated by the impulse, there is no need for rapid mixing. The mixture will also be at the required temperature.

2.5.1 Large perturbations

Large perturbations using flash or laser photolysis and shock tubes require a new equilibrium situation to be set up which is far from the initial equilibrium state. These methods are generally used in gas phase studies, and small perturbations are used for solutions, though there is nothing constraining the techniques in this way.

2.5.2 Flash photolysis

The reactants are premixed and reaction initiated by an intense flash of radiation lasting 10^{-4} to 10^{-6} s. This disrupts reactant molecules into atoms and radicals in very high concentrations. The high concentrations of the intermediates produced make the subsequent kinetic analysis very much easier, and many of these reactions are often not picked up in low intensity–low concentration photochemical reactions.

The first intense flash is followed by a series of secondary low intensity flashes which enable the identities and concentrations of the species to be found at various times, leading ultimately to rate constants for the reactions. Spectroscopic methods can easily measure intervals of 10^{-6} s, and lasers can take this down to 10^{-15} s.

The initial flash must be sufficiently intense to give the high concentrations of intermediates required, but the higher the intensity of the flash the longer is its duration. This means that very fast reactions occurring in a time less than or equal to the duration of the flash will be over by the time the flash is over. A compromise must be made, and intensities are used so that the flash lasts 10^{-4} to 10^{-6} s, and half lives down to 10^{-6} s are easy to follow. The secondary flashes are for analysis only and are of low intensity, so that it is easy to have a flash lasting only 10^{-6} s.

Flash photolysis greatly increases the number of intermediate reactions able to be studied. These are generally radical–radical reactions, in contrast to the radical–molecule reactions which predominate in low intensity photolysis.

The half-lives able to be studied by flash photolysis can be extended to shorter values in the following modification. Short interval secondary flashes are necessary if reaction times as short as 10^{-9} to 10^{-15} s are to be studied. This is done very ingeniously by splitting the beam (Figure 2.9). One part of the beam goes through the reaction mixture and causes photolysis; the other by-passes the reaction mixture, goes to a movable mirror and is reflected back to the reaction mixture through a fluorescent solution which converts the beam into continuous radiation in the visible. This is then used to analyse the mixture spectroscopically. The path lengths of the two parts of the beam are different. By altering the position of the mirror the time interval between the first flash and the analytical flash can be varied. For instance, a 30 cm distance corresponds to an interval of 10^{-9} s.

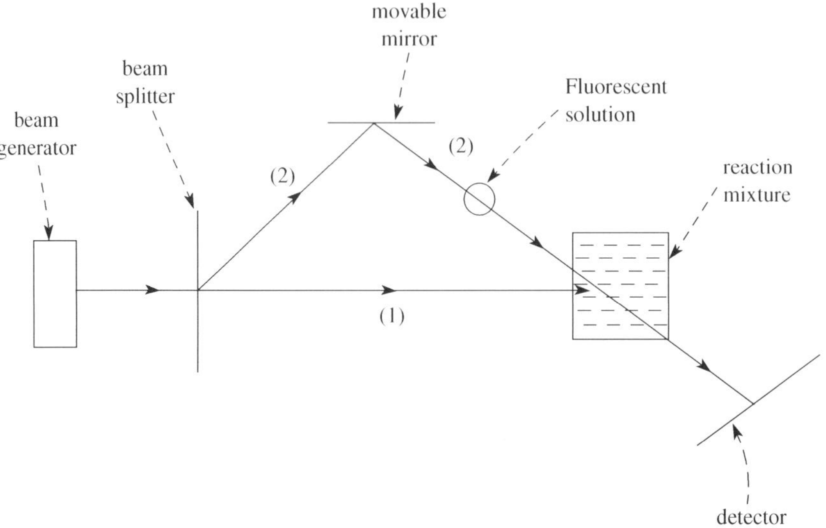

Figure 2.9 Schematic diagram for beam splitting in a flash photolysis experiment

> **Worked Problem 2.11**
>
> ***Question.*** Calculate the path-length differences which correspond to time intervals of 10^{-6}, 10^{-7}, 10^{-8}, 10^{-9}, 10^{-10}, 10^{-11} and 10^{-12} s between a photolysis and analytical flash. The velocity of light $= 2.998 \times 10^{10}$ cm s^{-1}. Comment on the values.
>
> ***Answer.*** $v = d/t \therefore d = v \times t$. Values for path-length differences are 29 980 cm, 2998 cm, 299.8 cm, 29.98 cm, 2.998 cm, 0.2998 cm and 0.029 98 cm.
>
> Time intervals of the order of 10^{-11} s and less cannot be achieved by this technique for following the rate. Pulsed lasers (Sections 2.1.4 and 2.5.3) have to be used.
>
> Likewise the technique cannot be used for time intervals of the order of 10^{-7} s and greater. Ordinary flash photolysis would be adequate.
>
> Only a narrow range of half-lives can be studied using this method.

2.5.3 Laser photolysis

With a laser, a beam of high intensity but short duration is now possible. The technique of beam splitting can also be used with lasers, though now high intensity pulsed lasers give a very wide range of pulses lasting from 10^{-9} to 10^{-15} s. In consequence most chemical and physical processes can be studied.

2.5.4 Pulsed radiolysis

X-radiation or electrons are used in an analogue of flash photolysis. Pulses of length 10^{-6} to 10^{-12} s are standard, but here the radiation causes ionization rather than disruption. The range of types of reaction which can now be studied using pulsed lasers and pulsed radiolysis is extensive.

2.5.5 Shock tubes

A diaphragm with a pressure difference over it of up to about 10^6 atm is ruptured, causing a shock wave which passes through a premixed reaction mixture. As the shock wave passes along the tube it heats the reaction mixture to 10^4–10^5 K. The shock wave reaches the observation point in a very short time (10^{-6}–10^{-8} s), and all of the reaction mixture up to this point is then at this very high temperature. This is zero time for the start of reaction. The tube is able to stay at the high temperature for around 10^{-3} s, and all the measurements must be made within that time. The very high temperature causes disruption of reactant molecules into radicals in very high concentrations, and these can subsequently react. At the same time the mixture is driven along the tube by a driver gas. This converts the system to a flow apparatus. As the mixture is driven past the observation point it will have been reacting for progressively longer times. Analysis at the observation point will give concentrations of the species in the mixture corresponding to these times. This often uses lasers at convenient time intervals ranging from 10^{-6} s to 10^{-15} s. When the driver gas reaches the observation point, the experiment is over. It generally takes 10^{-3} second between the arrival of shocked gas and driver gas, and this corresponds to the period of the high temperature.

Sometimes an alternative procedure places several observation points along the tube. These correspond to progressively longer times of reaction the further the observation point is from the diaphragm (Figure 2.10). The calculations are similar to those in Problem 2.9.

Shock waves combine features of flash photolysis with continuous flow. Though flash photolysis studies are more flexible and cover a wider range of reaction times, they are limited to photochemical reactions.

Shock tubes are used for gas reactions. When liquids are used the temperature rises are small, making the technique much less useful.

2.5.6 Small perturbations: temperature, pressure and electric field jumps

These single impulse displacements only alter the position of equilibrium provided certain conditions are met:

(a) temperature jump methods – reaction must have a significant ΔH;

(b) pressure jump methods – reaction must have a significant ΔV;

(c) electric field impulse – reaction must involve ionization.

Initial state before rupture of diaphragm:

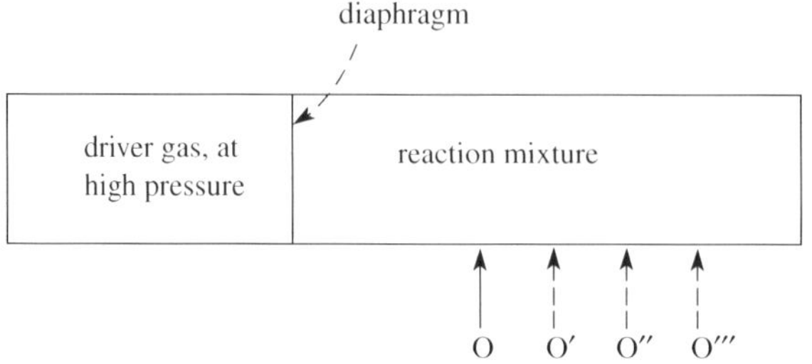

Two subsequent situations:

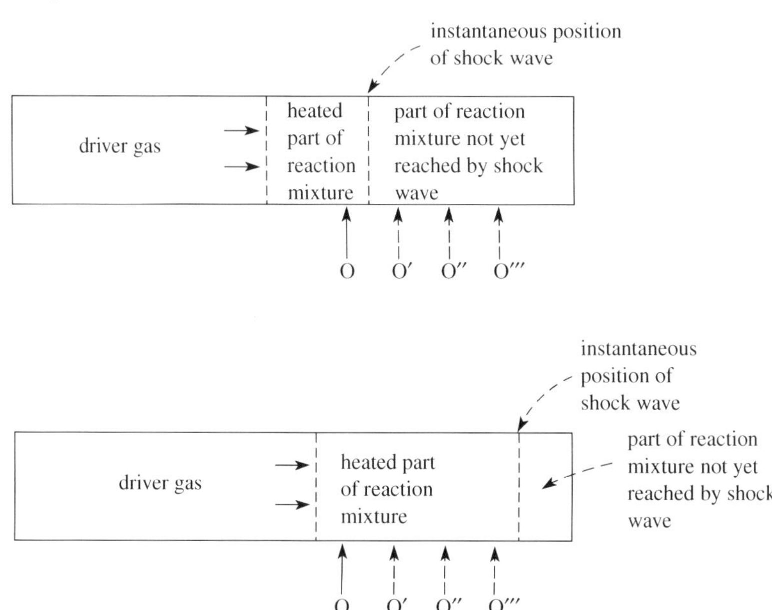

O: observation point

O′, O″, O‴: possible additional observation points

Figure 2.10 Schematic diagram of a shock tube apparatus

These techniques involve a sudden increase in temperature, a sudden increase in pressure and a sudden application of an electric field, respectively.

Table 2.1 summarizes the time taken for the impulse to make its effect, the time that the effect lasts and hence the time during which concentration measurements must be made.

For the mathematical analysis of the data to be valid, rates must be studied very close to equilibrium, and any displacement must be small (see Sections 3.21 and 3.21.1). If a single equilibrium is involved, then the approach to the new equilibrium position always simulates first order behaviour.

Table 2.1 Single impulse relaxation

Impulse	Time for jump to occur (s)	Change lasts (s)	Observations must be made in (s)
Temperature	10^{-6}	1	$1\text{–}10^{-6}$
Pressure	10^{-4}	50	$50\text{–}10^{-4}$
Electric	10^{-8}	2×10^{-4}	$2 \times 10^{-4}\text{–}10^{-8}$
Shock	10^{-6}	50	$50\text{–}10^{-6}$

The experimental data is concentration–time data, and integrated rate expressions give the observed rate constants. The technique can be extended to multi-step equilibria and to equilibrium reactions with consecutive steps. For multi-step equilibria, analysis results in a sum of exponentials, which may or may not reduce to first order.

2.6 Periodic Relaxation Techniques: Ultrasonics

This is the most powerful of the relaxation techniques, partly because it can handle multi-step equilibria easily, and often gives the relaxation times and the number of equilibria direct from the actual measurements (Section 3.10.2). It has been used to study physical processes such as energy transfer and ion solvation, as well as chemical reactions. This technique is used for reaction times of 1 to 10^{-10} s.

If passed through a reaction mixture, a sound wave causes a wave of alternating temperatures which calls for a corresponding *wave of new equilibrium positions to be set up*. If the concentrations of species involved in the equilibria can alter rapidly enough to enable *each new equilibrium position to be set up* as each new temperature is attained, then the concentrations and the equilibria can keep in phase with the periodic displacement. This occurs when the period of the sound wave is large compared with the relaxation time.

If reaction does not occur fast enough, then the concentration will not alter in phase with the periodic displacement, the *required equilibria for each temperature are not set up* and the concentrations remain approximately constant throughout. This occurs when the period of the sound wave is very small compared to the relaxation time.

These are two extremes. An intermediate situation, where the period of the sound wave is comparable to the relaxation time, occurs when reaction is fast enough to allow the concentrations to alter with time, but the change in concentration is out of phase with the sound wave. This shows up as an increase in the velocity of sound, or as a maximum in the absorption of sound by the reaction mixture as the frequency of the sound wave alters. This is the region where all the useful information is obtained.

If there is only one equilibrium, i.e. one relaxation time involved, there is a single point of inflection on the velocity–frequency diagram, or one maximum in the absorption of sound–frequency diagram (Figures 2.11 and 2.12). For more than one equilibrium then there will be a corresponding number of points of inflection, or maxima (Figures 2.13 and 2.14), provided the relaxation processes are resolved. These are easily and accurately measured. Sound waves are passed through the reaction mixture and the velocity of sound, or its absorption, is measured for a series of frequencies of the sound wave, and a graph drawn. From each point of inflection or maximum the relaxation time can be found directly and the rate constant calculated (Table 3.3). Solving the kinetics of complex equilibria is straightforward using relaxation times, in contrast to the increasing difficulty and eventual impossibility of using integrated rate equations and concentration data (Section 3.9).

If the relaxation processes are not resolved, best fit computer analysis is used.

Unfortunately, chemical identification of the relaxation process is by far the most difficult and ambiguous part of the analysis. There are no direct chemical observations. The magnitude of the relaxation time can often distinguish between physical processes and chemical reactions, but this does not reveal what the chemical reactions are.

Relaxation times of the order of 10^{-5} to 10^{-10} s are usually studied, giving first order rate constants lying between 10^5 and 10^{10} s^{-1}, though slower reactions can

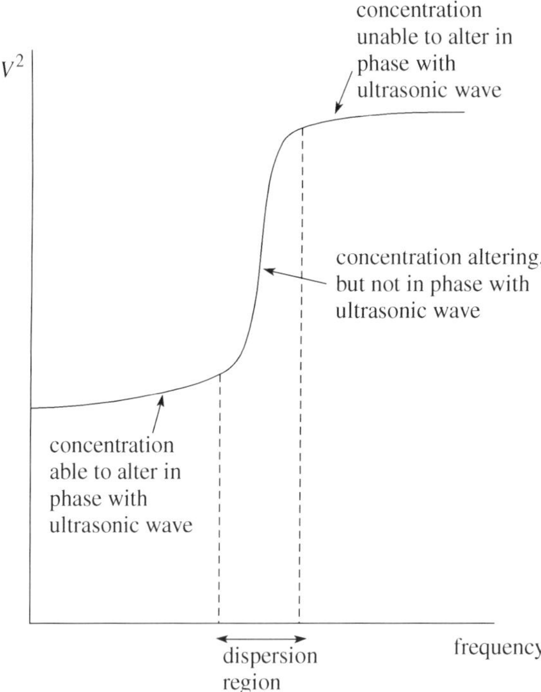

Figure 2.11 Periodic relaxation diagram showing the dependence of the velocity of an ultrasonic wave through a reaction mixture on its frequency

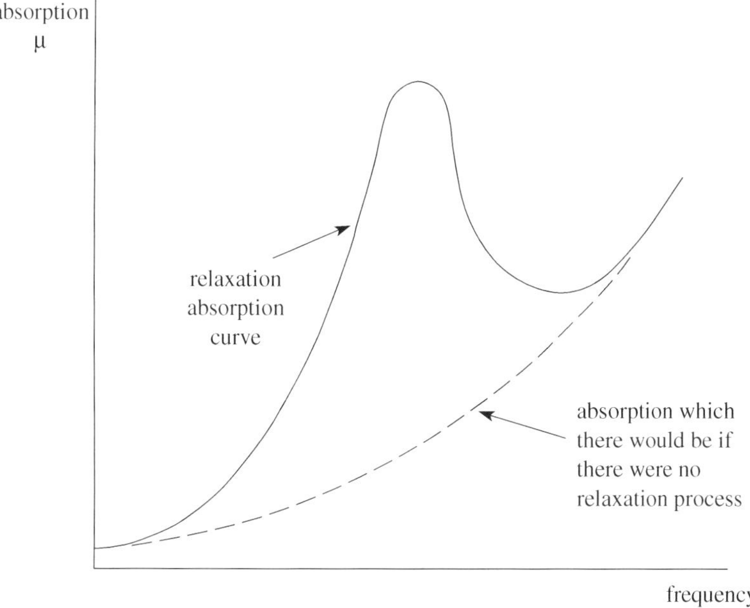

Figure 2.12 Periodic relaxation diagram showing the dependence of the absorption of an ultrasonic wave through a reaction mixture on its frequency

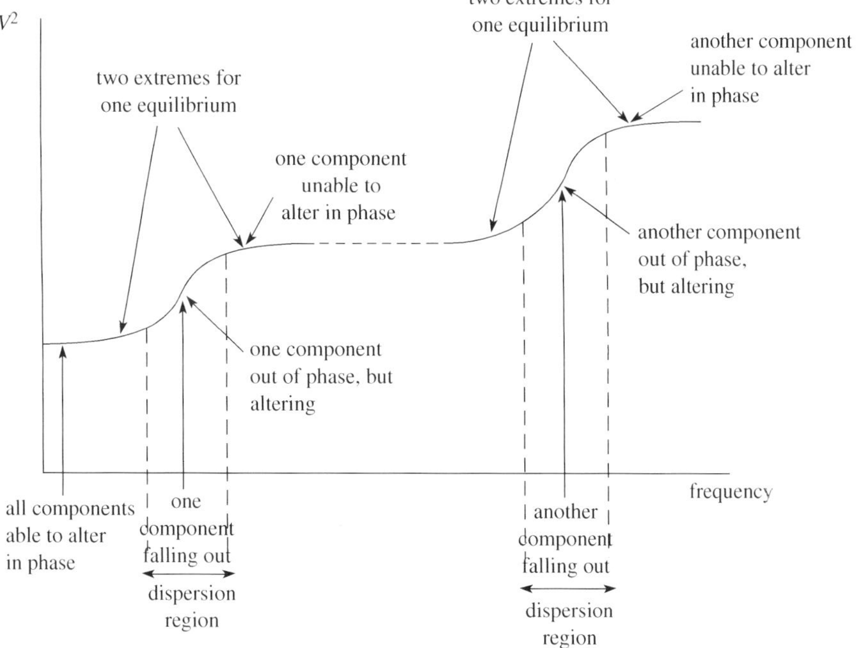

Figure 2.13 Periodic relaxation for a double equilibrium: diagram for velocity of wave

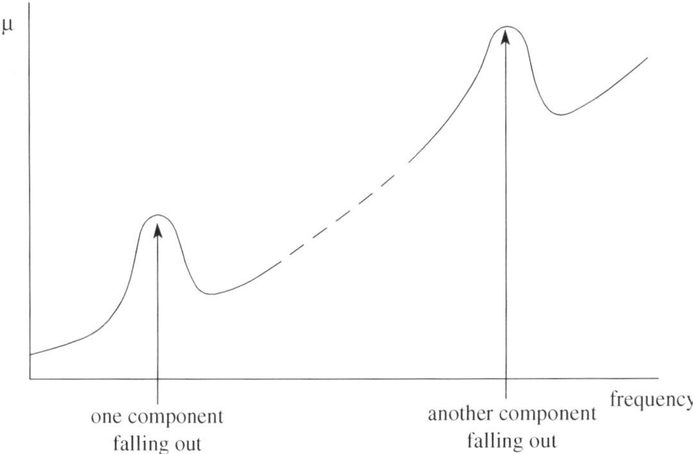

Figure 2.14 Periodic relaxation for a double equilibrium: diagram for absorption of wave

easily be studied. With appropriate low concentrations, second order processes up to diffusion-controlled rates can be examined.

2.7 Line Broadening in NMR and ESR Spectra

All spectroscopic lines have a natural line width, and this can be of great use to the kineticist. This natural line width is determined by the lifetime of the excited state of the molecule. If this is short the line width is broad, while longer lifetimes give more sharply defined lines. If reaction occurs, this can alter the lifetime of the excited state and so change the natural line width of the transition. A detailed spectroscopic analysis gives relations between the width of the line, the lifetime of the reacting species and the rate constant for the reaction. This has proved a very important tool, especially for reactions in solution such as proton transfers and acid/base ionization processes.

Further reading

Laidler K.J., *Chemical Kinetics*, 3rd edn, Harper and Row, New York, 1987.

Laidler K.J. and Meiser J.H., *Physical Chemistry*, 3rd edn, Houghton Mifflin, New York, 1999.

Alberty R.A. and Silbey R.J., *Physical Chemistry*, 2nd edn, Wiley, New York, 1996.

Logan S.R., *Fundamentals of Chemical Kinetics*, Addison-Wesley, Reading, MA, 1996.

* Nicholas J., *Chemical Kinetics*, Harper and Row, New York, 1976.
* Wilkinson F., *Chemical Kinetics and Reaction Mechanisms*, Van Nostrand Reinhold, New York, 1980.
* Out of print, but should be in university libraries.

Further problems

1. Suggest ways for determining the rate of the following reactions.

   - Bromination of $CH_3CH{=}CH_2$ in aqueous solution.
   - Reaction of $C_2H_5^\bullet$ with CH_3COCH_3 in the gas phase.
   - A polymerization reaction.
   - Iodination of CH_3COCH_3.
   - $2NO(g) + 2H_2(g) \rightarrow N_2(g) + 2H_2O(g)$
   - A reaction in solution with a half-life of 10^{-4} s.
   - $K^\bullet(g) + CH_3I(g) \rightarrow KI(g) + CH_3^\bullet(g)$
   - $C_6H_5N_2^+(aq) + H_2O(l) \rightarrow C_6H_5OH(aq) + H^+(aq) + N_2(g)$
   - $A + B \rightleftarrows C \rightleftarrows D \rightleftarrows E + F$
   - $S_2O_8^{2-}(aq) + 2I^-(aq) \rightarrow 2SO_4^{2-}(aq) + I_2(aq)$
   - A reaction with a half-life of 10^{-9} s.

2. The following table gives the absorbances for various concentrations of a solution. A 1 cm optical cell is used. Use this to calculate the molar absorption coefficient for the solute.

$\frac{\text{conc.}}{\text{mol dm}^{-3}}$	0.0045	0.0105	0.0140	0.0190
A	0.380	0.899	1.195	1.615

   The kinetics of a reaction involving this solute as reactant, in which only the reactant absorbs, gives the following data. Convert this data to [reactant] as a function of time.

A	1.420	1.139	0.904	0.724	0.584
$\frac{\text{time}}{\text{s}}$	0	100	200	300	400

3. Interpret the spectrum shown in Figure 2.15.

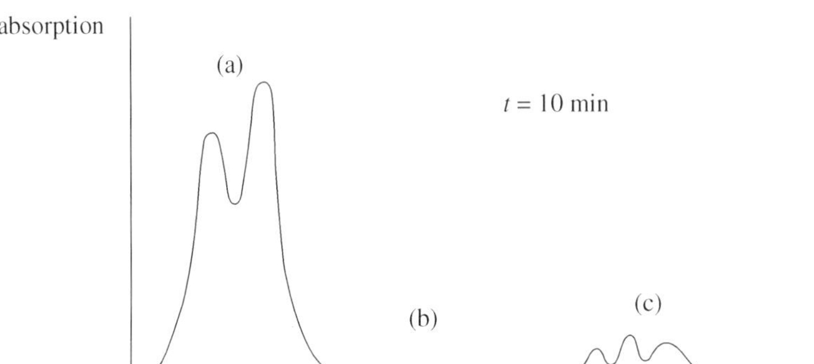
absorption
(a)
t = 10 min
(b)
(c)
frequency

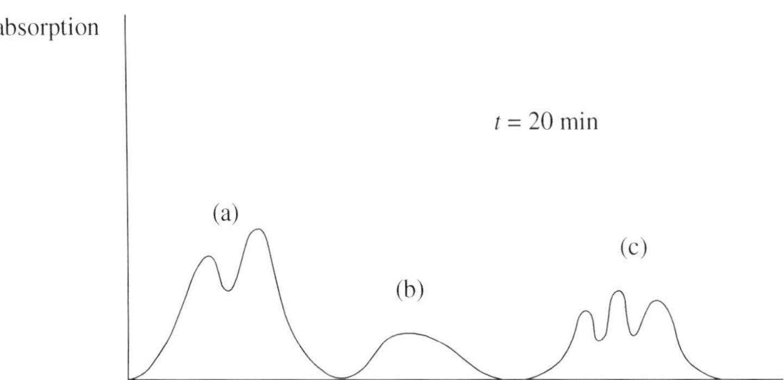
absorption
t = 20 min
(a)
(c)
(b)

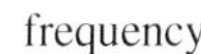
frequency

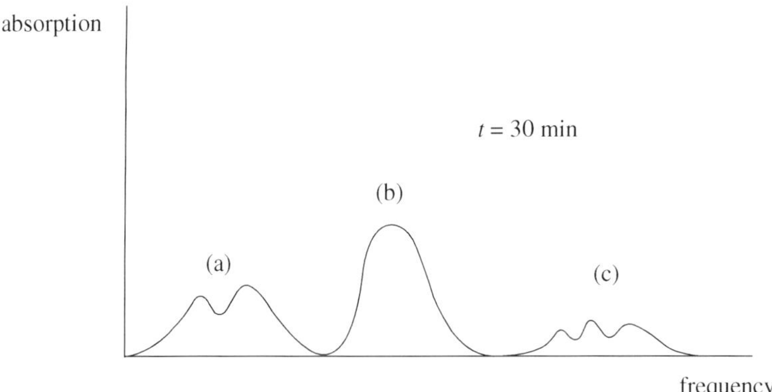
absorption
t = 30 min
(b)
(a)
(c)
frequency

Figure 2.15

4. The following is kinetic data. Plot graphs of [A] versus t, $\log_e$[A] versus t and $\frac{1}{[A]}$ versus t, and comment.

$\frac{\text{time}}{\text{s}}$	0	100	200	350	600	900
$\frac{[A]}{\text{mol dm}^{-3}}$	0.0200	0.0161	0.0137	0.0111	0.0083	0.0065

5. A kinetic experiment is carried out in a flow tube where the flow rate is 25 m s^{-1}. The detector is a spectrophotometer, and absorbances are measured at various distances along the tube. The reactant has a molar absorption coefficient, $\varepsilon =$ 1180 mol^{-1} dm^3 cm^{-1}. The absorbances are measured using a 1 cm optical cell. Use this information, and that in the following table, to calculate [reactant] as a function of time. Then plot a graph of [reactant] versus time and comment on its shape.

$\frac{\text{distance}}{\text{cm}}$	2.0	4.0	6.0	8.0	10.0
A	0.575	0.457	0.346	0.232	0.114

6. The conversion of ethylene oxide to ethane-1,2-diol in aqueous solution is catalysed by H_3O^+ and occurs with an decrease in volume.

$$\underset{\text{O}}{CH_2{-}CH_2}\text{(aq)} + H_2O\text{(l)} \longrightarrow CH_2OHCH_2OH\text{(aq)}$$

The reaction can be followed in a dilatometer, by watching the decrease in the height of the solution in the capillary tube.

In the following experiment, a 0.100 mol dm^{-3} solution of ethylene oxide was studied in the presence of a constant concentration of acid. The following data lists the heights in the dilatometer capillary as a function of time.

$\frac{\text{height}}{\text{cm}}$	27.72	27.08	26.45	25.88	25.34	23.55	22.83	22.20	18.30
$\frac{\text{time}}{\text{min}}$	0	20	40	60	80	160	200	240	∞

Using the heights at zero time and infinity, and the known initial and final concentrations of ethylene oxide, convert the above data into concentration/time data and draw the corresponding graph.

7. The base hydrolysis of acetylcholine can be conveniently followed by monitoring the drop in pH with time.

$$(CH_3)_3N^+CH_2CH_2OCOCH_3(aq) + OH^-(aq) \rightarrow (CH_3)_3N^+CH_2CH_2OH(aq) + CH_3COO^-(aq)$$

Use the following kinetic data to plot (a) a graph of pH versus time, (b) a graph of $[OH^-]$ versus time.

pH	12.81	12.50	12.40	12.30	12.20	12.10	12.00	11.90	11.80
pOH	1.19	1.50	1.60	1.70	1.80	1.90	2.00	2.10	2.20
$\frac{\text{time}}{\text{s}}$	0	35	45	57	70	80	90	100	114

8. The following dimerization has been studied by measuring the decrease in total pressure with time:

$$2A(g) \rightarrow A_2(g)$$

$\frac{p_{\text{total}}}{\text{mmHg}}$	632	591	533	484	453	432	416
$\frac{\text{time}}{\text{min}}$	0	10	30	60	90	120	150

Find the partial pressure of the monomer remaining as a function of time.

9. The base hydrolysis of ethyl ethanoate was studied at 25 °C in a conductance apparatus. Both reactants were at an initial concentration 5.00×10^{-3} mol dm^{-3}, and the conductivity, κ, was found at various times.

$\frac{\text{time}}{\text{min}}$	0	10	20	30	40	∞
$\frac{10^3\,\kappa}{\text{ohm}^{-1}\,\text{cm}^{-1}}$	1.235	1.041	0.925	0.842	0.787	0.443

Find $[OH^-]$ as a function of time.

3 The Kinetic Analysis of Experimental Data

In Chapter 2 methods of following a reaction with time were discussed, and techniques for probing changes in concentration of reactant remaining at various times were described. Practice was given at converting the experimental data into [reactant] left at times throughout the experiment. Graphs of [reactant] versus time were drawn, and all the kinetic quantities which are described in this book derive from such basic data.

The data from such experiments must now be manipulated to give quantitative information about each reaction.

Aims

By the end of this chapter you should be able to

- define the rate of reaction and discuss its dependence on reactant concentration,
- understand the meaning of order and rate constant and determine both from rate/concentration data,
- know the meaning of half-life,
- find the order and rate constant from concentration/time data using integrated rate expressions,
- understand and use the steady state treatment and
- appreciate the complexities involved in unravelling the kinetics of complex reactions.

An Introduction to Chemical Kinetics. Margaret Robson Wright
 ISBNs: 0-470-09058-8 (hbk) 0-470-09059-6 (pbk)

3.1 The Experimental Data

The first step in the analysis converts the experimental observations to *plots of [reactant] versus time*. Figures 3.1 and 3.2 illustrate some typical data.

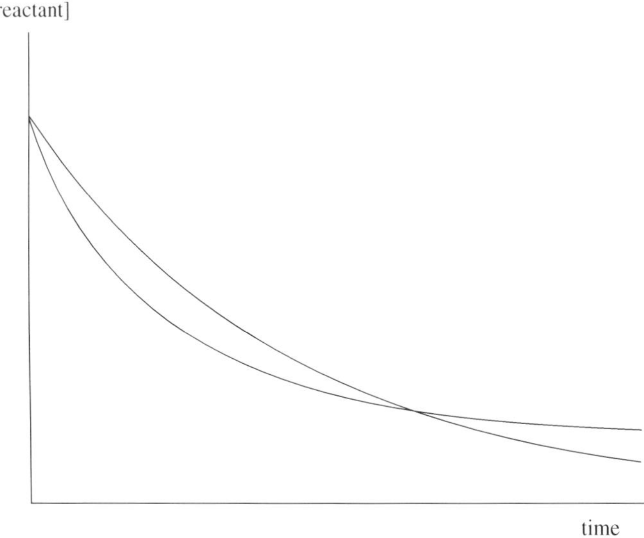

Figure 3.1 Graphs of [reactant] versus time showing curvature

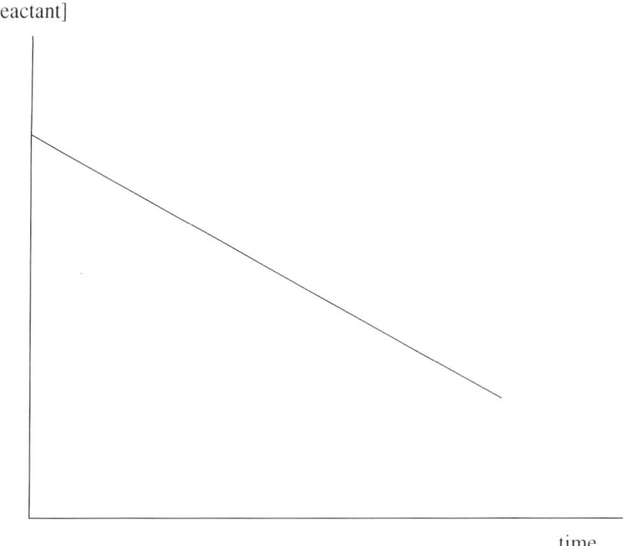

Figure 3.2 Graph of [reactant] versus time showing linearity

A simple qualitative description of how fast reaction occurs can be taken from a direct observation of how long it takes for a certain percentage reaction to occur. But in a quantitative analysis rate must be precisely defined, and once this has been done it becomes apparent how inadequate the loose definition of rate in terms of percentage reaction actually is.

The more precise meaning of the term *rate of reaction*, defined as how *fast* the [reactant] changes with time, can be illustrated on graphs of concentration versus time (see Figures 3.3(a), (b) and (c)).

The *average* rate over the time interval t_1 to t_2 when the concentration *decreases* from c_1 to c_2 is specified by

$$\frac{c_2 - c_1}{t_2 - t_1} = \text{gradient of line AB.}$$

However, this gives limited information. What is needed is the *actual rate* at a particular [reactant], called the '*instantaneous rate*'. This corresponds to situations where $c_2 - c_1 \rightarrow 0$, and $t_2 - t_1 \rightarrow 0$, and describes the gradient of a tangent to the curve at the particular concentration or time, e.g. the gradient of line EF, Figure 3.3(b), gives the instantaneous rate at c' and t'. The gradient of EF is *negative*, but

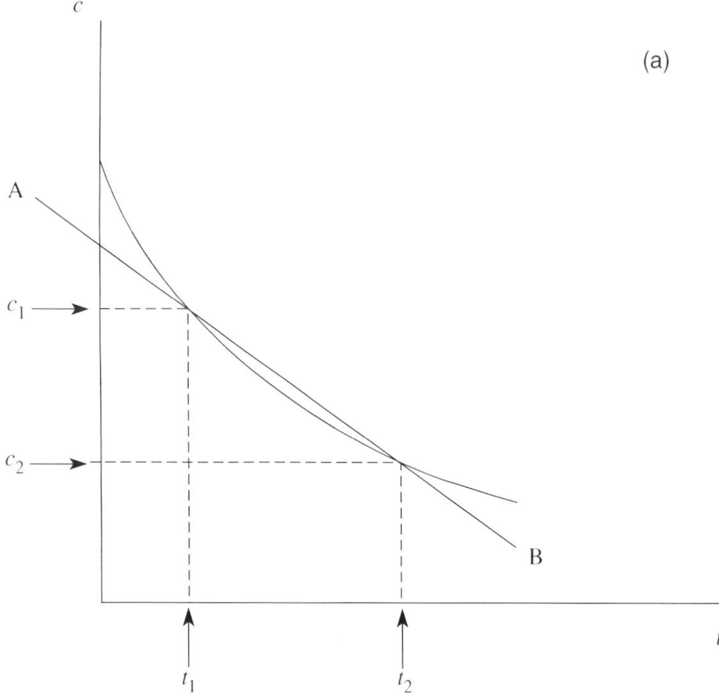

Figure 3.3 (a) Graph of concentration versus time, illustrating the meaning of average rate. (b) Graph of concentration versus time, illustrating the meaning of instantaneous rate. (c) Graph of concentration versus time, illustrating the meaning of initial rate

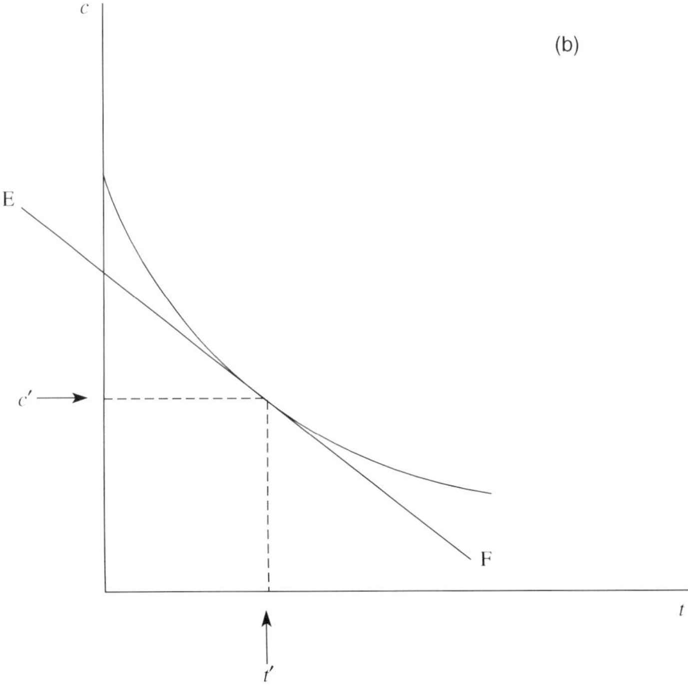

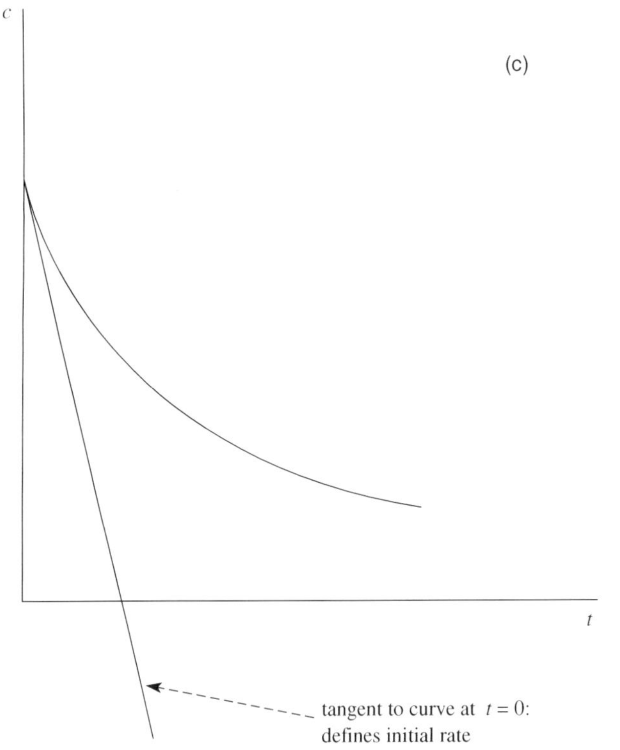

Figure 3.3 (*Continued*)

the *rate* of reaction is *defined* to be *positive*, and so the rate of reaction in terms of reactant = − the gradient. *It is important to include the minus sign.*

The *initial* rate is a very important quantity in kinetics, especially for complex reactions involving many steps where secondary reactions and products of reaction may affect the rate. The *initial rate* is the rate at the very start of reaction, Figure 3.3(c), and, with the exception of chain reactions, Section 6.9.1, is the line of *steepest gradient* giving the *maximum rate.*

The *units* of rate are given in terms of concentration per unit time or pressure per unit time, i.e.

$$\text{mol dm}^{-3}\ \text{time}^{-1},\ \text{or N m}^{-2}(\text{or atm, mmHg})\ \text{time}^{-1}.$$

3.2 Dependence of Rate on Concentration

In Figure 3.4, three tangents have been drawn. The magnitudes of the gradients lie in the order AB > CD > EF and these are at points (t_1,c_1), (t_2,c_2) and (t_3,c_3) respectively, i.e. as [reactant] decreases so does the rate.

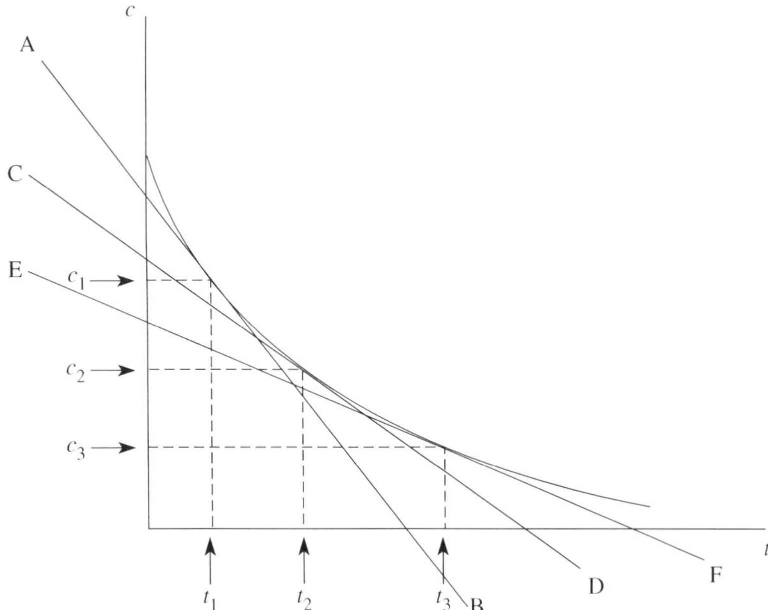

Figure 3.4 Graph of concentration versus time, illustrating the dependence of rate on [reactant]

The conclusion follows that the rate *depends* in some way on [*reactant*] *remaining*, i.e.

$$\text{rate} \propto [\text{reactant}]^n \tag{3.1}$$

where n is a number which states *exactly how* the rate depends on the [reactant]. It is called the *order*. From this,

$$\text{rate} = k\,[\text{reactant}]^n \tag{3.2}$$

where k is a constant of proportionality called the *rate constant*.

- If $n = 1$, the reaction is *first* order; if $n = 2$ the reaction is *second* order.
- If $n = 3/2$, the reaction is *three-halves* order; if $n = 0$, it is *zero* order.

If the reaction involves two substances A and B and

$$\text{rate} = k\,[\text{A}][\text{B}] \tag{3.3}$$

then the order is first with respect to A and first with respect to B, with the reaction being second order overall.

Likewise, if

$$\text{rate} = k\,[\text{P}]^2[\text{Q}] \tag{3.4}$$

then the reaction is second order in P, first order in Q and third order overall.

The equations

$$\text{rate} = k\,[\text{reactant}]^n \tag{3.5}$$

$$\text{rate} = k\,[\text{reactant}_1]^m[\text{reactant}_2]^n \tag{3.6}$$

are called the *rate expression*, or the *rate equation*. In general if the rate of reaction depends on the concentrations of several species, then

$$\text{rate} = k\,[\text{P}]^m[\text{Q}]^n\cdots \quad \text{and} \quad \text{order} = m + n + \cdots$$

and a kinetic analysis aims to find the value of m, n, ... and k at constant temperature for the reaction being studied.

3.3 Meaning of the Rate Expression

1. rate $= k$ [reactant] describes a *first order* reaction where

   - if the concentration increases by a factor of two, the rate also increases by a factor of two,

- if the concentration increases by a factor of six, the rate also increases by a factor of six.

2. rate $= k$ [reactant]^2 describes a *second order* reaction where

   - if the concentration increases by a factor of two, the rate increases by a factor of 2^2, i.e. four,
   - if the concentration increases by a factor of six, the rate increases by a factor of 6^2, i.e. 36.

3. rate $= k$[reactant]^0 = k describes a *zero order* reaction where

   - no matter how the concentration varies, the rate remains *constant*.

3.4 Units of the Rate Constant, *k*

These depend on the order of the reaction and are worked out as follows.

- First order reaction:

$$\text{rate} = k\,[\text{reactant}] \tag{3.7}$$

$$k = \text{rate}/[\text{reactant}] \tag{3.8}$$

Units of k are, therefore, $\dfrac{\text{mol dm}^{-3}\text{time}^{-1}}{\text{mol dm}^{-3}}$ i.e time^{-1}

- Second order reaction:

$$\text{rate} = k\,[\text{reactant}]^2 \tag{3.9}$$

$$k = \text{rate}/[\text{reactant}]^2 \tag{3.10}$$

Units of k are, therefore, $\dfrac{\text{mol dm}^{-3}\text{time}^{-1}}{\text{mol}^2\text{ dm}^{-6}}$ i.e. mol^{-1} dm^3 time^{-1}

- Zero order reaction:

$$\text{rate} = k \tag{3.11}$$

Units of k are, therefore, mol dm^{-3} time^{-1}.

Worked Problem 3.1

Question.

(a) The reaction between A + B is first order in A and second order in B. Give the rate expression, and then find the units of k (assume time in minutes).

(b) A reaction between P and Q is 3/2 order in P and order −1 in Q. Give the rate expression, and find the units of k (again assume minutes).

Answer.

(a) $\text{rate} = k[\text{A}][\text{B}]^2 \quad k = \dfrac{\text{rate}}{[\text{A}][\text{B}]^2} = \dfrac{\text{mol dm}^{-3}\,\text{min}^{-1}}{\text{mol dm}^{-3}\,\text{mol}^2\,\text{dm}^{-6}} = \text{mol}^{-2}\,\text{dm}^6\,\text{min}^{-1}$

(b) $\text{rate} = k\dfrac{[\text{P}]^{3/2}}{[\text{Q}]} \quad k = \dfrac{\text{rate}[\text{Q}]}{[\text{P}]^{3/2}} = \dfrac{\text{mol dm}^{-3}\text{min}^{-1}\,\text{mol dm}^{-3}}{\text{mol}^{3/2}\,\text{dm}^{-9/2}} = \text{mol}^{1/2}\,\text{dm}^{-3/2}\,\text{min}^{-1}$

3.5 The Significance of the Rate Constant as Opposed to the Rate

For all orders, except zero order, the rate of reaction varies with concentration and is useless as a means of either quantifying reactions or comparing them. In contrast, the rate constant is a constant at a given temperature and is independent of concentration. It can be used to quantify reactions and to compare them. Also, if the rate constant and the order are known, then it easy to calculate the rate for any concentration.

Worked Problem 3.2

Question.

(a) If $k = 5.7 \times 10^{-4}$ mol^{-1} dm^3 s^{-1} calculate the rate for a reaction which is first order in both A and B when [A] = 5.0×10^{-2} mol dm^{-3} and [B] = 2.0×10^{-2} mol dm^{-3}.

(b) If [A] had been 5.0×10^{-4} mol dm^{-3} and [B] had been 2.0×10^{-3} mol dm^{-3} what would the rate have been?

(c) What conclusion can be drawn from this?

Answer.

(a) Rate $= k[A][B]$

$= 5.7 \times 10^{-4}\,\text{mol}^{-1}\,\text{dm}^3\,\text{s}^{-1} \times 5.0 \times 10^{-2}\,\text{mol}\,\text{dm}^{-3} \times 2.0 \times 10^{-2}\,\text{mol}\,\text{dm}^{-3}$

$= 5.7 \times 10^{-7}\,\text{mol}\,\text{dm}^{-3}\,\text{s}^{-1}$

(b) Rate $= 5.7 \times 10^{-4}\,\text{mol}^{-1}\,\text{dm}^3\,\text{s}^{-1} \times 5.0 \times 10^{-4}\,\text{mol}\,\text{dm}^{-3} \times 2.0 \times 10^{-3}\,\text{mol}\,\text{dm}^{-3}$

$= 5.7 \times 10^{-10}\,\text{mol}\,\text{dm}^{-3}\,\text{s}^{-1}$

(c) In these two experiments the rate depends on the concentrations. The rate is not a characteristic of the reaction itself, but the rate constant is.

These two calculations show that the rate depends on the quite fortuitous concentration at which the rate is measured. To compare reactions using rates, all reactions would have to have the rate calculated at exactly the same concentrations. However, if the rate constants are tabulated for a given temperature, this is sufficient to define all reactions *absolutely*.

Worked Problem 3.3

Question. Base hydrolyses of amino-acid esters have two contributing reactions:

(a) OH^- reacting with the protonated ester, HE^+, and

(b) OH^- reacting with the unprotonated ester, E.

At 25 °C the rate constant for the protonated ester is 1550 $\text{mol}^{-1}\,\text{dm}^3\,\text{min}^{-1}$ and the rate constant for the unprotonated ester is 42 $\text{mol}^{-1}\,\text{dm}^3\,\text{min}^{-1}$. At pH = 9.30, $[OH^-] = 2.0 \times 10^{-5}\,\text{mol dm}^{-3}$, and if the total $[\text{ester}] = 2 \times 10^{-2}\,\text{mol dm}^{-3}$, then $[HE^+] = 5 \times 10^{-4}\,\text{mol dm}^{-3}$, and $[E] = 195 \times 10^{-4}\,\text{mol dm}^{-3}$.

(a) Calculate the contributions to the overall rate from the two reactions.

(b) What conclusions can be drawn?

Answer.

(a) Rate of protonated reaction

$= 1550\,\text{mol}^{-1}\,\text{dm}^3\,\text{min}^{-1} \times 5 \times 10^{-4}\,\text{mol dm}^{-3} \times 2.0 \times 10^{-5}\,\text{mol dm}^{-3}$

$= 1.55 \times 10^{-5}\,\text{mol dm}^{-3}\,\text{min}^{-1}$

Rate of unprotonated reaction

$= 42\ \text{mol}^{-1}\ \text{dm}^3\ \text{min}^{-1} \times 195 \times 10^{-4}\ \text{mol dm}^{-3} \times 2.0 \times 10^{-5}\ \text{mol dm}^{-3}$

$= 1.64 \times 10^{-5}\ \text{mol dm}^{-3}\ \text{min}^{-1}$

(b) Although $[HE^+] \ll [E]$ the contributions to the overall rate from the protonated and unprotonated forms of the ester are approximately equal. This is because $k_{HE^+} \gg k_E$.

This illustrates a general point in kinetics. It should *not be assumed* that, because a species is present in low concentrations, it does not contribute significantly to the rate.

3.6 Determining the Order and Rate Constant from Experimental Data

Sometimes it is easy to determine the order directly from the experimental data, though in general this does not happen.

Worked Problem 3.4

Question. Find the orders of the following reactions by *inspection*, then calculate the rate constants.

(a)

$\frac{10^2[\text{reactant}]}{\text{mol dm}^{-3}}$	1.0	2.0	4.0	12.0
$\frac{\text{rate}}{\text{mol dm}^{-3}\text{s}^{-1}}$	0.05	0.10	0.20	0.60

(b)

$\frac{10^3[\text{reactant}]}{\text{mol dm}^{-3}}$	4.0	8.0	16.0	32.0
$\frac{10^8\ \text{rate}}{\text{mol dm}^{-3}\text{s}^{-1}}$	3.0	12.0	48	192

(c)

$\frac{10^2[\text{reactant}]}{\text{mol dm}^{-3}}$	6.0	12.0	24.0	48.0
$\frac{10^4\ \text{rate}}{\text{mol dm}^{-3}\text{h}^{-1}}$	5.0	4.9	5.1	5.0

Answer.

(a) Determination of n:

- concentration up by a factor of two, rate up by a factor of two;
- concentration up by a factor of three, rate up by a factor of three.

Reaction is *first* order.

Determination of k:

rate $= k$[reactant] and $k =$ rate/[reactant]. *All* data should be used and the *average* taken, giving $k = 5.0\ \text{s}^{-1}$.

(b) Determination of n:

- concentration up by a factor of two, rate up by a factor of four, i.e. 2^2;
- concentration up by a factor of four, rate up by a factor of 16, i.e. 4^2.

Reaction is *second* order.

Determination of k:

rate $= k[\text{reactant}]^2$ and $k = \text{rate}/[\text{reactant}]^2$, giving an average value of $k = 1.88 \times 10^{-3}\ \text{mol}^{-1}\ \text{dm}^3\ \text{s}^{-1}$.

(c) Determination of n:

- altering the concentration, rate remains the same; reaction is *zero* order.

Determination of k:

$$\text{rate} = k[\text{reactant}]^0 = k \text{ and the average is } 5.0 \times 10^{-4}\ \text{mol dm}^{-3}\ \text{h}^{-1}$$

Pay particular attention to the units of k in these examples.

3.7 Systematic Ways of Finding the Order and Rate Constant from Rate/Concentration Data

Data are rarely as obvious as in the above examples, and the relation of rate to concentration often cannot be found by inspection; e.g., the orders of the following reactions are not obvious (Table 3.1).

Table 3.1 Rate/concentration data. (a) Reaction A; (b) Reaction B

(a)	$\dfrac{10^5\ \text{rate}}{\text{mol dm}^3\text{min}^{-1}}$	3.8	25.2	45	62.1
	$\dfrac{10^3\ \text{conc.}}{\text{mol dm}^{-3}}$	1.78	4.58	6.15	7.22
(b)	$\dfrac{10^3\ \text{rate}}{\text{mol dm}^3\text{min}^{-1}}$	2.0	4.2	7.8	17.5
	$\dfrac{10^2\ \text{conc.}}{\text{mol dm}^{-3}}$	1.0	1.6	2.5	4.4

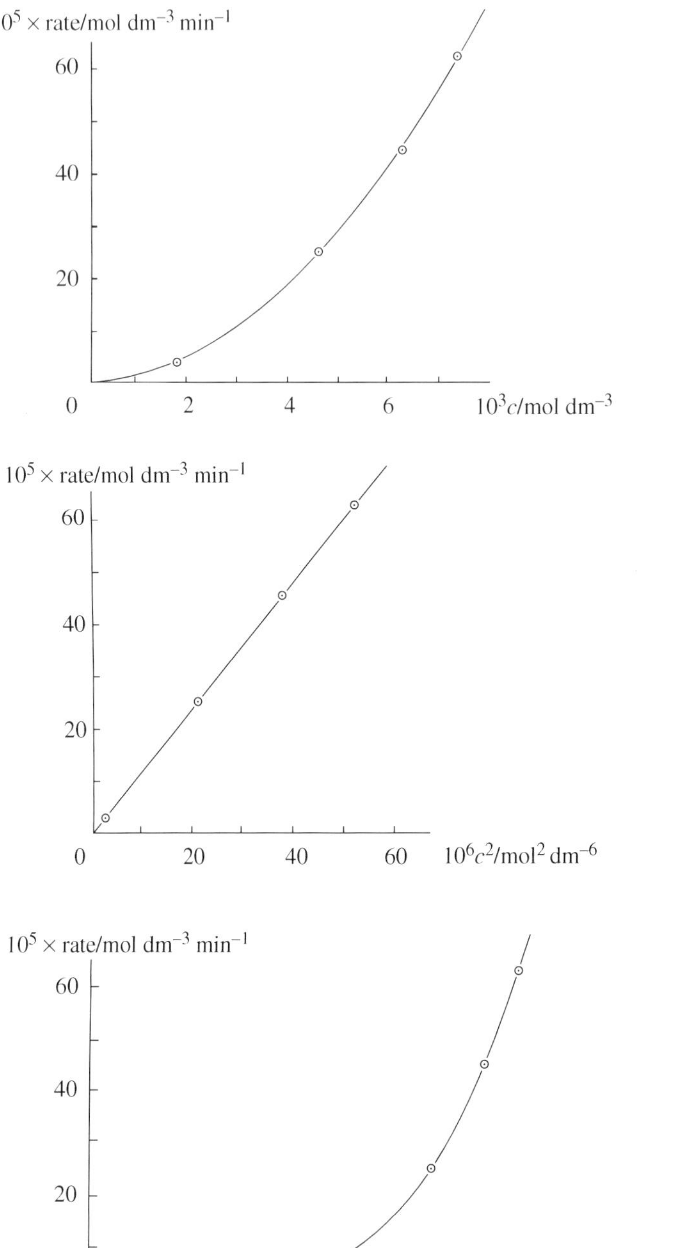

Figure 3.5 Determination of order: graphs of rate versus concentration, (concentration)^2 and (concentration)$^{1/2}$

3.7.1 A straightforward graphical method

In general, the aim is to find just how the rate depends on concentration, e.g. is it proportional to [reactant], $[\text{reactant}]^2$, $[\text{reactant}]^{1/2}$, $[\text{reactant}]^0$ etc.? If graphs of rate versus [reactant], $[\text{reactant}]^2$, $[\text{reactant}]^{1/2}$, $[\text{reactant}]^0$ etc. are drawn, then the one which is *linear* gives the order.

An interesting point. Why is it not necessary to draw the graph of rate versus $[\text{reactant}]^0$ to show that the reactions A and B in Table 3.1 are not zero order?

Answer. If the reaction were zero order then the rate would be independent of [reactant]. This is clearly not so.

Taking the first set of data for reaction A in Table 3.1, the graphs suggested above can be drawn. The graph of rate versus $[\text{reactant}]^2$ is the only one that is linear, and the reaction is second order (see Figure 3.5).

The value of k can be found from the graph. Since the rate $= k[\text{reactant}]^2$, a graph of rate versus $[\text{reactant}]^2$ is linear with slope $= k$, and intercept $=$ zero. The gradient of the line gives k with units $\text{mol}^{-1}\ \text{dm}^3\ \text{min}^{-1}$.

This method depends on *making guesses* as to which orders might fit the results, and hoping that the correct order is one of those chosen. In the above example only four guesses were made, and luckily they included the correct choice. If the order is complex this could be a 'hit or miss' procedure, but it illustrates the meaning of order clearly. However, if the four guesses made for the first set of data were tried with the second set of data, none would give a linear graph and other guesses would be needed.

3.7.2 log/log graphical procedures

These are completely systematic, and eliminate the necessity of making guesses as to possible orders. They give the order and rate constant direct from one graph.

$$\text{rate} = k\,[\text{reactant}]^n \tag{3.2}$$

Taking logarithms (to base 10 or base e) of both sides gives

$$\log \text{rate} = \log k + n \log[\text{reactant}] \tag{3.12}$$

and a plot of log rate versus log[reactant] should be linear with gradient $= n$, and intercept $= \log k$. Hence n and k can be found.

Worked Problem 3.5

Question. Use this method to find the order and rate constant for the second set of data, i.e. for reaction B in Table 3.1.

Answer.

$\log_e$ rate	−6.21	−5.47	−4.85	−4.05
$\log_e$ conc	−4.61	−4.14	−3.69	−3.12

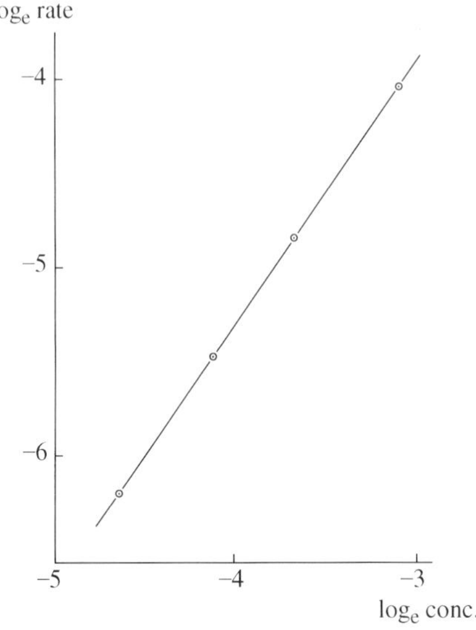

Figure 3.6 Determination of order: graph of $\log_e$ rate versus $\log_e$[reactant]

> Figure 3.6 shows that this reaction is 3/2 order (from the slope). The intercept is $\log_e k$. However, extrapolation is not sensible here, and $\log_e k$ and k must be calculated from $\log_e \text{rate} = \log_e k + n \log_e[\text{reactant}]$, possible since n is now known; $n = 3/2$, giving $k = 1.98\ \text{mol}^{-1/2}\ \text{dm}^{3/2}\ \text{min}^{-1}$. Note that in this case $\log_e$ has been used; it does not matter which base is used.

3.7.3 A systematic numerical procedure

Since rate $= k\,[\text{reactant}]^n$ a series of values of the rates at various [reactant] can give the order and rate constant as follows:

$$\frac{\text{rate in expt 2}}{\text{rate in expt 1}} = \frac{k}{k}\left(\frac{\text{conc in expt 2}}{\text{conc in expt 1}}\right)^n \qquad (3.13)$$

The k in the numerator and denominator cancel and each ratio can be evaluated, from which n can be found as

$$\text{ratio of rates} = \{\text{ratio of concs}\}^n \qquad (3.14)$$

If the value of n is not clear from a direct comparison of the ratios then a log/log plot will give a systematic procedure:

$$\log\{\text{ratio of rates}\} = n \ \log\{\text{ratio of concs}\} \tag{3.15}$$

Worked Problem 3.6

Question. Use the following data to find n and k in this way, and compare this method with methods 3.7.1 and 3.7.2.

$\dfrac{10^3 \text{ conc.}}{\text{mol dm}^{-3}}$	1.25	4.76	8.05
$\dfrac{10^6 \text{ rate}}{\text{mol dm}^{-3} \text{ min}^{-1}}$	5.15	72.5	200.8
experiment	1	2	3

Answer.

(a) Rate in expt 2/rate in expt 1 = (conc. in expt 2/conc. in expt 1)^n

$$\frac{7.25 \times 10^{-5}}{5.15 \times 10^{-6}} = \left(\frac{4.76 \times 10^{-3}}{1.25 \times 10^{-3}}\right)^n$$

$14.1 = (3.8)^n$, giving $n = 2$, or $\log_{10} 14.1 = n \log_{10} 3.8$, hence $n = 2$

(b) Rate in expt 3/rate in expt 1 gives

$$\frac{200.8 \times 10^{-6}}{5.15 \times 10^{-6}} = \left(\frac{8.05 \times 10^{-3}}{1.25 \times 10^{-3}}\right)^n$$

$$39.0 = (6.44)^n, \text{ giving } n = 2.$$

Likewise for experiments 3 and 2

$$\frac{200.8 \times 10^{-6}}{72.5 \times 10^{-6}} = \left(\frac{8.05 \times 10^{-3}}{4.76 \times 10^{-3}}\right)^n$$

$$2.77 = (1.69)^n, \text{ giving } n = 2.$$

Therefore the reaction is second order.

$$\text{rate} = k[\text{reactant}]^n \text{ from which } k = \text{rate}/[\text{reactant}]^2.$$

Experiment 1	Experiment 2	Experiment 3
$k = \dfrac{5.15 \times 10^{-6}}{(1.25 \times 10^{-3})^2}$ mol^{-1} dm^3 min^{-1} $= 3.3$ mol^{-1} dm^3 min^{-1}	$k = \dfrac{72.5 \times 10^{-6}}{(4.76 \times 10^{-3})^2}$ mol^{-1} dm^3 min^{-1} $= 3.2$ mol^{-1} dm^{-3} min^{-1}.	$k = \dfrac{200.8 \times 10^{-6}}{(8.05 \times 10^{-3})^2}$ mol^{-1} dm^3 min^{-1} $= 3.1$ mol^{-1} dm^3 min^{-1}

The average $k = 3.2$ mol^{-1} dm^3 min^{-1}.

The most straightforward method is the log/log procedure.

3.8 Drawbacks of the Rate/Concentration Methods of Analysis

All the methods so far described require finding the rate, and this requires drawing *tangents* to the graph of [reactant] versus time. It is difficult and tedious to draw *accurate* tangents, though with computer analysis both accuracy and speed of manipulation improve dramatically. Nonetheless, using rate/concentration data, especially initial rates, is often the best way to analyse complex kinetics where the rate may be complex and secondary reactions affect the kinetics. But for routine determination of the rate constant, once the order has been established, integrated rate equations are excellent.

3.9 Integrated Rate Expressions

These methods use calculus, and make use of the fact that a derivative at a point corresponds to the gradient of the tangent at that point (see Figure 3.7).

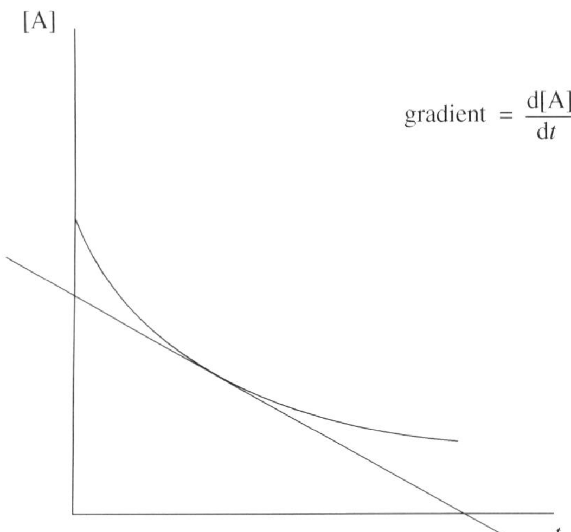

Figure 3.7 Definition of a derivative as a gradient of a tangent

The rate of reaction is given as the *negative* of the gradient of the tangent to the curve of [reactant] versus time, and the gradient *defines* the derivative: d[reactant]/dt. Using this, the rate expressions derived earlier can be formulated as *differential equations*, i.e. equations having a derivative present. Using [A] as a shorthand notation for [reactant] gives Table 3.2.

Table 3.2 Differential rate equations

First order:	rate $= k[A]$	or $-d[A]/dt = k[A]$
Second order:	rate $= k[A]^2$	or $-d[A]/dt = k[A]^2$
Zero order:	rate $= k[A]^0 = k$	or $-d[A]/dt = k\ [A]^0 = k$
3/2 order:	rate $= k[A]^{3/2}$	or $-d[A]/dt = k[A]^{3/2}$

Differential equations can be solved by integration, and this results in *linear* equations relating [reactant] and time, so that the order and rate constant can be found *directly* from *linear* plots.

In scientific work it is always much easier and more accurate to work with linear relationships. These have an easily defined gradient, and the intercept is accurately determinable. Also, it is much simpler to discriminate between systematic scatter and random scatter using linear plots rather than curves.

Integrated rate expressions can also be used to demonstrate that the *half-life* of a reaction varies systematically with [reactant], and so is *diagnostic* of the reaction *order.*

3.9.1 Half-lives

The half-life is the time taken for a given concentration to decrease to half of its value; e.g, if $[M] = 12 \times 10^{-3}$ mol dm^{-3} at the start of the reaction, then the first half-life is the time taken for [M] to drop to 6×10^{-3} mol dm^{-3}; the second half-life is the time taken for [M] to drop from 6×10^{-3} mol dm^{-3} to 3×10^{-3} mol dm^{-3}; the third half-life is the time taken for [M] to drop from 3×10^{-3} mol dm^{-3} to 1.5×10^{-3} mol dm^{-3}; and so on.

Worked Problem 3.7

Question. Draw graphs of [reactant] versus time for the following reactions. Find as many half-lives as possible, and state the dependence on concentration for each reaction.

Reaction A

$\frac{10^2 c}{\text{mol dm}^{-3}}$	1.000	0.840	0.724	0.506	0.390	0.335	0.225	0.180	0.145
$\frac{\text{time}}{\text{min}}$	0	20	40	100	150	200	350	450	600

Reaction B

$\frac{10^2 c}{\text{mol dm}^{-3}}$	1.000	0.835	0.580	0.375	0.172
$\frac{\text{time}}{\text{min}}$	0	40	100	150	200

Reaction C

$\frac{10^2 c}{\text{mol dm}^{-3}}$	1.000	0.810	0.620	0.480	0.370	0.285	0.225	0.150	0.060
$\frac{\text{time}}{\text{min}}$	0	50	100	150	200	250	300	400	600

Answer.

Reaction A. Successive $t_{1/2}$ occur at 100 min and 300 min, with the third being unobtainable from the graph. The successive $t_{1/2}$ values are 100 min, 200 min and ≫200 min, i.e. half-lives increase as the concentration decreases (see Figure 3.8(a)).

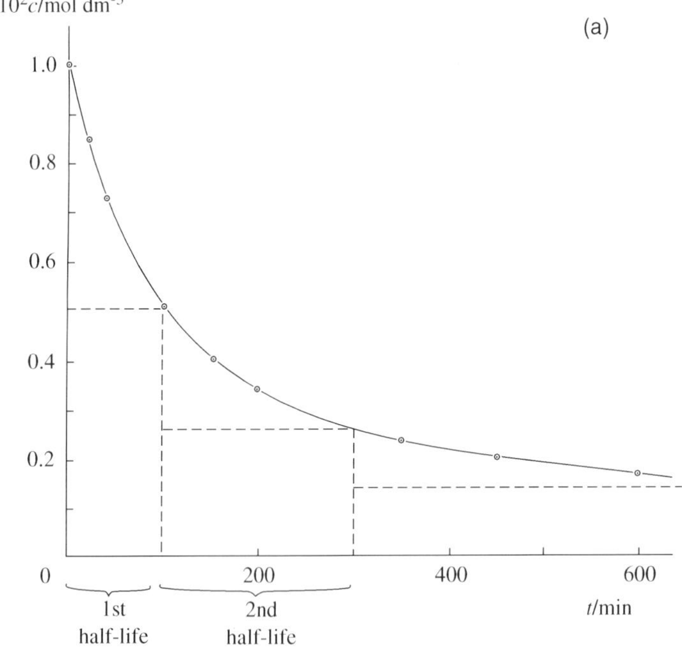

Figure 3.8 Graphs of concentration versus time showing dependence of half-life on concentration

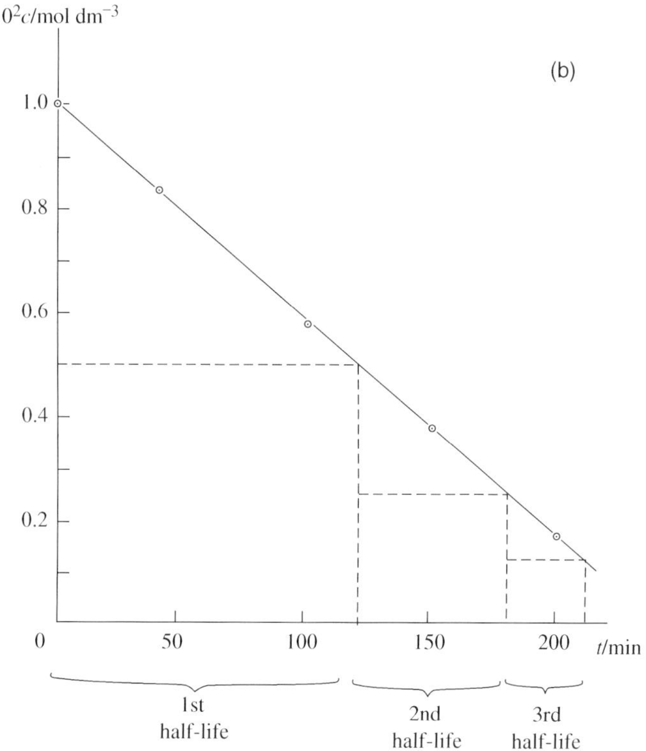

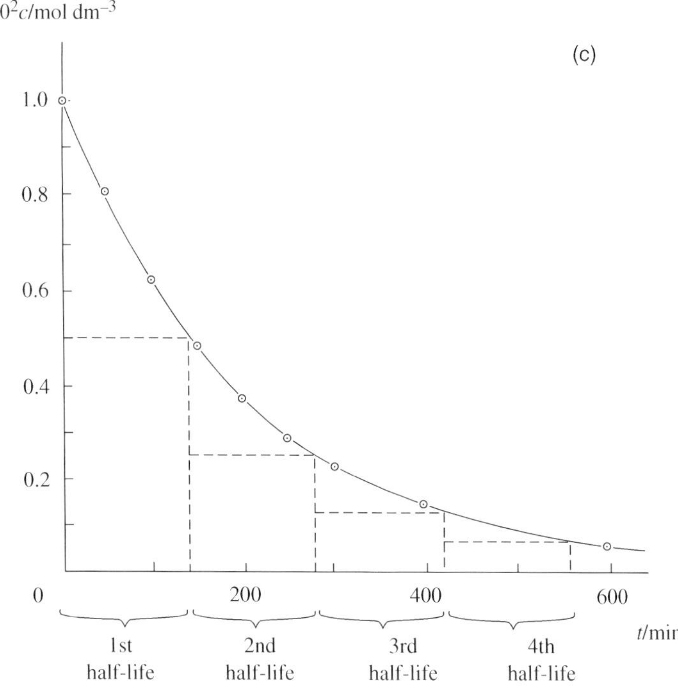

Figure 3.8 (*Continued*)

Reaction B. Successive $t_{1/2}$ occur at 120, 180 and 212 min. The successive $t_{1/2}$ values are 120, 60 and 32 min, i.e. half-lives decrease as the concentration decreases (see Figure 3.8(b)).

Reaction C. Successive $t_{1/2}$ occur at 140, 280, 420 and 560 min. The successive $t_{1/2}$ values are 140, 140, 140 and 140 min., i.e. the half-life is independent of concentration (see Figure 3.8(c)).

A very important point to note. Looking very carefully at the graphs for *reactions A* and *C* shows that although both are curves, they have different shapes. This is reflected in the behaviour of the half-lives and the equations describing the precise shapes of the graphs (Sections 3.10, 3.11 and 3.12).

3.10 First Order Reactions

$$\text{Rate of reaction} = k[\mathrm{A}] \tag{3.16}$$

$$-\frac{\mathrm{d}[\mathrm{A}]}{\mathrm{d}t} = k[\mathrm{A}] \tag{3.17}$$

This equation is put into a form so that a *standard integral* results: take all terms in one variable, i.e. [A], to one side and all terms in the other variable, i.e. t, to the other side.

$$-\frac{\mathrm{d}[\mathrm{A}]}{[\mathrm{A}]} = k\,\mathrm{d}t, \text{ or moving the negative sign } \frac{\mathrm{d}[\mathrm{A}]}{[\mathrm{A}]} = -k\,\mathrm{d}t \tag{3.18}$$

These can be integrated. Because k is a constant it can be taken outside the integral sign:

$$\int \frac{\mathrm{d}[\mathrm{A}]}{[\mathrm{A}]} = -k \int \mathrm{d}t \tag{3.19}$$

In kinetics the change in concentration with time is followed from the start of the reaction, $[\mathrm{A}]_0$ at $t = 0$ to $[\mathrm{A}]_t$ at time t. These are the *limits* between which the integral is taken.

$$\int_{[\mathrm{A}]_0}^{[\mathrm{A}]_t} \frac{\mathrm{d}[\mathrm{A}]}{[\mathrm{A}]} = -k \int_0^t \mathrm{d}t \tag{3.20}$$

These are both standard integrals: $$\int_{x_0}^{x_t} \frac{\mathrm{d}x}{x} = \log_e x_t - \log_e x_0 \tag{3.21}$$

$$\int_0^x \mathrm{d}x = x - 0 \tag{3.22}$$

$$\therefore \log_e[\mathrm{A}]_t - \log_e[\mathrm{A}]_0 = -kt \quad \text{or} \quad \log_e[\mathrm{A}]_t = \log_e[\mathrm{A}]_0 - kt \tag{3.23}$$

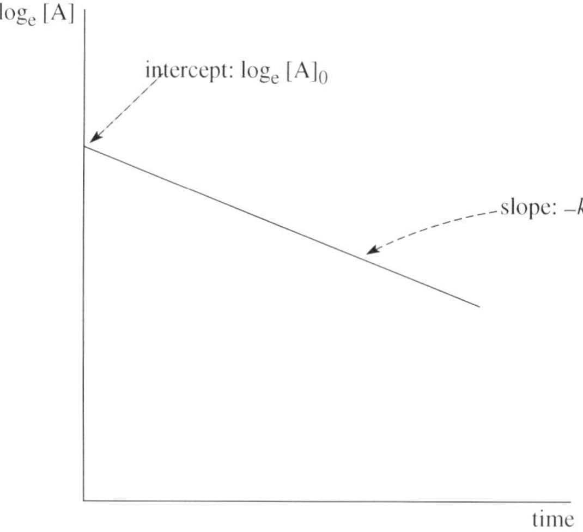

Figure 3.9 First order plot: log$_e$[A] versus time

This is the equation of a *straight* line: a plot of $\log_e[A]_t$ versus t should have slope $-k$ and intercept $\log_e[A]_0$ (see Figure 3.9).

If the experimental data *fit* first order kinetics, then a plot of $\log_e[A]_t$ versus t *should be linear*. If the plot is curved the data do not fit first order kinetics. The slope is negative and equals $-k$, hence k is found.

Equation (3.23) is also the equation of the experimental curve of [reactant] versus time. This can be confirmed by calculating values of $\log_e[A]_t$ and thence $[A]_t$ from equation (3.23), and plotting $[A]_t$ versus t. This should regenerate the experimental curve.

3.10.1 The half-life for a first order reaction

At the first half-life, $t_{1/2}$,

$$[A]_t = \tfrac{1}{2}[A]_0 \qquad (3.24)$$

$$\therefore\ kt_{1/2} = \log_e[A]_0 - \log_e\left(\tfrac{1}{2}[A]_0\right) = \log_e 2 \qquad (3.25)$$

$$t_{1/2} = \frac{\log_e 2}{k} \qquad (3.26)$$

The half-life is *independent* of the concentration, and takes the same value no matter what the extent of reaction.

Worked Problem 3.8

Question. Show that the decomposition of N_2O follows first order kinetics, and find the rate constant and half-life. Why cannot the methods of Section 3.7 be used here?

$\frac{10^3[N_2O]}{\text{mol dm}^{-3}}$	100	61	37	10
time/s	0	50	100	230

Answer. We need to find $\log_e[N_2O]$ to plot $\log_e[N_2O]_t$ versus t.

$\log_e[N_2O]_t$	−2.30	−2.80	−3.30	−4.61

The graph of $\log_e[N_2O]_t$ versus t is linear $\therefore$ reaction is first order.
Slope $= -1.00 \times 10^{-2}\ s^{-1}$, hence $k = 1.00 \times 10^{-2}\ s^{-1}$.

$$t_{1/2} = \frac{\log_e 2}{1.00 \times 10^{-2}\ s^{-1}} = 69\ s.$$

This value agrees with the value read directly from the experimental graph.
The data is concentration/time data and is *immediately adaptable* to integrated rate methods. To use the methods of Section 3.7 the data has to be rate/concentration data, which is not the manner in which it is presented in this problem.

3.10.2 An extra point about first order reactions

The half-life of a first order reaction remains constant throughout reaction and is independent of concentration. This applies to any fractional lifetime, though the half-life is the one most commonly used. The *relaxation time* is the other common fractional lifetime. The relaxation time is relevant *only* to *first order* reactions, and is a *fractional* lifetime which bears a very simple relation to the rate constant as a direct consequence of the exponential behaviour of first order reactions.

The first order integrated rate equation

$$\log_e[A]_t = \log_e[A]_0 - kt \tag{3.23}$$

can also be written in exponential form as

$$[A]_t = [A]_0 \exp(-kt) \tag{3.27}$$

Fractional lifetimes can be formulated as in Table 3.3.

Table 3.3 Fractional lives

Fractional life	Half-life: $t_{1/2}$	Quarter-life: $t_{1/4}$	Relaxation time: τ
$[A]_t = [A]_0$	by definition	by definition	by definition
$\exp(-kt)$	$[A]_t = (1/2)[A]_0$	$[A]_t = (1/4)[A]_0$	$[A]_t = (1/e)[A]_0$
integrated	$(1/2)[A]_0 = [A]_0\,e^{-kt_{1/2}}$	$(1/4)[A]_0 = [A]_0\,e^{-kt_{1/4}}$	$(1/e)[A]_0 = [A]_0 e^{-k\tau}$
rate equation	$1/2 = e^{-kt_{1/2}}$ *or**	$1/4 = e^{-kt_{1/4}}$	$e^{-1} = e^{-k\tau}$
	$2 = e^{+kt_{1/2}}$	*or** $4 = e^{+kt_{1/4}}$	*or** $e = e^{+k\tau}$
	$\log_e 2 = kt_{1/2}$	$\log_e 4 = kt_{1/4}$	†$\log_e e = k\tau$
	$t_{1/2} = \frac{\log_e 2}{k}$	$t_{1/4} = \frac{\log_e 4}{k}$	$\tau = \frac{1}{k}$

*Note the change of sign in the exponential.
†Note $\log_e e = 1$.

Figure 3.10 shows the data of Table 3.3 graphically. The relaxation time is easily read off from the graph, being the time at which $[A]_t = (1/e)[A]_0 = 0.3679[A]_0$ (see Table 3.3). At the relaxation time the reaction has gone to 63.2 per cent completion and still has 36.8 per cent to go.

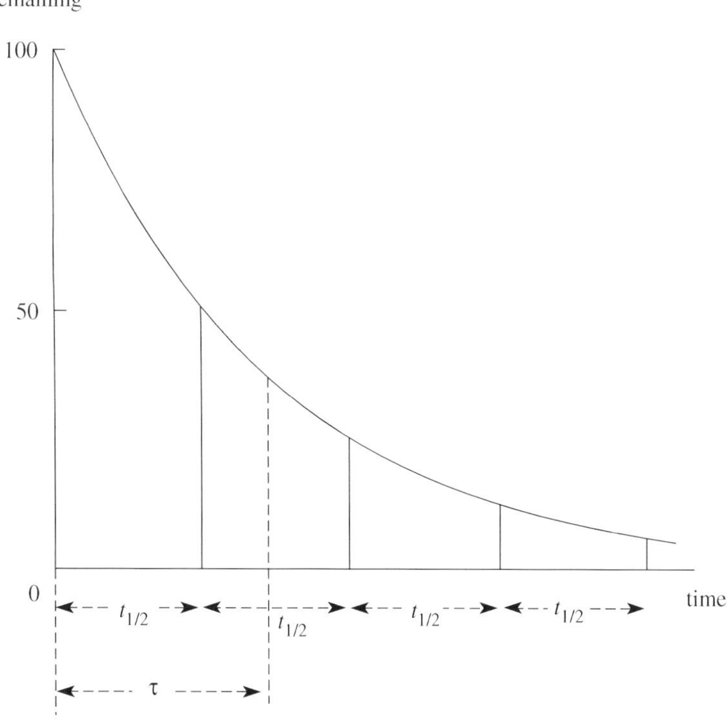

Figure 3.10 Half-lives for a first order reaction

Relaxation times become important when studying the approach to equilibrium either directly or as a result of a small perturbation from equilibrium. They also emerge directly from ultrasonic relaxation techniques for studying single and multiple equilibria (Section 2.6). However, *they only apply to first order or pseudo- first order processes* (Section 3.15).

3.11 Second Order Reactions

$$\text{Rate of reaction} = k[\mathrm{A}]^2 \tag{3.28}$$

$$-\frac{\mathrm{d}[\mathrm{A}]}{\mathrm{d}t} = k[\mathrm{A}]^2 \tag{3.29}$$

$$\text{Proceeding as before: } -\frac{\mathrm{d}[\mathrm{A}]}{[\mathrm{A}]^2} = k\,\mathrm{d}t \quad \text{or} \quad \frac{\mathrm{d}[\mathrm{A}]}{[\mathrm{A}]^2} = -k\,\mathrm{d}t \tag{3.30}$$

Again integrating between the limits $[\mathrm{A}]_0$ and $[\mathrm{A}]_t$, and $t = 0$ and t, gives

$$\int_{[\mathrm{A}]_0}^{[\mathrm{A}]_t} \frac{\mathrm{d}[\mathrm{A}]}{[\mathrm{A}]^2} = -k\int_0^t \mathrm{d}t \tag{3.31}$$

$$\text{These are again standard integrals: } \int_{x_0}^{x_t} \frac{\mathrm{d}x}{x^2} = -\frac{1}{x_t} + \frac{1}{x_0} \tag{3.32}$$

$$\int_0^x \mathrm{d}x = x - 0 \tag{3.22}$$

$$\therefore\ -\frac{1}{[\mathrm{A}]_t} + \frac{1}{[\mathrm{A}]_0} = -kt \quad \text{i.e.} \quad \frac{1}{[\mathrm{A}]_t} = \frac{1}{[\mathrm{A}]_0} + kt \tag{3.33}$$

This is the equation of a *straight* line: a plot of $1/[\mathrm{A}]_t$ versus t should be *linear* with slope equal to $+k$ and intercept equal to $1/[\mathrm{A}]_0$ (Figure 3.11).

If the experimental data *fit* second order kinetics, then a plot of $1/[\mathrm{A}]_t$ versus t *should be linear.* If the plot is curved, the data do not fit second order kinetics.

The slope is positive and equals $+k$, hence k is found.

Equation (3.33) is also the equation of the experimental curve of [reactant] versus time. This can be confirmed by calculating values of $1/[\mathrm{A}]_t$ and thence $[\mathrm{A}]_t$ from equation (3.33), and plotting $[\mathrm{A}]_t$ versus t, which should regenerate the experimental curve.

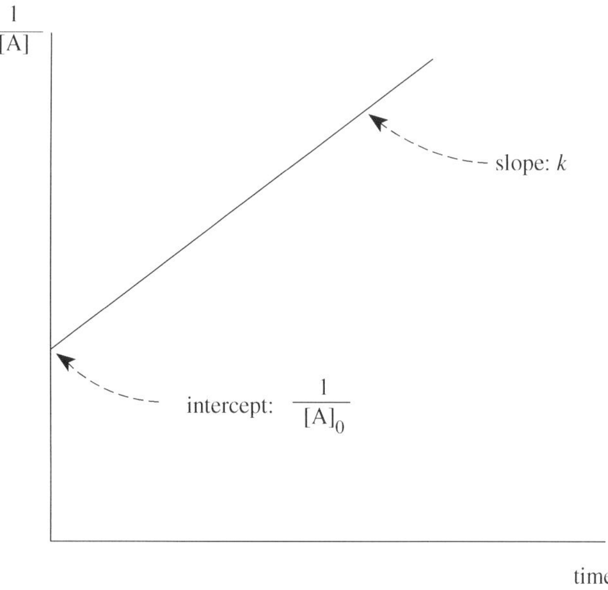

Figure 3.11 Second order plot: 1/[A] versus time

3.11.1 The half-life for a second order reaction

At the first half-life, $t_{1/2}$, $[A]_t = \frac{1}{2}[A]_0$ (3.24)

$$\therefore\ kt_{1/2} = \frac{1}{(1/2)[A]_0} - \frac{1}{[A]_0} = \frac{1}{[A]_0} \quad (3.34)$$

$$t_{1/2} = \frac{1}{k[A]_0} \quad (3.35)$$

The half-life *is proportional to* the reciprocal of the concentration, i.e. concentration^{-1}, and so the larger the concentration the smaller is the half-life. *The half-life increases as reaction proceeds.*

Worked Problem 3.9

Question. Show that the following reaction follows second order kinetics, and find the rate constant and first half-life.

$\dfrac{10^3[A]}{\text{mol dm}^{-3}}$	7.36	3.68	2.45	1.47	1.23
time/s	0	60	120	240	300

Answer. We need to find $1/[A]_t$ to plot $1/[A]$ versus t

$\frac{1/[A]}{mol^{-1}\ dm^3}$	136	272	408	680	813

The graph of $1/[A]$ versus t is linear ∴ reaction is second order.
The slope $= +2.26\ mol^{-1}\ dm^3\ s^{-1}$ and $k = +2.26\ mol^{-1}\ dm^3\ s^{-1}$.

$t_{1/2} = 1/(k[A]_0) = 1/(2.26\ mol^{-1}\ dm^3\ s^{-1} \times 7.36 \times 10^{-3}\ mol\ dm^{-3}) = 60.1\ s$

3.11.2 An extra point about second order reactions

The analyses given above apply to reactions of *one* molecular species, i.e $A \rightarrow$ products, or $A + A \rightarrow$ products. But what about reactions such as $A + B \rightarrow$ products, where *both* [A] and [B] decrease steadily with time and the rate is proportional to the concentrations of *both* A and B?

$$\text{Rate} = k[A][B] \tag{3.3}$$

The second order integrated rate equation quoted earlier only applies to this case *if* [A] = [B]. If unequal concentrations of A and B are used, the integrated rate equation becomes

$$\frac{1}{[A]_0 - [B]_0}\log_e\frac{[A]_t}{[B]_t} = \frac{1}{[A]_0 - [B]_0}\log_e\frac{[A]_0}{[B]_0} + kt \tag{3.36}$$

which is cumbersome to handle. For this reason it is much better to handle such second order reactions as *pseudo*-first order reactions (Section 3.15).

3.12 Zero Order Reaction

$$\text{Rate of reaction} = k[A]^0 = k \tag{3.37}$$

$$-\frac{d[A]}{dt} = k[A]^0 = k \tag{3.38}$$

Proceeding as before:

$$d[A] = -k\,dt \tag{3.39}$$

Integrating between the limits $[A]_0$ and $[A]_t$, and $t = 0$ and t, gives

$$\int_{[A]_0}^{[A]_t} d[A] = -k \int_0^t dt \tag{3.40}$$

Both of these are the same standard integral: $\int_0^x dx = x - 0$ (3.22)

$$\therefore \ [A]_t - [A]_0 = -kt \quad \text{i.e.} \quad [A]_t = [A]_0 - kt \tag{3.41}$$

This is the equation of a *straight* line: a plot of $[A]_t$ versus t should be *linear* with slope equal to $-k$ and intercept equal to $[A]_0$ (Figure 3.12).

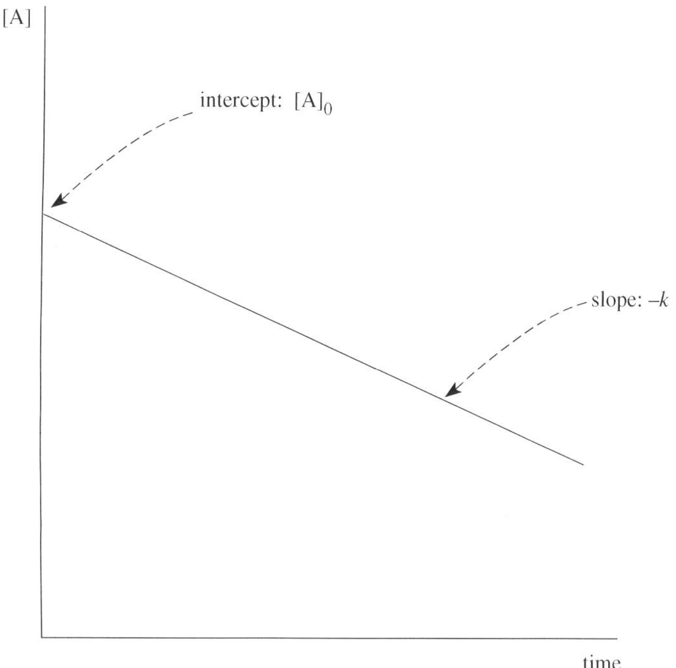

Figure 3.12 Zero order plot: [A] versus time

If the experimental data *fit* zero order kinetics, then a plot of $[A]_t$ versus t *should be linear.* If the plot is curved the data do not fit zero order kinetics.

The slope is negative and equals $-k$, hence k is found.

But note: the graph for the integrated rate expression *is the experimental graph* of [reactant] versus time. This only happens for a zero order reaction, and it is immediately obvious from the experimental plot that reaction is zero order.

3.12.1 The half-life for a zero order reaction

At the first half-life, $t_{1/2}$,

$$[A]_t = \tfrac{1}{2}[A]_0 \quad (3.24)$$

$$\therefore \; kt_{1/2} = [A]_0 - \tfrac{1}{2}[A]_0 = [A]_0/2 \quad (3.42)$$

$$t_{1/2} = [A]_0/2k \quad (3.43)$$

The half-life *is proportional* to the concentration, and so the larger the concentration the greater is the half-life. The *half-life decreases as the reaction proceeds.*

Worked Problem 3.10

Question. Show that the following reaction obeys zero order kinetics, and find the rate constant and first half-life.

$\frac{10^2 \times [A]}{\text{mol dm}^{-3}}$	13.80	11.05	8.43	5.69	3.00
$\frac{\text{time}}{\text{h}}$	0	10	20	30	40

Answer. We need a graph of [A] versus t. This graph is linear, $\therefore$ reaction is zero order with slope $= -2.7 \times 10^{-3}$ mol dm^{-3} h^{-1} and $k = 2.7 \times 10^{-3}$ mol dm^{-3} h^{-1}.

$t_{1/2} = [A]_0/2k = 13.80 \times 10^{-2}$ mol dm$^{-3}/(2 \times 2.7 \times 10^{-3}$ mol dm^{-3} h$^{-1}) = 25.5$ h.

The following problem illustrates the use which can be made of half-lives.

Worked Problem 3.11

Question. Look at the answer to problem 3.7 and infer the order of each reaction.

Answer.

Reaction A: half-life increases as concentration decreases, second order.

Reaction B: half-life decreases as concentration decreases, zero order.

Reaction C: half-life remains constant, first order.

3.13 Integrated Rate Expressions for Other Orders

These are summarized in Table 3.4.

Table 3.4 Integrated rate equations

Order	Differential equation	Integrated equation	Graph to be drawn	Half-life expression
0	$\frac{-d[A]}{dt} = k[A]^0 = k$	$[A]_t = [A]_0 - kt$	$[A]_t$ vs t slope $= -k$	$t_{1/2} = \frac{[A]_0}{2k}$
1	$\frac{-d[A]}{dt} = k[A]$	$\log_e[A]_t = \log_e[A]_0 - kt$	$\log_e[A]_t$ vs t slope $= -k$	$t_{1/2} = \frac{\log_e 2}{k}$
2	$\frac{-d[A]}{dt} = k[A]^2$	$\frac{1}{[A]_t} = \frac{1}{[A]_0} + kt$	$\frac{1}{[A]_t}$ vs t slope $= +k$	$t_{1/2} = \frac{1}{k[A]_0}$
3	$\frac{-d[A]}{dt} = k[A]^3$	$\frac{1}{[A]_t^2} = \frac{1}{[A]_0^2} + 2kt$	$\frac{1}{[A]_t^2}$ vs t slope $= +2k$	$t_{1/2} = \frac{3}{2k[A]_0^2}$
1/2	$\frac{-d[A]}{dt} = k[A]^{1/2}$	$[A]_t^{1/2} = [A]_0^{1/2} - \frac{1}{2}kt$	$[A]_t^{1/2}$ vs t slope $= -\frac{1}{2}k$	$t_{1/2} = \frac{(2-\sqrt{2})[A]_0^{1/2}}{k}$
3/2	$\frac{-d[A]}{dt} = k[A]^{3/2}$	$\frac{1}{[A]_t^{1/2}} = \frac{1}{[A]_0^{1/2}} + \frac{1}{2}kt$	$\frac{1}{[A]_t^{1/2}}$ vs t slope $= +\frac{1}{2}k$	$t_{1/2} = \frac{2(\sqrt{2}-1)}{k[A]_0^{1/2}}$

3.14 Main Features of Integrated Rate Equations

- Confirms a suspected order if the relevant plot is linear;
- a quick, easy and accurate way to determine k once the order is known;
- can be used to calculate the concentration at any time, or the time taken for the concentration to drop by a given amount;

But

- if the reaction is not followed over a long enough extent the graphs will appear to be linear for all orders – to distinguish orders it is *vital* to follow to *at least* 60 per cent reaction, though less than 60 per cent conversion is possible with *very* accurate data;
- if the rate is affected by [product] then the use of integrated rate expressions can give completely misleading conclusions.

Worked Problem 3.12

Question. Use the following data to demonstrate the impossibility of distinguishing between first, second and zero orders. What is the percentage of reaction for this data?

$\frac{p_{N_2O}}{\text{mmHg}}$	100.0	91.8	81.0	75.2	67.7	60.9
t/min	0	10	20	30	40	50

Answer. Graphs of p_{N_2O} versus t (zero order), $\log_e p_{N_2O}$ versus t (first order) and $1/p_{N_2O}$ versus t (second order) are all approximately linear, and no conclusive decision can be made (Figures 3.13(a)–(c)). The percentage reaction at 50 min is 39. To get an unambiguously linear plot using integrated rate expressions it is vital to take reaction to at least 60 per cent. If this had been done for this experiment, the first order plot would have been unambiguously linear while the other two would have been decidedly curved.

With highly accurate data, the extent of reaction needed could become less.

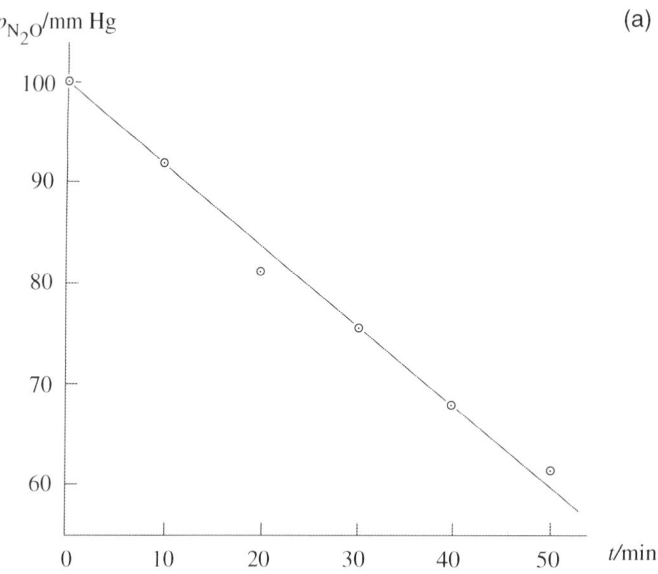

Figure 3.13 (a) Graph of $p(N_2O)$ versus time. (b) Graph of $\log_e p(N_2O)$ versus time. (c) Graph of $1/p(N_2O)$ versus time

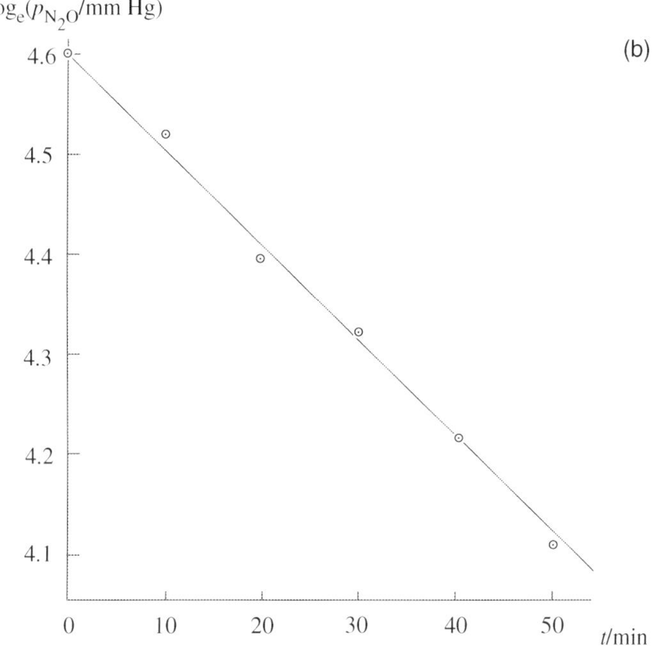

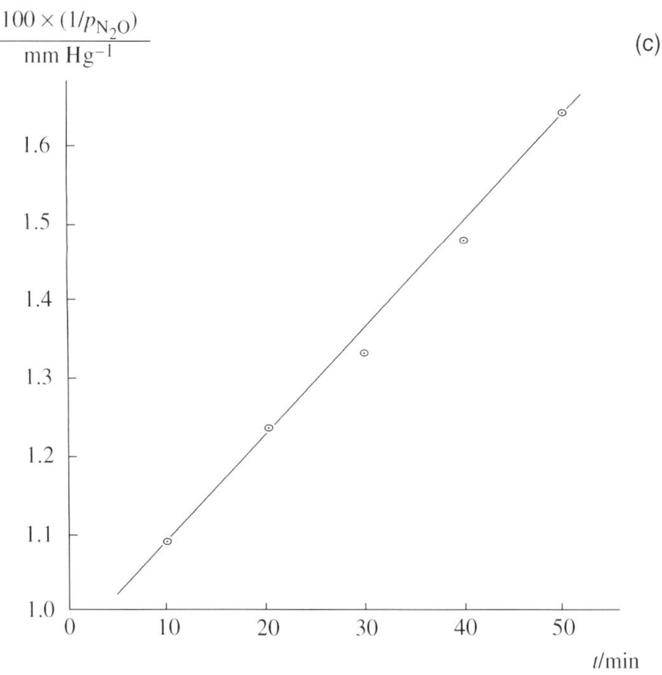

Figure 3.13 *(Continued)*

3.15 Pseudo-order Reactions

In general, the reaction A + B → products has rate equation

$$\text{Rate} = k[\text{A}]^{\alpha}[\text{B}]^{\beta} \tag{3.44}$$

where α is the order with respect to A and β is the order with respect to B. The total order is $n = \alpha + \beta$.

Neither (a) the differential method, based on rate/concentration data, nor (b) the integrated rate method, based on concentration/time data, is easily applicable.

The differential method gives

$$\log \text{rate} = \log k + \alpha \log[\text{A}] + \beta \log[\text{B}] \tag{3.45}$$

This is an equation in three unknowns, and the simple log/log plot is no longer possible.

The integrated rate method has also been shown (Section 3.11.2) to become cumbersome to use if the reactant concentrations are different.

Trick to overcome this difficulty. If the effect of [B] on the rate is kept constant, i.e. [B] is kept effectively constant, then the effect of [A] on the rate could be studied independently of [B], or *vice versa*. This is attained by keeping one of the concentrations in *large* excess over the other *throughout the reaction. What does this mean for the rate equation?*

$$\text{Rate} = k[\text{A}]^{\alpha}[\text{B}]^{\beta} \tag{3.44}$$

If [B] is in *excess* and *effectively* constant throughout reaction, then $[\text{B}]^{\beta}$ is also *effectively* constant throughout reaction, and can be taken with k to give rate $= k'\,[\text{A}]^{\alpha}$ where k' is called the pseudo-rate constant – meaning '*as if*'.

So, effectively,

$$\text{rate} = k'[\text{A}]^{\alpha} \tag{3.46}$$

where

$$k' = k[\text{B}]^{\beta} \tag{3.47}$$

and all the previous methods are immediately applicable, and k' and α can be found.

The value of β can then be found by repeating the experiment with B still in excess, but at various different values. The value of k' can be found for each constant excess [B], giving

$$k' = k[\text{B}]^{\beta} \tag{3.47}$$

$$\log k' = \log k + \beta \ \log[\text{B}] \tag{3.48}$$

and so a plot of $\log k'$ versus log[B] should be linear with slope $= \beta$ and intercept $= \log k$.

Worked Problem 3.13

Question. Show that the following fits the condition above.

Initially: $[B] = 0.150 \text{ mol dm}^{-3}$ $[A] = 1.0 \times 10^{-4} \text{ mol dm}^{-3}$

Hint. consider concentrations at 90 per cent reaction.

Answer. Let 90 per cent of A have reacted, i.e. $90/100 \times 1.0 \times 10^{-4}$ mol $\text{dm}^{-3} = 0.9 \times 10^{-4}$ mol dm^{-3}, leaving [A] at time $t = 0.1 \times 10^{-4}$ mol dm^{-3}. At time t, [B] has dropped by 0.9×10^{-4} mol dm^{-3}, leaving [B] at time $t = 1499.1 \times 10^{-4}$ mol $\text{dm}^{-3} \approx 0.150$ mol dm^{-3}. During most of the reaction [B] has remained *effectively* constant.

Worked Problem 3.14

Question. The acid-catalysed hydrolysis of an ester is first order in both H_3O^+ and ester. In an experiment $[H_3O^+]$ is kept in constant excess at 0.0100 mol dm^{-3}, and a first order $k' = 5.0 \times 10^{-5}\ \text{s}^{-1}$ is found. Calculate the true second order rate constant.

Answer.

$$\text{Rate} = k[\text{ester}][H_3O^+] = k'[\text{ester}] \tag{3.49}$$

where

$$k' = k[H_3O^+] \tag{3.50}$$

and k' is the *pseudo*-first order rate constant.

$$k = k'/[H_3O^+] = 5.0 \times 10^{-5}\text{s}^{-1}/0.0100 \text{ mol dm}^{-3} = 5.0 \times 10^{-3} \text{ mol}^{-1} \text{ dm}^3 \text{ s}^{-1}$$

3.15.1 Application of pseudo-order techniques to rate/concentration data

$$\text{Rate} = k[A]^{\alpha}[B]^{\beta} \tag{3.44}$$

which gives

$$\log \text{ rate} = \log k + \alpha \log[A] + \beta \log[B] \tag{3.45}$$

which cannot be solved easily. But rate/concentration data can still be easily handled by holding one reactant in excess, as is shown in the following example.

Worked Problem 3.15

Question. The rate of reaction between A and B depends on the concentrations of both A and B. In one experiment the following data were collected when the initial concentrations were

$$[\mathrm{A}] = 8 \times 10^{-4}\ \mathrm{mol\ dm^{-3}} \quad [\mathrm{B}] = 0.1\ \mathrm{mol\ dm^{-3}}$$

$\dfrac{\text{rate}}{\text{mol dm}^{-3}\text{min}^{-1}}$	8.0	4.5	2.0	0.5
$\dfrac{10^4[\mathrm{A}]}{\text{mol dm}^3}$	8.0	6.0	4.0	2.0

Find the order with respect to A, and the *pseudo*-rate constant k'. In a further series of experiments, with B still in large excess, the pseudo-rate constant k' varied with the concentration of B as

$\dfrac{[\mathrm{B}]}{\text{mol dm}^{-3}}$	0.2	0.3	0.4	0.6
$\dfrac{10^{-6}k'}{\text{mol}^{-1}\text{dm}^3\text{min}^{-1}}$	25.0	37.5	50.0	75.0

Find the order of the reaction with respect to B, and hence the overall order and the true rate constant.

Answer.

Dependence of rate on [A].

Concentration up by a factor of two, rate up by a factor of four, i.e. 2^2

concentration up by a factor of three, rate up by a factor of nine, i.e. 3^2

$\therefore$ reaction is second order in [A]: rate $= k'[\mathrm{A}]^2$ and $k' = \text{rate}/[\mathrm{A}]^2$

$k' = 8.0\ \mathrm{mol\ dm^{-3}\ min^{-1}}/(8.0 \times 10^{-4})^2\ \mathrm{mol^2\ dm^{-6}} = 12.5 \times 10^6\ \mathrm{mol^{-1}\ dm^3\ min^{-1}}$

Using all the data gives an average $k' = 12.5 \times 10^6\ \mathrm{mol^{-1}\ dm^3\ min^{-1}}$.

Dependence of rate on [B].

$$\text{Rate} = k[\mathrm{A}]^2[\mathrm{B}]^m = k'[\mathrm{A}]^2 \text{ when } [\mathrm{B}] \gg [\mathrm{A}] \ \therefore\ k' = k[\mathrm{B}]^m$$

A systematic procedure would now use a log–log plot to evaluate m and k. $\log k' = \log k + m \log[\mathrm{B}]$ and plot of $\log k'$ versus $\log[\mathrm{B}]$ is linear with slope $= m$ and intercept $= \log k$.

In this example the dependence of rate constant k' on [B] is obvious:
concentration up by a factor of two, rate constant up by a factor of two
concentration up by a factor of three, rate constant up by a factor of three
$\therefore$ reaction is first order in B i.e. rate $\propto$ [B] and rate $= k[A]^2$ [B]
k can be found either

1. graphically, by plotting k' versus [B] giving $k = 125 \times 10^6\ \text{mol}^{-2}\ \text{dm}^6\ \text{min}^{-1}$, or
2. numerically, since $k = k'/[B]$. Using data in the second column gives

$k = 37.5 \times 10^6\ \text{mol}^{-1}\ \text{dm}^3\ \text{min}^{-1}/0.3\ \text{mol dm}^{-3} = 125 \times 10^6\ \text{mol}^{-2}\ \text{dm}^6\ \text{min}^{-1}$.

All four sets of data must be used to get the average k.

3.16 Determination of the Product Concentration at Various Times

Sometimes it is much easier to determine the product concentration than the reactant concentration, and the data must be converted into concentrations of reactant left at various times, e.g.

$$S_2O_8^{2-}(aq) + 2I^-(aq) \rightarrow 2SO_4^{2-}(aq) + I_2(aq)$$

is easily followed by analysing for I_2, and from this $[S_2O_8^{2-}]$ remaining can be found.

At time t $[I_2]$ is y mol dm^{-3}. Knowing this the stoichiometry will give $[S_2O_8^{2-}]$.
x mol $S_2O_8^{2-}$ reacts to give x mol I_2 and $2x$ mol SO_4^{2-}.
$\therefore$ if $[I_2]$ is y mol dm^{-3}, then $[S_2O_8^{2-}]$ has decreased by y mol dm^{-3}.
If the initial concentration was $[S_2O_8^{2-}]_0$ at time, $t = 0$ then at time t, $[S_2O_8^{2-}]_t = [S_2O_8^{2-}]_0 - y$.

Sometimes it is not necessary to convert from product concentration to reactant concentration, illustrated in the next problem (Problem 3.16).

Worked Problem 3.16

Question. Trichloroethanoic acid is readily decarboxylated in aqueous solution:

$$CCl_3COOH(aq) \rightarrow CHCl_3(aq) + CO_2(g)$$

$\frac{\text{volume } CO_2}{\text{cm}^3}$	2.25	8.30	14.89	31.14	40.04
$\frac{\text{time}}{\text{min}}$	330	1200	2400	7760	∞

Show the reaction to be first order, and calculate the time taken for the initial concentration of CCl_3COOH to fall by 25 per cent.

Answer.

Volume of CO_2 produced when reaction is complete $\propto [CCl_3COOH]_0$

Volume of CO_2 produced at time $t \propto [CCl_3COOH]_{reacted}$

$[CCl_3COOH]$ left at time $t = [CCl_3COOH]_0 - [CCl_3COOH]_{reacted}$

$$\propto v_\infty - v_t$$

where v is the appropriate volume of CO_2.
First order plot: $\log_e[A]_t = \log_e[A]_0 - kt$.
The decrease in concentration of reactant $\propto$ volume of CO_2 evolved = constant $\times$ vol.
Hence plot $\log_e(v_\infty - v_t)$ versus t.

$\frac{v_t}{cm^3}$	2.25	8.30	14.89	31.14	40.04
$\frac{v_\infty - v_t}{cm^3}$	37.79	31.74	25.15	8.90	0
$\log_e(v_\infty - v_t)$	3.632	3.458	3.225	2.186	–
$\frac{time}{min}$	330	1200	2400	7760	∞

Plot of $\log_e(40.04 - v_t)$ versus t is linear and reaction is first order.

$$\text{Slope} = -1.94 \times 10^{-4}\,\text{min}^{-1} \therefore k = 1.94 \times 10^{-4}\,\text{min}^{-1}$$

When 25 per cent of the CCl_3COOH has reacted, 25 per cent of the final amount of CO_2 will have been formed, i.e. 10.01 cm^3.

$\therefore$ at this time $[CCl_3COOH] \propto (40.04 - 10.01)\,cm^3$.

$\log_e(40.04 - 10.01) = \log_e 40.04 - 1.94 \times 10^{-4}\,t$

$t = 1485$ min.

Worked Problem 3.17

Question. Why is it possible in this case that the *actual concentrations* of CCl_3COOH are not needed for the first order plot?

Answer. A logarithm has to be plotted against time and, since $\log_e$ conc. = $\log_e$ constant + $\log_e$ vol., then a plot of $\log_e$ vol. will be displaced up or down vertically from the $\log_e$ conc. by a constant amount equal to $\log_e$ constant. However, the slopes of the two graphs will be the same. Hence for a first order plot there is no need to obtain actual concentrations; a plot of the logarithm of a quantity proportional to the concentration will suffice. This has considerable value when physical methods of following reactions are being used.

3.17 Expressing the Rate in Terms of Reactants or Products for Non-simple Stoichiometry

- Reaction: $A \rightarrow 2B$

  x mol of A gives $2x$ mol of B
  $\therefore$ rate of production of B is twice that of removal of A

$$\text{i.e.}\quad +\mathrm{d}[B]/\mathrm{dt} = -2\mathrm{d}[A]/\mathrm{d}t \tag{3.51}$$

$$-\mathrm{d}[A]/\mathrm{d}t = +\tfrac{1}{2}\mathrm{d}[B]/\mathrm{d}t \tag{3.52}$$

- Reaction: $A + 2B \rightarrow 2C + D$

  x mol of A reacts with $2x$ mol of B to give $2x$ mol of C and x mol of D,
  $\therefore$ rate of removal of B is twice that of removal of A, and is twice the rate of production of D, but equal to the rate of production of C, *or* rate of removal of A is half the rate of removal of B, and rate of removal of A is half the rate of production of C and equal to the rate of production of D.

$$-2\,\mathrm{d}[A]/\mathrm{d}t = -\mathrm{d}[B]/\mathrm{d}t = +\mathrm{d}[C]/\mathrm{d}t = +2\,\mathrm{d}[D]/\mathrm{d}t \tag{3.53}$$

$$-\mathrm{d}[A]/\mathrm{d}t = -\tfrac{1}{2}\mathrm{d}[B]/\mathrm{d}t = +\tfrac{1}{2}\mathrm{d}[C]/\mathrm{d}t = +\mathrm{d}[D]/\mathrm{d}t \tag{3.54}$$

3.18 The Kinetic Analysis for Complex Reactions

Only a relatively few kinetic schemes for complex reactions have differential rate expressions which can be integrated straightforwardly. Examples are the following.

- $A \xrightarrow{k_1} P_1$
  $A \xrightarrow{k_2} P_2$

The differential equations describing this scheme are

$$-\mathrm{d}[\mathrm{A}]/\mathrm{d}t = k_1[\mathrm{A}] + k_2[\mathrm{A}] \tag{3.55}$$

$$+\mathrm{d}[\mathrm{P}_1]/\mathrm{d}t = k_1[\mathrm{A}] \tag{3.56}$$

$$+\mathrm{d}[\mathrm{P}_2]/\mathrm{d}t = k_2[\mathrm{A}] \tag{3.57}$$

On integration these give

$$[\mathrm{A}]_t = [\mathrm{A}]_0 \exp\{-(k_1 + k_2)t\} \tag{3.58}$$

$$[\mathrm{P}_1]_t = \frac{k_1}{k_1 + k_2}[\mathrm{A}]_0\{1 - \exp[-(k_1 + k_2)t]\} \tag{3.59}$$

$$[\mathrm{P}_2]_t = \frac{k_2}{k_1 + k_2}[\mathrm{A}]_0\{1 - \exp[-(k_1 + k_2)t]\} \tag{3.60}$$

and k_1 and k_2 can both be determined directly from the experimental data.

- $\mathrm{A} \xrightarrow{k_1} \mathrm{I} \xrightarrow{k_2} \mathrm{P}$

and this gives differential equations

$$-\mathrm{d}[\mathrm{A}]/\mathrm{d}t = k_1[\mathrm{A}] \tag{3.61}$$

$$+\mathrm{d}[\mathrm{P}]/\mathrm{d}t = k_2[\mathrm{I}] \tag{3.62}$$

$$+\mathrm{d}[\mathrm{I}]/\mathrm{d}t = k_1[\mathrm{A}] - k_2[\mathrm{I}] \tag{3.63}$$

which on integration give

$$[\mathrm{A}]_t = [\mathrm{A}]_0 \exp(-k_1 t) \tag{3.64}$$

$$[\mathrm{P}]_t = \frac{1}{k_2 - k_1}[\mathrm{A}]_0\{k_2[1 - \exp(-k_1 t)] - k_1[1 - \exp(-k_2 t)]\} \tag{3.65}$$

$$[\mathrm{I}]_t = \frac{k_1}{k_2 - k_1}[\mathrm{A}]_0\{\exp(-k_1 t) - \exp(-k_2 t)\} \tag{3.66}$$

Again the rate constants can be found directly from the experimental data, and it is also possible to draw graphs showing how [I] varies with time for assumed relative values of k_1 and k_2 (see Figures 3.14 and 3.15 below).

Other schemes that can be easily integrated include

- $\mathrm{A} \xrightarrow{k_1} \mathrm{I}_1 \xrightarrow{k_2} \mathrm{I}_2 \xrightarrow{k_3} \mathrm{P}$

- $A \xrightarrow{k_1} I$
  $I + X \xrightarrow{k_2} P + X$
- $A \underset{k-1}{\overset{k_1}{\rightleftarrows}} I \xrightarrow{k_2} P$

3.18.1 Relatively simple reactions which are mathematically complex

One problem for kineticists is that only a relatively slight increase in complexity of the kinetic scheme results in differential equations which cannot be integrated in a straightforward manner to give a manageable analytical expression. When this happens the differential equations have to be solved by either numerical integration or computer simulation. This is a mathematical limitation of the use of integrated rate expressions which is not apparent in the kinetic scheme. Typical schemes which are mathematically complex are

- $A + B \underset{k_{-1}}{\overset{k_1}{\rightleftarrows}} I \xrightarrow{k_2} P$
- $A + B \underset{k_{-1}}{\overset{k_1}{\rightleftarrows}} I_1 \underset{k_{-2}}{\overset{k_2}{\rightleftarrows}} I_2 \xrightarrow{k_3} P$

These mathematically complex kinetic schemes should be compared with the mathematically straightforward schemes:

- $A \underset{k_{-1}}{\overset{k_1}{\rightleftarrows}} I \xrightarrow{k_2} P$
- $A \underset{k_{-1}}{\overset{k_1}{\rightleftarrows}} I_1 \underset{k_{-2}}{\overset{k_2}{\rightleftarrows}} I_2 \xrightarrow{k_3} P$

These schemes are not much less complex chemically, but can be integrated easily. The important distinction between the two categories lies in the first reversible step. In the mathematically simple case the forward step is first order, while in the complex case it is second order. Schemes which are first order in all steps are mathematically simple and can be integrated in a straightforward manner. When there are second order steps, the situation is normally more complicated.

3.18.2 Analysis of the simple scheme $A \xrightarrow{k_1} I \xrightarrow{k_2} P$

This sort of analysis is very important in the formulation of the steady state approximation, developed to deal with kinetic schemes which are too complex mathematically to give simple explicit solutions by integration. Here the differential rate expression can be integrated. The differential and integrated rate equations are given in equations (3.61)–(3.66).

3.18.3 Two conceivable situations

Remember: reactions involving more than one step often have one step that is slow compared with the other steps, and the overall rate is limited by the rate of this step, and can be approximated to the rate of this slow step. When this occurs, the slow step is termed the *rate-determining step*. Not all reactions have a rate-determining step. Sometimes the individual reactions have comparable rates, and the overall rate cannot be approximated to the rate of any individual step.

For the scheme

$$A \xrightarrow{k_1} I \xrightarrow{k_2} P$$

there are two conceivable situations.

1. $k_2 \gg k_1$. This implies that the production of the intermediate I is the rate-determining step, or the slowest step, in the sequence. Once the intermediate is formed it is very rapidly converted to product P. Graphs of the consequent concentration dependences of A, I and P on time can be drawn (Figure 3.14). The magnitudes of [I] have been very grossly magnified to enable them to be placed on the graph.

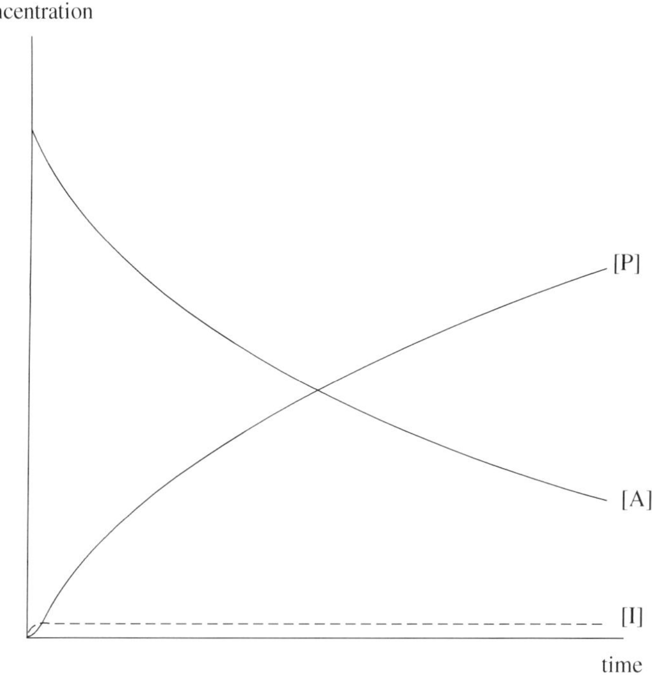

Figure 3.14 Graph of concentration versus time for reactant, product and intermediate in the reaction $A \xrightarrow{k_1} I \xrightarrow{k_2} P$ where $k_2 \gg k_1$

When $k_2 \gg k_1$ there is always a build-up of intermediate to a *very very low*, but *steady*, value of the concentration, which remains almost constant at this value until reaction is virtually complete, when it falls back to zero. The build-up to the steady plateau often occurs so very rapidly that it can only be followed using the specialized techniques of fast reactions. Determination of the very low concentrations of the intermediate is also difficult, and often cannot be achieved, though the modern laser-induced fluorescence methods are helping here.

2. $k_2 \ll k_1$. This implies that the intermediate is formed from A much faster than it is converted to P. The graph, Figure 3.15, shows that the intermediate concentration never builds up to a steady value, and that the magnitudes of the intermediate concentrations are always relatively large, and comparable to either [A] or [P] depending on the stage of reaction. Towards the end of reaction [I] tails off to zero. A similar result is obtained if $k_2 \approx k_1$.

These two situations show totally different behaviour, and the graphs demonstrate two points:

- in some reactions there is a build-up of the intermediate to an almost constant value; in others there is not;

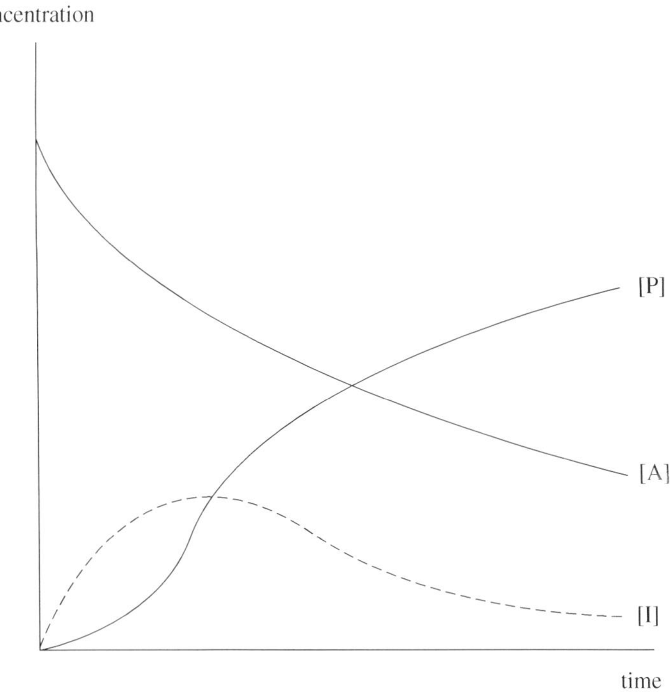

Figure 3.15 Graph of concentration versus time for reactant, product and intermediate in the reaction $A \xrightarrow{k_1} I \xrightarrow{k_2} P$ where $k_2 \ll k_1$

- the criterion for build-up to a plateau is that the intermediate must be present in very, very low concentrations. This implies that it must always be removed in a very fast reaction, where in this case $k_2 \gg k_1$. In general the intermediate can be removed in one reaction, or in a series of reactions, but these must be *fast*.

3.19 The Steady State Assumption

Reactions involving *intermediates* present in *very very low, and almost constant* concentrations, called *steady state concentrations*, are in the *steady state region.* As will be shown in Chapter 6, many chain and non-chain reactions involve such intermediates.

If intermediates are present in low steady concentrations this implies that

- the intermediates are *very reactive* and they *never build up* to significant concentrations throughout reaction;
- this, in turn, implies that the total rate of production of the intermediates is *balanced* by the total rate of their removal;
- the steady concentration over the time of reaction implies that the *rate of change* of the intermediate concentration with time is *approximately zero*, i.e.
- $+\mathrm{d[intermediate]}/\mathrm{d}t \approx 0$ (3.67)

Note: this is written as a rate of *formation* of the intermediate. This is the normal way of expressing the change in concentration of the intermediate with time.

- the rate of change of intermediate concentration with time can be expressed in terms of the rates of all steps producing the intermediate and the rates of all steps removing the intermediate;
- $+\mathrm{d[intermediate]}/\mathrm{d}t$ = sum of all the rates of production of the intermediate − sum of all the rates of removal of the intermediate = 0

3.19.1 Using this assumption

- This simplifies the subsequent analysis of the kinetic scheme since the relevant differential equations do not have to be solved. When several intermediates are present, the steady state treatment is an invaluable aid.
- For given mechanisms, it is possible to predict expressions for the steady state concentrations of the intermediates in terms of rate constants and reactant concentrations.

- The steady state method must never be applied to reactions where the intermediates are present in significant amounts.
- It must never be applied during the period of build-up to the steady state.
- It must never be applied to reactants or products, i.e. d[reactant]/dt and d[product]/dt are *never* to be equated to zero.
- Sometimes instead of a steady state concentration of intermediates there is an 'explosive' build-up.
- In some autocatalytic reactions where the products accelerate the reaction, see Section 6.14, the concentration of intermediates switches abruptly from a steady value to one which oscillates violently.

Worked Problem 3.18

Question. The following is a highly simplified mechanism for the decomposition of C_3H_8:

$$C_3H_8 \xrightarrow{k_1} CH_3^{\bullet} + C_2H_5^{\bullet}$$

$$CH_3^{\bullet} + C_3H_8 \xrightarrow{k_2} CH_4 + C_3H_7^{\bullet}$$

$$C_3H_7^{\bullet} \xrightarrow{k_3} C_2H_4 + CH_3^{\bullet}$$

$$CH_3^{\bullet} + CH_3^{\bullet} \xrightarrow{k_4} C_2H_6$$

Pick out the intermediates and formulate a steady state expression, +d[intermediate]/d$t \approx 0$, where possible. From these, find expressions for each steady state concentration, and then formulate the overall rate of reaction in terms of the rate of production of CH_4.

Answer.

- Intermediates are $CH_3^{\bullet}$, $C_2H_5^{\bullet}$ and $C_3H_7^{\bullet}$.
- $CH_3^{\bullet}$ is formed in steps 1 and 3 and removed in steps 2 and 4.
- $C_2H_5^{\bullet}$ is formed in step 1, but the step or steps removing $C_2H_5^{\bullet}$ are not given and so a steady state equation cannot be written.
- $C_3H_7^{\bullet}$ is formed in step 2 and removed in step 3.
- Steady state equations: these are always written in terms of *production* of the intermediate i.e. +ve for rates of production and −ve for rates of removal.
- $$+\mathrm{d}[CH_3^{\bullet}]/\mathrm{d}t = \underset{\text{(formation+ve)}}{\underset{\uparrow}{k_1[C_3H_8]}} - \underset{\text{(removal−ve)}}{\underset{\uparrow}{k_2[CH_3^{\bullet}][C_3H_8]}} + \underset{\text{(formation+ve)}}{\underset{\uparrow}{k_3[C_3H_7^{\bullet}]}} - \underset{\text{(removal−ve)}}{\underset{\uparrow}{2k_4[CH_3^{\bullet}]^2}} = 0 \quad (3.68)$$

The factor of two arises because the steady state equation is written in terms of the rate of production of $CH_3^\bullet$, and for every step 4, two $CH_3^\bullet$ are removed.

- $$+\mathrm{d}[C_3H_7^\bullet]/\mathrm{d}t = \underset{\substack{\uparrow \\ \text{(formation+ve)}}}{k_2[CH_3^\bullet][C_3H_8]} - \underset{\substack{\uparrow \\ \text{(removal−ve)}}}{k_3[C_3H_7^\bullet]} = 0 \qquad (3.69)$$

Both these equations are equations in two unknowns, and neither can be solved on its own. They are simultaneous equations, solved by adding (3.68) + (3.69).

- $k_1[C_3H_8] = 2k_4[CH_3^\bullet]^2 \qquad \therefore [CH_3^\bullet] = (k_1/2k_4)^{1/2}[C_3H_8]^{1/2}$ (3.70)
- Substituting into (3.69) gives $[C_3H_7^\bullet] = \frac{k_2}{k_3}\left(\frac{k_1}{2k_4}\right)^{1/2}[C_3H_8]^{3/2}$ (3.71)
- Rate of reaction $= k_2[CH_3^\bullet][C_3H_8] = k_2(k_1/2k_4)^{1/2}[C_3H_8]^{3/2}$ (3.72)

3.20 General Treatment for Solving Steady States

Although each mechanism requires an individual solution, a general treatment for solving steady states can be given.

- Identify *all* reactants, products and intermediates.
- Note in which *steps each* intermediate appears, i.e. which step or steps *produce* each intermediate, and which step or steps *remove* it.
- Formulate the *steady state equation* for *each* intermediate; be *careful* of +ves and −ves.
- Choose a *reactant* or *product* in terms of which the overall rate could be experimentally measured and in terms of which the rate, as predicted by the mechanism, can be written. Often alternative expressions will be possible. Choose the one with the *smallest* number of terms and the *smallest* number of steady state concentrations.
- Note the *intermediates* whose concentrations appear in the predicted overall rate expression, and realize that the steady state equations *must be solved* to give these concentrations.
- See whether the required steady state concentration can come from one steady state equation directly. If *not*, then try *adding* all the steady state equations. This may give the required intermediate concentration direct, or it may be necessary to *substitute* into *one or other* of the steady state equations. This manipulation is often the *tricky* part of the procedure, and *needs practice*.

- Substitution into the rate expression will give an expression involving ks and observable concentrations. This *predicted* rate is *equated* to the *experimental* rate, and the overall observed k can then be expressed in terms of the *individual* ks.
- Often a *rigorous* treatment can lead to a very *complex* expression, and approximations are required. These are decided on by use of other knowledge and/or sensible assumptions.
- One further relation which can be explicitly used is that

  total rate of formation of intermediates = *total rate of removal* of intermediates.
- This is a statement equivalent to the steady state statement that there is *no build-up of intermediates* throughout the reaction.
- If two steps in the sequence constitute a reversible reaction and *equilibrium is set up* and *maintained* throughout reaction, then formulation of the equilibrium constant, in terms of the ks for the forward and back reactions and the concentrations of the species involved, gives another relation between the rate constants.

$$\mathrm{A} + \mathrm{B} \underset{k_{-1}}{\overset{k_1}{\rightleftarrows}} \mathrm{C} + \mathrm{D}$$

$$K = \frac{k_1}{k_{-1}} = \left(\frac{[\mathrm{C}][\mathrm{D}]}{[\mathrm{A}][\mathrm{B}]}\right)_{\mathrm{eq}} \qquad \therefore \quad k_1[\mathrm{A}][\mathrm{B}] = k_{-1}[\mathrm{C}][\mathrm{D}] \tag{3.73}$$

There are other *special procedures* that are applicable to *chain reactions only*. These are given in Chapter 6.

Worked Problem 3.19

Question. In Chapter 1, the physical mechanism involved in elementary reactions was given as

$$\mathrm{A} + \mathrm{A} \xrightarrow{k_1} \mathrm{A}^* + \mathrm{A} \qquad \text{activation by binary collisions}$$

$$\mathrm{A}^* + \mathrm{A} \xrightarrow{k_{-1}} \mathrm{A} + \mathrm{A} \qquad \text{deactivation by binary collisions}$$

$$\mathrm{A}^* + b\mathrm{A} \xrightarrow{k_2} \text{products} \qquad \text{reaction step}$$

where $b = 0$ for unimolecular reaction steps

$b = 1$ for bimolecular reaction steps

$b = 2$ for termolecular reaction steps.

Carry out a steady state analysis to show that unimolecular reactions are first order at high pressures, moving to second order at low pressures. State the rate-determining steps in each region. In contrast, show that bimolecular reactions are second order at all pressures.

Answer.

1. For a unimolecular reaction

$$\mathrm{d}[\mathrm{A}^*]/\mathrm{d}t = k_1[\mathrm{A}]^2 - k_{-1}[\mathrm{A}^*][\mathrm{A}] - k_2[\mathrm{A}^*] = 0 \tag{3.74}$$

$$[\mathrm{A}^*] = \frac{k_1[\mathrm{A}]^2}{k_{-1}[\mathrm{A}] + k_2} \tag{3.75}$$

$$-\mathrm{d}[\mathrm{A}]/\mathrm{d}t = k_2[\mathrm{A}^*] = \frac{k_2 k_1[\mathrm{A}]^2}{k_{-1}[\mathrm{A}] + k_2} \tag{3.76}$$

which is neither first nor second order.

If $k_{-1}[\mathrm{A}] \gg k_2$, then

$$k_{-1}[\mathrm{A}][\mathrm{A}^*] \gg k_2[\mathrm{A}^*]$$

i.e. rate of deactivation $\gg$ rate of reaction
and the slow forward step is thus reaction

$$-\mathrm{d}[\mathrm{A}]/\mathrm{d}t = \frac{k_1 k_2[\mathrm{A}]}{k_{-1}} \tag{3.77}$$

The reaction is now first order, and this is more likely to occur at high pressures where [A] is large.

If $k_{-1}[\mathrm{A}] \ll k_2$, then

$$k_{-1}[\mathrm{A}][\mathrm{A}^*] \ll k_2[\mathrm{A}^*]$$

i.e. rate of deactivation $\ll$ rate of reaction

The rate of collision is thus rate determining, and the slow *forward* step will be the collisional rate of activation. The rate of collision is rate determining and the slow forward step is activation, step 1, and

$$-\mathrm{d}[\mathrm{A}]/\mathrm{d}t = k_1[\mathrm{A}]^2 \tag{3.78}$$

The reaction is now second order, and this is more likely to occur at low pressures when [A] is small.

2. For a bimolecular reaction

$$d[A^*]/dt = k_1[A]^2 - k_{-1}[A^*][A] - k_2[A^*][A] = 0 \quad (3.79)$$

$$[A^*] = \frac{k_1[A]^2}{k_{-1}[A] + k_2[A]} \quad (3.80)$$

$$-d[A]/dt = k_2[A^*][A] = \frac{k_2k_1[A]^3}{k_{-1}[A] + k_2[A]} = \frac{k_2k_1[A]^2}{k_{-1} + k_2} \quad (3.81)$$

and this is strictly second order. Nothing can be said about which forward step is rate determining.

3.21 Reversible Reactions

The simplest example is

$$A \underset{k_{-1}}{\overset{k_1}{\rightleftarrows}} B$$

At any time t,

$$[A]_{total} = [A]_{actual} + [B]_{actual} \quad (3.82)$$

During reaction the actual concentrations of A and B are constantly changing, but the above relation always holds.

When reaction reaches equilibrium, $[A]_{actual}$ and $[B]_{actual}$ remain constant with time and

$$K = \frac{k_1}{k_{-1}} = \left(\frac{[B]}{[A]}\right)_{eq} \quad (3.83)$$

Worked Problem 3.20

Question. Why is the situation at equilibrium, where $-d[A]_{eq}/dt = 0$, not a steady state?

Answer. In the steady state $+d[\text{intermediate}]/dt = 0$. The above equilibrium situation is not a steady state because the concentrations at equilibrium are significant and are not the very, very low concentrations of the steady state.

If reaction is followed by measuring the change in the concentration of A with time:

$$-\mathrm{d}[\mathrm{A}]_{\mathrm{actual}}/\mathrm{d}t = k_1[\mathrm{A}]_{\mathrm{actual}} - k_{-1}[\mathrm{B}]_{\mathrm{actual}} \tag{3.84}$$

$$= k_1[\mathrm{A}]_{\mathrm{actual}} - k_{-1}\{[\mathrm{A}]_{\mathrm{total}} - [\mathrm{A}]_{\mathrm{actual}}\} \tag{3.85}$$

$$= (k_1 + k_{-1})[\mathrm{A}]_{\mathrm{actual}} - k_{-1}[\mathrm{A}]_{\mathrm{total}} \tag{3.86}$$

Following standard integration procedures we end up with a non-standard integral, but this can be overcome by using the properties of the equilibrium position.

At equilibrium the actual concentrations of A and B are constant:

$$\text{i.e.} - \mathrm{d}[\mathrm{A}]_{\mathrm{eq}}/\mathrm{d}t = 0 \tag{3.87}$$

and so *at equilibrium and only at equilibrium*

$$-\mathrm{d}[\mathrm{A}]_{\mathrm{eq}}/\mathrm{d}t = (k_1 + k_{-1})[\mathrm{A}]_{\mathrm{actual}} - k_{-1}[\mathrm{A}]_{\mathrm{total}} = 0 \tag{3.88}$$

$$\therefore\ k_{-1}[\mathrm{A}]_{\mathrm{total}} = (k_1 + k_{-1})[\mathrm{A}]_{\mathrm{actual}},\ \text{with } [\mathrm{A}]_{\mathrm{actual}} = [\mathrm{A}]_{\mathrm{eq}} \tag{3.89}$$

Be careful: remember that this equation is only valid *at equilibrium.*

Substitution for $k_{-1}[\mathrm{A}]_{\mathrm{total}}$ into equation (3.86) gives

$$-\mathrm{d}[\mathrm{A}]_{\mathrm{actual}}/\mathrm{d}t = (k_1 + k_{-1})\{[\mathrm{A}]_{\mathrm{actual}} - [\mathrm{A}]_{\mathrm{eq}}\} \tag{3.90}$$

This equation holds *for all situations, not just* for equilibrium, and can be integrated to give

$$\log_e\{[\mathrm{A}]_{\mathrm{actual}} - [\mathrm{A}]_{\mathrm{eq}}\} = \log_e\{[\mathrm{A}]_0 - [\mathrm{A}]_{\mathrm{eq}}\} - (k_1 + k_{-1})t \tag{3.91}$$

This is a first order integrated rate equation and a plot of $\log_e\{[\mathrm{A}]_{\mathrm{actual}} - [\mathrm{A}]_{\mathrm{eq}}\}$ versus t has slope $= -(k_1 + k_{-1})$ and intercept $= \log_e\{[\mathrm{A}]_0 - [\mathrm{A}]_{\mathrm{eq}}\}$

$$\therefore\ k_{\mathrm{obs}} = k_1 + k_{-1} \tag{3.92}$$

and

$$K = k_1/k_{-1} \tag{3.93}$$

and so k_1 and k_{-1} can be found.

3.21.1 Extension to other equilibria

When this treatment is extended to other equilibria, non-standard integrals result and simple integrated rate expressions cannot be obtained. All these equilibria are non-first order in at least one direction, and all show complex kinetics at all stages away from the position of equilibrium. However, each one simulates first order behaviour *close to the position of equilibrium*, and so is well suited to being studied by single impulse techniques. Here the equilibrium is slightly displaced and the return to equilibrium is studied. The term describing such methods is 'the approach to equilibrium'. Generalizations of Equation (3.91) can be given

Worked Problem 3.21

Question. The isomerization of *cis*- and *trans*- 1-ethyl-2-methyl cyclopropane comes to equilibrium.

$$cis \underset{k_{-1}}{\overset{k_1}{\rightleftarrows}} trans$$

At 425.6 °C $[trans]_{eq}/[cis]_{eq} = 2.79$.
Throughout reaction the following data were obtained:

$\frac{[cis]}{\text{mol dm}^{-3}}$	0.016 79	0.014 06	0.011 02	0.008 92	0.007 75
$\frac{\text{time}}{\text{s}}$	0	400	1000	1600	2100

Find k_1 and k_{-1}.

Answer. This reaction is one where both the forward and back reactions are first order. Equation (3.91) can be used.

At all times: $[cis]_0 = [cis]_{actual} + [trans]_{actual} = 0.016\,79 \text{ mol dm}^{-3}$

At equilibrium: $[trans]_{eq} = 2.79[cis]_{eq}$

$\therefore\ 3.79\ [cis]_{eq} = 0.016\,79 \text{ mol dm}^{-3} \quad \therefore\ [cis]_{eq} = 4.43 \times 10^{-3} \text{ mol dm}^{-3}$

$\frac{[cis]_t}{\text{mol dm}^{-3}}$	0.016 79	0.014 06	0.011 02	0.008 92	0.007 75
$\left\{\frac{[cis]_t - [cis]_{eq}}{\text{mol dm}^{-3}}\right\}$	0.012 36	0.009 63	0.006 59	0.004 49	0.003 32
$\log_e\{[cis]_t - [cis]_{eq}\}$	−4.393	−4.643	−5.022	−5.406	−5.708
$\frac{\text{time}}{\text{s}}$	0	400	1000	1600	2100

A plot of $\log_e\{[cis]_t - [cis]_{eq}\}$ versus time is linear with slope $= -6.25 \times 10^{-4} \text{ s}^{-1}$

$$\therefore\ k_1 + k_{-1} = 6.25 \times 10^{-4} \text{ s}^{-1} \qquad K = k_1/k_{-1} = 2.79$$

giving $k_1 = 4.60 \times 10^{-4} \text{ s}^{-1}$ and $k_{-1} = 1.65 \times 10^{-4} \text{ s}^{-1}$

3.22 Pre-equilibria

These involve a reversible reaction followed by one or more other reactions. Analysis shows that if the equilibrium is not established very rapidly then the kinetics are complex and are not of simple order. Special computer techniques are required to analyse the results.

If the equilibrium is established very rapidly and is maintained throughout reaction, analysis becomes straightforward. This can be illustrated by the reaction

$$\mathrm{A} \underset{k_{-1}}{\overset{k_1}{\rightleftarrows}} \mathrm{B} \xrightarrow{k_2} \mathrm{C}$$

Under these conditions

- the step $\mathrm{B} \xrightarrow{k_2} \mathrm{C}$ will be rate determining, and the reaction reduces effectively to $\mathrm{B} \rightarrow \mathrm{C}$;
- so that the mechanistic rate $= k_2[\mathrm{B}]_{\text{actual}}$; (3.94)
- mathematical analysis shows that for the step $\mathrm{B} \xrightarrow{k_2} \mathrm{C}$ to be slow in comparison to the fast steps $\mathrm{A} \xrightarrow{k_1} \mathrm{B}$ and $\mathrm{B} \xrightarrow{k_{-1}} \mathrm{A}$ it is sufficient that k_1 *or* k_{-1} is large *and* $k_1 + k_{-1} \gg k_2$;
- although the equilibrium is maintained throughout the reaction both $[\mathrm{A}]_{\text{actual}}$ and $[\mathrm{B}]_{\text{actual}}$ are varying, but the ratio

$$\frac{[B]_{\text{actual}} \text{ at time } t}{[A]_{\text{actual}} \text{ at time } t} = \text{constant} = K \tag{3.95}$$

- the above holds for most reactions studied. If the pre-equilibrium is not set up fast and maintained, special techniques of analysis are required.

Reactions involving pre-equilibria are discussed in Chapter 8.

3.23 Dependence of Rate on Temperature

Most reactions show an increase in rate as the temperature increases, studied by measuring the rate constant at various temperatures. Table 3.5 gives rate constants for bromination of propanone in acidified aqueous Br_2 solution.

Table 3.5 shows the dramatic increase in rate constant over a range of a few degrees. If the rate constant is plotted against temperature an exponential curve is obtained (Figure 3.16). However, to go further requires data over a much larger range of temperature.

Table 3.5 Effect of temperature on rate constant

$\frac{\text{temp}}{°\text{C}}$	20	24	25	27	29
$\frac{10^6 k}{\text{mol}^{-1}\text{dm}^3\text{s}^{-1}}$	1.16	2.02	2.79	5.59	8.60

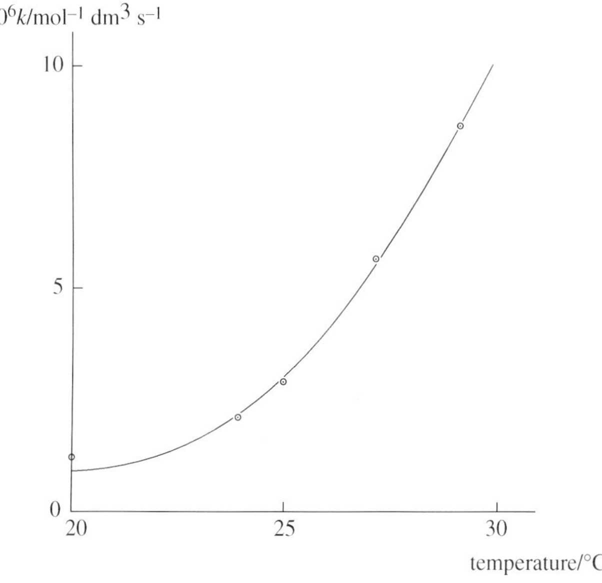

Figure 3.16 Graph of k versus temperature for the bromination of propanone in acidified aqueous solution

The dependence of the rate constant on temperature for the conversion of thiourea to ammonium thiocyanate in aqueous solution is given in Table 3.6.

Again, a plot of k versus T shows an exponential rise. If $\log_e k$ is plotted against $1/T$, *where T is in kelvin*, a straight line of negative slope is obtained. This behaviour is summarized in the Arrhenius equation:

$$k = A \exp(-E_A/RT) \tag{3.96}$$

where A is the pre-exponential or A factor and E_A is the activation energy for reaction. This implies that

$$\log_e k = \log_e A - E_A/RT \tag{3.97}$$

Table 3.6 Effect of temperature on rate constant

$\frac{10^8 k}{\text{s}^{-1}}$	4.53	59.3	197	613
$\frac{\text{temp.}}{°\text{C}}$	90	110	120	130

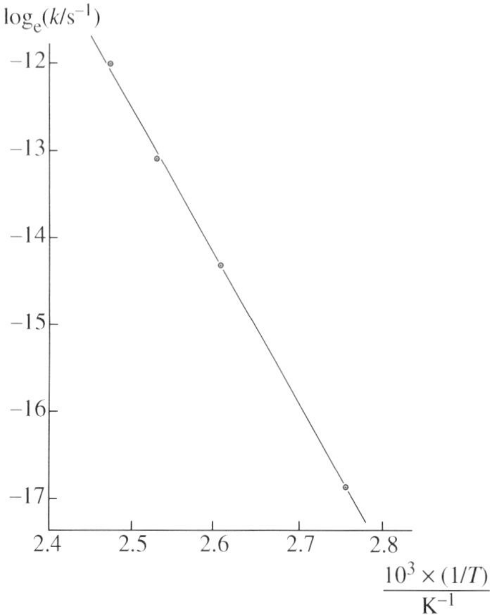

Figure 3.17 Arrhenius plot: $\log_e k$ versus $1/T$

and that a graph of $\log_e k$ versus $1/T$ should be linear with slope $= -E_A/RT$ and intercept $= \log_e A$ (Figure 3.17).

Note that the temperature must always be given in kelvin.

Most reactions obey the Arrhenius equation, but sometimes highly accurate and precise measurements give a discernible curvature, implying that either A or E_A or both are varying with temperature.

Worked Problem 3.22

Question. Use the data in Table 3.6 to find the activation energy and the A factor for the reaction quoted.

Answer.

$\frac{10^8 k}{s^{-1}}$	4.53	59.3	197	613
$\log_e k$	−16.910	−14.338	−13.137	−12.002
$\frac{\text{temp.}}{°C}$	90	110	120	130
$\frac{T}{K}$	363.15	383.15	393.15	403.15
$\frac{10^3 \times \left(\frac{1}{T}\right)}{K^{-1}}$	2.754	2.610	2.544	2.480

Note: the temperature must be in kelvin.

A plot of $\log_e k$ versus $1/T$ is linear with slope $= -17.85 \times 10^3$ K

$$\therefore \quad -E_A/R = -17.85 \times 10^3 \text{ K} \quad \text{and} \quad E_A = 8.314 \times 17.85 \times 10^3 \text{ J mol}^{-1} \text{ K}^{-1} \text{ K} = 148 \text{ kJ mol}^{-1}$$

A has to be found by calculation; extrapolation is impossible. Take one temperature, its corresponding rate constant and the graphical value of E_A, and substitute into the Arrhenius equation, e.g.

$$k = 4.53 \times 10^{-8} \text{ s}^{-1} \text{ at } T = 363.15 \text{ K}$$

$$E_A = 148 \times 10^3 \text{ J mol}^{-1} \qquad R = 8.314 \text{ J mol}^{-1} \text{ K}^{-1}$$

$$4.53 \times 10^{-8} \text{ s}^{-1} = A \exp(-148 \times 10^3/8.314 \times 363.15)$$

$$= 5.14 \times 10^{-22} A$$

$$A = 8.8 \times 10^{13} \text{ s}^{-1}$$

watch consistency of units: $R = 8.314$ J mol^{-1} K^{-1}; $\therefore$ E_A must be in J mol^{-1}.

Further reading

Laidler K.J., *Chemical Kinetics*, 3rd edn, Harper and Row, New York, 1987.

Laidler K.J. and Meiser J.H., *Physical Chemistry,* 3rd edn, Houghton Mifflin, New York, 1999.

Alberty R.A. and Silbey R.J., *Physical Chemistry*, 2nd edn, Wiley, New York, 1996.

Logan S.R., *Fundamentals of Chemical Kinetics*, Addison-Wesley, Reading, MA, 1996.

*Nicholas J., *Chemical Kinetics*, Harper and Row, New York, 1976.

*Wilkinson F., *Chemical Kinetics and Reaction Mechanisms*, Van Nostrand Reinhold, New York, 1980.

*Out of print, but should be in University Libraries.

Further problems

1. (a) The following kinetic data were found for the reaction

$$A \rightarrow B + C$$

	Expt 1	Expt 2	Expt 3	Expt 4
$\frac{10^2[A]}{\text{mol dm}^{-3}}$	0.50	1.78	7.22	19.9
$\frac{10^5 \text{ rate}}{\text{mol dm}^{-3} \text{min}^{-1}}$	0.30	3.80	62.2	475

Explaining your reasoning, find the order of the reaction and hence the rate constant.

(b) Show that your conclusions are consistent with the following data for the same reaction at the same temperature.

$\frac{10^3[A]}{\text{mol dm}^{-3}}$	10	8.05	6.75	5.0
$\frac{\text{time}}{\text{min}}$	0	200	400	800

2. The following data refer to the vapour phase decomposition of ethylene oxide at 414 °C:

$$C_2H_4O(g) \rightarrow CH_4(g) + CO(g)$$

$\frac{\text{time}}{\text{min}}$	0	9	13	18	50	100
$\frac{p_{\text{total}}}{\text{mmHg}}$	116.5	129.4	134.7	141.3	172.1	201.2

Find the order and rate constant for this reaction.
What further information would be needed to determine the activation energy?

3. From the following measurements of the initial rates of a reaction involving reactants A, B and C, estimate the order of reaction with respect to each reactant and the value of the rate constant. $[A]_0$ etc refer to the initial concentrations.

Expt	1	2	3	4
$\frac{10^5 \times \text{initial rate}}{\text{mol dm}^{-3}\,\text{s}^{-1}}$	2.5	3.75	10.0	2.5
$\frac{[A]_0}{\text{mol dm}^{-3}}$	0.010	0.010	0.010	0.020
$\frac{[B]_0}{\text{mol dm}^{-3}}$	0.005	0.005	0.010	0.005
$\frac{[C]_0}{\text{mol dm}^{-3}}$	0.010	0.015	0.010	0.010

4. The thermal decomposition of nitryl chloride obeys the stoichiometric equation

$$2NO_2Cl(g) \rightarrow 2NO_2(g) + Cl_2(g)$$

(a) Use the following data to show that the reaction obeys first order kinetics. Find the time required for the partial pressure of NO_2Cl to fall by

60 per cent of its initial value.

$\frac{\text{time}}{\text{s}}$	0	300	900	1500	2000
$\frac{p_{\text{total}}}{\text{mmHg}}$	19.9	21.7	24.4	26.3	27.3

(b) Does the following data give concordant results with (a)?

$\frac{10^3 \times \text{initial rate}}{\text{mmHg s}^{-1}}$	0.96	2.23	4.06	10.4
$\frac{\text{initial } p_{NO_2Cl}}{\text{mmHg}}$	1.45	3.39	6.17	15.9

The initial rates refer to the decrease in partial pressure of NO_2Cl

$$2NO_2Cl(g) \rightarrow 2NO_2(g) + Cl_2(g)$$

5. For a second order reaction with stoichiometry $2A \rightarrow A_2$, the total pressure at 400 K varied with time as follows:

$\frac{p_{\text{total}}}{\text{mmHg}}$	645	561	476	436	412	390
$\frac{\text{time}}{\text{s}}$	0	50	150	250	350	500

Determine the rate constant and the half-life. What prediction can be made about the dependence of $t_{1/2}$ on concentration? Verify this by calculating the first and second $t_{1/2}$.

6. The reaction

$$N_2O(g) \rightarrow N_2(g) + 1/2O_2(g)$$

is catalysed by $Cl_2(g)$ and the rate of reaction is found to depend on the partial pressures of both the reactant N_2O and the catalyst Cl_2.
The following initial rates were found for various partial pressures of N_2O and Cl_2.

	Expt 1	Expt 2	Expt 3
$\frac{p(N_2O)}{\text{mmHg}}$	300	150	300
$\frac{p(Cl_2)}{\text{mmHg}}$	40	40	10
$\frac{\text{initial rate}}{\text{mmHg s}^{-1}}$	3.0	1.5	1.5

Determine the order with respect to N_2O and Cl_2.

7. The reaction between liquid toluene and dissolved chlorine at 170 °C is catalysed by acids. With a catalyst concentration of 0.117 mol dm^{-3}, the reaction was followed by analysing for chlorine.

$\frac{\text{time}}{\text{min}}$	0	3	6	10	14
$\frac{[Cl_2]}{\text{mol dm}^{-3}}$	0.0775	0.0565	0.0410	0.0265	0.0175

Show that the reaction is first order in chlorine, and find the rate constant. Use the following results to determine the order with respect to catalyst. The values of k' quoted were found as above.

$\frac{[\text{catalyst}]}{\text{mol dm}^{-3}}$	0.050	0.075	0.100	0.153	0.200
$\frac{k'}{\text{min}^{-1}}$	0.0278	0.053	0.090	0.172	0.270

4 Theories of Chemical Reactions

Developments in theoretical chemical kinetics have made dramatic progress in the last few decades and these have been complemented by developments in experimental techniques, particularly molecular beam experiments and modern spectroscopic techniques. Many of these advances are modifications and developments of the original basic theories. It is essential that the concepts behind these older ideas are fully understood before moving on to the recent ideas.

Aims

By the end of this chapter you should be able to

- understand simple collision theory and be aware of its defects,
- appreciate that molecular beam ideas are much more physically realistic than those of simple collision theory,
- have an awareness of the power of modified collision theory,
- list the properties of potential energy surfaces relevant to transition state theory,
- follow an outline of the theoretical derivation of transition state theory in partition function form, and in thermodynamic form,
- compare predictions of collision theory, transition state theory and experiment and
- understand the fundamental concepts underlying unimolecular theory.

An Introduction to Chemical Kinetics. Margaret Robson Wright
 ISBNs: 0-470-09058-8 (hbk) 0-470-09059-6 (pbk)

4.1 Collision Theory

Simple collision theory assumes reaction occurs when molecules, with energy greater than a critical minimum, collide. Calculation of two quantities, the *total rate* of collision of reactant molecules and the *fraction* of molecules which have at least the critical energy, gives an equation to compare with the experimental Arrhenius equation:

$$k = A\exp(-E_A/RT) \tag{4.1}$$

The *main features* of collision theory are

- molecules are hard spheres, i.e. there are no *inter*molecular interactions,
- vibrational and rotational structures of reactants and products are ignored,
- the activated complex plays no part in the theory, and
- redistribution of energy on reaction is ignored.

Molecular beams, chemiluminescence and laser-induced fluorescence experiments show the theory in its simple form to be fundamentally flawed, with internal states of reactants and products and the redistribution of energy on reaction being of fundamental importance.

Molecular beam experiments study reactive and non-reactive collisions between molecules by observing the deflection of each molecule from its original path as a result of the collision. By studying the amount of scattering into various angles, much useful kinetic information can be found.

Chemiluminesence is emission of radiation from the products of a chemical reaction, and is normally associated with exothermic reactions where the products are in excited vibrational, rotational and sometimes electronic states. The particular excited states which are produced immediately after reaction has occurred can be studied and kinetic information obtained. Such studies also throw doubts on the validity of assuming that these states play no part.

Laser-induced fluorescence has been discussed in Chapter 2. Lasers excite molecules to excited levels from which they can lose energy by emission of radiation. This technique has been of major importance in the molecular beam experiments which force modification of the original collision theory outlined below.

4.1.1 Definition of a collision in simple collision theory

For reaction between unlike molecules, $A + B \rightarrow products$, only collisions between A and B can lead to reaction, and the appropriate total rate of collision is for collisions between A and B only.

Molecules A and B approach with relative velocity v_{AB}, represented as B being stationary and A approaching with velocity v_{AB}.

$$\text{A} \quad \bullet \quad \xrightarrow{v_{AB}} \quad \bullet \quad \text{B(stationary)}$$

The relative velocity describes the relative motion of two moving objects:

$$v_{AB} = v_A - v_B \tag{4.2}$$

If A and B are moving in the same direction, both velocities have the same sign, and the relative velocity is the difference in the magnitudes of the velocities.

If A and B are moving in opposite directions, the velocities of approach have different signs and the relative velocity is the sum of the magnitudes of the velocities.

A *collision only occurs* if the line representing the approach of A (radius r_A) to B (radius r_B) lies at a lateral distance *less than* $r_A + r_B$ from the centre of B, but the molecules *will not collide* if the centre of B lies at a distance *greater than* $r_A + r_B$ (Figure 4.1). A disc around B with radius equal to $r_A + r_B$ has cross-section $\pi(r_A + r_B)^2$. If an A molecule *hits this disc* then *collision will occur.*

The distance, b, on the diagram is called the impact parameter (Figure 4.2), and the magnitude of b defines whether collision is possible.

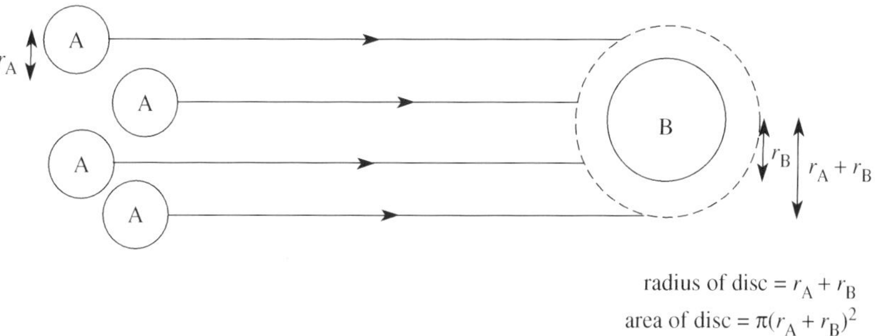

Figure 4.1 Definition of a collision

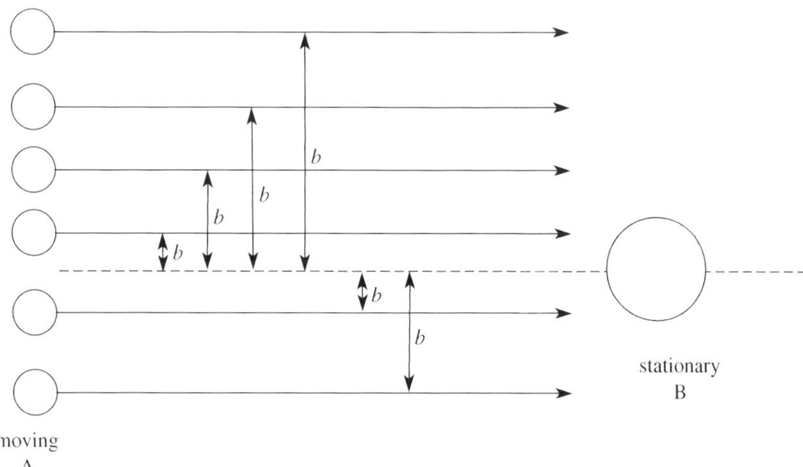

Figure 4.2 Definition of the impact parameter

- If $b \leq r_A + r_B$ collisions are possible.
- If $b > r_A + r_B$ collision is impossible.
- If $b = 0$ collision is head-on.
- If $b = r_A + r_B$ collision is *just* successful.

4.1.2 Formulation of the total collision rate

Encounters with impact parameter $b < r_A + r_B$ count towards the collision rate; those with $b > r_A + r_B$ do not. The total number of collisions per unit time per unit volume between A and B can be found from kinetic theory, and this *collision rate* is often called the *collision frequency*, $\mathcal{Z}_{AB}$. This quantity is proportional to the numbers of molecules of A and B per unit volume, n_A and n_B:

$$\mathcal{Z}_{AB} \propto n_A n_B \tag{4.3}$$

$$\mathcal{Z}_{AB} = Z_{AB} n_A n_B \tag{4.4}$$

Z_{AB} is the constant of proportionality, called the *collision number*, and this can be found from kinetic theory:

$$Z_{AB} = (8\pi kT/\mu)^{1/2}(r_A + r_B)^2 \tag{4.5}$$

where μ is the reduced mass, explained as follows.

Certain combinations of masses turn up in mechanical arguments when discussing systems of more than one object. Such combinations are called the reduced mass for the system. One such combination is the reduced mass, μ, for the approach of two molecules to each other, given by

$$\frac{1}{\mu} = \frac{1}{m_A} + \frac{1}{m_B} = \frac{m_A + m_B}{m_A m_B} \tag{4.6}$$

$$\mu = \frac{m_A m_B}{m_A + m_B} \tag{4.7}$$

The calculated collision rate was found to be *very much greater* than observed rates, suggesting that some factor limits the *effectiveness* of collisions in causing *reaction*; *viz*., only collisions which have at least a critical energy can result in reaction.

In simple collision theory, for a bimolecular reaction this critical energy, ε_0, is the *kinetic energy of relative translational motion along the line of centres* of the colliding molecules, loosely described as the 'violence' of the collision.

The fraction of molecules with energy at least a certain critical amount ε_0 is given by the Maxwell–Boltzmann distribution. This is a statistical mechanical deduction relating to the fraction of molecules which have energy at least a certain quantity. The kinetic energy of relative motion along the line of centres of two colliding particles involves $\frac{1}{2}mv^2$ for each; i.e., there are two terms involving a square, here v_A^2 and v_B^2.

This is called an energy in two squared terms. The Maxwell–Boltzmann expression for such a situation is

$$\frac{n}{n_{\text{total}}} = \exp\left(\frac{-\varepsilon}{kT}\right) \tag{4.8}$$

where the fraction refers to the number of molecules with energy at least ε in two squared terms.

In more complex situations, where e.g. rotations and vibrations are involved, there will be squared terms relating to each rotational and each vibrational mode. Each rotational mode involves the square of the angular velocity and contributes one squared term. Each vibrational mode involves a term from the potential energy and a term for the kinetic energy of vibration, and contributes two squared terms if the vibration is harmonic. If there are a total of $2s$ squared terms, then this is called energy in $2s$ squared terms. The Maxwell–Boltzmann distribution is correspondingly more complex (Section 4.5.8).

The fraction of molecules with energy at least ε_0 along the line of centres is thus a simple exponential term, $\exp(-\varepsilon_0/kT)$, so that the total *rate of reaction* is equal to the total rate of collisions *modified* by this exponential energy term. This is the total rate of *successful collisions*:

$$\text{calculated rate of reaction} = Z_{AB} n_A n_B \exp(-\varepsilon_0/kT) \tag{4.9}$$

$$= (8\pi kT/\mu)^{1/2}(r_A + r_B)^2 n_A n_B \exp(-\varepsilon_0/kT) \tag{4.10}$$

This rate of reaction can also be given as $k_{\text{calc}} n_A n_B$, giving for the rate constant:

$$k_{\text{calc}} = Z_{AB} \exp(-\varepsilon_0/kT) \tag{4.11}$$

$$= (8\pi kT/\mu)^{1/2}(r_A + r_B)^2 \exp(-\varepsilon_0/kT) \tag{4.12}$$

The term in $(r_A + r_B)^2$ in Z_{AB} can be replaced by the collision diameter, d_{AB}^2, or the cross sectional area, $\sigma = \pi(r_A + r_B)^2$.
Note. The area of a circle is πr^2, and the area of one surface of a disc is also πr^2.

These equations, as quoted, are in *molecular units*, and when comparison is made with the experimental Arrhenius equation, equation (4.11) must be converted into *molar units*.

Conversion to molar units:

μ per molecule must be converted to an equivalent quantity per mole;

k in J K^{-1} must be converted to R in J K^{-1} mol^{-1};

ε_0 in J must be converted into E_A in kJ mol^{-1};

n_A and n_B, as number of molecules per unit volume, into concentrations in mol dm^{-3};

or *vice versa* depending on how the values are given initially.

Worked Problem 4.1

Question. Taking Z, the collision number, to be 4×10^{10} mol^{-1} dm^3 s^{-1}, calculate the rate of collision, Ƶ, for two reactants both at a concentration of 2.5×10^{-2} mol dm^{-3}. Compare the result with an experimental rate of reaction of 1.4×10^{-6} mol dm^{-3} s^{-1}. Suggest a reason for any discrepancy.

Answer.

$$\begin{aligned} \text{Rate of collision} &= Z \times [\mathrm{A}][\mathrm{B}] \qquad (4.13) \\ &= 4 \times 10^{10} \times 2.5 \times 10^{-2} \times 2.5 \times 10^{-2}\ \mathrm{mol\ dm^{-3}\ s^{-1}} \\ &= 2.5 \times 10^{7}\ \mathrm{mol\ dm^{-3}\ s^{-1}}. \\ \text{Observed rate of reaction} &= 1.4 \times 10^{-6}\ \mathrm{mol\ dm^{-3}\ s^{-1}}. \\ \text{Observed rate/rate of collision} &= \frac{1.4 \times 10^{-6}\ \mathrm{mol\ dm^{-3}\ s^{-1}}}{2.5 \times 10^{7}\ \mathrm{mol\ dm^{-3}\ s^{-1}}} \\ &= 5.6 \times 10^{-14} \end{aligned}$$

Observed rate $\ll$ collision rate, suggesting that only a few collisions lead to reaction.

Problem 4.1 indicates that there is a large factor limiting the effectiveness of collisions leading to reaction. As shown in Section 4.1.2, this is the requirement for a certain critical energy, where for bimolecular reactions the activation energy is the kinetic energy of relative translational motion along the line of centres of the colliding molecules. When molecules collide they will not react unless this energy is at least the critical value, the activation energy. The rate of reaction depends crucially on the magnitude of this *activation energy*. The Arrhenius equation, (4.1), can be used to demonstrate the dramatic effect which the magnitude of E_{A} has on the rate constant and hence on the rate.

Worked Problem 4.2

Question. Use the Arrhenius equation to calculate the rate constants for the following two reactions which have the same magnitude for the A factor, $A = 5 \times 10^{10}$ mol^{-1} dm^3 s^{-1}. The temperature is 300 K.

(a) A reaction with activation energy of $+30$ kJ mol^{-1}.

(b) A reaction with activation energy of $+100$ kJ mol^{-1}.

Answer.

$$k = A \exp(-E_{\mathrm{A}}/RT) \qquad (4.1)$$

(a) $k = 5 \times 10^{10} \exp\left(\frac{-30 \times 10^3}{8.314 \times 300}\right) = 5 \times 10^{10}\,e^{-12.0} = 5 \times 10^{10} \times 6.0 \times 10^{-6}$
$= 3 \times 10^5 \text{ mol}^{-1} \text{ dm}^3 \text{ s}^{-1}$

(b) $k = 5 \times 10^{10} \exp\left(\frac{-100 \times 10^3}{8.314 \times 300}\right) = 5 \times 10^{10}\,e^{-40.1} = 5 \times 10^{10} \times 3.9 \times 10^{-18}$
$= 1.9 \times 10^{-7} \text{ mol}^{-1} \text{ dm}^3 \text{ s}^{-1}$

The reaction with the lower activation energy is faster by a factor of 1.5×10^{12}, demonstrating that the rate of a reaction is highly dependent on the magnitude of the activation energy.

There are a number of other variables in the collision theory expression for the rate constant and rate of reaction. These are considered explicitly in the following worked problems.

Worked Problem 4.3

Question.

(a) Calculate the *effect of temperature on Z* for a reaction if the temperature is increased from 500 K to 800 K.

(b) Calculate the *effect of temperature on the exponential term* for a reaction for which the activation energy is +45 kJ mol^{-1} if the temperature is increased from 500 K to 800 K.

(c) Calculate the *effect if $r_A + r_B$ is increased by a factor of two.*

(d) Calculate the *effect if μ, the reduced mass, is increased by a factor of two.*

(e) Calculate the *effect on the exponential term* if the activation energy is 35 kJ mol^{-1} when the temperature is 500 K.

(k, the Boltzmann constant, $= 1.380\,658 \times 10^{-23}$ J K^{-1}.)

Answer.

(a)
$$Z_{AB} = \left(\frac{8\pi kT}{\mu}\right)^{1/2} (r_A + r_B)^2 \tag{4.5}$$

If μ, r_A and r_B are kept constant then $Z \propto T^{1/2}$:

$$\frac{Z_{800}}{Z_{500}} = \left(\frac{800}{500}\right)^{1/2} = 1.265$$

(b) The activation energy given is the *observed* value, +45 kJ mol^{-1}. ε_0 in the exponential term is in molecular units.

Be careful of units here.

$$E_A = +45\,\text{kJ mol}^{-1} \text{ giving } \varepsilon_0 = \frac{45 \times 10^3}{6.022 \times 10^{23}} = 7.47 \times 10^{-20}\ \text{J}.$$

$$T = 500\,\text{K} \qquad \exp\left(\frac{-7.47 \times 10^{-20}\,\text{J}}{1.380\,658 \times 10^{-23}\text{J K}^{-1} \times 500\,\text{K}}\right) = \text{e}^{-10.82} = 2.00 \times 10^{-5}$$

$$T = 800\,\text{K} \qquad \exp\left(\frac{-7.47 \times 10^{-20}\,\text{J}}{1.380\,658 \times 10^{-23}\text{J K}^{-1} \times 800\,\text{K}}\right) = \text{e}^{-6.76} = 1.16 \times 10^{-3}$$

$$\text{Ratio:} \quad \frac{\exp_{800}}{\exp_{500}} = \frac{1.16 \times 10^{-3}}{2.00 \times 10^{-5}} = 57.8$$

Note the difference in the ratios for Z and the exponential term. Temperature has a *much larger* effect on the exponential term than on Z.

(c) $Z_{AB} \propto (r_A + r_B)^2$. If $r_A + r_B$ increases by a factor of two, then Z_{AB} increases by a factor of four.

(d) $Z_{AB} \propto 1/\mu^{1/2}$. If μ increases by a factor of two, then Z decreases by a factor of $1/2^{1/2} = 0.707$.

For the same fractional change, altering μ has a smaller effect than altering the size.

(e) $E_A = +35$ kJ mol^{-1} corresponds to

$$\varepsilon_0 = \frac{35 \times 10^3}{6.022 \times 10^{23}} = 5.81 \times 10^{-20}\ \text{J}$$

$$\exp\left(\frac{-\varepsilon_0}{kT}\right) = \exp\left(\frac{-5.81 \times 10^{-20}}{1.380\,658 \times 10^{-23}\ \text{J K}^{-1} \times 500\ \text{K}}\right) = \text{e}^{-8.42} = 2.21 \times 10^{-4}$$

A comparison of this exponential term for $E_A = +35$ kJ mol^{-1} with that for $E_A = +45$ kJ mol^{-1} gives $\frac{2.21 \times 10^{-4}}{2.00 \times 10^{-5}} = 11.1$. This is a large increase.

The above question demonstrates that the major effect on the rate constant is when temperature or activation energy change, rather than when mass or size, vary. These effects are examined more closely in Problem 4.4 below.

Worked problem 4.4

Question. The initial temperature is 300 K. The activation energy is +25 kJ mol^{-1}.

(a) Calculate the effect on the rate constant if the temperature is increased by a factor of two, $r_A + r_B$, μ and activation energy being kept constant.

(b) Calculate the effect on the rate constant if the activation energy is increased by a factor of two, $r_A + r_B$, μ and T being kept constant.

(c) Calculate the effect on the rate constant if the activation energy is decreased by a factor of two, $r_A + r_B$, μ and T being kept constant.

Answer.

$$k = \left(\frac{8\pi kT}{\mu}\right)^{1/2} (r_A + r_B)^2 \exp\left(\frac{-\varepsilon_0}{kT}\right) \tag{4.12}$$

$$Z = \left(\frac{8\pi kT}{\mu}\right)^{1/2} (r_A + r_B)^2 \tag{4.5}$$

(a) The effect of temperature on the rate constant comes from two sources; the exponential term and Z.

(i) Z increases by a factor of $(600/300)^{1/2} = \sqrt{2} = 1.414$.

(ii) The exponential term is $\exp(-\varepsilon_0/kT)$. E_A is given in kJ mol^{-1}. Units must be consistent, i.e. either molecular or molar.

At 300 K:

$$\text{molecular,}\ \exp\left(\frac{-\varepsilon_0}{kT}\right) = \exp\left(\frac{-25 \times 10^3\ \text{J}}{6.022 \times 10^{23} \times 1.380\ 658 \times 10^{-23}\ \text{J K}^{-1} \times 300\ \text{K}}\right)$$
$$= e^{-10.02} = 4.4 \times 10^{-5}$$

At 600 K:

$$\text{molecular,}\ \exp\left(\frac{-\varepsilon_0}{kT}\right) = \exp\left(\frac{-25 \times 10^3\ \text{J}}{6.022 \times 10^{23} \times 1.380\ 658 \times 10^{-23}\ \text{J K}^{-1} \times 600\ \text{K}}\right)$$
$$= e^{-5.01} = 6.7 \times 10^{-3}$$

$$\exp_{600}/\exp_{300} = \frac{6.7 \times 10^{-3}}{4.4 \times 10^{-5}} = 150$$

Doubling the temperature has a much greater effect on the exponential term than on Z. The total effect is an increase by a factor of $1.414 \times 150 = 212$.

(b) The exponential term is $\exp(-\varepsilon_0/kT)$. E_A is given in kJ mol^{-1}. Units must be consistent, i.e. either molecular or molar.

$$\text{Molecular,}\quad \exp\left(\frac{-\varepsilon_0}{kT}\right) = \exp\left(\frac{-25 \times 10^3\ \text{J}}{6.022 \times 10^{23} \times 1.380\,658 \times 10^{-23}\ \text{J K}^{-1} \times 300\ \text{K}}\right)$$
$$= e^{-10.02} = 4.4 \times 10^{-5}$$

Activation energy doubled gives

$$\text{molecular,}\quad \exp\left(\frac{-\varepsilon_0}{kT}\right) = \exp\left(\frac{-50 \times 10^3\ \text{J}}{6.022 \times 10^{23} \times 1.380\,658 \times 10^{-23}\ \text{J K}^{-1} \times 300\ \text{K}}\right)$$
$$= e^{-20.05} = 2.0 \times 10^{-9}$$

The effect of doubling the activation energy is to decrease the exponential by a factor of $2.0 \times 10^{-9}/4.4 \times 10^{-5} = 4.5 \times 10^{-5}$. This is a very large decrease.

(c) *Activation energy halved gives*

$$\text{molecular,}\quad \exp\left(\frac{-\varepsilon_0}{kT}\right) = \exp\left(\frac{-12.5 \times 10^3\ \text{J}}{6.022 \times 10^{23} \times 1.380\,658 \times 10^{-23}\ \text{J K}^{-1} \times 300\ \text{K}}\right)$$
$$= e^{-5.01} = 6.7 \times 10^{-3}$$

The effect of halving the activation energy is to increase the exponential by a factor of $6.7 \times 10^{-3}/4.4 \times 10^{-5} = 150$. This is a large change.

Note that, provided the units are consistent throughout, molecular and molar forms of the exponential give the same results:

$$\text{molecular,}\quad \exp\left(\frac{-\varepsilon_0}{kT}\right) = \exp\left(\frac{-25 \times 10^3\ \text{J}}{6.022 \times 10^{23} \times 1.380\,658 \times 10^{-23}\ \text{J K}^{-1} \times 300\ \text{K}}\right)$$
$$= e^{-10.02} = 4.4 \times 10^{-5}$$

$$\text{molar,}\quad \exp\left(\frac{-25 \times 10^3\ \text{J mol}^{-1}}{8.314\ \text{J mol}^{-1}\ \text{K}^{-1} \times 300\ \text{K}}\right) = e^{-10.02} = 4.4 \times 10^{-5}$$

4.1.3 The *p* factor

Equation (4.11) can be compared with the experimental Arrhenius equation (4.1), care being taken with the units of k and R in the exponential term.

$$k_{\text{calc}} = Z_{\text{AB}} \exp(-\varepsilon_0/kT) \tag{4.11}$$

$$k = A \exp(-E_A/RT) \tag{4.1}$$

For some reactions Z_{AB} and A have similar values, but there are also many reactions for which they differ by many powers of ten. This requires postulating another *factor, 'p', which limits the theoretical rate of reaction*, and whose magnitude can be found by comparison with experiment.

$$\frac{k_{calc}}{k_{obs}} = \frac{pZ_{AB}\exp(-\varepsilon_0/kT)}{A\exp(-E_A/RT)} \tag{4.14}$$

Watching units of k and R, and equating ε_0 with a *molecular* E_A, makes $k_{calc} = k_{obs}$ if

$$p = A/Z_{AB} \tag{4.15}$$

Worked Problem 4.5

Question.

(a) The following gives values of observed A factors to compare with a typical Z value of 4×10^{10} mol^{-1} dm^3 s^{-1} for the same temperature. Calculate the p factor for each reaction, and comment on its magnitudes.

Reaction	A/mol^{-1} dm^3 s^{-1}
$NO_2^\bullet + F_2 \rightarrow NO_2F + F^\bullet$	1.6×10^9
$2ClO^\bullet \rightarrow Cl_2 + O_2$	6×10^7
$2NOCl \rightarrow 2NO + Cl_2$	1×10^{10}
$O_3 + C_3H_8 \rightarrow C_3H_7O^\bullet + HO_2^\bullet$	1×10^6
$CH_3^\bullet + C_6H_5CH_3 \rightarrow CH_4 + C_6H_5CH_2^\bullet$	1×10^7
butadiene + propenal $\rightarrow$ 1,2,3,6-tetrahydrobenzaldehyde	1.5×10^6

(b) Suggest whether the collision rate would increase or decrease for the following reactions:

(i) a reaction between ions of like charge;
(ii) a reaction between ions of unlike charge.

Answer.

(a)

Reaction	p
$NO_2^\bullet + F_2 \rightarrow NO_2F + F^\bullet$	4×10^{-2}
$2ClO^\bullet \rightarrow Cl_2 + O_2$	1.5×10^{-3}
$2NOCl \rightarrow 2NO + Cl_2$	2.5×10^{-1}
$O_3 + C_3H_8 \rightarrow C_3H_7O^\bullet + HO_2^\bullet$	2.5×10^{-5}
$CH_3^\bullet + C_6H_5CH_3 \rightarrow CH_4 + C_6H_5CH_2^\bullet$	2.5×10^{-4}
butadiene + propenal $\rightarrow$ 1,2,3,6-tetrahydrobenzaldehyde	3.8×10^{-5}

As the complexity of the reactants increases the p factor decreases, and the discrepancy with collision theory increases. This reflects the inadequacy of simple collision theory, especially the neglect of the internal structures of the reactants, and intermolecular interactions.

(b) The latter effect becomes particularly important when considering reactions between charged species:

(i) there will be strong attractive forces between the reactants and this would be expected to increase the collision rate;

(ii) there will be strong repulsive forces leading to a decrease in collision rate.

This 'p' factor has also been interpreted as a preference for a certain direction, or angle of approach, of the reacting molecules. Molecular beam experiments show some reactions to have a quite decided preference for a specific direction and/or angle of approach.

The 'p' factor could also be a consequence of the physically naïve hard sphere model used. This model requires that there be no intermolecular interactions either at close distances or at a distance. This has been shown not to be the case, with evidence coming from many aspects of physical chemistry. Also, vibrations and rotations *do* affect reaction, as is shown by experiments with molecular beams.

4.1.4 Reaction between like molecules

The rate of reaction can be calculated in a similar way, making sure that each A molecule is not counted twice – each A molecule can be considered as both 'hitting' and being 'hit by' another A molecule.

$$k = Z_{\mathrm{AA}}(-\varepsilon_0/kT) \tag{4.16}$$

where

$$Z_{\mathrm{AA}} = (32\pi kT/\mu)^{1/2} r_{\mathrm{A}}^2 \tag{4.17}$$

4.2 Modified Collision Theory

4.2.1 A new definition of a collision

Here a *collision* is seen in terms of the *deflection* of the molecules from their original paths as they approach each other. Molecular beam experiments are ideally suited to study collisions. Two beams of molecules are shot at 90° to each other. Several things can happen at the point of intersection of the beams.

- Both A and B molecules can continue on *undeflected* along the original direction of the beam. They are *unscattered.*
- A and B molecules can interact and as a result be *deflected* from their original trajectories. They are *scattered.*

The beam experiments detect scattered and unscattered A and B molecules. Any A or B *deflected* and *scattered* must have *collided.*

In simple collision theory, if the impact parameter $b \leq r_A + r_B$, then collision will occur; if $b > r_A + r_B$ collision cannot occur. In molecular beam studies a collision is still defined in terms of a distance apart of the trajectories, but this is no longer $r_A + r_B$, but is a distance b_{max}. If the impact parameter $b \leq b_{max}$, collision occurs, and the *numerical value of* b_{max} is found from scattering experiments and *is closely related to the minimum angle of scattering able to be detected.*

The impact parameter is defined as before (Figure 4.2 and Section 4.1.1).

Within the range of impact parameters from $b = 0$ to $b = \infty$ there is a critical range of values which result in collision.

- If $b = 0$ there will be a head-on collision
- If $b < b_{max}$ there will be a collision
- If $b > b_{max}$ there can be no collision
- If $b = b_{max}$ the encounter has *just* been successful as a collision.

With this definition of a collision the cross sectional area is now

$$\sigma = \pi b_{max}^2 \tag{4.18}$$

and this cross section is closely related to the rate constant. Each impact parameter is associated with a particular value of the angle of scattering: $b = 0$ corresponds to a 'head-on' collision which results in totally backward scattering at 180°, whereas if b is just less than b_{max} interactions will be small, resulting in a very small angle of scattering. Intermediate situations, resulting in stronger interactions, lead to scattering between 0 and 180°. If $b > b_{max}$ there will be no collision and no scattering (Figure 4.3).

The crucial parameter is thus b_{max}, and its magnitude can be found from experiment.

4.2.2 Reactive collisions

When *reaction* occurs then successful collisions are detected from the *scattering patterns of products*. These give an effective $b_{max(R)}$, and cross section for reaction,

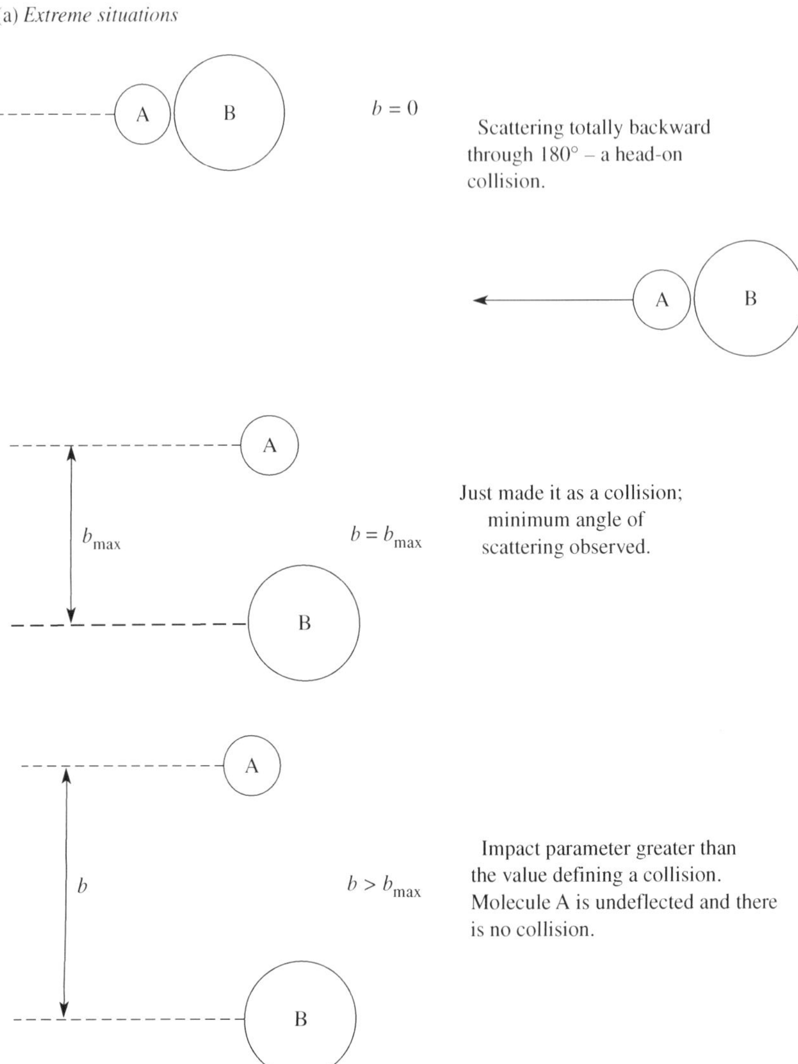

Figure 4.3 The relation between impact parameter and angle of scattering

$\sigma_{(R)}$, from which the related rate constant for reaction can be found. If the impact parameter for the two molecules is greater than $b_{max(R)}$, reaction does not occur and there can be no scattering of products.

4.2.3 Contour diagrams for scattering of products of a reaction

The *amount of scattering into various angles* and the *distribution of velocities* of the products can be represented on a *contour diagram* where the initial direction of

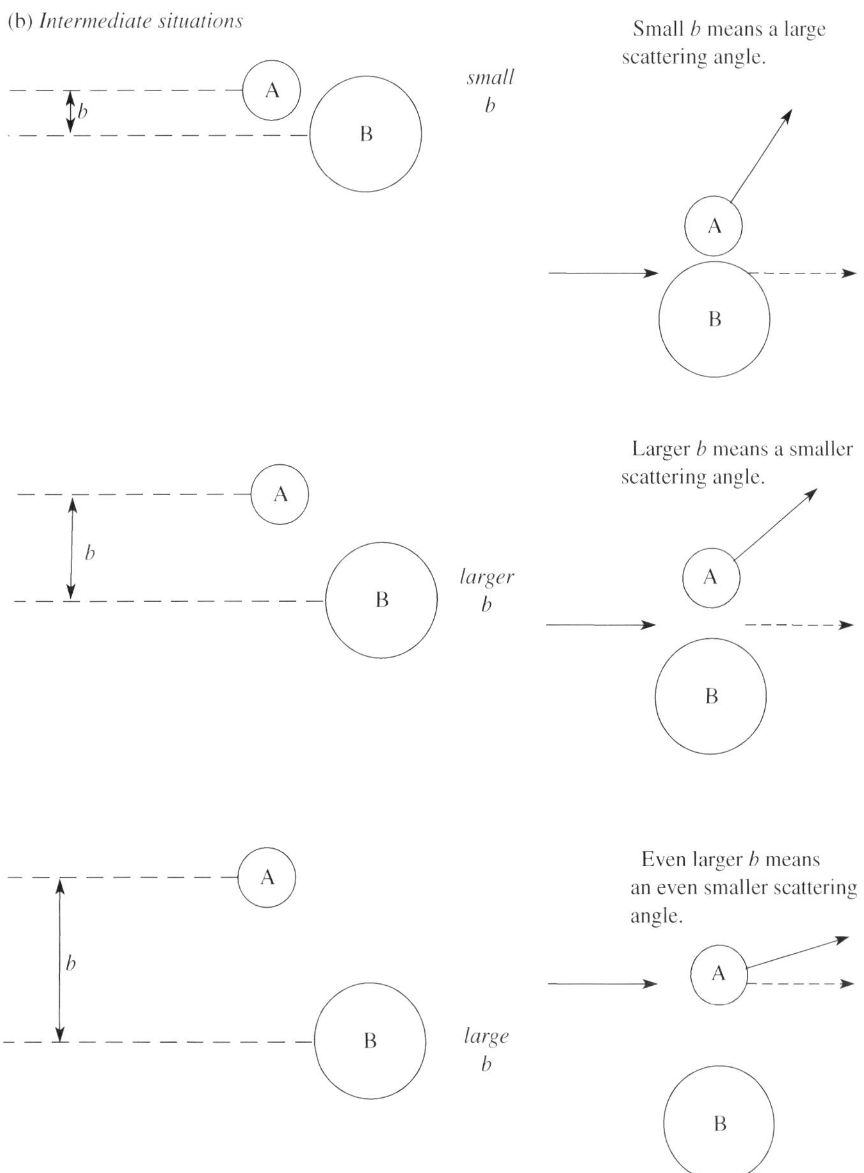

Figure 4.3 (*Continued*)

approach of reactants is from left to right. Contours join up regions of *equal scattering*, and the diagram shows immediately the distribution of scattering of products. *Concentric circles* represent regions where the scattered products have *equal velocities*, and from this and the contours the most probable velocity of the scattered products is found in the region of maximum scattering (Figure 4.4).

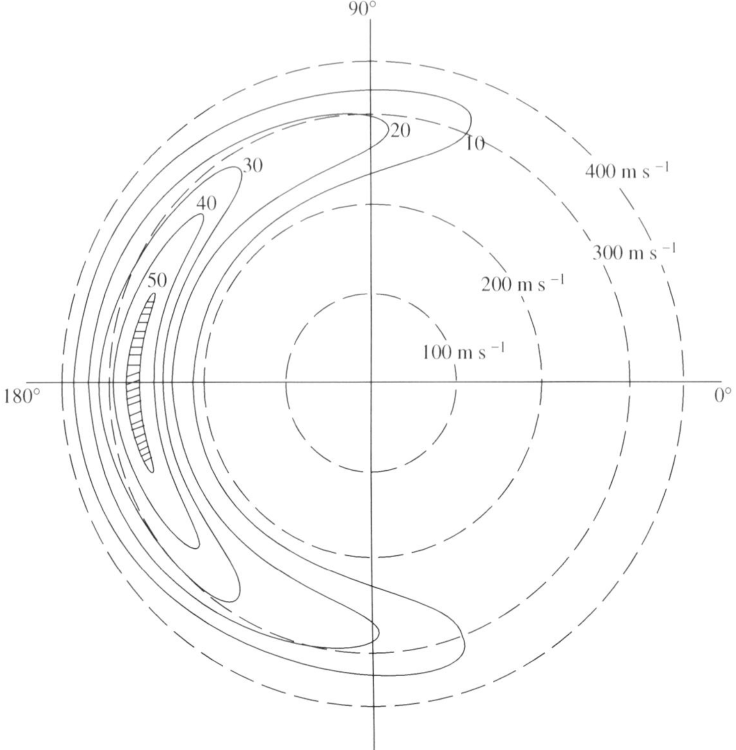

Figure 4.4 Molecular beam contour diagram showing scattering and velocities

Worked Problem 4.6

Question. Interpret the following molecular beam contour diagrams for reactions (see Figures 4.5–4.8). Indicate the regions of most intense scattering, the most likely angle of scattering, the most likely velocity of the scattered products and the region of least scattering.

Answer.

Figure	Most intense scattering	Most likely angle	Most likely velocity	Least scattering
4.5	contour 25	150° and 210°	300 m s^{-1}	1st/4th quadrants
4.6	contour 30	90° and 270°	200 m s^{-1}	angles close to 0 and 180°
4.7	contour 40	0°	80 m s^{-1}	2nd/3rd quadrants
4.8	contour 15	60° and 300°	100 m s^{-1}	2nd/3rd quadrants

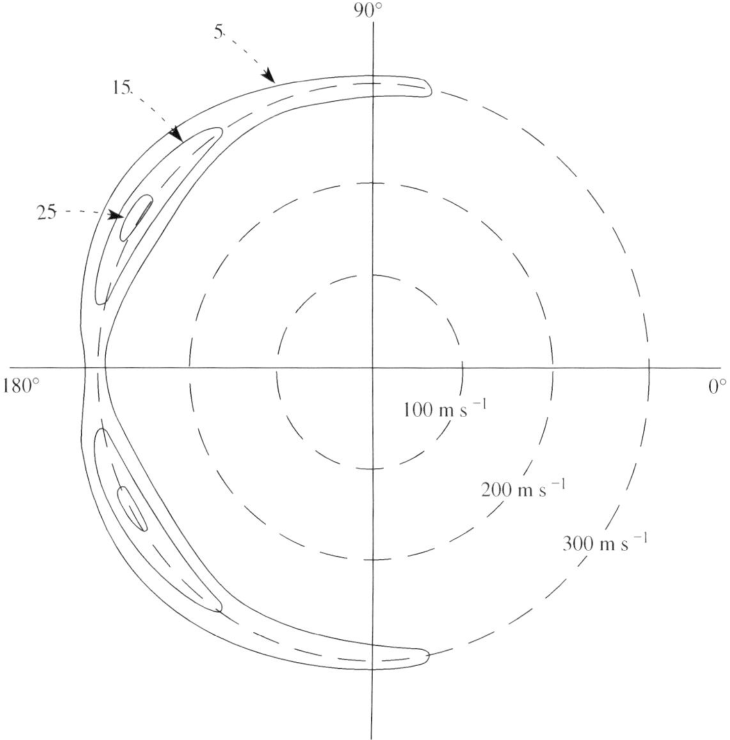

Figure 4.5 Molecular beam contour diagram

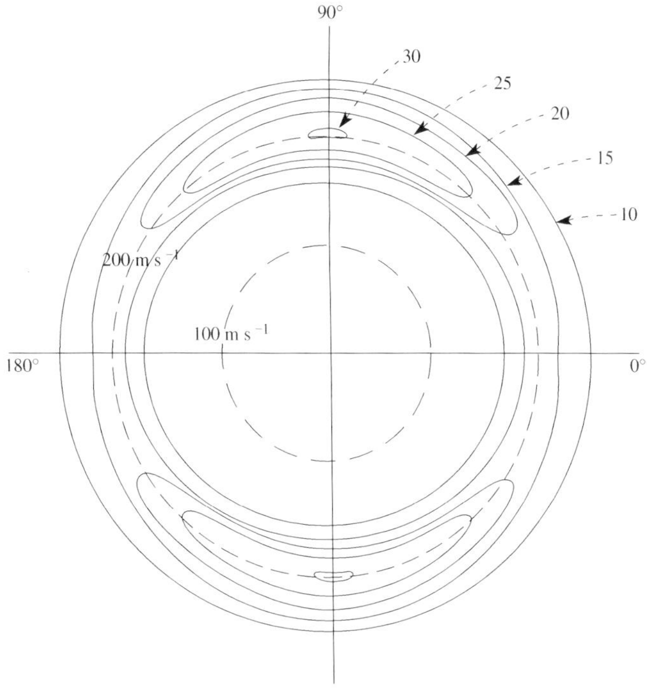

Figure 4.6 Molecular beam contour diagram

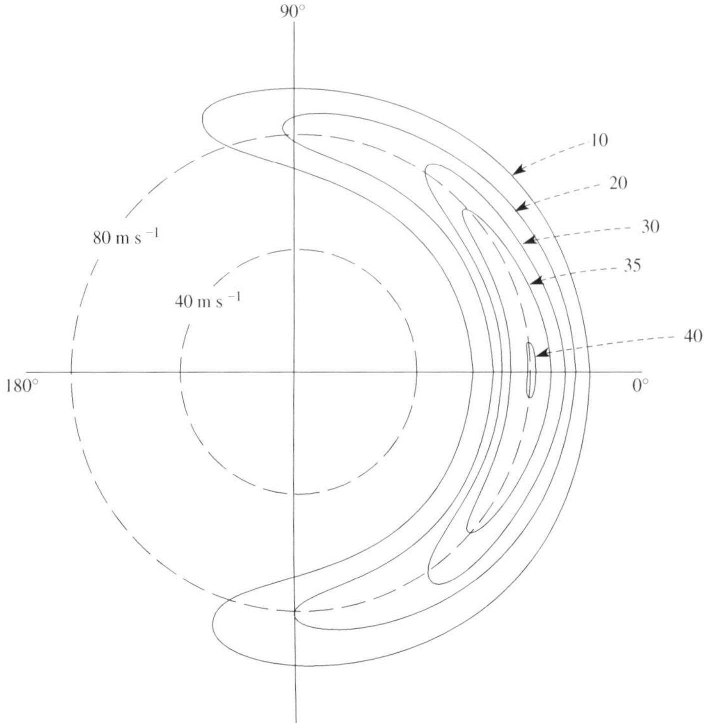

Figure 4.7 Molecular beam contour diagram

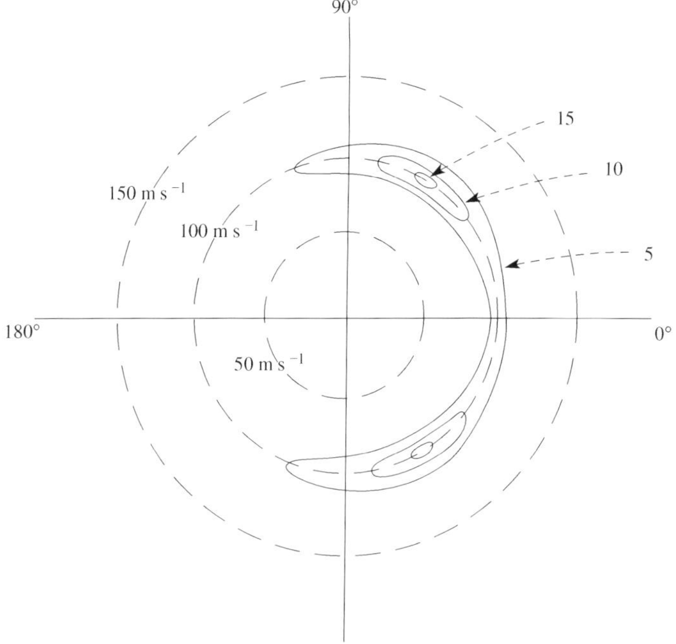

Figure 4.8 Molecular beam contour diagram

4.2.4 Forward scattering: the stripping or grazing mechanism

A typical contour diagram for a reaction e.g. A + BC → AB + C showing *forward* scattering is given in Figure 4.9.

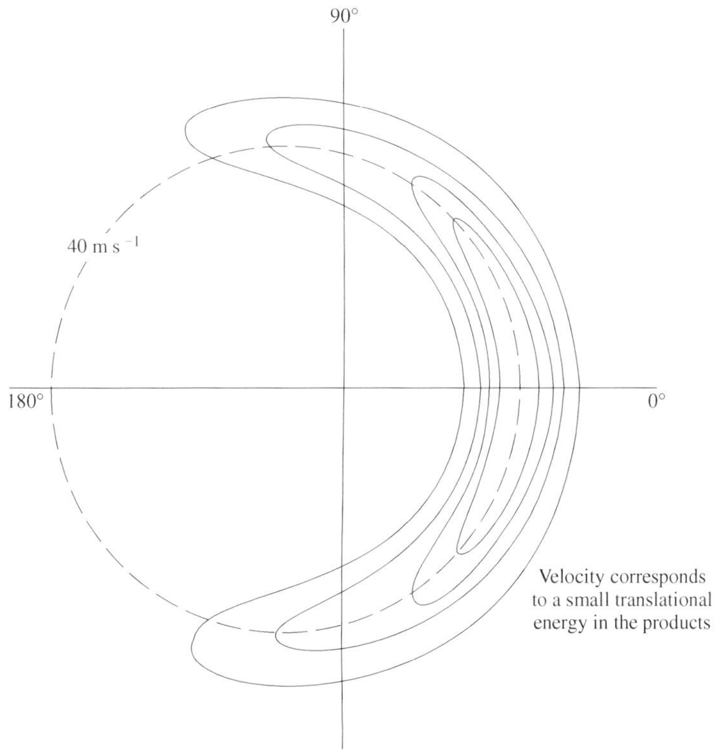

Figure 4.9 Molecular beam contour diagram showing forward scattering

Product AB has forward scattering with maximum very close to 0° so that AB emerges from the reactive collision not much deflected from the direction of A before collision. Other angles of forward scattering are of course possible. C is also not much deflected by the reactive collision and continues on with only slight deflection from the original direction of BC (Figure 4.10).

Reactions showing *forward* scattering have *large impact parameters* and *large cross sections*. The mechanism is called '*stripping*', since A simply pulls off B from BC in passing and at a distance, leaving C to continue almost in the same direction. A, having made a successful 'grab' at B, continues on as AB with little change in direction. The lack of deflection shows that the interactions are so strong that they are significant even at a distance, thereby removing the need for close contact. Obviously there are gradations. The *most probable velocity* of the scattered products is low,

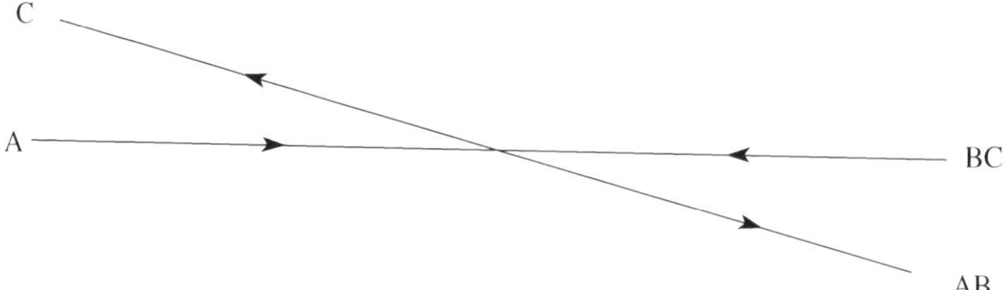

Figure 4.10 Deflection as a result of collision

showing that *translational energy* of the products is *low* so that most of the energy of reaction in this exothermic reaction goes into *vibrational excitation* of products. This ties in with the potential energy surface for reaction (Section 5.5).

4.2.5 Backward scattering: the rebound mechanism

A typical contour diagram for reaction, e.g. $P + MN \rightarrow PM + N$, is given in Figure 4.11.

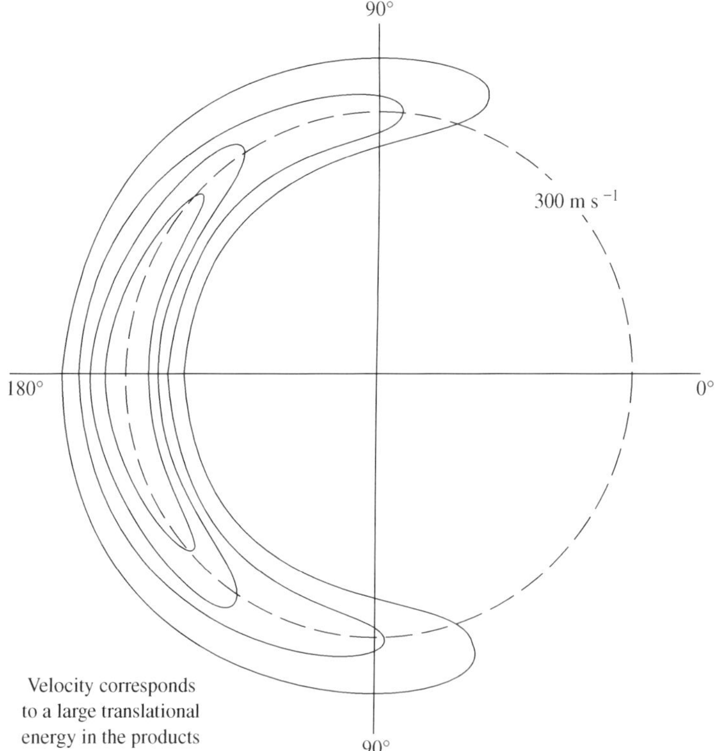

Figure 4.11 Molecular beam contour diagram showing backward scattering

Product PM has backward scattering with maximum very close to 180°, so that PM emerges from the reactive collision rebounding almost totally backwards from the direction in which P was originally moving. Likewise, N rebounds backward from the original direction of MN.

Reactions showing *backward* scattering have *small impact parameters* and *cross sections*. The mechanism is called '*rebound*' because P hits MN head on, pulls M away from N and, under the influence of the repulsive forces between M and N, PM moves almost totally backwards. N also rebounds backwards from PM under the same repulsive forces, and returns only slightly deflected from its original path (Figure 4.12).

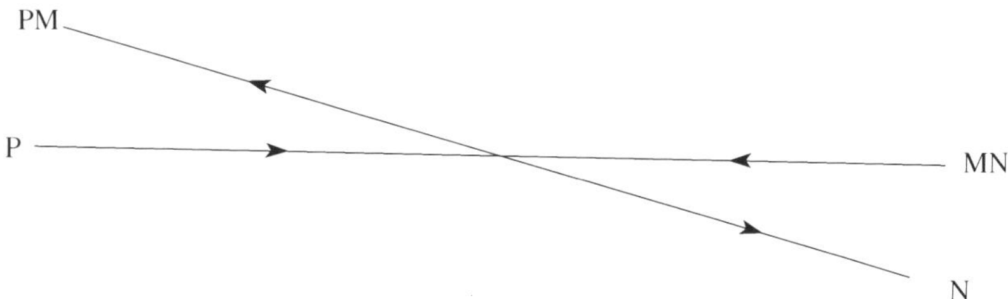

Figure 4.12 Deflection as a result of collision

The *most probable velocity* is high, showing that the *translational energy* of the products is *high* and that they are separating under short range repulsions. The *vibrational energy* of the products of backward scattering is consequently *low*. This is totally different from the 'stripping' mechanism, where the reactive collision takes place under a very strong attractive interaction and energy is channelled into vibration. Again the results tie up with the potential energy surface for a reaction occurring with low impact parameter (Section 5.6.1).

4.2.6 Scattering diagrams for long-lived complexes

When the *lifetime* of the collision is of the order of *several rotations* a *collision complex* is formed, and the contour diagram can be approximately *symmetric* with respect to 90° (Figures 4.13 and 4.14).

In this type of reaction, the interactions keep the two reactants in close contact with each other for long enough for them to 'forget' the directions in which they would have parted had they been able to do so before rotation. They thus move apart in random directions.

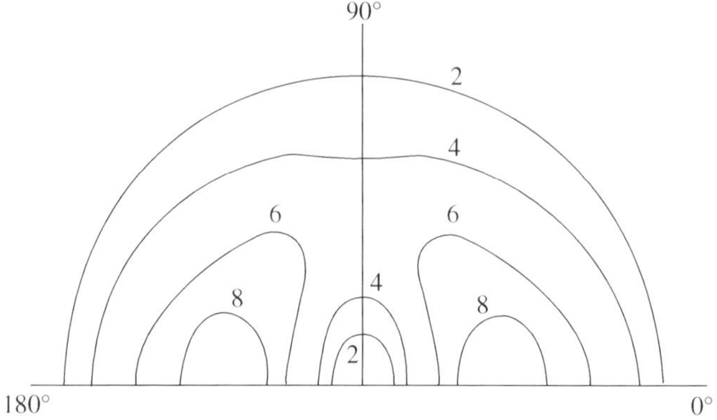

Figure 4.13 Molecular beam contour diagram showing symmetric scattering

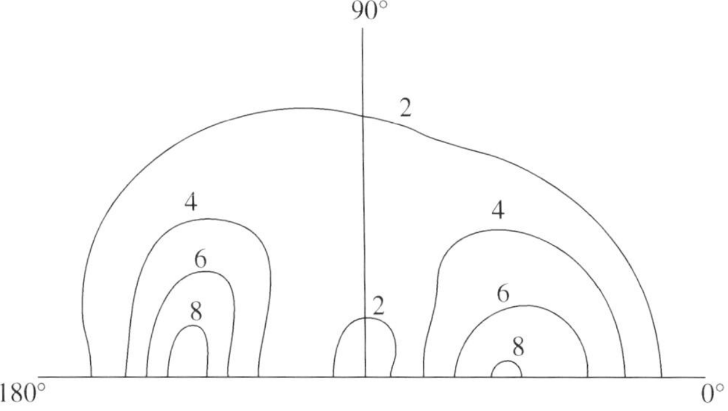

Figure 4.14 Molecular beam contour diagram showing symmetric scattering

Worked Problem 4.7

Question. Interpret the following diagrams (Figures 4.15–4.17).

Answer.

Figure	Type of scattering	Symmetry	Type of reaction	Cross section	Velocity in product	Product energy
4.15	backward	asymmetric	rebound	small	high	translation high
4.16	mixed	symmetric	collision complex	varied	mixed	mixed
4.17	forward	asymmetric	stripping	large	low	vibration high

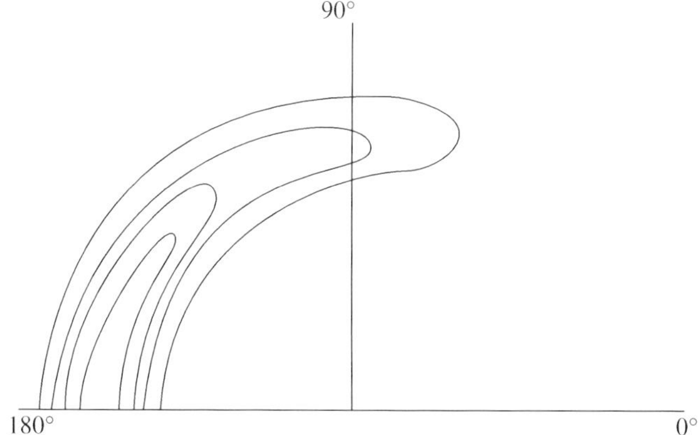

Figure 4.15 Molecular beam contour diagram

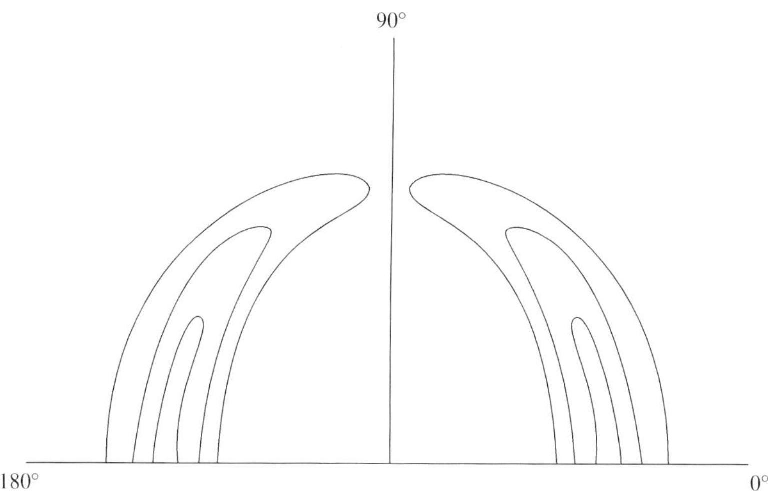

Figure 4.16 Molecular beam contour diagram

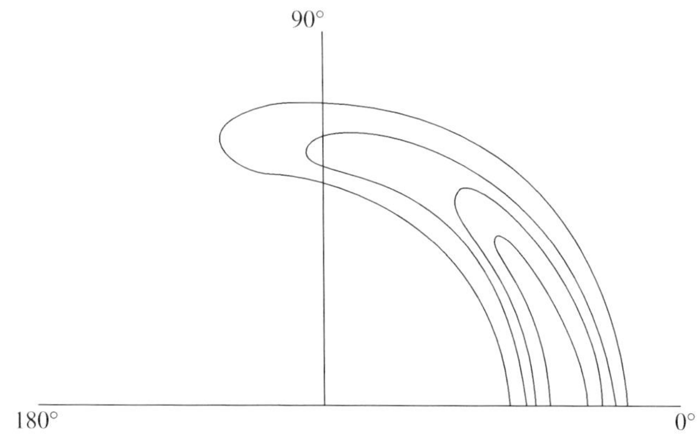

Figure 4.17 Molecular beam contour diagram

The first three columns, type of scattering, symmetry and type of reaction, give the inferences made from the actual diagrams. The final three columns, cross section, velocity in product and product energy, are then further deductions which come from the type of reaction which has been inferred.

Modified collision theory along with molecular beam studies has greatly expanded the knowledge of reaction at the microscopic level for many simple elementary reactions.

4.3 Transition State Theory

Transition state theory describes the changes of geometrical configuration which occur when suitably activated molecules, having the required critical energy, react. It gives a detailed account of the absolute rate of reaction. The physical steps of energy transfer in activation and deactivation are outside the remit of the theory.

4.3.1 Transition state theory, configuration and potential energy

The reacting molecules and products are described in terms of all the atoms in a single 'reaction unit' made up of all the reacting species and products, e.g. the reaction

$$CH_3^\bullet + C_2H_6 \rightarrow CH_4 + C_2H_5^\bullet$$

is described in terms of the unit

```
H    H     H
 \    \   /
H-C---C-C-H
 /    /\   \
H    H  H   H
```

"reaction unit"

rather than as the independent species

$$CH_3^\bullet \qquad C_2H_6 \qquad CH_4 \qquad C_2H_5^\bullet$$

When the geometrical configuration changes, the potential energy, PE, of the 'reaction unit' also changes, and a PE surface summarizes these changes.

For the *general* reaction e.g. $A + B \rightarrow C + D$, where A, B, C and D are polyatomic molecules, the PE surface is n-dimensional. Fortunately, all the basic properties of an n-dimensional surface of relevance to kinetics can be exemplified by the simpler three-dimensional surface. This describes the reaction e.g. $X + YZ \rightarrow XY + Z$, where

X, Y and Z are atoms, and where a linear approach of X to YZ and recession of Z is assumed, implying a linear configuration for the 'reaction unit':

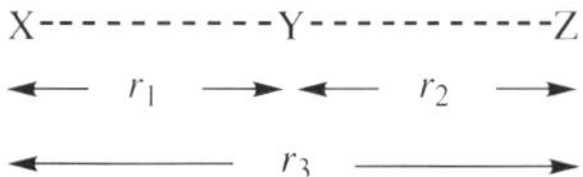

in which all configurations can be described by the distances r_1 and r_2, with $r_3 = r_1 + r_2$.

There is no interaction between X and Y or between X and Z when X is at large distances from YZ, and the PE is that for YZ at its equilibrium internuclear distance. As r_1 decreases, attractive interactions are set up, these being different at different distances. A quantum mechanical calculation shows how the PE increases as the distance r_1 decreases. Interactions between X and Y gradually become comparable to these between Y and Z.

Finally the interactions between X and Y become greater than those between Y and Z. The configurations reached as Z recedes from XY result in progressively decreasing potential energies. When r_2 is very large the PE is virtually that for XY at its equilibrium internuclear distance.

These configurational changes occur at constant total energy, resulting in an interconversion of kinetic and potential energy.

The calculations are summarized in the form of a table such as

r_1	r_2	PE
–	–	–
–	–	–

or on a three-dimensional PE surface (Figure 4.18).

This shows the PE for all possible configurations. The total energy is given as a horizontal plane through the surface, and the kinetic energy for any configuration is the vertical distance between this plane and the point representing the configuration.

4.3.2 Properties of the potential energy surface relevant to transition state theory

- The PE surface (Figure 4.18) is unique to a given reaction. All units (X---Y---Z) with the same configuration have the same PE, and correspond to the same point on the surface, irrespective of their total energy. One point could represent a unit (X---Y---Z) with enough energy to react, or one having insufficient energy for reaction.

- An understanding of what is implied by a Morse curve is necessary here, in order to understand the following point. A diatomic molecule has one normal mode of vibration. If the vibration behaves as a harmonic oscillator, the PE is proportional to the square of the displacement from the equilibrium internuclear distance. The

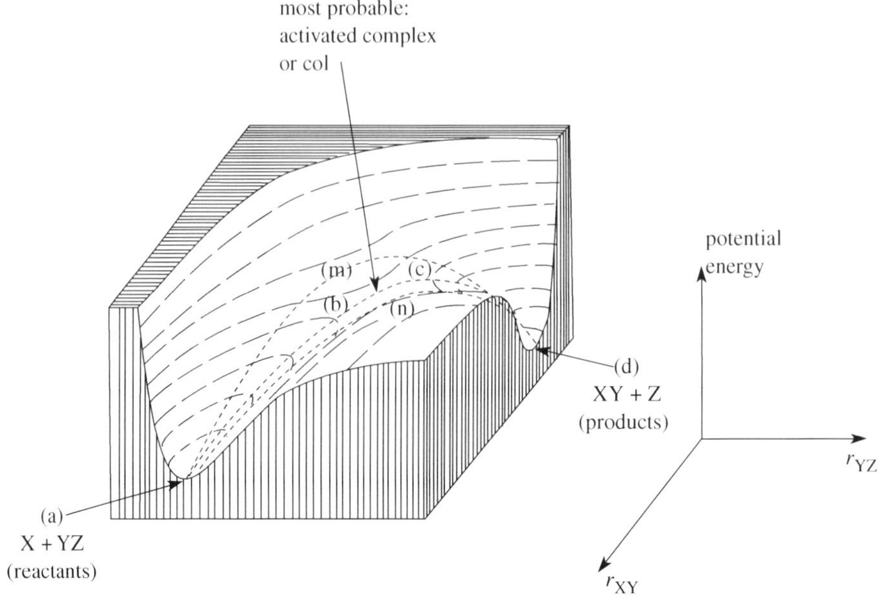

Figure 4.18 A 3-D potential energy surface

equation describing this motion is a parabola where PE is plotted against distance. The energy levels are equally spaced.

If the motion is not that of a harmonic oscillator, the PE is no longer proportional to the square of the displacement, and the equation describing the motion is now a curve with a plateau approaching a horizontal PE asymptote. This curve is the Morse curve. The asymptote represents the dissociation limit and the energy levels become closer together as the PE increases (Figure 4.19).

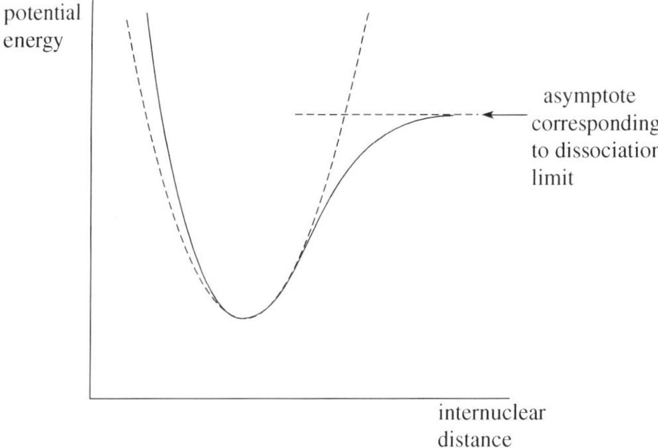

Figure 4.19 Morse curve for anharmonic vibration (the broken curve is a parabola corresponding to harmonic vibration)

- Point *a* corresponds to X at large distances from YZ. Interaction energy between X and YZ is virtually zero, and the side view of the model shows the typical Morse PE curve for the diatomic molecule YZ. Point *d* represents products, giving the PE for molecule XY with Z at a large distance, the side view giving the Morse curve for XY.

  During reaction the unit (X---Y---Z) changes from configuration *a* to configuration *d*. Any continuous path linking point *a* to point *d* is a physically possible route, and the route chosen by the vast majority of the units (X----Y----Z) will be the most probable one.

- Moving away from *a* or *d* in *all* directions gives an *increase* in PE, and on *every geometrically possible route* between *a* and *d* there will be some point with a *maximum* PE. Reaction requires the unit (X----Y----Z) to pass through some *configuration of maximum PE*. For instance paths *a*, *m*, *d* or *a*, *n*, *d* are possible routes.

- The Maxwell–Boltzmann distribution defines the most probable route. Here the Maxwell–Boltzmann distribution is being used in terms of a *probability*, rather than a fraction of molecules with energy at least a certain critical energy. The probability that a molecule has a given potential energy at any point on the surface is proportional to $\exp(-\varepsilon/kT)$, where ε is the PE at the point.

  Probability decreases as the PE increases, and the two most probable states are *a* the reactants, and *d* the products, while the lowest lying PE maximum is the most probable one. Reaction involves configurational changes along the most probable route. This is *the sequence of lowest lying configurations* which go through the *lowest lying PE maximum*. It is often called the minimum energy route or path, or the reaction coordinate, i.e. route *a*, *b*, *c*, *d*. Most 'units' (X----Y----Z) follow this route up the entrance valley through the critical maximum and down the exit valley. The theory can allow for other routes by using a weighting factor for each route calculated by the Maxwell–Boltzmann distribution.

- The configuration of maximum PE is called *the activated complex; the transition state; or the critical configuration;* and the unit (X----Y---Z) *must* attain this configuration before reaction can take place. Possessing the critical energy is not sufficient; the fundamental requirement is attainment of this critical configuration.

- The vertical distance between *a* and the critical configuration gives the necessary PE for reaction, and the minimum total energy needed to attain the critical PE.

- Any change in the dimensions of the activated complex along the reaction coordinate leads to a decrease in PE, behaviour not found in molecules. The Morse curve shown earlier (Figure 4.19) demonstrates that for a molecule displacement from the equilibrium position always leads to an increase in PE. The activated complex should be regarded *as merely a critical configuration*. Changes in dimensions in all other directions result in an increase in PE, behaviour typical of vibration about an equilibrium internuclear distance.

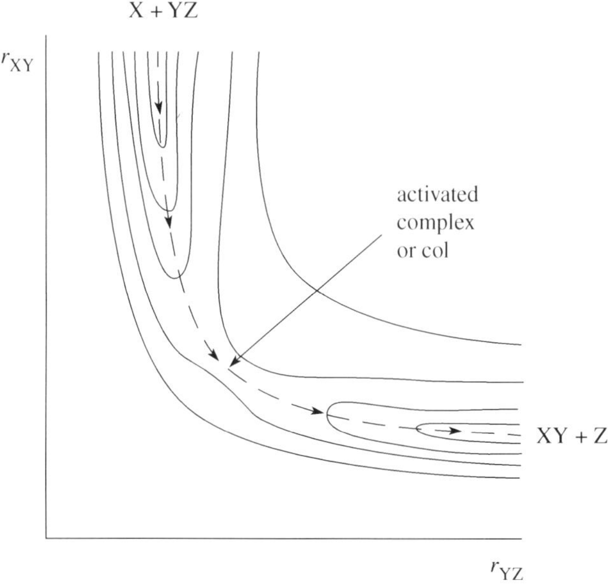

Figure 4.20 The potential energy contour diagram corresponding to figure 4.18

- Values of r_1 and r_2 for configurations of the same PE give a contour diagram of curves of constant PE (Figure 4.20).The activated complex is again at the col or the saddle point.

- Following the reaction path on the PE surface gives a PE profile showing how the PE changes as the reaction entity (X----Y----Z) changes configuration along the reaction coordinate (Figure 4.21).

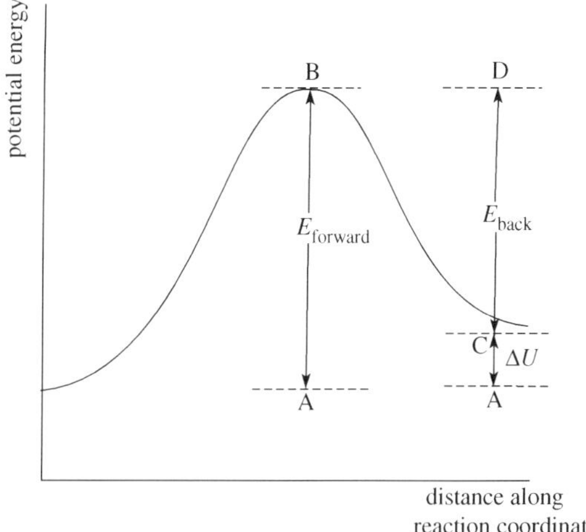

Figure 4.21 The potential energy profile corresponding to figure 4.18

- The activated complex lies at the top of the PE profile. The vertical distance AB corresponds to the critical energy required for reaction. The vertical distance CD corresponds to the critical energy for the back reaction. AC gives the PE change between reactants and products, approximately equal to ΔU^{θ} of reaction.
- The PE profile shows very clearly the *decrease* in PE resulting from any change in dimensions of the activated complex *along the reaction coordinate.*

Worked Problem 4.8

Question. Figures 4.22–4.24 show three PE profiles. Classify these as exothermic, endothermic or thermoneutral in the forward and back reactions.

Answer.

Figure	Forward reaction	Back reaction
4.22	thermoneutral	thermoneutral
4.23	*endo*thermal	*exo*thermal
4.24	*exo*thermal	*endo*thermal

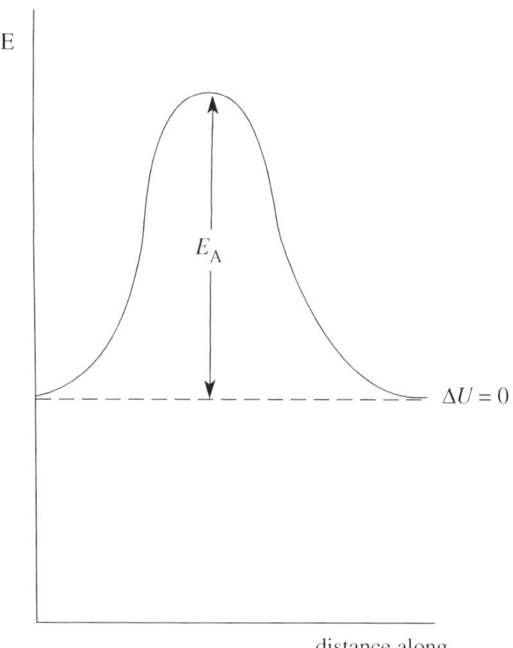

Figure 4.22 A potential energy profile

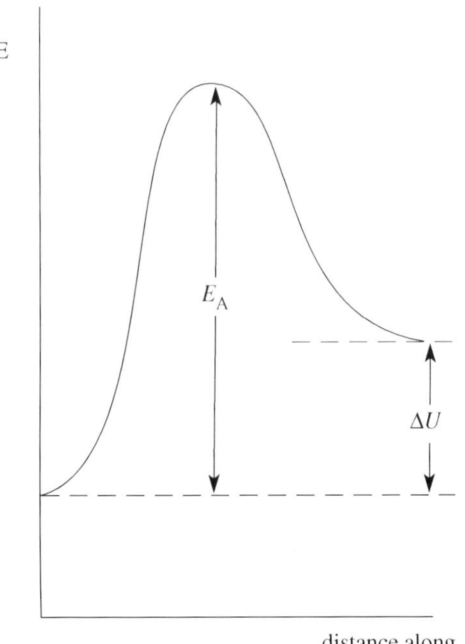

Figure 4.23 A potential energy profile

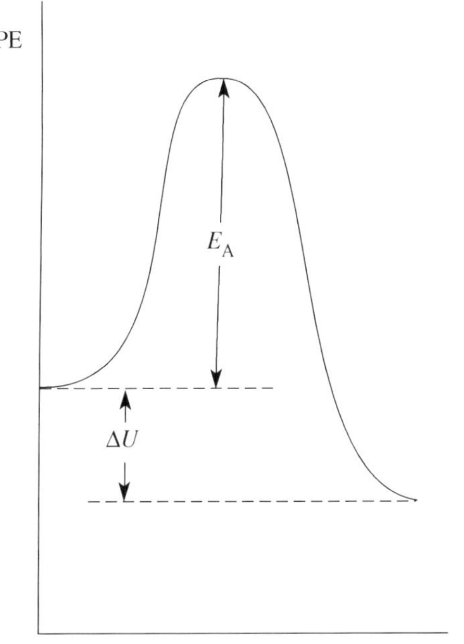

Figure 4.24 A potential energy profile

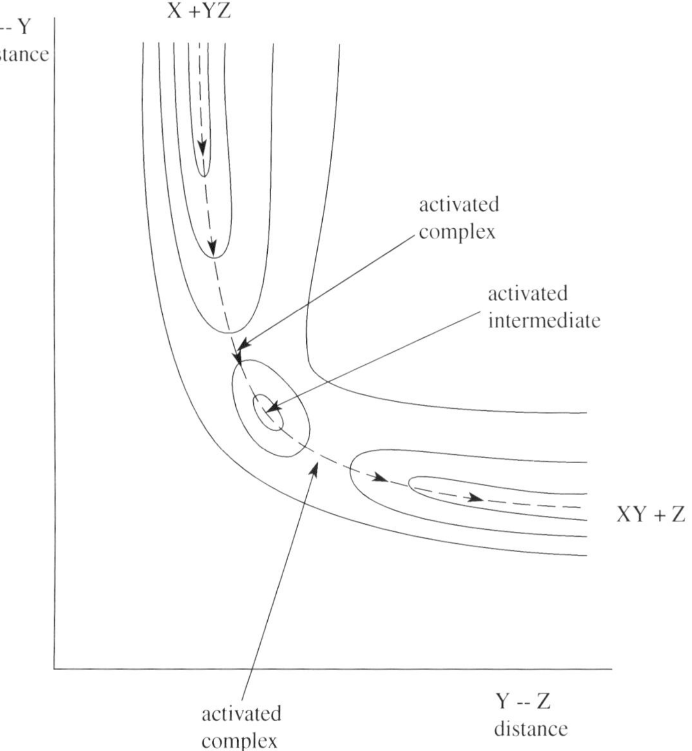

Figure 4.25 A potential energy contour diagram showing an activated intermediate

- An *activated intermediate* can be formed at any stage along the reaction pathway. It appears on the *surface* as a *well*, on the *contour* diagram as a *series of contours*, and on the *profile* as a *minimum* (Figures 4.25 and 4.26).

  For the *activated intermediate* displacements in *all* directions, *even along the reaction coordinate*, give *increases* in potential energy, in sharp contrast to the *decrease along the reaction coordinate* found for the *activated complex*. An intermediate is simply a normal but high energy molecule, in contrast to the critical arrangement of atoms of the activated complex.

  Note: it is very important always to be aware of this very crucial distinction.

- Activated or energized molecules have energy equal to, or greater than, the critical energy. They can be anywhere on the surface, but can only become activated complexes by rearranging the relative positions of the atoms in the reaction unit (X----Y----Z) until the critical configuration is reached.

- Molecules with less than the critical energy also appear on the surface, but these can only alter in a manner allowed by their total energy. For these molecules the possibilities of alteration in configuration are limited. In particular, the critical configuration can never be reached. Quantum mechanical tunnelling allows the

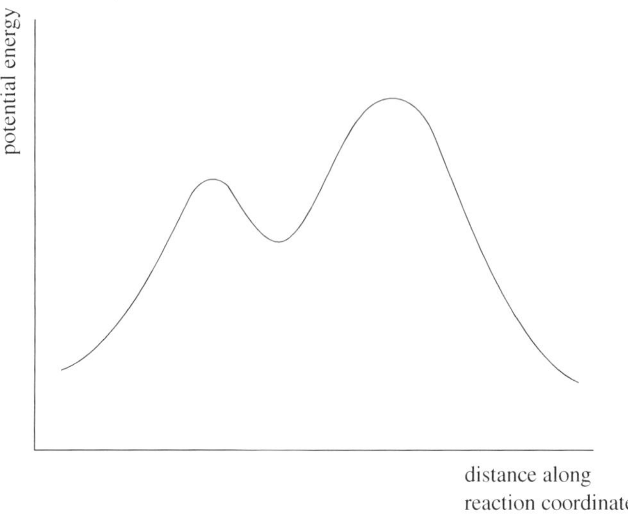

Figure 4.26 A potential energy profile showing an activated intermediate

possibility that products can be reached without the reactants having the critical energy. Tunnelling appears as a possibility when the reaction process is treated quantum mechanically. Reaction treated classically requires the unit (X----Y---- Z) to pass over the top of the col, i.e. through the critical configuration. But, if movement over the surface is quantized, there is a finite probability that the unit (X- - -Y- - -Z) can move from a point on the reactant side of the barrier to another point on the product side without passing over the barrier. This is tunnelling, and reaction has occurred without the reactants having the critical energy. The result is that the transmission coefficient (Section 4.3.3) is greater than unity. Tunnelling is more likely for reactions of light mass species, such as H, H^+, H^- or electrons. This is a refinement on basic transition state theory not dealt with here.

- The process of reaction is often described as 'moving over the PE surface'. This is *not* a movement in space, but only an internal movement of atoms with respect to each other. Likewise, phrases such as 'surmounting a PE barrier' or 'getting over the energy hump' and 'climbing up and over the col or saddle point' are *descriptive* only, and the term 'the rate of passage through the critical configuration' simply means a rate of change of configuration.

- The critical configuration occupies a flat non-zero area, corresponding to a flat non-zero length along the reaction coordinate. The potential energy is virtually constant over that part of the surface defining the activated complex, with zero force acting on the critical configuration. This results in *a constant rate of change in configuration through the critical configuration*. This describes a free translation along the reaction coordinate. Once the system reaches configurations with decreasing PE, the motion ceases to be a free translation.

- It is vitally important to understand what a free translation means, and to be aware of the important consequences of motion through the critical configuration being a free translation. The force acting on a system is

$$\text{force} = -\frac{\mathrm{d(PE)}}{\mathrm{d}r} \tag{4.19}$$

  If the PE is constant, then $\mathrm{d(PE)}/\mathrm{d}r$ is zero, and the force is zero and there is no acceleration in the motion of the reaction unit as it passes through the critical configuration, i.e. through the region of the surface where the PE is constant. If there is no acceleration, then the reaction unit must pass over the barrier at a constant rate. This rate can be calculated from kinetic theory.

  If the PE is not constant, then $\mathrm{d(PE)}/\mathrm{d}r$ is non-zero, and the rate of change of configuration over that region of the PE surface will vary depending on the value of $\mathrm{d(PE)}/\mathrm{d}r$. This happens at all points other than the critical configuration.

- The concentration of activated complexes is the *total* number of units (X----Y----Z) per unit volume which are in the critical configuration, no matter what their past history is.

- The rate of reaction is the total number of units (X----Y----Z) per unit volume which pass through the critical configuration per unit time *from reactant valley to product valley.*

Worked Problem 4.9

Question. Figure 4.27 gives a potential energy contour diagram. What can be said about configurations along the line AB? Identify the activated complex and activated intermediate in the diagram.

Answer. Along AB, $r_{AB} = r_{BC}$. The activated complex lies at the saddle-point lying on AB. The PE contour diagram is symmetrical about the critical configuration. The activated intermediate lies in the well of the elliptical contours and at configurations where $r_{BC} > r_{AB}$. There is a further activated complex as marked along the line PQ.

4.3.3 An outline of arguments involved in the derivation of the rate equation

This involves calculating the rate at which the 'reaction unit' passes through the critical configuration. Since the PE is constant over the region of the surface occupied by the critical configuration, the rate at which it moves through the critical

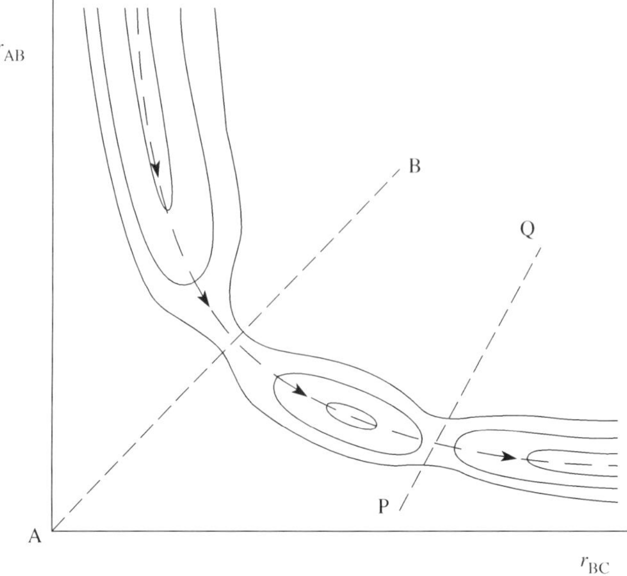

Figure 4.27 A potential energy contour diagram

configuration is constant. This rate can be calculated from kinetic theory. The concentration of activated complexes must also be calculated. This is done using *equilibrium statistical mechanics*, and by assuming that activated complexes are in equilibrium with the reactants, which is fully justified if $E_0 > kT$.

The statistical mechanical contribution to transition state theory uses *partition functions*. These are statistical mechanical quantities made up from translational, rotational, vibrational and electronic terms, though the electronic terms can normally be ignored if the reaction occurs in the ground state throughout.

Calculation of partition functions requires spectroscopic quantities for the rotational and vibrational partition functions. The quantities required are moments of inertia, rotational symmetry numbers and fundamental vibration frequencies for all normal modes of vibration. The translational terms require the mass of the molecule. All terms depend on temperature. Calculation of partition functions is routine for species for which a detailed spectroscopic analysis has been made.

Two equations emerge:

- the statistical mechanical expression for the rate constant for reaction;
- the equilibrium constant for formation of the activated complex from reactants.

$$k = \kappa \frac{kT}{h} \frac{Q^{\neq *}}{Q_X Q_{YZ}} \exp\left(\frac{-E_0}{RT}\right) \tag{4.20}$$

$$K^{\neq *} = \frac{Q^{\neq *}}{Q_X Q_{YZ}} \exp\left(\frac{-E_0}{RT}\right) \tag{4.21}$$

κ is the transmission coefficient and allows for the possibility that not all units (X----Y---Z) in the critical configuration may actually react, k and h are the Boltzmann and Planck constants, E_0 is the critical energy and can be found from the PE surface, $Q^{\neq *}$ is the partition function per unit volume for the activated complex *with one term missing* and Q_X and Q_{YZ} are *ordinary complete* partition functions per unit volume for the reactants. Like $Q^{\neq *}$, $K^{\neq *}$ is a quantity with one term missing. This is explained in the brief derivation given below.

Equation (4.20) is the *working* equation of transition state theory in statistical mechanical form.

Brief derivation of transition state theory

Reaction occurs when the reaction unit attains the critical configuration, becomes an activated complex, and passes on to product. The critical configuration occurs at the maximum on the potential energy profile and extends over a length δ along the reaction coordinate. The activated complex spends a time τ in the critical configuration, and from this the rate of passage through the critical configuration from left to right is given as

$$v = \frac{\delta}{\tau}$$

where the magnitudes of both δ and τ are unspecified.

If $N^{\neq}$ is the total number of activated complexes, then the number which pass through the critical configuration per unit time per unit volume is $N^{\neq}v/V\delta$, and so the rate of reaction can be given as

$$\frac{\kappa N^{\neq} v}{V\delta} = \frac{\kappa c^{\neq} v}{\delta} \tag{4.22}$$

where κ is the transmission coefficient which accounts for the possibility that not all activated complexes reach product configuration, and $c^{\neq}$ is the concentration of activated complexes expressed as a *number* of activated complexes per unit volume.

v can be found from kinetic theory, and is $v = \left(kT/2\pi\mu^{\neq}\right)^{1/2}$, giving

$$\text{rate of reaction} = \frac{\kappa c^{\neq}}{\delta}\left(\frac{kT}{2\pi\mu^{\neq}}\right)^{1/2} \tag{4.23}$$

where $\mu^{\neq}$ is the reduced mass for the change of configuration, calculable from the potential energy surface, but generally left unspecified.

The rate of reaction can also be given as $kc_X c_{YZ}$ so that the calculated rate constant is given by

$$k = \kappa \frac{c^{\neq}}{c_X c_{YZ}} \frac{1}{\delta}\left(\frac{kT}{2\pi\mu^{\neq}}\right)^{1/2} \tag{4.24}$$

There is no easy way to calculate $c^{\neq}/c_{X}c_{YZ}$ except by assuming equilibrium between reactants and activated complexes when equilibrium statistical mechanics can be used to calculate $c^{\neq}/c_{X}c_{YZ}$:

$$K^{\neq} = \left(\frac{c^{\neq}}{c_{X}c_{YZ}}\right) = \frac{Q^{\neq}}{Q_{X}Q_{YZ}}\exp\left(\frac{-\Delta U_0}{RT}\right) \tag{4.25}$$

where ΔU_0 is ΔU at absolute zero for the process of forming activated complexes from reactants, and is equal to E_0, the activation energy per mole found from the potential energy surface. The Q are complete partition functions per unit volume.

Substitution into the equation for k gives

$$k = \kappa\frac{Q^{\neq}}{Q_{X}Q_{YZ}}\frac{1}{\delta}\left(\frac{kT}{2\pi\mu^{\neq}}\right)^{1/2}\exp\left(\frac{-E_0}{RT}\right) \tag{4.26}$$

This equation contains the awkward terms δ and $\mu^{\neq}$ for which an assignment of magnitude would not be an easy task, and which conventionally is not made.

There is, however, a very convenient way out of this problem. The activated complex has been shown to have a free translation along the reaction coordinate over the distance δ. Statistical mechanics can furnish an expression for this quantity: the partition function for a free translation over a distance δ is $\dfrac{(2\pi\mu^{\neq}kT)^{1/2}\delta}{h}$, and this contains the awkward terms δ and $\mu^{\neq}$. If this term is factorized out of the overall total molecular partition function for the activated complex $Q^{\neq}$, then $Q^{\neq}$ can be written as

$$\frac{(2\pi\mu^{\neq}kT)^{1/2}\delta}{h}Q^{\neq *} \tag{4.27}$$

where $Q^{\neq *}$ is the partition function with the term for the free translation factorized out; the * summarizes the fact that this is a partition function with one term struck out.

This now results in an expression where cancellation of the awkward terms occurs:

$$k = \frac{Q^{\neq *}}{Q_{X}Q_{YZ}}\frac{\kappa}{\delta}\frac{(2\pi\mu^{\neq}kT)^{1/2}\delta}{h}\left(\frac{kT}{2\pi\mu^{\neq}}\right)^{1/2}\exp\left(\frac{-E_0}{RT}\right) \tag{4.28}$$

$$= \frac{\kappa kT}{h}\frac{Q^{\neq *}}{Q_{X}Q_{YZ}}\exp\left(\frac{-E_0}{RT}\right) \tag{4.20}$$

which is the fundamental equation of transition state theory in statistical mechanical form.

Calculation of the partition functions for *reactants* is straightforward, but the partition function for the *activated complex* needs explanation. The activated complex has been shown to have the *unique feature of a free translation along the reaction coordinate over the distance occupied by the activated complex*. The statistical mechanical quantity for this free translation has already been *factorized out* from the total partition function for the activated complex in the derivation. This has been done simply because doing so allows cancellation of some awkward terms in the derivation of the rate constant equation. This is why the symbol has * appeared along with the symbol $\neq$, this latter indicating that the process is one of forming the activated complex, often very loosely termed activation. $Q^{\neq *}$ is now a partition function per unit volume for the activated complex but with *one crucial term missing* from it, i.e. the term for the free translation. This is more fully explained in the section below.

4.3.4 Use of the statistical mechanical form of transition state theory

E_0 can be calculated from the PE surface, and if $Q^{\neq *}$, Q_X and Q_{YZ} can be calculated then a completely theoretical calculation would give k, the absolute rate constant for the given reaction.

Calculation of partition functions for *reactants* is a routine matter if a complete spectroscopic analysis is available.

The partition function calculation for the *activated complex* is more tricky. If the surface is known the quantities quoted above can be found from the dimensions and the curvature of the surface around the critical configuration. If this is not possible, then estimates of $Q^{\neq *}$ can be made by analogy with a molecule of similar structure. *The vital feature is that the free translation along the reaction coordinate has already been accounted for and must not be included in the calculation*, and so the activated complex has *one degree of vibrational freedom less* than that for a molecule with the same number of atoms.

The statistical mechanical form of transition theory makes considerable use of the concepts of translational, rotational and vibrational degrees of freedom. For a system made up of N atoms, $3N$ coordinates are required to completely specify the positions of all the atoms. Three coordinates are required to specify translational motion in space. For a linear molecule, two coordinates are required to specify rotation of the molecule as a whole, while for a non-linear molecule three coordinates are required. These correspond to two and three rotational modes respectively. This leaves $3N - 5$ (linear) or $3N - 6$ (non-linear) coordinates to specify vibrations in the molecule. If the vibrations can be approximated to harmonic oscillators, these will correspond to *normal modes of vibration*.

The number of coordinates needed to specify each type of motion is called the number of degrees of freedom; e.g., non-linear molecules have three translational degrees of freedom, three rotational degrees of freedom and $3N - 6$ vibrational degrees of freedom.

4.3.5 Comparisons with collision theory and experimental data

Worked Problem 4.10

Question. The transition state theory expression

$$k = \kappa \frac{kT}{h} \frac{Q^{\neq *}}{Q_X Q_{YZ}} \exp\left(\frac{-E_0}{RT}\right) \qquad (4.20)$$

can be compared with collision theory, and with the Arrhenius equation. Which terms can be compared with each other?

Answer. It is assumed that the Arrhenius activation energy E_A will be at least approximately equal to E_0.

$$\text{Collision theory}: \quad k = pZ \exp\left(\frac{-E_A}{RT}\right) \qquad (4.29)$$

$$\text{Arrhenius equation}: \quad k = A \exp\left(\frac{-E_A}{RT}\right) \qquad (4.1)$$

$\kappa \frac{kT}{h} \frac{Q^{\neq *}}{Q_X Q_{YZ}}$ can be compared with pZ and A

$\frac{Q^{\neq *}}{Q_X Q_{YZ}}$ is related to p

$\kappa \frac{kT}{h}$ is related to Z

Table 4.1 shows the remarkably good predictions made by transition state theory, and can be compared with the much poorer predictive value of collision theory, Table 4.2, demonstrating the considerable superiority of transition state theory.

Table 4.1 Comparison of experimental A factors with pre-exponential factors calculated from transition state theory

Reaction	$\log_{10} A_{observed}$ A/mol^{-1} dm^3 s^{-1}	$\log_{10} A_{calculated}$ A/mol^{-1} dm^3 s^{-1}
$F_2 + ClO_2^{\bullet} \rightarrow FClO_2 + F^{\bullet}$	7.5	7.9
$NO_2 + CO \rightarrow NO + CO_2$	10.1	9.8
$2NO_2 \rightarrow 2NO + O_2$	9.3	9.7
$2NOCl \rightarrow 2NO + Cl_2$	10.0	8.6
$NO_2 + F_2 \rightarrow NO_2F + F^{\bullet}$	9.2	8.1
$H^{\bullet} + CH_4 \rightarrow H_2 + CH_3^{\bullet}$	10.0	10.3
$CH_3^{\bullet} + C_2H_6 \rightarrow CH_4 + C_2H_5^{\bullet}$	9.3	8.0
$H^{\bullet} + H_2 \rightarrow H_2 + H^{\bullet}$	10.7	10.9

Table 4.2 Comparison of experimental A factors with collision numbers, Z, calculated from collision theory

Reaction	$\log_{10} A_{observed}$ $A/\text{mol}^{-1}\ \text{dm}^3\ \text{s}^{-1}$	$\log_{10} Z_{calculated}$ $Z/\text{mol}^{-1}\ \text{dm}^3\ \text{s}^{-1}$
$NO_2 + F_2 \rightarrow NO_2F + F^\bullet$	9.2	10.8
$2NO_2 \rightarrow 2NO + O_2$	9.3	10.6
$2NOCl \rightarrow 2NO + Cl_2$	10.0	10.8
$F_2 + ClO_2^\bullet \rightarrow FClO_2 + F^\bullet$	7.5	10.4
$2ClO^\bullet \rightarrow Cl_2 + O_2$	7.8	10.4

Worked Problem 4.11

Question. Why is collision theory inferior to transition state theory?

Answer. This is mainly because transition state theory considers the internal structure, rotational and vibrational behaviour of reactants and activated complex, explicitly, while collision theory assumes them to be hard spheres and the internal structures are ignored.

Qualitative comparisons can be made using rough order of magnitude contributions made to $\kappa(kT/h)Q^{\neq *}/Q_X Q_{YZ}$ from translation, rotation and vibration, and such calculations again demonstrate the importance of the internal motions of the molecules in determining predicted A factors.

Worked Problem 4.12

Question. Given that the total number of degrees of freedom for a polyatomic molecule is $3N$, calculate the number of vibrational modes open to (a) an atom, (b) a diatomic molecule, (c) a non-linear polyatomic molecule with N atoms and (d) a non-linear activated complex with N atoms.

Answer.

(a) an atom has three translations, but no rotations and no vibrations;

(b) a diatomic molecule has three translations, two rotations and $(3 \times 2) - 3 - 2 = 1$ vibration;

(c) a non-linear molecule has three translations, three rotations and $\therefore\ 3N - 6$ vibrations;

(d) a non-linear activated complex has three translations, three rotations, one internal free translation and $\therefore\ 3N - 7$ vibrations.

Note. Compare the answers to points (c) and (d). The activated complex has *one vibration less* open to it compared with the polyatomic molecule with the same number of atoms.

It is also possible to look at the changes in types of degrees of freedom, and changes in internal motions on forming the activated complex. Considerable insight into trends in A factors, and interpretations of these in terms of changes in internal motions, can be made using rough order of magnitude calculations without having recourse to detailed accurate values for the partition functions. The approximations made are acceptable because contributions from the internal motions do not vary much with mass, moment of inertia or vibrational frequency.

Typical values of these contributions:

- for each translation: 10^9 to 10^{10} dm^{-1}
- for each rotation: 10 to 10^2
- for each vibration: 1 to 10

Worked Problem 4.13

Question. At 298 K, and assuming $\kappa = 1$ and $\kappa kT/h = 6.2 \times 10^{12}$ s^{-1}, and using typical values (in the middle of the ranges above) for the contributions from translation, rotation and vibration,

for each translation: 5×10^9 dm^{-1}

for each rotation: 50

for each vibration 5

calculate the predicted A factor for the following reaction types:

(a) atom + atom $\rightarrow$ activated complex;

(b) atom + linear molecule $\rightarrow$ non-linear activated complex;

(c) two non-linear molecules $\rightarrow$ non-linear activated complex.

Answer.

(a) Each atom has three translations; the activated complex has three translations, two rotations, one internal free translation and $\therefore (3 \times 2) - 6$ vibrations i.e. zero. The contribution from the internal translation has already been incorporated in the derivation, so that

$$\text{contributions to } \frac{Q^{\neq *}}{Q_X Q_{YZ}} \text{ are } \frac{\text{3trans, 2rots, 0vib}}{\text{3trans, 3trans}} = \frac{\text{2rots}}{\text{3trans}}$$

$$\text{value of } \kappa\frac{kT}{h}\frac{Q^{\neq *}}{Q_X Q_{YZ}} = \frac{6.2\times 10^{12}\times 50\times 50}{5\times 10^9\times 5\times 10^9\times 5\times 10^9}\text{dm}^3\ \text{s}^{-1}$$
$$= 1.2\times 10^{-13}\ \text{dm}^3\ \text{s}^{-1}$$

(b) An atom has three translations; a linear molecule has three translations, two rotations and $3N - 5$ vibrations; the non-linear activated complex has three translations, three rotations, one internal free translation and $\therefore\ 3(N+1) - 7$ vibrations i.e. $3N - 4$ vibrations.

$$\text{Contributions to } \frac{Q^{\neq *}}{Q_X Q_{YZ}} \text{ are } \frac{\text{3trans, 3rots, }(3N-4)\text{vibs}}{\text{3trans, 3trans, 2rots, }(3N-5)\text{vib}} = \frac{\text{1rot, 1vib}}{\text{3trans}}$$

$$\text{value of } \kappa\frac{kT}{h}\frac{Q^{\neq *}}{Q_X Q_{YZ}} = \frac{6.2\times 10^{12}\times 50\times 5}{5\times 10^9\times 5\times 10^9\times 5\times 10^9}\text{dm}^3\ \text{s}^{-1}$$
$$= 1.2\times 10^{-14}\ \text{dm}^3\ \text{s}^{-1}$$

(c) A non-linear molecule, A, has three translations, three rotations and $3N_A - 6$ vibrations. Likewise non-linear molecule B has three translations, three rotations and $3N_B - 6$ vibrations. Total for reactants = six translations, six rotations and $3N_A + 3N_B - 12$ vibrations.

The non-linear activated complex has three translations, three rotations, one internal free translation and $3N_A + 3N_B - 7$ vibrations.

$$\text{Contributions to } \frac{Q^{\neq *}}{Q_X Q_{YZ}} \text{ are } \frac{\text{3trans, 3rots, }(3N_A + 3N_B - 7)\text{vibs}}{\text{3trans, 3trans, 3rots, 3rots, }(3N_A + 3N_B - 12)\text{vibs}}$$
$$= \frac{\text{5vib}}{\text{3trans, 3rots}}$$

$$\kappa\frac{kT}{h}\frac{Q^{\neq *}}{Q_X Q_{YZ}} = \frac{6.2\times 10^{12}\times 5\times 5\times 5\times 5\times 5}{5\times 10^9\times 5\times 10^9\times 5\times 10^9\times 50\times 50\times 50}\text{dm}^3\ \text{s}^{-1}$$
$$= 1.2\times 10^{-18}\ \text{dm}^3\ \text{s}^{-1}$$

The results from Problem 4.13 are instructive. They are

(a) atom + atom $\rightarrow$ activated complex, $A = 1.2\times 10^{-13}\ \text{dm}^3\ \text{s}^{-1}$;

(b) atom + linear molecule $\rightarrow$ non-linear activated complex, $A = 1.2\times 10^{-14}\ \text{dm}^3\ \text{s}^{-1}$;

(c) two non-linear molecules $\rightarrow$ non-linear activated complex, $A = 1.2\times 10^{-18}\ \text{dm}^3\ \text{s}^{-1}$.

These values of A are in molecular units, and cannot be compared *directly* with a typical value of Z. This is because partition functions are molecular quantities and the contributions to $\kappa(kT/h)Q^{\neq *}/Q_X Q_{YZ}$ are in molecular units. Multiplying by

$6.022 \times 10^{23}\ \text{mol}^{-1}$ converts to molar units, giving the following values:

reaction (a) $A = 7 \times 10^{10}\ \text{mol}^{-1}\ \text{dm}^3\ \text{s}^{-1}$

reaction (b) $A = 7 \times 10^{9}\ \text{mol}^{-1}\ \text{dm}^3\ \text{s}^{-1}$

reaction (c) $A = 7 \times 10^{5}\ \text{mol}^{-1}\ \text{dm}^3\ \text{s}^{-1}$

Taking the typical value for Z of $4 \times 10^{10}\ \text{mol}^{-1}\ \text{dm}^3\ \text{s}^{-1}$ and using $p = A_{\text{calc}}/Z$ gives the values of 'p' below:

reaction (a) $p \approx 1$

reaction (b) $p \approx 10^{-1}$

reaction (c) $p \approx 10^{-5}$

No great reliance can be placed on the *actual* numerical values because *approximate* values were used for the contributions from translation, rotation and vibration. However, the powers of ten are significant, as is the trend with degree of complexity of the reactants.

Table 4.3 shows that even this approximate version of transition state theory gives a satisfactory interpretation of the experimental 'p' factors. As the complexity of the reactants increases so p decreases.

Table 4.3 Typical values of p calculated from approximate transition state theory and the average values of some experiments giving observed p factors

Reaction	p_{calc}	p_{obs}
$Br^\bullet + H_2 \rightarrow HBr + H^\bullet$	8×10^{-1}	5×10^{-1}
$H^\bullet + C_2H_6 \rightarrow H_2 + C_2H_5^\bullet$	8×10^{-2}	13×10^{-2}
$CH_3^\bullet + CH_3COCH_3 \rightarrow CH_4 + {}^\bullet CH_2COCH_3$	6×10^{-4}	2×10^{-3}
$CH_3^\bullet + HC(CH_3)_3 \rightarrow CH_4 + {}^\bullet C(CH_3)_3$	5×10^{-5}	3×10^{-4}

4.4 Thermodynamic Formulations of Transition State Theory

There are some situations, such as reactions in solution, where the partition function form is not immediately useful, whereas a thermodynamic formulation is more immediately applicable.

The thermodynamic quantities for equilibrium which are taken over by transition state theory are

$$K = \text{quotient of concentrations at equilibrium} \tag{4.30}$$

$$\Delta G^\theta = -RT \log_e K \tag{4.31}$$

$$\Delta G^{\theta} = \Delta H^{\theta} - T\Delta S^{\theta} \tag{4.32}$$

$$K = \exp\left(\frac{-\Delta G^{\theta}}{RT}\right) = \exp\left(\frac{-\Delta H^{\theta}}{RT}\right)\exp\left(\frac{+\Delta S^{\theta}}{R}\right) \tag{4.33}$$

$$\mathrm{d}\Delta G^{\theta} = \Delta V^{\theta}\mathrm{d}p - \Delta S^{\theta}\mathrm{d}T \tag{4.34}$$

$$\left(\frac{\partial \Delta G^{\theta}}{\partial T}\right)_p = -\Delta S^{\theta} \tag{4.35}$$

$$\left(\frac{\partial \Delta G^{\theta}}{\partial p}\right)_T = \Delta V^{\theta} \textit{ but only for reactions in solutions, } \text{Section 7.6.3} \tag{4.36}$$

$$\left(\frac{\partial \log_e K}{\partial T}\right)_p = +\frac{\Delta H^{\theta}}{RT^2} \tag{4.37}$$

$$\left(\frac{\partial \log_e K}{\partial p}\right)_T = -\frac{\Delta V^{\theta}}{RT} \textit{ but only for reactions in solution, } \text{Section 7.6.3} \tag{4.38}$$

Note. These last four equations involve partial derivatives. If an equation involves more than two variables, then normally the effect of varying one of the quantities on a second quantity will be studied while the third is held constant, e.g.

$$\mathrm{d}\Delta G^{\theta} = \Delta V^{\theta}\mathrm{d}p - \Delta S^{\theta}\mathrm{d}T \tag{4.34}$$

Here a small change in ΔG^{θ} could be a result of a small change in p and/or a small change in T. When this happens and the first quantity, ΔG^{θ}, is plotted against the second, p, while the third, T, is held constant, then the slope of the graph is called a *partial derivative*, denoted by the symbol $\left(\partial \Delta G^{\theta}/\partial p\right)_T$. The presence of the ∂s indicates a partial derivative, and the subscript tells what the quantity is which is being held constant.

The equilibrium constant for formation of the activated complex from reactants has been given (Equation (4.25)), and substitution into the rate expression, Equation (4.20), and cancellation of terms gives

$$k = \frac{\kappa kT}{h} K^{\neq *} \tag{4.39}$$

By applying standard thermodynamic functions to the equilibrium, an expression for the rate constant in terms of $\Delta G^{\neq *}$, $\Delta H^{\neq *}$ and $\Delta S^{\neq *}$ can be given:

$$k = \frac{\kappa kT}{h}\exp\left(\frac{-\Delta G^{\neq *}}{RT}\right) \tag{4.40}$$

$$= \frac{\kappa kT}{h}\exp\left(\frac{-\Delta H^{\neq *}}{RT}\right)\exp\left(\frac{+\Delta S^{\neq *}}{R}\right) \tag{4.41}$$

where $\Delta S^{\neq *}$ is the entropy of activation, $\Delta G^{\neq *}$ is the free energy of activation, and $\Delta H^{\neq *}$ is the enthalpy of activation. Here '*of activation*' means '*for the process of forming the activated complex*'.

The dependence of the rate constant on pressure allows $\Delta V^{\neq *}$, the volume of activation, to be found for reactions in solution, see Section 4.4.1.

The thermodynamic functions for activation are standard values, with the standard state normally taken to be unit concentration. These all have one term missing, see Sections 4.3.3 and 4.3.4.

4.4.1 Determination of thermodynamic functions for activation

Detailed calculations using partition functions, calculated from spectroscopic data for reactants and the PE surface for the activated complex, allows $K^{\neq *}$, $\Delta G^{\neq *}$, $\Delta H^{\neq *}$, $\Delta S^{\neq *}$ and $\Delta V^{\neq *}$ to be found. Experimental data can give experimental values to compare direct with the theoretical ones. Equation (4.41) gives

$$\log_e \frac{k}{T} = \log_e \frac{\kappa k}{h} + \frac{\Delta S^{\neq *}}{R} - \frac{\Delta H^{\neq *}}{RT} \tag{4.42}$$

A plot of $\log_e(k/T)$ versus $1/T$ has slope $= -\Delta H^{\neq *}/R$ and intercept $= \log_e(\kappa k/h) + \Delta S^{\neq *}/R$. Since $\kappa kT/h$ is known, κ being generally taken to be unity, $\Delta S^{\neq *}$ can be found.

Similarly, the effect of pressure on the rate constant gives $\Delta V^{\neq *}$ for reactions in solution. Equation (4.40) gives

$$\left(\frac{\partial \log_e k}{\partial p}\right)_T = -\frac{1}{RT}\left(\frac{\partial \Delta G^{\neq *}}{\partial p}\right)_{\mathrm{T}} = -\frac{\Delta V^{\neq *}}{RT} \tag{4.43}$$

A plot of $\log_e k$ versus p has slope $= -\Delta V^{\neq *}/RT$.

Note: the partial derivatives here, $(\partial \log_e k/\partial p)_T$ and $\left(\partial \Delta G^{\neq *}/\ \partial p\right)_T$ where k varies with p at constant T, and $\Delta G^{\neq *}$ varies with p at constant T.

4.4.2 Comparison of collision theory, the partition function form and the thermodynamic form of transition state theory

Worked Problem 4.14

Question. Write out the equations for the statistical mechanical form, the thermodynamic form, collision theory and the Arrhenius theory. What are the equivalent terms?

Answer.

$$k = \kappa \frac{kT}{h} \frac{Q^{\neq *}}{Q_{\mathrm{X}} Q_{\mathrm{YZ}}} \exp\left(\frac{-E_0}{RT}\right) \tag{4.20}$$

$$k = \frac{\kappa k T}{h} \exp\left(\frac{-\Delta H^{\neq *}}{RT}\right) \exp\left(\frac{+\Delta S^{\neq *}}{R}\right) \tag{4.41}$$

$$k = pZ \exp\left(\frac{-E_A}{RT}\right) \tag{4.29}$$

$$k = A \exp\left(\frac{-E_A}{RT}\right) \tag{4.1}$$

If the exponential terms E_0/RT, $\Delta H^{\neq *}/RT$ and E_A/RT can be correlated with each other and if $\kappa kT/h$ is related to Z, then A, pZ, $\kappa(kT/h)Q^{\neq *}/Q_X Q_{YZ}$ and $(\kappa kT/h)\exp(+\Delta S^{\neq *}/R)$ can be correlated, while p, $Q^{\neq *}/Q_X Q_{YZ}$ and $\Delta S^{\neq *}/R$ are related to each other.

Since $\kappa kT/h$ is a constant for a given T, A factors explicitly reflect any trend in $\Delta S^{\neq *}$ with reactant type (see Table 4.4). For these reactions $\Delta S^{\neq *}$ is taken to reflect the complexity of reacting molecules and activated complexes. $\text{Log}_{10} A$ and $\Delta S^{\neq *}$ parallel each other, and decrease as the complexity of the reacting molecules increases.

Table 4.4 Values of $\log_{10} A$ and $\Delta S^{\neq *}$ associated with Figure 4.28

Reaction	T/K	$\log_{10}\left(\frac{A}{\text{mol}^{-1}\,\text{dm}^3\,\text{s}^{-1}}\right)$	$\Delta S^{\neq *}$/J K^{-1} mol^{-1}
$H^\bullet + D_2 \rightarrow HD + D^\bullet$	600	10.7	−62
$OH^\bullet + CH_4 \rightarrow H_2O + CH_3^\bullet$	300	9.5	−80
$CH_3^\bullet + H_2 \rightarrow CH_4 + H^\bullet$	600	9.5	−85
$CH_3^\bullet + C_2H_6 \rightarrow CH_4 + C_2H_5^\bullet$	650	9.3	−90
$CH_3^\bullet + (CH_3)_3CH \rightarrow CH_4 + (CH_3)_3C^\bullet$	650	8.5	−105

Ideally this data should be collected at the same temperature for all reactions. Not doing so increases the scatter in the graph.

Worked Problem 4.15

Question. Comment on the results given in Table 4.4. Plot $\log_{10} A$ versus $\Delta S^{\neq *}$ and comment.

Answer. As molecular complexity increases, $\Delta S^{\neq *}$ becomes increasingly more negative and so $\exp(\frac{\Delta S^{\neq *}}{R})$ becomes increasingly smaller. This correlates with the A factors quoted.

A graph of $\log_{10} A$ versus $\Delta S^{\neq *}$ is approximately linear (Figure 4.28). This is expected since the same physical phenomenon manifests itself in each.

However, note the comment in the table.

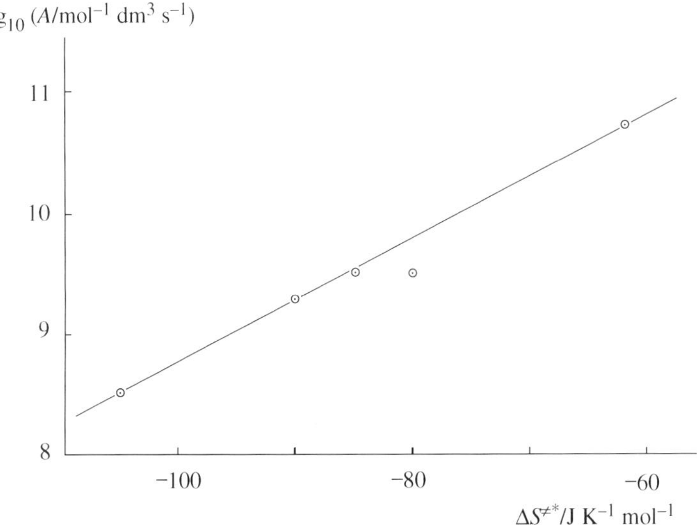

Figure 4.28 A graph of $\log_{10} A$ versus $\Delta S^{\neq *}$

4.4.3 Typical approximate values of contributions entering the sign and magnitude of $\Delta S^{\neq *}$

Translational terms are based here on a standard concentration of 1 mol dm^{-3}. All values depend somewhat on temperature.

Translation: ca 125 to 175 J mol^{-1} K^{-1}

ca 40 to 60 J mol^{-1} K^{-1} per degree of freedom

Rotation: ca 50 to 100 J mol^{-1} K^{-1} for linear molecules

ca 75 to 150 J mol^{-1} K^{-1} for non-linear molecules

ca 25 to 50 J mol^{-1} K^{-1} per degree of freedom

Vibration: per normal mode: from nearly zero, for high frequencies at normal temperatures, up to ca 15 J mol^{-1} K^{-1} for low frequencies at higher temperatures

Worked Problem 4.16

Question. Assuming a contribution to $\Delta S^{\neq *}$ of

50 J mol^{-1} K^{-1} for each of three translational degrees of freedom,

40 J mol^{-1} K^{-1} per each rotational degree of freedom and

10 J mol^{-1} K^{-1} per normal mode of vibration;

calculate $\Delta S^{\neq *}$ for the following reaction,

$$A + B \leftrightarrows (A{-}{-}{-}{-}B)^{\neq} \rightarrow \text{products}$$

A, B and the activated complex are non-linear with N_A, N_B and $N_A + N_B$ atoms respectively.

Answer.

A has three translations, three rotations and $3N_A - 6$ vibrations.
B has three translations, three rotations and $3N_B - 6$ vibrations.
$(A{-}{-}{-}B)^{\neq}$ has three translations, three rotations, one internal free translation and $(3N_A + 3N_B - 7)$ vibrations. Reactants have a total of six translations, six rotations and $(3N_A + 3N_B - 12)$ vibrations. The activated complex has a total of three translations, three rotations, one internal free translation and $(3N_A + 3N_B - 7)$ vibrations.

On forming the activated complex, we *lose* three translations and three rotations and *gain* 5 vibrations. The activated complex gains one internal free translation, but this has already been taken care of in the derivation of the rate expression and does not appear in the $\exp(+\Delta S^{\neq *}/R)$ term. Quantifying this gives translation, *lose* 150 J mol^{-1} K^{-1}; rotation, *lose* 120 J mol^{-1} K^{-1}; vibration, *gain* 50 J mol^{-1} K^{-1}; *net loss* is 220 J mol^{-1} K^{-1}.

Hence for this reaction using the rough and ready magnitudes quoted, $\Delta S^{\neq *} = -220$ J mol^{-1} K^{-1}. Since $\Delta S^{\neq *}/R$ correlates with the p factor the large negative value for $\Delta S^{\neq *}$ corresponds to a low p factor.

This is, of course, only a very descriptive calculation.

4.5 Unimolecular Theory

In unimolecular reactions one single reactant molecule passes into the activated complex and reacts. Experimental observations show that, *within any given experiment*, unimolecular reactions are strictly first order irrespective of the initial concentration or pressure (Figure 4.29).

At *high pressures* the observed first order rate constant is *strictly independent of pressure*, but if experiments are carried out at *low or intermediate pressures* then the first order rate constant *depends on pressure*, and the reaction moves from strict first order kinetics at high pressures to second order at low pressures. At pressures intermediate between these two limits, the reaction shows complex kinetics with no simple order. This requires explanation, see below and Problem 4.17.

When studying reactions in the gas phase, the kinetics have often been studied by following the change in pressure with time. This is particularly so with unimolecular reactions, and kineticists tend to discuss the experimental results for these reactions in

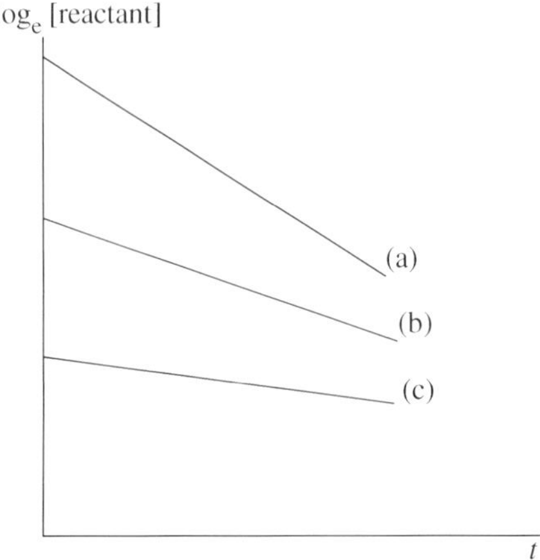

Figure 4.29 First order plots for a unimolecular reaction at three different initial pressures of reactant. (a) Highest pressure. (b) Intermediate pressure. (c) Lowest pressure

terms of pressures. When the theoretical aspects are being discussed, the number of molecules per unit volume is used. It is common to move between the two frames of reference without explaining this, and this juxtaposition of the frames of reference will become obvious when the steady state equations are written in terms of concentrations, and the discussion thereafter is in terms of pressures. There is no genuine discrepancy here, since at constant volume and temperature, $p \propto n$.

$$pV = nRT \tag{4.44}$$

and

$$p = \frac{n}{V}RT = cRT \tag{4.45}$$

Worked Problem 4.17

Question. The following data give experimental results for a unimolecular reaction, A → B, carried out at (a) high pressures and (b) low pressures. Show that both experiments follow strictly first order kinetics and determine the rate constants. Comment on the relative values for the two experiments.

(a)	$\frac{10^{-4}p_A}{\text{N m}^{-2}}$	10.08	8.02	6.03	4.49	3.30	2.51
	$\frac{\text{time}}{\text{min}}$	0	10	22	35	48	60

(b) $\frac{p_A}{N\,m^{-2}}$	1470	1290	1120	850	660	505
$\frac{\text{time}}{\text{min}}$	0	10	20	40	60	80

Answer. First order: $\log_e p_t = \log_e p_0 - kt$. Plot $\log_e p_t$ versus t. If first order, graph is linear with slope $= -k$.

(a) $\log_e p_t$	11.521	11.292	11.007	10.712	10.404	10.131
$\frac{\text{time}}{\text{min}}$	0	10	22	35	48	60

Graph is linear with slope $= -0.0234\ \text{min}^{-1}$ giving $k = 0.0234\ \text{min}^{-1}$.

(b) $\log_e p_t$	7.293	7.162	7.021	6.745	6.492	6.225
$\frac{\text{time}}{\text{min}}$	0	10	20	40	60	80

Graph is linear with slope $= -0.0134\ \text{min}^{-1}$, giving $k = 0.0134\ \text{min}^{-1}$.

Both sets of data show first order kinetics within each experiment. The k values have different magnitudes, confirming the dependence of k_{obs}^{1st} on pressure.

The crucial step in the development of unimolecular theory was the postulate of a time lag between the activation and reaction steps in the master mechanism for all elementary reactions given in Chapter 1. During this time an activated molecule can either be deactivated in a deactivating energy transfer collision, or it can alter configuration to reach the critical configuration and react. All elementary reactions involve three steps, two energy transfer steps and one reaction step, and for unimolecular reactions

$$A + A \xrightarrow{k_1} A^* + A \qquad \text{energy transfer activation}$$

$$A^* + A \xrightarrow{k_{-1}} A + A \qquad \text{energy transfer deactivation}$$

$$A^* \xrightarrow{k_2} \text{products} \qquad \text{reaction}$$

A steady state analysis gives (Problem 3.19)

$$\frac{d[A^*]}{dt} = k_1[A]^2 - k_{-1}[A][A^*] - k_2[A^*] = 0 \tag{4.46}$$

$$[A^*] = \frac{k_1[\mathrm{A}]^2}{k_{-1}[\mathrm{A}] + k_2} \tag{4.47}$$

$$\text{rate} = k_2[\mathrm{A}^*] \tag{4.48}$$

$$-\frac{\mathrm{d}[\mathrm{A}]}{\mathrm{d}t} = \frac{k_1 k_2[\mathrm{A}]^2}{k_{-1}[\mathrm{A}] + k_2} \tag{4.49}$$

which is neither first order nor second order.

Since each given experiment is strictly first order within the experiment, this rate (Equation (4.49)) can also be expressed in terms of an observed first order rate constant, $k_{\mathrm{obs}}^{1\mathrm{st}}$:

$$\text{rate} = k_{\mathrm{obs}}^{1\mathrm{st}}[\mathrm{A}] \tag{4.50}$$

$$\therefore k_{\mathrm{obs}}^{1\mathrm{st}} = \frac{k_1 k_2[\mathrm{A}]}{k_{-1}[\mathrm{A}] + k_2} \tag{4.51}$$

which predicts that the first order rate constant found from individual experiments depends on pressure as observed. Equation (4.49) shows that first order kinetics result when $k_{-1}[\mathrm{A}] \gg \mathrm{k}_2$ and the reaction step in the sequence is rate determining. Second order kinetics result when $k_{-1}[\mathrm{A}] \ll k_2$ and activation is rate determining, while at intermediate pressures Equation (4.49) is predicted to hold.

Note the use of pressure and concentration terminology here. It is standard practice to use both terms, even when discussing experimental results as is done in Problem 4.18.

4.5.1 Manipulation of experimental results

In general, over the full range of pressures

$$k_{\mathrm{obs}}^{1\mathrm{st}} = \frac{k_1 k_2[\mathrm{A}]}{k_{-1}[\mathrm{A}] + k_2} \tag{4.51}$$

At high pressures $k_{\mathrm{obs}}^{1\mathrm{st}}$ is strictly constant and independent of pressure, and

$$(k_{\mathrm{obs}}^{1\mathrm{st}})_{\text{high pressures}} = k_\infty = \frac{k_1 k_2}{k_{-1}} \tag{4.52}$$

and at low pressures $k_{\mathrm{obs}}^{1\mathrm{st}}$ becomes dependent on pressure, and

$$k_{\mathrm{obs}}^{1\mathrm{st}} = k_1[\mathrm{A}]. \tag{4.53}$$

(a) When the rate of deactivation = the rate of the reaction step

$$k_{-1}[\mathrm{A}^*][\mathrm{A}] = k_2[\mathrm{A}^*] \tag{4.54}$$

$$\therefore\ k_2 = k_{-1}[\mathrm{A}] \tag{4.55}$$

When this condition is met, *and only then*, Equation (4.51) reduces to

$$k_{\mathrm{obs}}^{\mathrm{1st}} = \frac{k_1 k_2}{2k_{-1}} = \frac{k_\infty}{2} \tag{4.56}$$

The pressure at which the observed first order rate constant has fallen to one-half of its high pressure value corresponds to

$$k_2 = k_{-1}[\mathrm{A}]_{1/2} \tag{4.57}$$

or put otherwise

$$[\mathrm{A}]_{1/2} = \frac{k_2}{k_{-1}} \tag{4.58}$$

$[\mathrm{A}]_{1/2}$ can be found from experiment, and if k_{-1} is taken to be λZ, then

$$k_2 = \lambda Z[\mathrm{A}]_{1/2} \tag{4.59}$$

where Z is the collision number and λ is the collision efficiency.

Once k_2 is found from experiment in this way, then k_1 can be found from the experimental value of k_∞:

$$k_1 = \frac{k_\infty k_{-1}}{k_2} = \frac{k_\infty \lambda Z}{k_2} \tag{4.60}$$

(b) The experimental results can be presented graphically in a plot of $1/k_{\mathrm{obs}}^{\mathrm{1st}}$ versus $1/[\mathrm{A}]$. Rearrangement of Equation (4.51) gives

$$\frac{1}{k_{\mathrm{obs}}^{\mathrm{1st}}} = \frac{k_{-1}}{k_1 k_2} + \frac{1}{k_1[\mathrm{A}]} \tag{4.61}$$

This will only be linear if k_1, k_{-1} and k_2 are constants, in which case the graph will have slope $= 1/k_1$ and intercept $= k_{-1}/k_1 k_2$. This gives an alternative route to k_2, again provided k_{-1} is known.

If, however, k_1 and k_2 are not single valued, then the graph will be a curve. Experimental data do, in fact, demonstrate conclusively that the graph is a curve. Section 4.5.2 discusses the physical significance of the constancy or otherwise of these rate constants.

Worked Problem 4.18

Question. The decomposition of cyclopropane to propene is a unimolecular reaction. It can be studied over a range of pressures. The following data gives values of $k_{\mathrm{obs}}^{\mathrm{1st}}$ over a range of concentrations. Plot $k_{\mathrm{obs}}^{\mathrm{1st}}$ versus concentration and comment on the shape of the graph.

Note: $pV = nRT$, hence pressure is proportional to concentration.

$\frac{10^4 c}{\text{mol dm}^{-3}}$	0.120	0.605	1.27	2.31	7.13	159
$\frac{10^5 k_{\text{obs}}^{\text{1st}}}{\text{s}^{-1}}$	8.6	15.4	20.0	22.3	28.2	38.7

Answer. The graph of $k_{\text{obs}}^{\text{1st}}$ versus concentration is a very decided curve, showing the predicted fall-off in $k_{\text{obs}}^{\text{1st}}$ at low pressures.

At high pressures, where $k_{\text{obs}}^{\text{1st}}$ is strictly constant and independent of pressure, Equation (4.51) reduces to

$$(k_{\text{obs}}^{\text{1st}})_{\text{high pressures}} = k_{\infty} = \frac{k_1 k_2}{k_{-1}} \tag{4.52}$$

where $k_{\infty} = 38.7 \times 10^{-5}\ \text{s}^{-1}$, and at low pressures $k_{\text{obs}}^{\text{1st}}$ becomes proportional to pressure, so that

$$k_{\text{obs}}^{\text{1st}} = k_1[\text{A}] \tag{4.53}$$

Values of k_1, k_{-1} and k_2 can be found from experimental data, as illustrated in Problem 4.19.

Worked Problem 4.19

Question.

(a) From the data in Problem 4.18 find $[\text{A}]_{1/2}$ and show this to equal k_2/k_{-1}.

(b) Find k_2 and then k_1, taking $Z - 5 \times 10^{10}\ \text{mol}^{-1}\ \text{dm}^3\ \text{s}^{-1}$, and $\lambda = 1$.

(c) Why can k_{-1} be taken to equal a typical value of Z, the collision number?

Answer. The pressure at which the observed first order rate constant has fallen to one-half of its high pressure value corresponds to

$$k_2 = k_{-1}[\text{A}]_{1/2} \tag{4.57}$$

or put otherwise

$$[\text{A}]_{1/2} = \frac{k_2}{k_{-1}} \tag{4.58}$$

If k_{-1} is taken to be λZ, then

$$k_2 = \lambda Z[\mathrm{A}]_{1/2} \tag{4.59}$$

From Problem 4.18 $k_\infty = 38.7 \times 10^{-5}\ \mathrm{s}^{-1}$ and $[\mathrm{A}]_{1/2} = 1.2 \times 10^{-4}\ \mathrm{mol\ dm^{-3}}$, giving $k_2 = 5 \times 10^{10}\ \mathrm{mol^{-1}\ dm^3\ s^{-1}} \times 1.2 \times 10^{-4}\ \mathrm{mol\ dm^{-3}} = 6 \times 10^6\ \mathrm{s^{-1}}$. Equation (4.52) can be rearranged to give

$$k_1 = \frac{k_\infty k_{-1}}{k_2} = \frac{k_\infty \lambda Z}{k_2} = \frac{38.7 \times 10^{-5}\ \mathrm{s^{-1}} \times 5 \times 10^{10}\ \mathrm{mol^{-1}\ dm^3\ s^{-1}}}{6 \times 10^6\ \mathrm{s^{-1}}}$$
$$= 3.2\ \mathrm{mol^{-1}\ dm^3\ s^{-1}}$$

where $\lambda = 1$.

It is likely that only one collision is needed to deactivate an activated molecule, and so k_{-1} can be equated to the collision number.

4.5.2 Physical significance of the constancy or otherwise of k_1, k_{-1} and k_2

The analysis of the experimental results given in Problem 4.19 assumes that k_1, k_{-1} and k_2 are constants. k_1 and k_{-1} describe the energy transfers of activation and deactivation. The assumption that these rate constants are single valued means that no distinction is drawn between the various excited vibrational levels corresponding to activated molecules. A unique value for k_2 implies that a very highly activated molecule decomposes at the same rate as a molecule with just the critical energy. Likewise a unique value for k_{-1} implies that deactivation from a highly excited molecule occurs at the same rate as that for a molecule with just the critical energy. These assumptions can be tested by experimental data in the following way.

Equation (4.51) can be rearranged to give

$$\frac{1}{k_{\mathrm{obs}}^{\mathrm{1st}}} = \frac{k_{-1}}{k_1 k_2} + \frac{1}{k_1[\mathrm{A}]} \tag{4.61}$$

If k_1, k_{-1} and k_2 are constants, i.e. are independent of the energy levels involved, see Sections 4.5.5, 4.5.7 and 4.5.10, the graph of $1/k_{\mathrm{obs}}^{\mathrm{1st}}$ versus $1/[\mathrm{A}]$ should be linear with slope $= 1/k_1$ and intercept $= k_{-1}/k_1 k_2$. This would provide an alternative route to k_1 and k_2, provided an assumed value is given to k_{-1}. If the graph is not linear, this means that some or all of the rate constants are not single valued. Experiments on energy transfer and molecular beams indicate that it is a reasonable approximation to assume that k_{-1} is single valued, but that it is incorrect to assume k_1 and k_2 to be single valued. Problem 4.20 illustrates this and allows a decision to be made.

Worked Problem 4.20

Question. Take the values of k_{obs}^{1st} and [A] given in Problem 4.18 and use them to decide whether k_1, k_{-1} and k_2 are constants.

Answer.

$\dfrac{10^{-4} \times \frac{1}{[A]}}{\text{mol}^{-1}\ \text{dm}^{-3}}$	8.3	1.65	0.79	0.433	0.140	0.006
$\dfrac{10^{-3}\ \frac{1}{k_{obs}^{1st}}}{\text{s}}$	11.6	6.5	5.0	4.5	3.55	2.58

Figure 4.30 gives a plot of $1/k_{obs}^{1st}$ versus 1/[A], and demonstrates conclusively that at least one of the rate constants is not single valued.

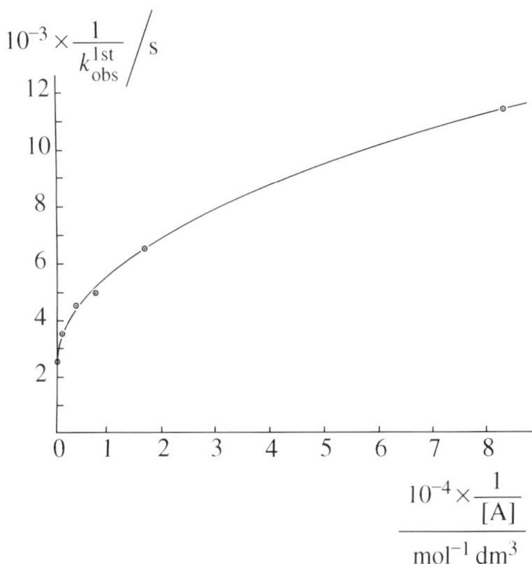

Figure 4.30 Graph of $1/k_{obs}^{1st}$ versus 1/[A] showing departure from linearity

4.5.3 Physical significance of the critical energy in unimolecular reactions

In bimolecular reactions where *two molecules* with at least the critical energy between them move into the critical configuration and react, the critical energy is assumed to have been accumulated in the energy of relative translational motion along the line of centres of the colliding molecules. But is this a sensible criterion for the critical energy in a unimolecular reaction, where the *single* activated molecule moves into the critical configuration and reacts? The answer is an unambiguous 'no'. All unimolecular theories consider vibrational energy to be the principal factor contributing to the critical energy which enables the activated molecule to split up. It is easily visualized that the over-extension of a bond or bond angle as a result of vigorous vibration of the molecule as a whole could result in reaction.

Worked Problem 4.21

Question. Why is rotational energy not considered to be as important as vibrational energy in a unimolecular reaction? In what circumstances might it be important?

Answer. The maximum number of rotational modes is three. In nearly all polyatomic molecules there are considerably more vibrational modes, e.g. CH_4 has nine. It is only with very small molecules that rotation might play a significant role, e.g. in the decompositions of O_3 and N_2O. The non-linear O_3 has three rotations and three vibrations, while the linear N_2O has two rotations and four vibrations.

4.5.4 Physical significance of the rate constants k_1, k_{-1} and k_2

These can be described in terms of a simple model and a complex model. For both, it is assumed that the critical energy corresponds to vibrational energy.

4.5.5 The simple model: that of Lindemann

- *The rate constant for activation, k_1.* There is *one* rate constant describing activation to *all* vibrational levels above the critical energy, ε_0, so there is no discrimination between these high energy levels. The critical energy is assumed to be accumulated in one vibrational mode.

This model makes a simple distinction between molecules without the critical energy, and those with at least the critical energy, i.e. activated or energized molecules. It takes no account of the fact that activated molecules actually do have vibrational levels of differing energies; and that the question 'is it as easy to activate a molecule to the first vibrational level above the critical as it is to activate the molecule to a high vibrational level?' could be asked. Intuitively this seems unlikely. As shown in the analysis of Problem 4.20 this assumption is unlikely to be correct.

- *The rate constant for deactivation, k_{-1}.* In all the major theories k_{-1} takes a unique value and describes deactivation from all energy levels above the critical to any level below the critical. The magnitude is often taken to be λZ, where Z is the collision number and λ is a collision efficiency that lies between zero and unity.

Again this implies that no distinction is made between differing energy levels above the critical. Energy transfer experiments show that this is a better approximation than the corresponding activation assumption, in so far as one collision is often all that is required for deactivation no matter what excited level is involved. In contrast, many collisions are required for activation.

- *The rate constant for reaction, k_2.* This is related to the time lag between the activating collision and the moment of reaction. For a first order process (see Table 3.3) $k = 1/\tau$, where τ is the *mean lifetime* of the activated molecule. In this simple model this time lag is *independent* of how much energy the activated molecule A^* has above the critical. This means that on average a very highly activated molecule lasts as long as one with energy just above the critical before reaction occurs. There is one k_2, which describes reaction from all energy levels above the critical energy, ε_0.

Intuitively, it would seem that a vibrationally very excited molecule is more likely to fly apart than is a molecule with a low vibrational energy.

As seen in Problem 4.20 it is unlikely that the assumptions given above are all correct. Experiment shows that at least one of k_1, k_{-1} and k_2 depends on the vibrational energy.

4.5.6 Quantifying the simple model

For the high pressure region of a unimolecular reaction where the reaction step is rate determining, statistical mechanics allows a calculation of the fraction of molecules which can have an energy at least the critical, ε_0, in one vibrational mode. One vibrational mode has energy made up of kinetic energy of vibration and potential energy of vibration. Each of these energies has one squared term in the equation specifying it, which gives two squared terms in total for one vibrational mode. For the explanation of energy in two squared terms and the Maxwell–Boltzmann distribution see Section 4.1.2.

The Maxwell–Boltzmann distribution gives the relevant fraction in two squared terms as equal to $\exp(-\varepsilon_0/kT)$, which gives

$$\frac{k_1}{k_{-1}} = \exp\left(\frac{-\varepsilon_0}{kT}\right) \tag{4.62}$$

- If k_{-1} is taken as λZ then

$$k_1 = \lambda Z \exp\left(\frac{-\varepsilon_0}{kT}\right) \tag{4.63}$$

- The value of k_2 can be found from k_1/k_{-1}, and the high pressure rate constant k_∞:

$$(k_{\text{obs}}^{\text{1st}})_{\text{high pressures}} = k_\infty = \frac{k_1 k_2}{k_{-1}} \tag{4.52}$$

$$= k_2 \exp\left(\frac{-\varepsilon_0}{kT}\right) \tag{4.64}$$

- Since k_2 is a constant, the dependence of k_∞ on temperature should show strict Arrhenius behaviour, giving a linear plot of $\log_e k_\infty$ versus $1/T$.

- The dependence of k_{obs}^{1st} on pressure is given as

$$\frac{1}{k_{obs}^{1st}} = \frac{k_{-1}}{k_1 k_2} + \frac{1}{k_1[A]} \tag{4.61}$$

Using the values of k_1, k_{-1}, and k_2 found as above the dependence of k^{1st} on pressure can be found.

When these calculated values were compared with experiment,

- k_1(observed) $\gg k_1$(calculated),
- the falling off of the rate constant $(k_{obs}^{1st})_{high\ pressures}$ with decrease in pressure occurs at a much lower pressure than predicted and
- the graph of $1/k_{obs}^{1st}$ versus $1/[A]$ is quite decidedly curved.

These observations show that the simple model is inadequate, even though the observed effect of temperature on k_∞ is as predicted.

4.5.7 A more complex model: that of Hinshelwood

- *The rate constant for activation, k_1.* This theory considers a more efficient activation process to bring the calculated k_1 more in line with the observed value. Energy can be accumulated in s normal modes, generally taken to be vibrational modes, with the value of s not specified. Likewise, the details of how the normal modes are involved, the number of modes involved and how the energy is distributed into the normal modes are not specified. The theory does not consider how the activated molecule moves into the activated complex. It only states that the activated molecule which is formed lasts until, by an internal redistribution of energy among the s modes, the relevant geometrical arrangement of the atoms in the molecule is that corresponding to reaction.
- *The rate constant for deactivation, k_{-1}.* This is given as equal to λZ.
- *The rate constant for the reaction step, k_2.* Again k_2 is assumed to be a constant, which does not depend on the energy above the critical value from which any given activated molecule reacts.

4.5.8 Quantifying Hinshelwood's theory

Energy is accumulated in s vibrational modes, i.e. in $2s$ squared terms, and the Maxwell–Boltzmann distribution gives to a first approximation

$$\frac{k_1}{k_{-1}} = \frac{(\varepsilon_0/kT)^{s-1}}{(s-1)!}\exp(-\varepsilon_0/kT) \tag{4.65}$$

$$k_1 = \frac{\lambda Z(\varepsilon_0/kT)^{s-1}}{(s-1)!}\exp(-\varepsilon_0/kT) \tag{4.66}$$

Note: The symbol $n!$ (read as n factorial) means the product of all the positive integers from 1 to n, e.g. $4! = 4 \times 3 \times 2 \times 1 = 24$.

It is important to be aware of the difference in algebraic form between the Maxwell–Boltzmann distribution for two squared terms (Section 4.1.2) and that for $2s$ squared terms. Also note well that $2s$ squared terms are used. This is because one vibrational mode has two squared terms associated with it, and so if activation energy is accumulated in s vibrational modes then it will be associated with $2s$ squared terms. Physically, it may be easier to visualise energy going into normal modes of vibration than to think of its going into squared terms (Section 4.1.2).

A calculated value of $k_{\text{obs}}^{\text{1st}}$ can be found from Equation (4.51), which when rearranged gives

$$k_{\text{obs}}^{\text{1st}} = \frac{k_2 k_1/k_{-1}}{1 + k_2/k_{-1}[\text{A}]} \tag{4.67}$$

Substituting from Equation (4.65) for k_1/k_{-1} gives

$$k_{\text{obs}}^{\text{1st}} = \frac{k_2}{1 + k_2/k_{-1}[\text{A}]} \frac{1}{(s-1)!} \left(\frac{\varepsilon_0}{kT}\right)^{s-1} \exp\left(\frac{-\varepsilon_0}{kT}\right) \tag{4.68}$$

where k_2 is single valued and constant, in keeping with the model of assuming that the lifetime of an activated molecule is independent of how much energy it possesses above the critical.

This equation requires explicit values of k_{-1}, k_2, ε_0 and s.

A first approximate value of k_2 can be found from the pressure at which the observed first order rate constant has fallen to half the constant high pressure value (Problem 4.19), and this first approximate value can be used to obtain a first approximate value for k_1. Thereafter k_2, ε_0 and s are found by successive iterations and best fit of experiment with theory.

The high pressure rate constant obtained when the reaction step in the sequence is rate determining can be calculated using equation (4.52),

$$k_\infty = \frac{k_1 k_2}{k_{-1}}$$

giving

$$k_\infty = k_2 \frac{1}{(s-1)!} \left(\frac{\varepsilon_0}{kT}\right)^{s-1} \exp\left(\frac{-\varepsilon_0}{kT}\right) \tag{4.69}$$

where again best fit values for k_2, ε_0 and s are used. It is found that best fits occur when s is approximately equal to half the total number of normal modes of vibration.

Even although k_2 is a constant this equation predicts a non-linear Arrhenius plot.

4.5.9 Critique of Hinshelwood's theory

- Activation into s vibrational modes gives a larger value for k_1 compared with the value for activation into one vibrational mode and is automatically a more efficient activation process. This represents a considerable advance on the simple model.
- The simple model predicts strict linearity of an Arrhenius plot, whereas the Hinshelwood model predicts a curved Arrhenius graph of $\log_e k_\infty$ versus $1/T$. Linear plots are observed.
- Allowing activation into s normal modes gives a considerable improvement in predicting the pressure at which a fall-off in the values of the calculated first order rate constant is found, i.e. Equation (4.68) is a better fit than Equation (4.61).
- The most damaging criticism of the theory is that it does not account for the very decided curvature observed in the plots of $1/k_{\mathrm{obs}}^{\mathrm{1st}}$ versus $1/[\mathrm{A}]$.

Worked Problem 4.22

Question. Calculate values for k_1, assuming $s = 1$, 3 and 6. Take Z to be 5×10^{10} $\mathrm{mol^{-1}\ dm^3\ s^{-1}}$, $T = 300$ K and $E_A = +60$ kJ $\mathrm{mol^{-1}}$.

Answer. The theoretical equations are in molecular units. Both Z and E_A are given in molar units. *But note* $(-E_A/RT) = (-\varepsilon_0/kT)$, and in this case E_A, given in kJ $\mathrm{mol^{-1}}$, does not have to be converted to a molecular quantity.

(a) $s = 1$

$$\exp(-\varepsilon_0/kT) = \exp\left(\frac{-60 \times 10^3}{8.3145 \times 300}\right) = \exp(-24.05) = 3.6 \times 10^{-11}$$

$$\begin{aligned} k_1 &= Z\exp(-\varepsilon_0/kT) = 5 \times 10^{10} \times 3.6 \times 10^{-11}\ \mathrm{mol^{-1}\ dm^3\ s^{-1}} \\ &= 1.8\ \mathrm{mol^{-1}\ dm^3\ s^{-1}} \end{aligned}$$

(b) $s = 3$

$$\frac{(\varepsilon_0/kT)^{s-1}}{(s-1)!}\exp(-\varepsilon_0/kT) = \frac{(24.05)^2}{2!} \times 3.6 \times 10^{-11} = 1.04 \times 10^{-8}$$

$$\begin{aligned} k_1 &= Z\frac{(\varepsilon_0/kT)^{s-1}}{(s-1)!}\exp(-\varepsilon_0/kT) \\ &= 5 \times 10^{10} \times 1.04 \times 10^{-8}\ \mathrm{mol^{-1}\ dm^3\ s^{-1}} \\ &= 520\ \mathrm{mol^{-1}\ dm^3\ s^{-1}} \end{aligned}$$

(c) $s = 6$

$$\frac{(\varepsilon_0/kT)^{s-1}}{(s-1)!}\exp(-\varepsilon_0/kT) = \frac{(24.05)^5}{5!} \times 3.6 \times 10^{-11} = 2.4 \times 10^{-6}$$

$$\begin{aligned} k_1 &= Z\frac{(\varepsilon_0/kT)^{s-1}}{(s-1)!}\exp(-\varepsilon_0/kT) \\ &= 5 \times 10^{10} \times 2.4 \times 10^{-6}\ \text{mol}^{-1}\ \text{dm}^3\ \text{s}^{-1} \\ &= 1.2 \times 10^5\ \text{mol}^{-1}\ \text{dm}^3\ \text{s}^{-1} \end{aligned}$$

From these values it is clear that allowing accumulation of energy into more than one vibrational mode dramatically increases the rate of activation. A value of $s = 6$ corresponds to a non-linear molecule with four atoms. If larger molecules were considered, s would be much larger and the increase in k_1 even more dramatic.

4.5.10 An even more complex model: that of Kassel

- *The rate constant for activation, k_1.* As with Hinshelwood, the critical energy is accumulated in s vibrational terms, but here s is the *total* number of vibrational modes open to the molecule. k_1 is then allowed to take different values dependent on how much energy above the critical energy, ε_0, the activated molecule possesses. It is easier to produce an activated molecule with energy slightly above the critical than it is to produce an activated molecule with energy far in excess of ε_0. The k_1 in the first instance would be higher than k_1 for the second situation.

- *The rate constant for deactivation, k_{-1}.* Even although activation occurs at a different rate depending on how much energy above the critical the molecule possesses in its vibrational modes, deactivation from all the levels above the critical is assumed to occur after one collision, i.e. $k_{-1} = \lambda Z$ with $\lambda = 1$.

- *The rate constant for the reaction step, k_2.* Hinshelwood assumed that the lifetimes of the activated molecules were the same no matter how much energy the molecule possessed in excess of the critical. Kassel assumes that a highly activated molecule will react faster than one which has just the critical energy. A series of k_2 values is considered, with the magnitude of each dependent on the energy value from which reaction is taking place.

- In the Hinshelwood theory the time lag corresponds to the time taken for the activated molecule to rearrange configuration into the critical configuration of the activated complex. The Kassel theory deals explicitly with this process, and imposes a *much more severe restriction* than does Hinshelwood. Before an activated molecule can react there must be a flow of energy at least ε_0 into a critical normal mode. Once this has happened the activated molecule has the correct distribution of vibrational energy for it to be able to react. The time lag is the time taken for this to happen.

4.5.11 Critique of the Kassel theory

The derivation is much more complicated than those outlined previously, especially when the quantum version is considered. Both versions are a vast improvement on the previous theories, and predict that $1/k^{1st}$ versus $1/[A]$ should be a curve, and that the plot of $\log_e k_\infty$ versus $1/T$ should be linear. Both predictions are verified by experiment.

More advanced theories continue to make further improvements.

4.5.12 Energy transfer in the activation step

The change over from first order kinetics, when the reaction step is rate determining, to second order kinetics, when activation is rate determining, depends on the total pressure, the complexity of the reactant and to a lesser extent on the temperature.

- *Effect of pressure.* This has been dealt with in detail on a quantitative basis. But on a physical basis it is more likely that activation is faster than the reaction step at high pressures where there is a greater chance of collisions, simply because there are more molecules around. At low pressures there is less chance of collisions occurring, again simply because there are fewer molecules around, and so activation is now slower than reaction.

Remember: activation is bimolecular, while reaction is unimolecular.

- *Complexity of the reacting molecule.* This reflects the number of atoms in the molecule, which determines the number of normal modes of vibration open to the molecule. The critical energy is accumulated in the normal modes of vibration, and the more complex the molecule the greater the number of modes over which *this* energy can be spread. Reaction is a consequence of a bond breaking and only some of the normal modes will contribute to the behaviour of this bond. The larger the total number of normal modes, i.e. the greater the complexity, the more difficult it will be for energy to flow into the critical coordinate, and the greater will be the time lag between activation and reaction. The reaction step is thus likely to be rate determining if the molecule is complex and first order kinetics is observed.

  If the molecule is small, there are fewer normal modes over which the critical energy can be spread, and it is much more likely that the required critical energy can flow into the critical coordinate and the molecule decompose. The smaller the reactant molecule, the more likely it is that activation will be rate determining. There will be a gradation as complexity increases and the final result may be determined by the pressure.

- *The temperature*. This has a much smaller effect, and is closely associated with the increase in the fraction of high energy molecules at higher temperatures. And so, at higher temperatures reaction is more likely to be faster than activation, which is thus more likely to be rate determining at high temperatures than at low. Second order kinetics, with activation rate determining, is thus more likely as the temperature increases.

These considerations are highly relevant to unimolecular steps in complex reactions, and to recombinations of atoms and radicals. These recombinations are the reverse of decompositions. If decomposition is in the high pressure first order region, with reaction rate determining, then the reverse recombination will be second order with reaction rate determining. If the decomposition is in the low pressure second order region, with activation rate determining, then the reverse combination will be third order and removal of energy will be rate determining.

These arguments are relevant to aspects of Chapter 6.

4.6 The Slater Theory

Slater's theory treats, in detail, the changing positions of the atoms in the molecule as the molecule vibrates, and studies these different arrangements of the atoms as a function of time. All the interatomic distances, bond lengths and bond angles are calculated as a function of time for a specified vibrational energy and energy distribution of the activated molecule. Since molecular vibrations vary periodically, then the bond lengths, bond angles and atomic distances will also change periodically with time. The changing atomic positions depend critically on the vibrational energy of the activated molecule, and the way in which this energy is distributed among the normal modes of vibration.

Fixing the magnitude of the energy, and the energy distribution, results in an inevitable and totally unique variation of the positions of the atoms with respect to each other, and reaction can only occur if certain atomic dimensions are achieved. What they are depends on the particular reaction, and the theory first decides what the critical configuration shall be, and then focuses attention on a critical aspect of the critical configuration, such as a bond length, a bond angle or some combination of both. The theory sets out the physical conditions and requirements which have to be met before it is posssible for the activated molecule to reach the critical dimensions.

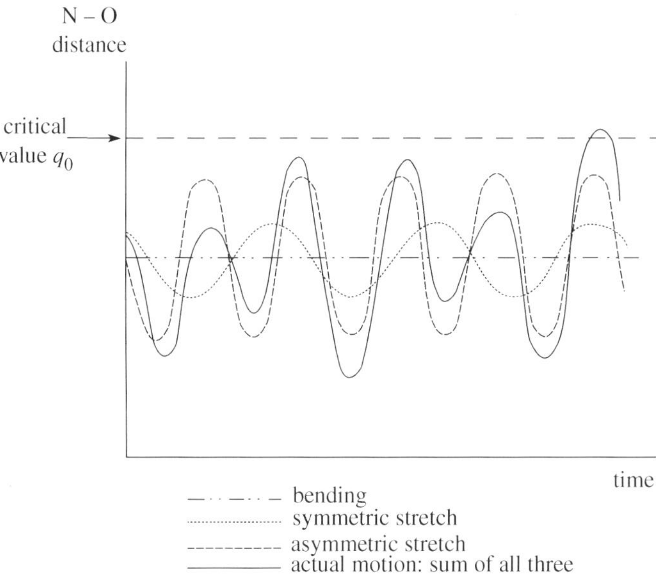

Figure 4.31 Summing of normal modes of vibration for N_2O

Figure 4.31 shows the sine waves for the three normal modes of vibration for the linear N_2O molecule. For the unimolecular decomposition $N_2O \rightarrow N_2 + O^{\bullet}$ it is assumed that reaction occurs with the breaking of the N–O bond after it reaches a certain critical extension. The diagram shows the summing of these three sine waves and the effect which that has on the extension of the N–O bond with time. The diagram shows the value of the critical extension which must be reached before the bond will break. The vibrational energy, and the distribution of this energy among the normal modes can be seen to be such that the bond *does* reach the critical value, and breaks.

The average number of times per unit time that the combination of sine waves, calculated for the summed normal modes, exceeds the critical extension can be found, and this gives a value for the first order high pressure rate constant.

Slater's theory assumes that the normal modes behave as harmonic oscillators, which requires that there be no flow of energy between the normal modes once the molecule is suitably activated, and so the energy distribution remains fixed between collisions. But spectroscopy shows that energy can flow around a molecule, and allowing for such a flow between collisions vastly improves the theory. Like Kassel's theory a fully quantum theory would be superior.

The equation for the high pressure rate constant predicts strict Arrhenius behaviour, and also a non-linear plot for $1/k_{obs}$ versus $1/[A]$, both in agreement with experiment.

However, the beauty of Slater's theory is that it gives a very clear description of the process of reaction, which is easy to visualize; that is, 'the molecule splits up if some bond(s), bond angle(s) or a combination of both is extended too far'.

Further reading

Laidler K.J., *Chemical Kinetics*, 3rd edn, Harper and Row, New York, 1987.
Laidler K.J. and Meiser J.H., *Physical Chemistry*, 3rd edn, Houghton Mifflin, New York, 1999.
Alberty R.A. and Silbey R.J., *Physical Chemistry*, 2nd edn, Wiley, New York, 1996.
Logan S.R., *Fundamentals of Chemical Kinetics*, Addison-Wesley, Reading, MA, 1996.
* Nicholas J., *Chemical Kinetics*, Harper and Row, New York, 1976.
* Wilkinson F., *Chemical Kinetics and Reaction Mechanisms*, Van Nostrand Reinhold, New York, 1980.
Robson Wright M., *Fundamental Chemical Kinetics*, Horwood, Chichester, 1999.
* Laidler K.J., *Theories of Chemical Reaction Rates*, McGraw-Hill, New York, 1969.
* Glasstone S., Laidler K.J. and Eyring H., *Theory of Rate Processes*, McGraw-Hill, New York, 1941.
* Out of print, but should be in university libraries.

Further problems

1. Find the change in the number of degrees of freedom of each type on forming the activated complex for the following reactions:
   - atom + non-linear molecule to non-linear AC
   - linear molecule + non-linear molecule to non-linear AC
   - two linear molecules to linear AC
   - two linear moleculcs to non-linear AC
   - non-linear molecule to non-linear AC.
2. How can $\Delta H^{\neq *}$ and $\Delta S^{\neq *}$ be found for a gas phase reaction? Why is it that a $\Delta V^{\neq *}$ value cannot be found for gas phase reactions?
3. Use transition state theory to calculate approximate $\Delta S^{\neq *}$ values for the following reactions:

$$Br^{\bullet} + H_2 \rightarrow HBr + H^{\bullet}$$
$$CH_3^{\bullet} + H_2 \rightarrow CH_4 + H^{\bullet}$$
$$F_2 + ClO_2^{\bullet} \rightarrow FClO_2 + F^{\bullet}$$
$$C_4H_6 + C_2H_4 \rightarrow \text{cyclo} - C_6H_{10}$$

   Comment on the values obtained.

The following are rough values for the contributions to the entropy for

translation:	$50\ \text{J mol}^{-1}\ \text{K}^{-1}$ per degree of freedom
rotation:	$40\ \text{J mol}^{-1}\ \text{K}^{-1}$ per degree of freedom
vibration:	$10\ \text{J mol}^{-1}\ \text{K}^{-1}$ per degree of freedom.

4. Taking the collision diameter to be 400 pm, calculate the collision number, Z, for collisions between the molecules $CH_2{=}CH{-}CH{=}CH_2$ and $CH_2{=}CH{-}CHO$ at 500 K, and from this find the pre-exponential factor, A.

The following are rough values for the contributions to the entropy from

translation:	$40\ \text{J mol}^{-1}\ \text{K}^{-1}$ per degree of freedom
rotation:	$25\ \text{J mol}^{-1}\ \text{K}^{-1}$ per degree of freedom
vibration:	$10\ \text{J mol}^{-1}\ \text{K}^{-1}$ per degree of freedom.

Using this, make a rough estimate of the transition state theory pre-exponential factor.
(The experimental A factor $= 1.6 \times 10^6\ \text{mol}^{-1}\ \text{dm}^3\ \text{s}^{-1}$.)

5. Use the following data for a unimolecular decomposition to determine k_1 and k_2 which appear in the simple Lindemann mechanism; assume that k_{-1} has a value of $5.0 \times 10^{10}\ \text{mol}^{-1}\ \text{dm}^3\ \text{s}^{-1}$. From this determine the mean lifetime of the activated molecule. Comment on the results.
(Proceed by plotting $1/k_{obs}^{1st}$ versus $1/[A]$.)

$\dfrac{10^6[A]}{\text{mol dm}^{-3}}$	10	15	20	40
$\dfrac{k_{obs}^{1st}}{\text{s}^{-1}}$	0.333	0.391	0.426	0.495

The following values extend the data to higher concentrations:

$\dfrac{10^6[A]}{\text{mol dm}^{-3}}$	100	200	400
$\dfrac{k_{obs}^{1st}}{\text{s}^{-1}}$	0.625	0.781	0.952

Comment on the plot of $1/k_{obs}$ versus $1/[A]$ which is obtained if all the values from both tables are included.

6. Predict the temperature dependence of A for the bimolecular reactions given in question 3.

$$Br^{\bullet} + H_2 \rightarrow HBr + H^{\bullet}$$
$$CH_3^{\bullet} + H_2 \rightarrow CH_4 + H^{\bullet}$$
$$F_2 + ClO_2^{\bullet} \rightarrow FClO_2 + F^{\bullet}$$
$$C_4H_6 + C_2H_4 \rightarrow \text{cyclo-}C_6H_{10}$$

(a) the translational partition function for each degree of freedom is proportional to $T^{1/2}$;

(b) the rotational partition function for each degree of freedom is proportional to $T^{1/2}$;

(c) the vibrational partition function for each degree of freedom is independent of temperature at low temperatures, but is proportional to T^1 at high temperatures.

7. The following are values of entropies of activation for some unimolecular reactions, together with values of ΔS^{θ} for the overall reaction:

Reaction	$\dfrac{\Delta S^{\neq *}}{\text{J mol}^{-1}\ K^{-1}}$	$\dfrac{\Delta S^{\theta}}{\text{J mol}^{-1}\ K^{-1}}$
cyclo-propane → propane	39	29.5
cyclo-butane → $2C_2H_4$	42	146.9
CH_2—CH_2 (bridged by O) → CH_3CHO	21	21.8
$CH_3NC \rightarrow CH_3CN$	4	−3.3
cyclo-$C_4F_8 \rightarrow 2C_2F_4$	49	172.9

Compare the $\Delta S^{\neq *}$ values with the corresponding ΔS^{θ}, and comment on the comparisons. Find the $\Delta S^{\neq *}$ values for the reverse reactions, and comment on the values obtained.

The values of ΔS^{θ} given in the table are based on a standard *concentration*, as distinct from the normal values based on a standard *pressure*. This has been done because the values of $\Delta S^{\neq *}$ are based on a standard *concentration*.

5 Potential Energy Surfaces

The calculation of the PE surface is basically quantum mechanical. Accurate surfaces are used to show how the topography of the surface affects the 'reaction unit' as it changes configuration across the surface. Predictions can be made, and these can be tested by molecular beams, spectroscopic techniques and chemiluminescence.

Aims

By the end of this chapter you should be able to

- distinguish between symmetrical, early and late barriers
- correlate the relative distances between atoms in the 'reaction entity' at the critical configuration with the type of potential energy surface
- list the properties of early barriers and correlate them with molecular beam results
- list the properties of late barriers and correlate them with molecular beam results and
- predict the dominant type of energy present in products and which type of energy promotes reaction.

5.1 The Symmetrical Potential Energy Barrier

It is possible to use the 3-D surface with its corresponding 2-D contour diagram and potential energy profile to discuss the general reaction $A + BC \rightarrow AB + C$ where A, AB, BC and C are all polyatomic molecules.

The progress of the reaction unit A----B----C along the reaction coordinate, or minimum energy path, is given in terms of relative values of r_{AB} and r_{BC}. Figures 5.1 and 5.2 have been given in terms of a symmetrical barrier where the activated complex lies symmetrically with respect to both the entrance and exit valleys.

An Introduction to Chemical Kinetics. Margaret Robson Wright
 ISBNs: 0-470-09058-8 (hbk) 0-470-09059-6 (pbk)

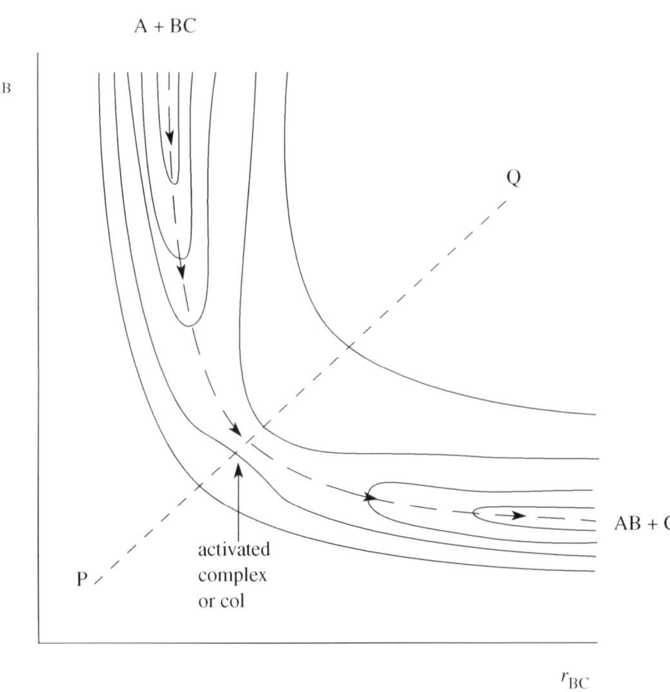

Figure 5.1 A potential energy contour diagram for a symmetrical barrier

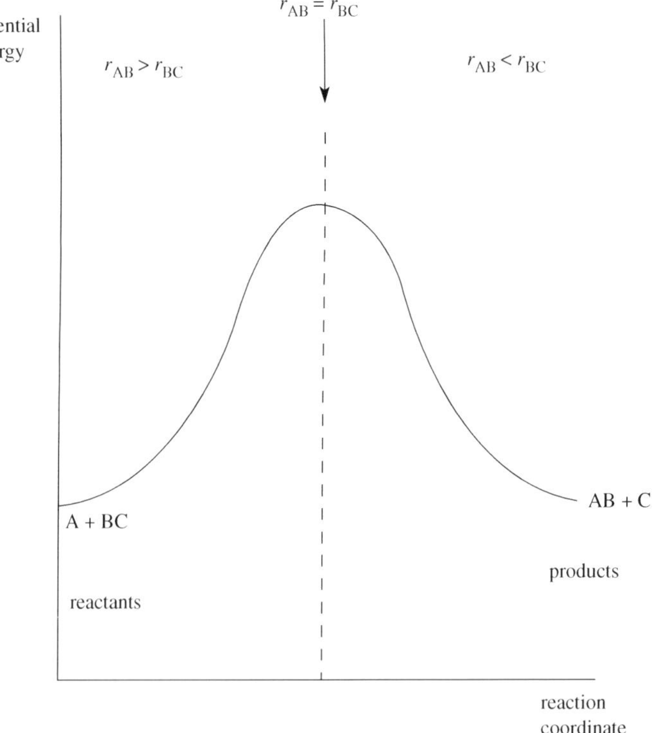

Figure 5.2 A potential energy profile for a symmetrical barrier

In the entrance valley r_{AB} is large and decreases towards the col. At the entrance to the exit valley r_{BC} is large, and decreases towards the col. At the col $r_{AB} = r_{BC}$, and this holds for all points along the line, PQ, as drawn. On the profile the activated complex lies at a maximum where $r_{AB} = r_{BC}$.

Other situations can occur where the activated complex lies in the entrance valley, an early barrier, or in the exit valley, a late barrier.

5.2 The Early Barrier

The activated complex and PE barrier are in the entrance valley, where $r_{AB} > r_{BC}$, corresponding to a configuration A-------B---C for the activated complex. This 3D surface is called an early or attractive surface with an early or attractive barrier (Figure 5.3).

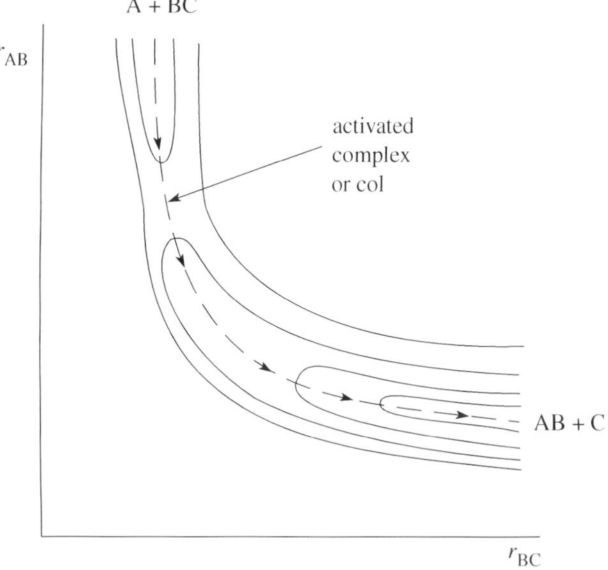

Figure 5.3 A potential energy contour diagram for an early barrier

The profile has the activated complex to the left of the line representing configurations where $r_{AB} = r_{BC}$ (Figure 5.4).

5.3 The Late Barrier

The activated complex and PE barrier are in the exit valley, where $r_{AB} < r_{BC}$, corresponding to a configuration A---B-------C for the activated complex. The 3-D

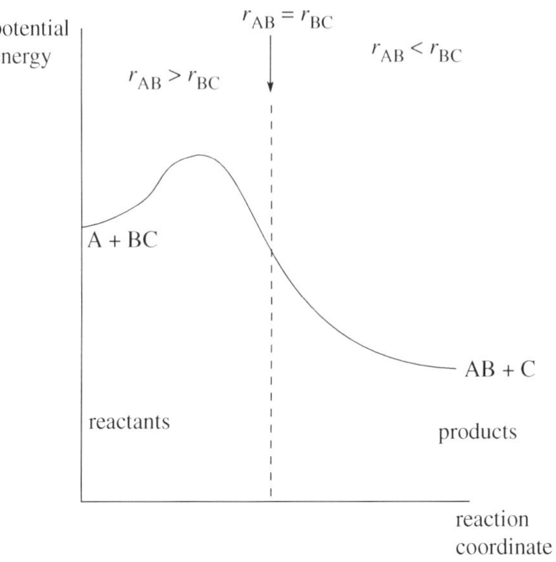

Figure 5.4 A potential energy profile for an early barrier

surface is called a late or repulsive surface, with a late or repulsive barrier (Figure 5.5).

The profile has the activated complex lying to the right of the line representing configurations where $r_{AB} = r_{BC}$ (Figure 5.6).

5.4 Types of Elementary Reaction Studied

Typical reactions that have been studied include

(a) $K^\bullet + Br_2 \rightarrow KBr + Br^\bullet$

(b) $H^\bullet + D_2 \rightarrow HD + D^\bullet$

(c) $K^\bullet + HCl \rightarrow KCl + H^\bullet$

(d) $D^\bullet + HBr \rightarrow DBr + H^\bullet$

(e) $Cl^\bullet + HBr \rightarrow HCl + Br^\bullet$

(f) $H^\bullet + HI \rightarrow H_2 + I^\bullet$

(g) $Br^\bullet + H_2 \rightarrow HBr + H^\bullet$

(h) $H^\bullet + Br_2 \rightarrow HBr + Br^\bullet$.

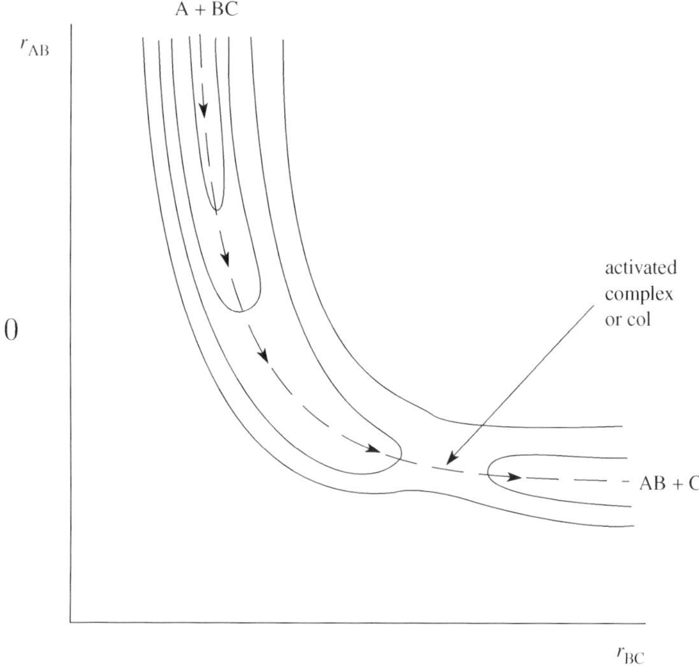

Figure 5.5 A potential energy contour diagram for a late barrier

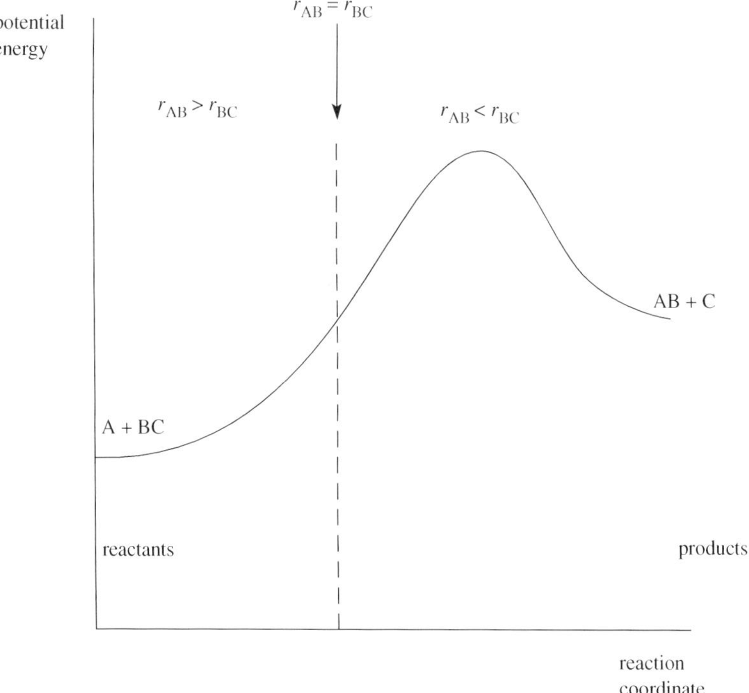

Figure 5.6 A potential energy profile for a late barrier

These are classified as a *heavy atom* approaching a molecule, or as a *light atom* approaching a molecule. Light and heavy are relative terms, and depend on the reaction in which the atoms are taking part, e.g. in the following two reactions $Na^{\bullet}$ can be light or heavy.

$$Na^{\bullet} + I_2 \rightarrow NaI + I^{\bullet}$$

$$Na^{\bullet} + H_2 \rightarrow NaH + H^{\bullet}$$

In the first reaction $Na^{\bullet}$ is *light* with respect to $I^{\bullet}$, but in the second is *heavy* with respect to $H^{\bullet}$.

Predictions depend on whether a *heavy* or a *light* atom is attacking a molecule, and on whether the overall reaction is *exo-* or *endo*thermic, or thermoneutral.

Worked Problem 5.1

Question. In the above examples which of the attacking atoms are heavy, and which are light? Do the same for the reverse reactions.

Answer.

Forward reaction	*Reverse reaction*
(a) $K^{\bullet}$ is heavy	$Br^{\bullet}$ is heavy
(b) $H^{\bullet}$ is light	$D^{\bullet}$ is heavy relative to $H^{\bullet}$, but is still a light atom. This is an intermediate situation.
(c) $K^{\bullet}$ is heavy	$H^{\bullet}$ is light
(d) $D^{\bullet}$ is light	$H^{\bullet}$ is light
(e) $Cl^{\bullet}$ is heavy	$Br^{\bullet}$ is heavy
(f) $H^{\bullet}$ is light	$I^{\bullet}$ is heavy
(g) $Br^{\bullet}$ is heavy	$H^{\bullet}$ is light
(h) $H^{\bullet}$ is light	$Br^{\bullet}$ is heavy

Predictions can be accurately tested by modern molecular beam, chemiluminescence and spectroscopic experiments. The more important aspects of these predictions are summarized below.

5.5 General Features of Early Potential Energy Barriers for Exothermic Reactions

Exothermic reactions release energy, and initially this energy resides in the products. The conclusions are independent of whether the attacking atom, A, is light or heavy in the general reaction, A + BC → AB + C, and are given in terms of

- the reaction unit A---B---C,
- the bonds A–B and B–C and
- the distances r_{AB} and r_{BC}.

On attractive surfaces (Figure 5.7), the unit (A---B---C) reaches the barrier or col before the bond B–C has altered much, and starts to move into the exit valley with A

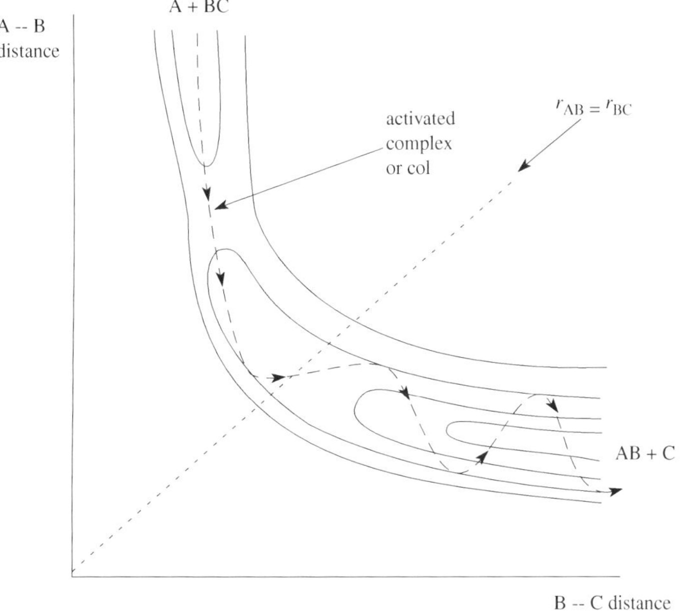

Figure 5.7 A potential energy contour diagram showing energy disposal for a reaction with an early barrier

still far from BC, but with the BC distance still close to its equilibrium internuclear distance. The barrier lies in the entrance valley, and the col corresponds to the configuration

$$\text{A-------B---C} \qquad r_{AB} \gg r_{BC}$$

In the exit valley C separates from B when A is still far from B. Even when C has separated from B by a very considerable extent, the AB distance remains large compared to its equilibrium internuclear distance. The reaction energy is released as the new bond is forming. Repulsion between B and C cause B and C to recoil from each other as they separate, and B is propelled towards the approaching A *as the AB bond is forming*.

$$\overset{\rightarrow}{\text{A}}\text{------}\overset{\leftarrow}{\text{B}}\text{---}\overset{\rightarrow}{\text{C}}$$

The motion set up can be thought of as extension of bond B–C during vibration, and inward approach of A and B towards each other in the vibration of bond A–B. Separation of B from C sets up vibrations in the bond A–B, which is vibrationally highly excited. The fraction of vibrational energy in the products is much higher than that of translational energy, giving predominantly vibrational energy in the products. This occurs when there is a curved route from the entrance of the exit valley to the critical configuration.

Reactions on attractive surfaces with early barriers are promoted by high translational energy in the reactants, with vibrational energy playing a minor role. Selective enhancement by translational energy is easiest when there is a straight run up the entrance valley to the critical configuration.

Vibrational energy is less important, because high vibrational energy would encourage the 'reaction unit' to hit the barrier wall perpendicular to the reaction coordinate at the end of the straight run up the entrance valley, from which it would be bounced back down the entrance valley.

Molecular beam experiments confirm these predictions and extend the results.

- Attractive surfaces are normally associated with the forward scattering of stripping reactions, where A approaches BC and from a distance attaches itself to B, and continues on undeflected.
- This results in large impact parameters and cross sections where A is still at large distances when BC splits. The intermolecular attractive forces between A and B must be strong.
- Typical reactions with an early barrier studied by molecular beams include those of the alkali metals with halogens and other simple molecules.

$$Na^{\bullet} + Cl_2 \rightarrow NaCl + Cl^{\bullet}$$
$$Cs^{\bullet} + I_2 \rightarrow CsI + I^{\bullet}$$
$$Na^{\bullet} + PCl_3 \rightarrow NaCl + PCl_2^{\bullet}$$

5.6 General Features of Late Potential Energy Surfaces for Exothermic Reactions

Discussion of repulsive surfaces splits into two categories:

- the minimum energy path, the 'traditional reaction coordinate', is the most probable route (Figure 5.8) when the attacking atom is light;
- a 'cutting the corner' trajectory is the most probable route (Figure 5.9) when the attacking atom is heavy.

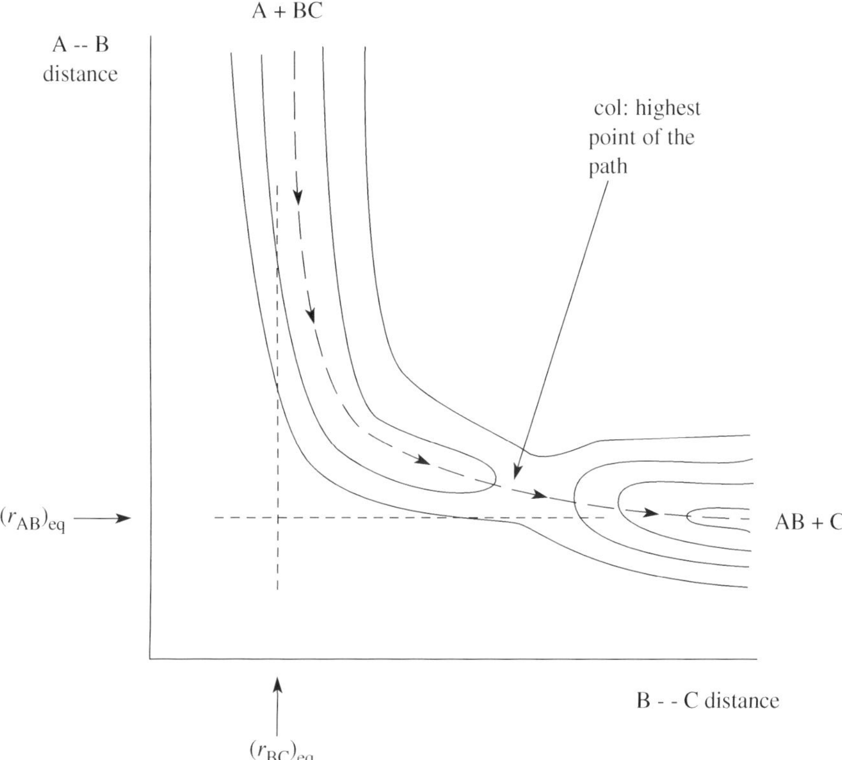

Figure 5.8 A potential energy contour diagram for a reaction with a late barrier, following the traditional reaction coordinate

For the minimum energy path, r_{AB} for the critical configuration at the col is close to $(r_{AB})_{eq}$, but r_{BC} is much larger than $(r_{BC})_{eq}$.

The 'cutting the corner' trajectory is displaced from the minimum energy path, and the activated complex configuration is displaced upwards with r_{AB} at the highest point now larger than $(r_{AB})_{eq}$.

5.6.1 General features of late potential energy surfaces where the attacking atom is light

The 'traditional' reaction coordinate (Figure 5.8) is followed. The reaction unit A---B---C reaches the col when $r_{AB} \approx (r_{AB})_{eq}$, i.e. it is close to its equilibrium internuclear distance, and the BC bond has extended to a considerable extent.

A----B-------C $\qquad r_{BC} \gg r_{AB}$

This places the barrier in the exit valley. In the exit valley, the AB bond is virtually formed before B and C have separated much, and energy release occurs while there is increasing separation of products, i.e. AB and C.

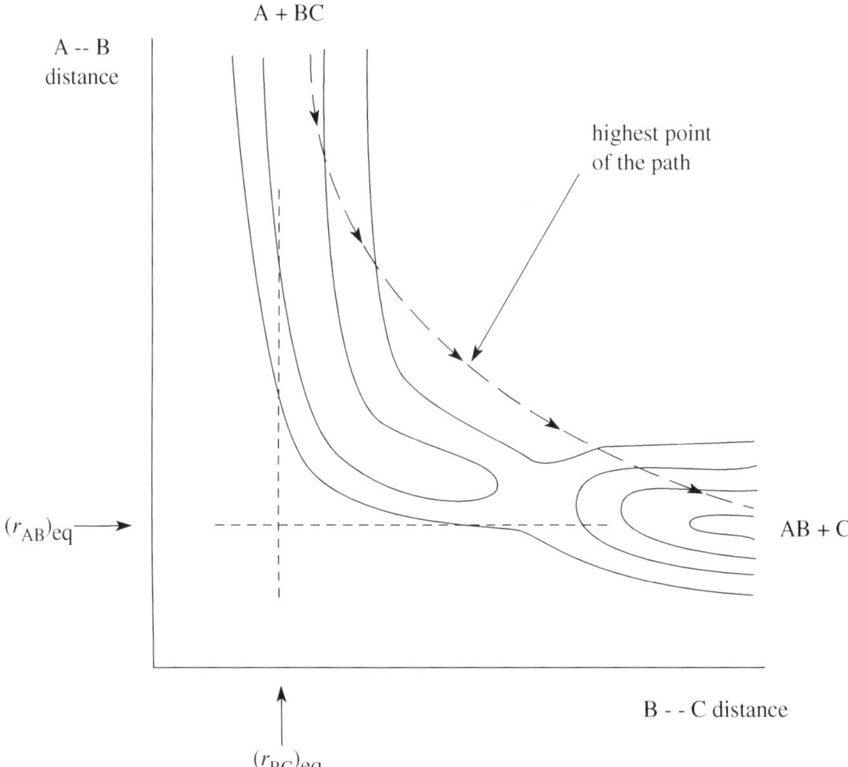

Figure 5.9 A potential energy contour diagram for a reaction with a late barrier, following a 'cutting the corner' trajectory

AB moves away from C as a single entity, and there is only minimal recoil of B into A, giving minimal vibrational energy in the products. There is a high proportion of translational energy in the products as predicted (Figure 5.10).

Reactions on these repulsive surfaces with late barriers are promoted by high vibrational energy in the reactants, with translational energy playing a minor role. Selective enhancement by vibration is best when there is a curved route up the entrance valley to the critical configuration. Vibration can enable the 'reaction unit' (A----B----C) to get round the bend easily. Translational energy is less important for surmounting the repulsive late barrier, because high translational energies would cause the 'reaction entity' (A----B----C) to hit the side walls of the entrance valley and be reflected back down the entrance valley.

Molecular beam experiments again confirm and extend these conclusions.

- Repulsive surfaces are associated with the backward scattering of a rebound mechanism, in which A collides with BC in a head-on collision and AB rebounds backwards.

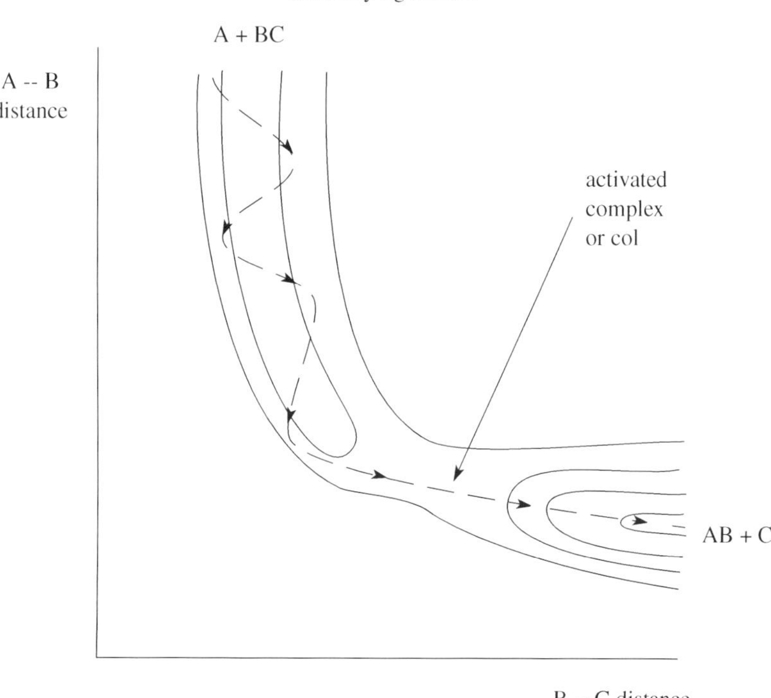

Figure 5.10 A potential energy contour diagram for a light atom attacking in a reaction with a late barrier, showing energy requirements for reaction and energy disposal in products

- This results in the small cross sections and small impact parameters expected when AB has formed as C splits off and separates from B under short range repulsion.
- Typical reactions studied in this way are between halogen atoms and HCl and HF

$$Br^{\bullet} + HCl \rightarrow HBr + Cl^{\bullet} \qquad Cl^{\bullet} + HF \rightarrow HCl + F^{\bullet}$$

along with the extensively studied reaction

$$K^{\bullet} + CH_3I \rightarrow KI + CH_3^{\bullet}$$

5.6.2 General features of late potential energy surfaces for exothermic reactions where the attacking atom is heavy

The behaviour for such reactions is very different from that predicted above for reactions where the attacking atom is light. Both vibration and translation are effective in promoting reaction, though vibrational energy is the more important. When a heavy atom is attacking, translational energy is still dominant in the products, but a much larger fraction of the energy is vibrational compared with the situations when light atoms attack. This is interpreted in terms of the reaction unit 'cutting the corner' (Figure 5.9), which is an alternative way of saying that the most probable route is now no longer the minimum energy path.

When B and C start to separate and move apart they do so before the AB bond is at its equilibrium internuclear distance, i.e. bond A–B is forming while B–C is breaking

$$\overset{\rightarrow}{\text{A}}\text{-------}\overset{\leftarrow}{\text{B}}\text{---}\overset{\rightarrow}{\text{C}}$$

AB is not moving apart from C as a single entity, and to some extent B is being propelled towards A while the A–B bond is forming, giving some vibrational release instead of the very dominant translational energy release found if a light atom attacks. This is the case *despite* the barrier being located in the exit valley. Both translation and vibration will appear in the products (Figure 5.11), with most vibration appearing as the 'cutting the corner' trajectory is progressively moved vertically upwards.

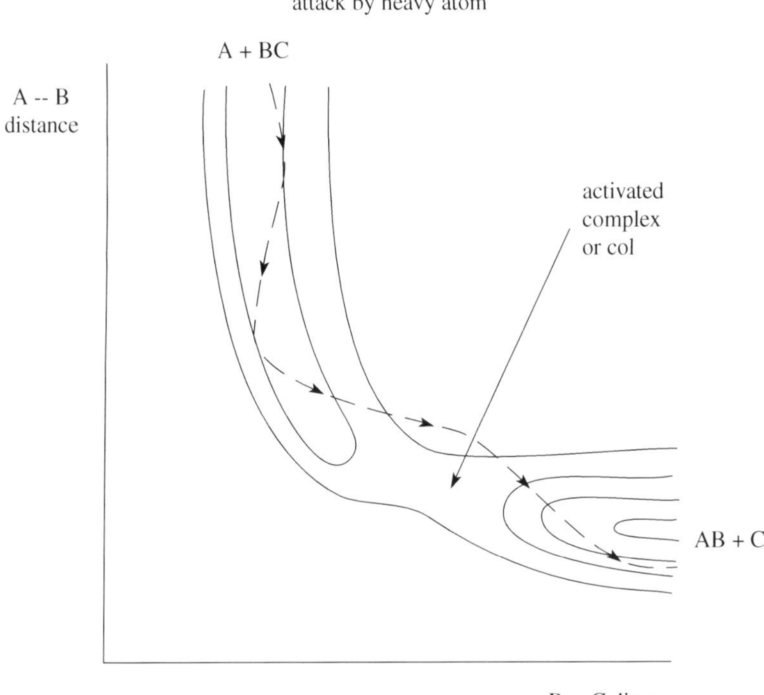

Figure 5.11 A potential energy contour diagram for a heavy atom attacking in a reaction with a late barrier, showing energy requirements for reaction and energy disposal in products

These conclusions are confirmed by chemiluminescence and molecular beam experiments for some reactions with highly repulsive surfaces, but which, nonetheless, give both vibration and translation in the products.

Worked Problem 5.2

Question. Interpret the molecular beam diagram in Figure 5.12.

Answer. Figure 5.12 shows forward scattering, and so is associated with an early barrier. This means that in the reaction $A + BC \rightarrow AB + C$, A is not much deflected from its original direction, and C continues with little deflection from the original direction of BC. This is a stripping reaction associated with large cross sections and impact parameters and with strong interactions between A and BC. Reaction is enhanced by translational energy in reactants, and vibrational energy is predominant in the products.

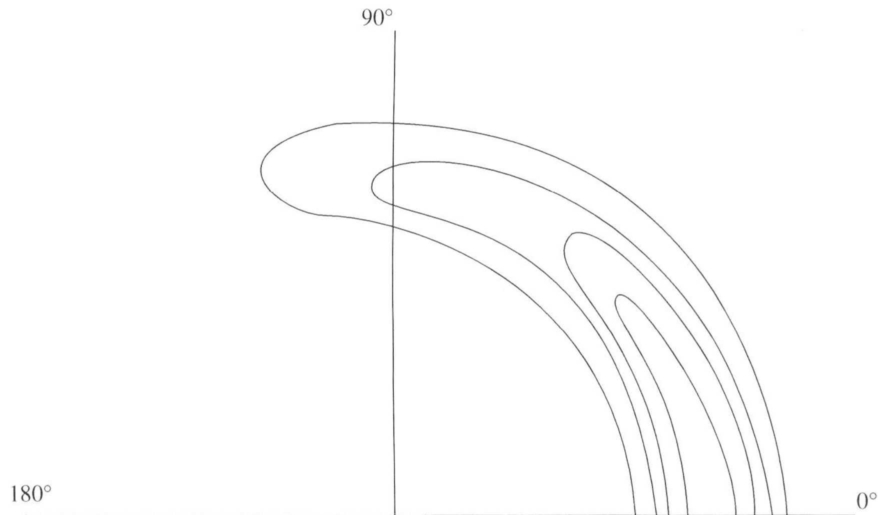

Figure 5.12 A molecular beam contour diagram

5.7 Endothermic Reactions

An endothermic reaction is the exact reverse of the corresponding exothermic reaction. An attractive surface for the exothermic reaction becomes a repulsive surface for the endothermic reaction, while the repulsive surface for the exothermic reaction becomes an attractive surface for the endothermic reaction. Likewise, if an exothermic reaction is enhanced by translational energy and has predominantly vibrational energy in the products, then the reverse endothermic reaction will be promoted by vibrational energy, and will have translational energy dominant in the products. Recognition of the mass types involved is imperative.

Worked Problem 5.3

Question. List the properties of the endothermic reaction

$$A + BC \rightarrow AB + C$$

corresponding to a reaction on an attractive early exothermic surface.

Answer. The endothermic reaction is the reverse of an exothermic reaction. The endothermic reaction will have

(a) a repulsive surface with a late barrier, so that the activated complex has a configuration

A---B------C

where the bond A–B is close to its equilibrium internuclear distance and bond B–C has yet to break;

(b) the late barrier means that the bond A–B is formed before the repulsion energy of B–C is released, so that translational energy is predominant in the products. Reaction will be enhanced by vibrational energy in reactants. These predictions are in keeping with the curved route up the entrance valley and the straight run down the exit valley from the activated complex.

5.8 Reactions with a Collision Complex and a Potential Energy Well

Sometimes there is a basin or well somewhere along the reaction coordinate on the surface and a minimum on the profile (Figures 5.13 and 5.14).

If the well is in the entrance valley the 'reaction unit' can often need vibrational energy, but if it is in the exit valley translational energy is often more effective. Such features are indicative of a collision complex which lasts sufficiently long for many vibrations and some rotations to occur. The lifetime of a collision complex is long

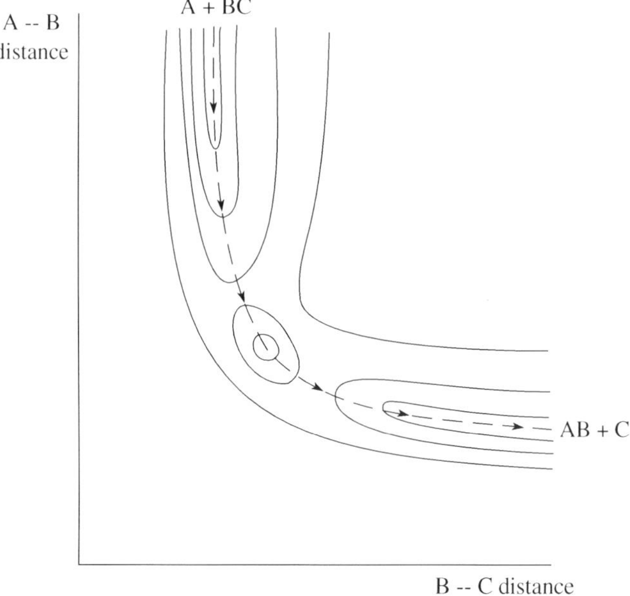

Figure 5.13 A potential energy contour diagram showing a collision complex

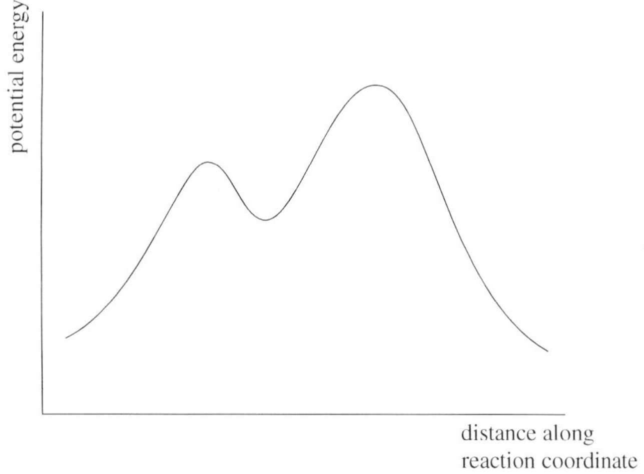

Figure 5.14 A potential energy profile showing a collision complex

enough to allow equilibration between vibrational and translational energy, so that the products separate with comparable amounts of both.

Again molecular beams give conclusive evidence as to whether a collision complex is involved. For such reactions the molecular beam contour diagram will be symmetric. The interactions keep the two reactants in close contact for long enough for them to 'forget' the directions in which they would have parted had they been able to do so before rotation. They thus part in random directions (Figure 5.15).

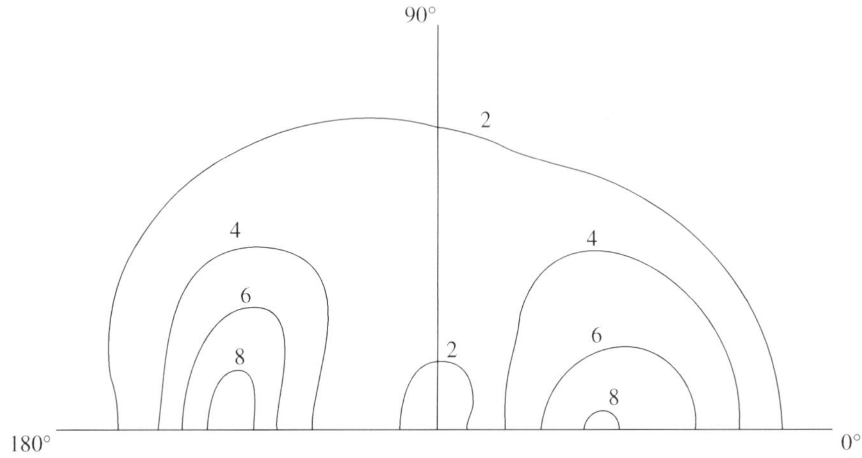

Figure 5.15 A molecular beam contour diagram showing a collision complex

Further reading

Laidler K.J., *Chemical Kinetics*, 3rd edn, Harper and Row, New York, 1987.

Laidler K.J. and Meiser J.H., *Physical Chemistry,* 3rd edn, Houghton Mifflin, New York, 1999.

Alberty R.A. and Silbey R.J., *Physical Chemistry*, 2nd edn, Wiley, New York, 1996.

Logan S.R., *Fundamentals of Chemical Kinetics*, Addison-Wesley, Reading, MA, 1996.

*Nicholas J., *Chemical Kinetics*, Harper and Row, New York, 1976.

*Wilkinson F., *Chemical Kinetics and Reaction Mechanisms*, Van Nostrand Reinhold, New York, 1980.

Robson Wright M., *Fundamental Chemical Kinetics*, Horwood, Chichester, 1999.

*Laidler K.J., *Theories of Chemical Reaction Rates*, McGraw-Hill, New York, 1969.

*Glasstone S., Laidler K.J. and Eyring H., *Theory of Rate Processes*, McGraw-Hill, New York, 1941.

*Out of print, but should be available in university libraries.

Further problems

1. In the reaction $P + MN \rightarrow PM + N$, the equilibrium internuclear distance, r_{eq}, for the bond M–N is 0.20 nm, while the M–N distance in the activated complex is 0.25 nm, and the P–M distance is 0.50 nm. The reaction is exothermic.

   - Predict the position of the activated complex on the potential energy profile, and draw out a schematic profile, indicating the relative values of $\Delta H^{\neq *}$ for the forward and back reactions.

- What are the energy requirements for reaction? Explain the energy disposal in the products. Predict the molecular beam contour diagram.
- Predict the behaviour of the back reaction.

2. (a) Interpret the following molecular beam diagram.

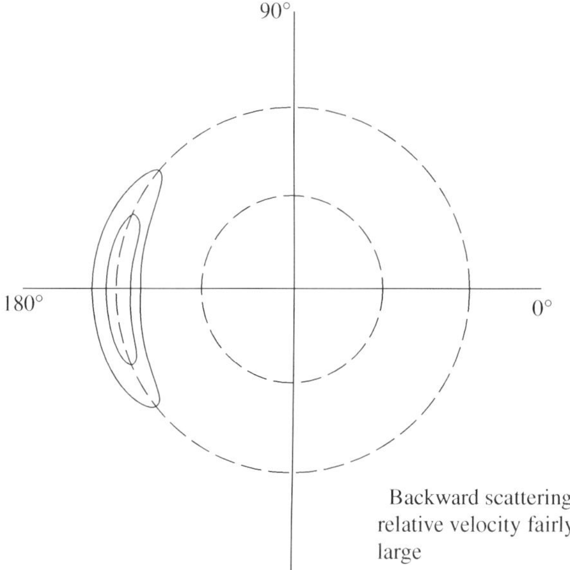

(b) Discuss the potential energy diagram shown below.

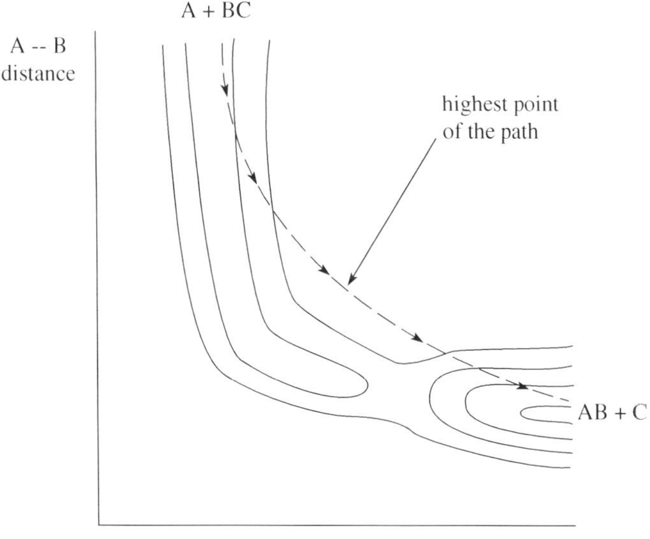

3. A given elementary reaction can be studied using molecular beam methods. It is found that reaction becomes faster as the vibrational ladder is climbed, i.e reaction

from $v = 1$ is slower than reaction from $v = 2$, which, in turn, is slower than that from $v = 3$. What can be said about the reaction?

4. The following reactions were studied using molecular beams. Suggest interpretations of these facts, and explain the reasoning.
   (a) In the reaction

$$Ca + NO_2 \rightarrow CaO + NO$$

   the cross section is large and there is predominantly forward scattering.

   (b) In the reaction

$$CH_3^{\bullet} + Br_2 \rightarrow CH_3Br + Br^{\bullet}$$

   the cross section is small and backward scattering is observed.

   (c) In the reaction

$$Cl^{\bullet} + CH_2 = CHBr \rightarrow CH_2 = CHCl + Br^{\bullet}$$

   forward and backward scattering are roughly equal in amount.

   (d) In the reaction of electronically excited Ca

$$Ca^* + O_2 \rightarrow CaO + O^{\bullet}$$

   the products are mainly in the ground vibrational level, $v = 0$, with only a very low population in other higher vibrational levels.

   (e) The reaction

$$O^{\bullet} + H_2 \rightarrow OH^{\bullet} + H^{\bullet}$$

   has a much greater cross section for $H_2(v = 1)$ than for $H_2(v = 0)$.

5. A reaction has a cross sectional area of 0.20 nm^2. The radii of the two reactants are respectively 80 and 190 pm. It is found that the translational energy of products in the region of greatest scattering is 42.5 kJ mol^{-1} while the product is predominantly in the ground vibrational state. The product is a diatomic molecule with $\nu_0 = 3 \times 10^{13} s^{-1}$. Vibrational energy $= (v + 1/2)h\nu_0$.

$$(h = 6.626 \times 10^{-34} \text{J s.})$$

   Deduce what you can about
   (a) the physical mechanism of the reaction,

   (b) the energy requirements of reaction, and energy disposal of the products,

   (c) the type of scattering that would be observed and

   (d) the PE contour and profile diagrams.

6 Complex Reactions in the Gas Phase

Aims

This chapter discusses complex reactions and focuses on the ***chemical*** mechanism. By the end of this chapter you should be able to

- recognize and classify the large variety of complex reactions
- distinguish between chain and non-chain mechanisms
- deduce mechanisms from experimental observations
- apply the steady state treatment
- recognize that there can be kinetically equivalent mechanisms
- recognize the distinctions between steady state and pre-equilibrium treatments
- quote the characteristics of chain reactions and their kinetic features
- appreciate the significance of third bodies in complex mechanisms.
- understand the special features of surface termination
- be aware of the features of branched chain reactions and the relation between branched chains and explosions, and
- be aware of the features of degenerate branching and the relation between degenerate branching and mild explosions

An Introduction to Chemical Kinetics. Margaret Robson Wright
 ISBNs: 0-470-09058-8 (hbk) 0-470-09059-6 (pbk)

6.1 Elementary and Complex Reactions

Elementary reactions occur via a single chemical step. If reaction occurs mechanistically as

$$A + B \rightarrow \text{products}$$

with

$$\text{mechanistic rate} = k[A][B] \tag{6.1}$$

then the reaction is bimolecular and second order.

There are only a few elementary bimolecular reactions in the gas phase, e.g. the reaction

$$NO + O_3 \rightarrow NO_2 + O_2$$

is believed to occur in one step involving the simultaneous coming together of NO and O_3 to form an activated complex, which changes configuration to give NO_2 and O_2 with

$$\text{rate} = k[NO][O_3] \tag{6.2}$$

Many of the bimolecular reactions for which extensive data are available occur as individual steps in reactions involving radicals, e.g.

$$CH_3^\bullet + C_2H_6 \rightarrow CH_4 + C_2H_5^\bullet$$
$$C_2H_5^\bullet + C_2H_5^\bullet \rightarrow C_4H_{10}$$

Reactions between light molecules have been extensively studied in the last two decades, generally by molecular beam techniques (see Chapter 4, Section 4.2), and these have allowed detailed testing of the predictions made from calculated potential energy surfaces. There are three typical mechanisms for gas phase reactions.

The stripping reactions showing forward scattering and large cross sections are typified by reactions such as

$$Na^\bullet + Cl_2 \rightarrow NaCl + Cl^\bullet$$
$$Na^\bullet + PCl_3 \rightarrow NaCl + PCl_2^\bullet$$

The rebound mechanism showing backward scattering and small cross sections is typified by

$$Br^\bullet + HCl \rightarrow HBr + Cl^\bullet$$
$$K^\bullet + CH_3I \rightarrow KI + CH_3^\bullet$$

Reactions involving collision complexes are sometimes found. Here an intermediate is produced, which lasts long enough for it to perform several rotations, and this results in symmetrical scattering. Typical reactions are

$$Cs^\bullet + SF_6 \rightarrow CsF + SF_5^\bullet$$
$$Cl^\bullet + CH_2{=}CHBr \rightarrow CH_2{=}CHCl + Br^\bullet$$

Ion–molecule reactions have also proved fertile ground for both theoretical studies and experiment. Here there are mainly two typical ion–molecule mechanisms: stripping reactions such as

$$N_2^{+\bullet} + H_2 \rightarrow N_2H^+ + H^\bullet$$
$$N_2^{+\bullet} + CH_4 \rightarrow N_2H^+ + CH_3^\bullet$$

and collision complexes for the more complex ion–molecule reactions such as

$$C_2H_4^{+\bullet} + C_2H_4 \rightarrow C_3H_5^{+\bullet} + CH_3^\bullet$$

Reaction can also be a simple breakdown of one molecule as

$$A \rightarrow \text{products}$$

with

$$\text{mechanistic rate} = k[A] \tag{6.3}$$

Such reactions are typified by the unimolecular decomposition of cyclopropane to give propene, where one molecule of reactant moves into the critical configuration and thence to products, with

$$\text{rate} = k[C_3H_6] \tag{6.4}$$

Like the bimolecular reactions, unimolecular reactions are often found as individual steps in complex reactions. These include the unimolecular breakdown of molecules into radicals often found as first initiation steps and propagation steps in chain reactions, e.g.

$$C_2H_6 \rightarrow CH_3^\bullet + CH_3^\bullet$$
$$CH_3CO^\bullet \rightarrow CH_3^\bullet + CO$$

Complex reactions proceed in several elementary chemical steps, and have a variety of mechanisms. But each elementary step in every mechanism also involves a *common physical* mechanism, with steps of activation and deactivation by binary collisions followed by the reaction step (Chapter 1 and Section 4.5). This can be important when discussing individual steps in, e.g., chain reactions, Section 6.9. This physical mechanism is

$$A + A \xrightarrow{k_1} A^* + A \qquad \text{activation by collision}$$

$$A^* + A \xrightarrow{k_{-1}} A + A \qquad \text{deactivation by collision}$$

$$A^* + b\,A \xrightarrow{k_2} \text{products} \qquad \text{reaction step}$$

$b = 0$, reaction is unimolecular;

$b = 1$, reaction is bimolecular;

$b = 2$, reaction is termolecular.

6.2 Intermediates in Complex Reactions

Complex reactions invariably involve *intermediates* which are formed in some steps, removed in others, and have a wide range of lifetimes. Longer lifetimes can result in build-up to significant intermediate concentrations during reaction, but these intermediates must also be sufficiently reactive to allow the subsequent reactions to occur. Intermediates can also be so short lived that they are removed almost as soon as they are formed, resulting in very, very low steady state concentrations. The lifetimes of intermediates and their concentrations have profound effects on the analysis of the kinetics of the reactions in which they occur.

Reminder. When steady state conditions prevail, the intermediates are highly reactive, and the total rate of their production is virtually balanced by their total rate of removal by reaction. They are present in very, very small and steady concentrations, and $d[I]_{ss}/dt = 0$.

Reactions involving intermediates are classified as non-chain or chain. A chain reaction is a special type of complex reaction where the distinguishing feature is the presence of propagation steps. Here one step removes an intermediate or chain carrier to form a second intermediate, also a chain carrier. This second chain carrier reacts to regenerate the first chain carrier and the characteristic cycle of a chain is set up, and continues until all the reactant is used up (see Section 6.9).

Although complex reactions can be classified as non-chain and chain, the type of experimental data collected and the manner in which it is analysed is common to both. The ultimate aim is to produce a mechanism, to determine the rate expression and to find the rate constants, activation energies and *A*-factors for all of the individual steps.

The following problem illustrates the variety of types of reaction which can occur.

Worked Problem 6.1

Question. Classify the following reactions in terms of the type of mechanism, i.e. elementary, complex etc:

(a) $ICl + H_2 \xrightarrow{\text{slow}} HI + HCl$

$HI + ICl \xrightarrow{\text{fast}} HCl + I_2$

(b) cyclo-$C_4H_8 \rightarrow 2\,C_2H_4$

(c) $^{238}U \rightarrow {}^{234}Th + \alpha$

$^{234}Th \rightarrow {}^{234}Pa + \beta$

$^{234}Pa \rightarrow {}^{234}U + \beta$

$^{234}U \rightarrow \ldots$

(d) $O_2^+ + NO \rightarrow O_2 + NO^+$

(e) $N_2O_4 \rightleftharpoons 2\,NO_2$

(f) $O_3 + M \rightarrow O_2 + O^\bullet + M$

$O_2 + O^\bullet + M \rightarrow O_3 + M$

$O_3 + O^\bullet \rightarrow 2\,O_2$

(g) *cis*-di-deuterocyclopropane → *trans*-di-deuterocyclopropane
cis-di-deuterocyclopropane → di-deuteropropenes

(h) $C_2H_6 \rightarrow 2\,CH_3^\bullet$

(i) $(CH_3)_3COOC(CH_3)_3 \rightarrow 2(CH_3)_3CO^\bullet$

$(CH_3)_3CO^\bullet \rightarrow CH_3COCH_3 + CH_3^\bullet$

$CH_3^\bullet + CH_3^\bullet \rightarrow C_2H_6$

(j) $NO + O_2 \xrightarrow{k_1} NO_3$

$NO_3 \xrightarrow{k_{-1}} NO + O_2$

$NO_3 + NO \xrightarrow{k_2} 2\,NO_2$

(k) $N_2O_5 \rightleftharpoons NO_2 + NO_3$

$NO_2 + NO_3 \rightarrow NO_2 + O_2 + NO$

$NO + N_2O_5 \rightarrow 3\,NO_2$

(l) $C_2H_5^\bullet + CH_3COCH_3 \rightarrow C_2H_6 + {}^\bullet CH_2COCH_3$

(m) $Br_2 + M \rightarrow 2\,Br^\bullet + M$

$Br^\bullet + CH_4 \rightarrow CH_3^\bullet + HBr$

$CH_3^\bullet + Br_2 \rightarrow CH_3Br + Br^\bullet$

$HBr + CH_3^\bullet \rightarrow CH_4 + Br^\bullet$

$2\,Br^\bullet + M \rightarrow Br_2 + M$

(n) $Kr^+ + H_2 \rightarrow KrH^+ + H^\bullet$

Answer.

(a) Two step consecutive.

(b) One step elementary.

(c) Many step consecutive.

(d) One step elementary.

(e) Reversible reaction.

(f) Three step complex with the second step being the reverse of the first.

(g) Two step parallel.

(h) Elementary step in a complex reaction.

(i) Three step complex.

(j) Three step complex with the second step being the reverse of the first.

(k) Four step complex with the second step being the reverse of the first.

(l) Elementary step in a complex reaction.

(m) Five step complex with two reversible steps, i.e. steps 1 and 5 are the reverse of each other, likewise for steps 2 and 4. This is a chain reaction. $Br^{\bullet}$ and $CH_3^{\bullet}$ are recycled.

(n) Elementary reaction.

6.3 Experimental Data

The following is a summary of the type of data involved in the study of complex reactions.

- Kinetic and non-kinetic information is required.
- Detection of products and intermediates is essential, and determination of their concentration throughout reaction allows products and intermediates to be classified as present in major, minor, trace or very trace amounts. This is relatively easy for products. Though easily detected, the very low concentrations of highly reactive intermediates can be difficult to determine accurately (Sections 2.1.4 and 2.1.5). Nonetheless, it is still possible to distinguish between very low steady state concentration intermediates and higher non-steady state concentrations. This has important consequences when unravelling the kinetics.
- Photochemically initiated reactions and quantum yields can yield vital clues in finding the mechanism; see Section 6.8.
- Thermochemical data, and the effect of temperature on relative yields can also help in determining mechanism.
- Kinetic studies, often from initial rates, give orders, experimental rate expressions, rate constants and activation parameters.

- The proposed mechanism must fit all the experimental facts, and the mechanistic rate expression must fit the experimental one. If these fit, the proposed mechanism is a possible or highly plausible one, but this does not *prove* the mechanism to be the correct one. Sometimes more than one mechanism can fit, called 'kinetically equivalent', and a distinction between them can only be made on non-kinetic evidence (see Section 6.6).
- The derivation of the mechanistic rate expression is considerably simplified if the steady state treatment can be used (Sections 3.19, 3.19.1 and 3.20). When intermediate concentrations are not sufficiently low and constant, the steady state approximation is no longer valid. Numerical integration by computer of the differential equations involved in the analysis, or computer simulation, may have to be used.

6.4 Mechanistic Analysis of Complex Non-chain Reactions

The methods and techniques described above will be illustrated by the following worked examples.

Worked Problem 6.2

This problem illustrates how to deduce a mechanism from basically *non-kinetic data.*

Question. The following results were obtained from a study of the *photochemical* decomposition of propanone, CH_3COCH_3:

$$CH_3COCH_3 \rightarrow C_2H_6 + CO$$

1. The quantum yield of CO is unity above 120 °C, but is less than unity below 120 °C.
2. The major products are C_2H_6 and CO, but CH_4 is also found in substantial amounts. $CH_3COCOCH_3$ is also formed below 120 °C.
3. The minor products are $CH_3COCH_2CH_3$, $CH_3COCH_2CH_2COCH_3$ and CH_2CO.
4. Spectroscopy, esr and mass spectrometry detect $CH_3^\bullet$, $CH_3COCH_2^\bullet$ and $CH_3CO^\bullet$.
5. Addition of O_2 and NO reduces yields of CH_4 and C_2H_6 to zero.

Deduce a mechanism that fits these observations.

Answer.

1. *Type of process.* Detection of $CH_3^\bullet$, $CH_3COCH_2^\bullet$ and $CH_3CO^\bullet$ demonstrates a free radical process, while the quantum yield of the product CO indicates a non-chain mechanism (see Section 6.8).

2. *Possible primary steps.* A rule of thumb suggests that C–C bonds break more readily than C–O bonds, which in turn break more easily than C–H bonds, suggesting two possible first steps. This rule of thumb is a summary of a vast amount of experimental data, including kinetic studies and bond dissociation measurements.

$$CH_3COCH_3 \rightarrow 2\,CH_3^\bullet + CO \qquad \text{(A)}$$

$$CH_3COCH_3 \rightarrow CH_3^\bullet + CH_3CO^\bullet \qquad \text{(B)}$$

$CH_3COCOCH_3$ is formed from recombination of two $CH_3CO^\bullet$ radicals, and is only found below 120 °C, suggesting step (B) as a possible primary step for these temperatures. This is consistent with the fact that $CH_3CO^\bullet$ is relatively stable below 120 °C and can undergo reactions other than decomposition to CO and $CH_3^\bullet$, i.e. recombination can occur.

At higher temperatures butane-2,3-dione, $CH_3COCOCH_3$, is no longer produced, and this is consistent with the known very rapid decomposition of the radical, $CH_3CO^\bullet$, above 120 °C,

$$CH_3CO^\bullet \rightarrow CH_3^\bullet + CO$$

which suggests that (A) is the dominant primary process above 120 °C.

These inferences are also consistent with the observed quantum yields.

(a) Above 120 °C a quantum yield of unity for CO fits

$$CH_3COCH_3 \rightarrow 2\,CH_3^\bullet + CO \qquad \text{(A)}$$

But it also fits

$$CH_3COCH_3 \rightarrow CH_3^\bullet + CH_3CO^\bullet \qquad \text{(B)}$$

since the very rapid decomposition of $CH_3CO^\bullet$ also results in one molecule of CO produced per molecule of CH_3COCH_3 decomposed.

(b) Below 120 °C the quantum yield of CO decreases, consistent with the emergence of the competitive reaction

$$CH_3CO^\bullet + CH_3CO^\bullet \rightarrow CH_3COCOCH_3$$

becoming possible, which will reduce the yield of CO per given amount of CH_3COCH_3 decomposing.

3. *Reactions of the $CH_3^\bullet$ radical produced by the primary processes.*

   (a) Radicals undergo H abstraction reactions readily, and also combine with other radicals.

$$CH_3^\bullet + CH_3COCH_3 \rightarrow CH_4 + {}^\bullet CH_2COCH_3$$

   This would account for the substantial yield of CH_4 and the presence of ${}^\bullet CH_2COCH_3$.

   (b) $CH_3^\bullet$ radicals recombine easily.

$$CH_3^\bullet + CH_3^\bullet + M \rightarrow C_2H_6 + M$$

   M is a third body necessary to remove energy from the newly formed excited product (see Section 6.12.4).
   But C_2H_6 is a major product, and so reaction (b) must occur more readily than the H abstraction reaction (a) producing CH_4.

4. *Reactions of the ${}^\bullet CH_2COCH_3$ radical produced by H abstraction by $CH_3^\bullet$.* Since CH_4 is produced in substantial amounts relative to the minor products, reasonable amounts of the ${}^\bullet CH_2COCH_3$ radical, the other product of the H abstraction, must also be formed, and be the source of the minor products formed by recombination reactions and a decomposition reaction.

$$CH_3^\bullet + {}^\bullet CH_2COCH_3 \rightarrow CH_3CH_2COCH_3$$
$$CH_3COCH_2^\bullet + {}^\bullet CH_2COCH_3 \rightarrow CH_3COCH_2CH_2COCH_3$$
$$CH_3COCH_2^\bullet \rightarrow CH_3^\bullet + CH_2CO$$

   All the minor products can therefore be accounted for by the presence of the ${}^\bullet CH_2COCH_3$ radical.

5. *Effect of O_2 and NO.* These are radical inhibitors and scavenge the $CH_3^\bullet$ radicals, removing the source of C_2H_6, CH_4 and the minor products. O_2 has two unpaired electrons and NO is an odd electron molecule. This means that these unpaired electrons can pair up with the unpaired electron on radicals to produce a single bond. This removes the radical characteristics, which are dependent on the presence of at least one unpaired electron. The yields will then drop to zero. This is observed and confirms the central role of $CH_3^\bullet$ in the mechanism.

6. *A mechanism that fits these observations.* The quantum yields have indicated that this reaction is not a chain reaction, and the mechanism proposed must reflect this.

$$CH_3COCH_3 \xrightarrow{8\mu} CH_3CO^\bullet + CH_3^\bullet$$

$$CH_3CO^\bullet \xrightarrow{k_2} CH_3^\bullet + CO \qquad \text{above } 120\ ^\circ\text{C}$$

$$CH_3^\bullet + CH_3COCH_3 \xrightarrow{k_3} CH_4 + {}^\bullet CH_2COCH_3$$

$$CH_3^\bullet + CH_3^\bullet \xrightarrow{k_4} C_2H_6$$

$$CH_3^\bullet + {}^\bullet CH_2COCH_3 \xrightarrow{k_5} CH_3CH_2COCH_3$$

$${}^\bullet CH_2COCH_3 + {}^\bullet CH_2COCH_3 \xrightarrow{k_6} CH_3COCH_2CH_2COCH_3 \qquad \text{very low}$$

$${}^\bullet CH_2COCH_3 \xrightarrow{k_7} CH_2CO + CH_3^\bullet$$

$$CH_3CO^\bullet + CH_3CO^\bullet \xrightarrow{k_8} CH_3COCOCH_3 \qquad \text{below } 120\ ^\circ\text{C}$$

$$CH_3^\bullet + CH_3CO^\bullet \xrightarrow{k_9} CH_3COCH_3 \qquad \text{below } 120\ ^\circ\text{C}$$

Although ${}^\bullet CH_2COCH_3$ is formed from $CH_3^\bullet$ and CH_3COCH_3, there is no subsequent major reaction of ${}^\bullet CH_2COCH_3$ to regenerate $CH_3^\bullet$, and so the mechanism proposed is not a chain reaction. Step 7 does regenerate $CH_3^\bullet$, but it is not a major step in this photochemical reaction.

This problem illustrates the ways in which the kineticist uses non-kinetic data to infer a plausible mechanism. However, the mechanism must also fit the observed kinetic facts. This is achieved by carrying out a steady state treatment on the proposed mechanism and then comparing the result with the observed rate expression.

6.5 Kinetic Analysis of a Postulated Mechanism: Use of the Steady State Treatment

The principles underlying this technique are set out in Sections 3.19 and 3.20. The steady state analysis must only be used in situations where the intermediates are present in *very, very low steady concentrations.* The following three problems explain how to carry out the steady state treatment.

Worked Problem 6.3

This is a straightforward example of moving from mechanism to rate expression.

Question. Decomposition of di-2-methylpropan-2-yl peroxide produces propanone and ethane:

$$(CH_3)_3COOC(CH_3)_3 \rightarrow 2\,CH_3COCH_3 + C_2H_6$$

and the generally accepted mechanism is

$$(CH_3)_3COOC(CH_3)_3 \xrightarrow{k_1} 2(CH_3)_3CO^\bullet$$
$$(CH_3)_3CO^\bullet \xrightarrow{k_2} CH_3COCH_3 + CH_3^\bullet$$
$$CH_3^\bullet + CH_3^\bullet \xrightarrow{k_3} C_2H_6$$

1. Explain why this is not a chain reaction.
2. Using the steady state assumption show that the reaction is first order throughout, even though it occurs in three consecutive steps. Formulate the rate of reaction in terms of production of C_2H_6.
3. Use of the steady state treatment requires that the radicals $CH_3^\bullet$ and $(CH_3)_3CO^\bullet$ are present in very, very low concentrations. Explain how this reaction can still be a standard source of $CH_3^\bullet$.

Answer.

1. The intermediates are $CH_3^\bullet$ and $(CH_3)_3CO^\bullet$. Although the $(CH_3)_3CO^\bullet$ splits up to form $CH_3^\bullet$ this *does not regenerate* $(CH_3)_3CO^\bullet$, and so this cannot be a chain reaction.
2. Steady states on the intermediates:

$$\frac{d[(CH_3)_3CO^\bullet]}{dt} = 2k_1[(CH_3)_3COOC(CH_3)_3] - k_2[(CH_3)_3CO^\bullet] = 0 \qquad (6.5)$$

rate of production ∴ +ve   rate of removal ∴ − ve

Note the factor of two: for each step 1, two $(CH_3)_3CO^\bullet$ are formed, and the rate of step 1 is given in terms of production of $(CH_3)_3CO^\bullet$

$$\therefore \quad [(CH_3)_3CO^\bullet] = \frac{2k_1}{k_2}[(CH_3)_3COOC(CH_3)_3] \qquad (6.6)$$

$$\frac{d[CH_3^\bullet]}{dt} = \underset{\substack{\uparrow \\ \text{rate of production } \therefore +\text{ve}}}{k_2[(CH_3)_3CO^\bullet]} - \underset{\substack{\uparrow \\ \text{rate of removal } \therefore -\text{ve}}}{2k_3[CH_3^\bullet]^2} = 0 \tag{6.7}$$

Again a factor of two appears because two $CH_3^\bullet$ radicals are removed for each act of recombination.

This is an equation in two unknowns and, unlike equation (6.5), cannot be solved. $[CH_3^\bullet]$ can be found by either of two methods.

(a) Use the standard procedure of adding the steady state equations, giving

$$[CH_3^\bullet] = \left(\frac{k_1}{k_3}\right)^{1/2} [(CH_3)_3COOC(CH_3)_3]^{1/2} \tag{6.8}$$

(b) Substitute for $[(CH_3)_3CO^\bullet]$ from (6.6) into (6.7), giving

$$[CH_3^\bullet] = \left(\frac{k_1}{k_3}\right)^{1/2} [(CH_3)_3COOC(CH_3)_3]^{1/2} \tag{6.9}$$

which is the same result, confirming the correctness of the algebra.

Rate of production of C_2H_6:

$$\frac{d[C_2H_6]}{dt} = k_3[CH_3^\bullet]^2 \tag{6.10}$$

$$= \frac{k_3k_1}{k_3}[(CH_3)_3COOC(CH_3)_3] \tag{6.11}$$

$$= k_1[(CH_3)_3COOC(CH_3)_3] \tag{6.12}$$

which predicts the reaction to be first order in reactant, as is observed experimentally, confirming that the mechanism fits the observed kinetics. *Note*: the rate of step 3 can be expressed in terms of

$$-\frac{d[CH_3^\bullet]}{dt} \quad \text{or} \quad +\frac{d[C_2H_6]}{dt}$$

For each step of reaction, two $CH_3^\bullet$ radicals are removed, and one C_2H_6 is formed.

$$\therefore \quad \text{rate at which 3 removes } CH_3^\bullet = 2k_3[CH_3^\bullet]^2 \tag{6.13}$$

and

$$\text{rate at which } C_2H_6 \text{ is formed} = +\frac{d[C_2H_6]}{dt} = k_3[CH_3^\bullet]^2 \tag{6.14}$$

so that

$$\text{rate at which 3 removes } CH_3^{\bullet} = +2\frac{d[C_2H_6]}{dt} \tag{6.15}$$

It is very important to understand these distinctions.

3. When the reaction is allowed to proceed without disturbance the $CH_3^{\bullet}$ radicals are present in steady state concentrations. However, if they are removed continuously from the reaction vessel, the reaction never gets to the steady state.

6.5.1 A further example where disentangling of the kinetic data is necessary

Another classic example of a complex reaction is the decomposition of N_2O_5 which shows first order kinetics, but with the first order rate constant decreasing in value as the pressure is lowered. Superficially, this could be taken as evidence of a typical unimolecular decomposition. However, even a first glance at the stoichiometry of the reaction should suggest that it is unlikely that there is a simple one step breakdown of N_2O_5 into the products.

$$N_2O_5(g) \rightarrow 2\,NO_2(g) + \tfrac{1}{2}O_2(g)$$

A complex reaction mechanism is likely, and all the experimental facts can be accounted for by the mechanism:

$$\begin{aligned} N_2O_5 &\xrightarrow{k_1} NO_2 + NO_3^{\bullet} \\ NO_2 + NO_3^{\bullet} &\xrightarrow{k_{-1}} N_2O_5 \\ NO_2 + NO_3^{\bullet} &\xrightarrow{k_2} NO_2 + NO^{\bullet} + O_2 \\ NO^{\bullet} + N_2O_5 &\xrightarrow{k_3} 3\,NO_2 \end{aligned}$$

This mechanism has a reversible unimolecular decomposition as a first step. As will be shown later, when unimolecular steps are involved in chain reactions, this can cause a change in order or a change in the value of the rate constant if the pressure is lowered.

This reaction is important in its own right, but it is also important in showing that all aspects of kinetics must be at one's fingertips when interpreting the kinetic data and proposing a mechanism. To understand the arguments leading to the interpretation of the data for this reaction it is essential that the basics of unimolecular theory, Section 4.5, are understood. It is also essential that, when a predicted rate expression is deduced which has a denominator made up of one or more terms, the physical significance of each term be understood and the relative values of each term assessed. The same situation arises in Worked Problems 6.4 and 6.5.

The intermediates are $NO_3^\bullet$ and $NO^\bullet$, and the steady state equations are

$$\frac{d[NO_3^\bullet]}{dt} = k_1[N_2O_5] - k_{-1}[NO_2][NO_3^\bullet] - k_2[NO_2][NO_3^\bullet] = 0 \qquad (6.16)$$

$$\therefore \quad [NO_3^\bullet] = \frac{k_1[N_2O_5]}{(k_{-1} + k_2)[NO_2]} \qquad (6.17)$$

$$\frac{d[NO^\bullet]}{dt} = k_2[NO_2][NO_3^\bullet] - k_3[NO^\bullet][N_2O_5] = 0 \qquad (6.18)$$

$$\therefore \quad [NO^\bullet] = \frac{k_2[NO_2][NO_3^\bullet]}{k_3[N_2O_5]} \qquad (6.19)$$

The rate of reaction can be given in three ways:

(a) the rate of removal of N_2O_5, expressed as three terms requiring the concentrations of both intermediates,

(b) the rate of production of NO_2, expressed as five terms, but with two cancelling out, and again involving both intermediates

(c) the rate of production of O_2, which is a one term expression involving only the concentration of $NO_3^\bullet$ which has been found.

The simplest one is chosen.

$$\frac{d[O_2]}{dt} = k_2[NO_2][NO_3^\bullet] = \frac{k_1 k_2[N_2O_5]}{(k_{-1} + k_2)} \qquad (6.20)$$

which is first order in N_2O_5 as found experimentally, with the observed first order rate constant, k_{obs}, given as

$$k_{obs} = \frac{k_1 k_2}{k_{-1} + k_2} \qquad (6.21)$$

Observation of a fall-off in the observed first order rate constant at low pressures is a consequence of the kinetics *mimicking* unimolecular behaviour; see Section 4.5 and Problems 4.17 and 4.18. At low pressures the unimolecular decomposition step, step 1, becomes second order because activation has become the rate determining step in the decomposition (see Section 4.5).

$$N_2O_5 + M \rightarrow N_2O_5^* + M$$

where $N_2O_5^*$ is an activated molecule with sufficient energy to react, and M is the species from which this energy has been transferred by collision. The rate of step 1 now becomes

$$\text{rate} = k_1'[N_2O_5][M] \qquad (6.22)$$

where k_1' is a second order rate constant, such that k_1 is replaced by $k_1'[M]$.

If the decomposition step has moved to having activation as rate determining at low pressures, then the reverse step of recombination must also have energy transfer as rate determining, i.e. it must become third order at low pressures.

The rate of the reverse recombination step becomes

$$\text{rate} = k'_{-1}[\text{NO}_2][\text{NO}_3^\bullet][\text{M}] \qquad (6.23)$$

where M is the third body required for the energy transfer step, and k'_{-1} is a third order rate constant, such that k_{-1} is replaced by $k'_{-1}[\text{M}]$.

Under conditions of low pressures

$$k_{\text{obs}} = \frac{k_1 k_2}{k_{-1} + k_3}$$

becomes

$$k_{\text{obs}} = \frac{k'_1[\text{M}]k_2}{k'_{-1}[\text{M}] + k_2} \qquad (6.24)$$

This, however, has no simple order, and under these conditions each of the terms in the denominator must be examined.

The term in k'_{-1} relates to the recombination reaction converting NO_2 and $NO_3^\bullet$ back to N_2O_5, and this reaction is in competition with the removal of NO_2 and $NO_3^\bullet$ to produce NO_2, $NO^\bullet$ and O_2. At low pressures there are fewer molecules present and the recombination reaction requiring the simultaneous presence of three species becomes progressively more difficult as the pressure is lowered, compared with reaction step 2, which only requires the simultaneous presence of two molecules. This means that

$$k'_{-1}[\text{NO}_2][\text{NO}_3^\bullet][\text{M}] \ll k_2[\text{NO}_2][\text{NO}_3^\bullet] \qquad (6.25)$$

$$\therefore\ k'_{-1}[\text{M}] \ll k_2 \qquad (6.26)$$

and so at low pressures the expression for k_{obs}

$$k_{\text{obs}} = \frac{k'_1[\text{M}]k_2}{k'_{-1}[\text{M}] + k_2} \qquad (6.24)$$

becomes

$$k_{\text{obs}} = \frac{k'_1[\text{M}]k_2}{k_2} = k'_1[\text{M}] \qquad (6.27)$$

The observed rate is second order as can be shown by the following argument.

$$\text{Observed rate} = k_{\text{obs}}[\text{N}_2\text{O}_5] \qquad (6.28)$$

and k_{obs} has been shown to be equal to $k_1'[M]$, Equation (6.27)

$$\therefore \quad \text{rate} = k_1'[M][N_2O_5] \tag{6.29}$$

which becomes second order as observed, manifested by the calculated first order rate constant progressively decreasing.

Remember that within any given experiment the reaction is strictly first order, see Problems 4.17 and 4.18, but the reaction moving to second order conditions will be shown up as the decreasing value of the first order constant.

6.6 Kinetically Equivalent Mechanisms

It is absolutely imperative to be aware that more than one mechanism may fit the experimental observations and the observed kinetics, and that an exact fit of prediction with experiment does not *prove* a given mechanism to be the correct one.

This is one of the fundamental tenets of the scientific method, *viz.*, that it is not possible to *prove* a hypothesis using experimental data, whereas it is possible to *refute* a hypothesis by reference to experimental data. All that can be said is that the data fits a given hypothesis, so that the hypothesis is a plausible one. The possibility remains open that, in the future, experimental evidence may show that the hypothesis is not justified.

The following problem illustrates this.

Worked Problem 6.4

Question. The reaction between nitric oxide (nitrogen monoxide) and oxygen gives nitrogen dioxide according to the stoichiometric equation.

$$2\,NO(g) + O_2(g) \rightarrow 2\,NO_2(g)$$

This reaction is second order in NO and first order in O_2. The following mechanisms can be proposed.

Mechanism A

$$2\,NO \xrightarrow{k_1} N_2O_2$$

$$N_2O_2 \xrightarrow{k_{-1}} 2\,NO$$

$$N_2O_2 + O_2 \xrightarrow{k_2} 2\,NO_2$$

Mechanism B

$$\mathrm{NO} + \mathrm{O_2} \xrightarrow{k_1} \mathrm{NO_3}$$
$$\mathrm{NO_3} \xrightarrow{k_{-1}} \mathrm{NO} + \mathrm{O_2}$$
$$\mathrm{NO_3} + \mathrm{NO} \xrightarrow{k_2} 2\,\mathrm{NO_2}$$

1. Assuming the steady state approximation, deduce the rate expressions for each mechanism.
2. Under what conditions will these mechanisms fit the experimental data and be kinetically equivalent? *Hint:* consider the denominator and the mathematical and chemical significance of each term.
3. What do the above analyses imply about the likelihood of the reversible steps being at equilibrium?
4. A further possibility is that the reaction occurs as an elementary termolecular reaction: which mechanism(s) is the more likely?

Answer.

1. *Applying the steady state treatment.*

   (a) *Mechanism A.* The intermediate is N_2O_2, and the rate of reaction is given in terms of step 2:

$$+\frac{\mathrm{d}[\mathrm{N_2O_2}]}{\mathrm{d}t} = k_1[\mathrm{NO}]^2 - k_{-1}[\mathrm{N_2O_2}] - k_2[\mathrm{N_2O_2}][\mathrm{O_2}] = 0 \tag{6.30}$$

$$\therefore\ [\mathrm{N_2O_2}] = \frac{k_1[\mathrm{NO}]^2}{k_{-1} + k_2[\mathrm{O_2}]} \tag{6.31}$$

$$+\frac{\mathrm{d}[\mathrm{NO_2}]}{\mathrm{d}t} = 2\,k_2[\mathrm{N_2O_2}][\mathrm{O_2}] \tag{6.32}$$

$$= \frac{2\,k_1k_2[\mathrm{NO}]^2[\mathrm{O_2}]}{k_{-1} + k_2[\mathrm{O_2}]} \tag{6.33}$$

   The factor of two is included, since two molecules of NO_2 are formed per step of reaction.

   (b) *Mechanism B.* The intermediate is now NO_3, but the rate of reaction is still given by the rate of step 2.

$$+\frac{d[NO_3]}{dt} = k_1[NO][O_2] - k_{-1}[NO_3] - k_2[NO_3][NO] = 0 \qquad (6.34)$$

$$\therefore \quad [NO_3] = \frac{k_1[NO][O_2]}{k_{-1} + k_2[NO]} \qquad (6.35)$$

$$+\frac{d[NO_2]}{dt} = 2\,k_2[NO_3][NO] \qquad (6.36)$$

$$= \frac{2\,k_1k_2[NO]^2[O_2]}{k_{-1} + k_2[NO]} \qquad (6.37)$$

Note the factor of two.

These two mechanisms give the same type of algebraic expression, but the actual expressions are different and neither fits the experimental rate expression:

$$\text{observed rate} = k_{obs}[NO]^2[O_2] \qquad (6.38)$$

2. However, *if in mechanism A* $k_2[O_2]$ can be ignored with respect to k_{-1} then

$$k_{-1} \gg k_2[O_2] \qquad (6.39)$$

and

$$k_{-1}[N_2O_2] \gg k_2[N_2O_2][O_2] \qquad (6.40)$$

which implies that

rate of reverse step $\gg$ the rate of reaction step

and most of the intermediate is removed by the reverse step and very, very little by reaction, suggesting that reaction is unlikely to disturb the equilibrium, and

If in mechanism B $k_2[NO]$ can be ignored with respect to k_{-1} then

$$k_{-1} \gg k_2[NO] \qquad (6.41)$$

and

$$k_{-1}[NO_3] \gg k_2[NO][NO_3] \qquad (6.42)$$

which implies that

rate of reverse step $\gg$ rate of reaction step

and most of the intermediate is removed by the reverse step and very, very little by reaction, suggesting that reaction is unlikely to disturb the equilibrium.

If these approximations are made, then both expressions reduce to

$$\frac{2k_1k_2[\mathrm{NO}]^2[\mathrm{O_2}]}{k_{-1}} \tag{6.43}$$

compatible with the observed rate expression above, with

$$k_{\mathrm{obs}} = \frac{2\,k_1k_2}{k_{-1}} \tag{6.44}$$

3. Hence, if these conditions hold, both mechanisms reduce to the *same rate expression* and must be *kinetically equivalent*. Non-kinetic evidence is required to distinguish them.

   As shown, these approximations imply that for both mechanisms reaction is unlikely to disturb the equilibrium.

   This conclusion can be demonstrated by carrying out a kinetic analysis *assuming* that the reversible reaction is *at equilibrium* throughout the reaction. This type of analysis is valid *only* if there is *equilibrium* so far as the *reversible reaction* is concerned.

   Assuming mechanism A:

$$K = \frac{[\mathrm{N_2O_2}]_{\mathrm{equil}}}{[\mathrm{NO}]^2_{\mathrm{equil}}} \tag{6.45}$$

$$\therefore \quad [\mathrm{N_2O_2}]_{\mathrm{equil}} = K[\mathrm{NO}]^2_{\mathrm{equil}} \tag{6.46}$$

$$+\frac{\mathrm{d}[\mathrm{NO_2}]}{\mathrm{d}t} = 2\,k_2[\mathrm{N_2O_2}][\mathrm{O_2}] \tag{6.32}$$

$$= 2\,k_2K[\mathrm{NO}]^2[\mathrm{O_2}] \tag{6.47}$$

   which fits the observed kinetics and the steady state analysis *with the approximation* $k_{-1} \gg k_2[\mathrm{O_2}]$. A similar pre-equilibrium analysis on mechanism B will yield the same result.

4. If the reaction were a termolecular elementary reaction

$$\mathrm{NO} + \mathrm{NO} + \mathrm{O_2} \rightarrow 2\,\mathrm{NO_2}$$

   then the predicted kinetics would be

$$\frac{\mathrm{d}[\mathrm{NO_2}]}{\mathrm{d}t} = 2\,k[\mathrm{NO}]^2[\mathrm{O_2}] \tag{6.48}$$

   kinetically equivalent to the previous two complex mechanisms.

Either of the two complex mechanisms would be more likely than the termolecular reaction, since the former only require two molecules to come together simultaneously in any given step. The termolecular mechanism would require the much more unlikely situation of a three body collision.

6.7 A Comparison of Steady State Procedures and Equilibrium Conditions in the Reversible Reaction

Many complex reactions involve a reversible step followed by one or more other steps. If the intermediate(s) are present in *steady state* concentrations, then the steady state analysis will give a general procedure for deducing the rate expression. If the intermediate(s) are present in *equilibrium* concentrations, an equilibrium analysis is appropriate. But if the intermediate(s) are not in either of these two categories, analysis becomes very complex (Sections 3.22 and 8.4). A comparison of the predicted and observed rate expressions can give considerable insight, e.g. Problem 6.4 above, and Problem 6.5 below.

Worked Problem 6.5

This problem focuses on the difference between steady state methods and the assumption of a pre-equilibrium for a reaction where the first two steps are reversible.

Question. The following mechanism is common to both inorganic and organic reactions:

$$\mathrm{RX} \xrightarrow{k_1} \mathrm{R}^+ + \mathrm{X}^-$$

$$\mathrm{R}^+ + \mathrm{X}^- \xrightarrow{k_{-1}} \mathrm{RX}$$

$$\mathrm{R}^+ + \mathrm{Y}^- \xrightarrow{k_2} \mathrm{RY}$$

- Carry out a steady state treatment, and find the conditions under which this reaction would be

  (a) first order in both RX and Y^- but inverse first order in X^-,

  (b) simple first order in RX.

- Now assume that the reversible reaction is at equilibrium, and deduce the rate expression.

- By comparing these two treatments, find the conditions under which the pre-equilibrium would be set up.

Answer.

- The steady state treatment

$$\frac{\mathrm{d}[\mathrm{R}^+]}{\mathrm{d}t} = k_1[\mathrm{RX}] - k_{-1}[\mathrm{R}^+][\mathrm{X}^-] - k_2[\mathrm{R}^+][\mathrm{Y}^-] = 0 \qquad (6.49)$$

$$[\mathrm{R}^+] = \frac{k_1[\mathrm{RX}]}{k_{-1}[\mathrm{X}^-] + k_2[\mathrm{Y}^-]} \qquad (6.50)$$

$$+\frac{\mathrm{d}[\mathrm{RY}]}{\mathrm{d}t} = k_2[\mathrm{R}^+][\mathrm{Y}^-] \qquad (6.51)$$

$$= \frac{k_1 k_2[\mathrm{RX}][\mathrm{Y}^-]}{k_{-1}[\mathrm{X}^-] + k_2[\mathrm{Y}^-]} \qquad (6.52)$$

Under steady state conditions, the reaction has complex kinetics with no simple order. As in the previous problem, there is a denominator made up of two terms, and again both terms should be considered in relation to each other. Doing so,

(a) If

$$k_{-1}[X^-] \gg k_2[Y^-] \tag{6.53}$$

then

$$k_{-1}[R^+][X^-] \gg k_2[R^+][Y^-] \tag{6.54}$$

which implies that

rate of reverse step $\gg$ rate of reaction step

and most of the intermediate is removed by the reverse step rather than by reaction.

The rate expression, Equation (6.52), then reduces to

$$\frac{d[RY]}{dt} = \frac{k_1 k_2[RX][Y^-]}{k_{-1}[X^-]} \tag{6.55}$$

with reaction first order in RX and Y^- and inverse first order in X^-.

(b) If

$$k_{-1}[X^-] \ll k_2[Y^-] \tag{6.56}$$

then

$$k_{-1}[R^+][X^-] \ll k_2[R^+][Y^-] \tag{6.57}$$

which implies that

rate of reverse step $\ll$ rate of reaction step.

Most of the intermediate is removed by reaction.

The rate expression, equation (6.52) then reduces to

$$\frac{d[RY]}{dt} = k_1[RX] \tag{6.58}$$

with reaction simple first order in reactant.

- If the reversible reaction is at equilibrium throughout reaction, then

$$K = \left(\frac{[R^+][X^-]}{[RX]}\right)_{\text{equil}} \tag{6.59}$$

$$\therefore \quad [R^+]_{\text{equil}} = \frac{K[RX]_{\text{equil}}}{[X^-]_{\text{equil}}} \tag{6.60}$$

$$\frac{d[RY]}{dt} = k_2[R^+]_{\text{equil}}[Y^-] \tag{6.61}$$

$$= \frac{k_2 K[RX]_{\text{equil}}[Y^-]}{[X^-]_{\text{equil}}} \tag{6.62}$$

$$= \frac{k_2 k_1[RX]_{\text{equil}}[Y^-]}{k_{-1}[X^-]_{\text{equil}}} \tag{6.63}$$

This corresponds to the steady state analysis (a), where most of the intermediate is removed by the reverse reaction and very little by reaction. It is only under these conditions that the pre-equilibrium is likely to be set up and maintained throughout reaction; see Section 8.4.

Note. It is very instructive to compare this analysis with that for the reaction of NO with O_2. Although these two reactions are totally different chemically, they analyse kinetically to be of the same algebraic form with the same chemical implications. This is a very common situation in kinetics, and there are many other rate processes which exhibit this algebraic form with a denominator. Other algebraic forms likewise can result on analysis for widely disparate reactions.

6.8 The Use of Photochemistry in Disentangling Complex Mechanisms

If a reaction can be initiated photochemically, a further route to investigating the mechanism and evaluating the rate constants becomes possible.

Radiation can be used to initiate a reaction (see Section 2.2.4). When molecules absorb radiation they are excited and often split up into radicals. Normally one molecule absorbs one quantum of radiation. This enables the rate of formation of the radicals produced as a result of absorption to be found from a measure of the radiation absorbed. These radicals may then react in a sequence of reactions, whereby more reactant is removed *independently* of the initial breakdown under the influence of radiation. The quantum yield is defined as the number of reactant molecules transformed per quantum absorbed, and gives a measure of how many molecules of reactant eventually react as a result of the initial first breakdown. If the quantum yield is large, this is conclusive evidence of a chain reaction, where many reactant molecules decompose per quantum absorbed.

The identity of the radicals formed from the initially excited molecule can be studied spectroscopically. If conventional radiation sources are used, the radicals will be formed in steady state concentrations and their rates of formation and removal cannot be measured. If, however, flash or laser photolysis is used the radicals are formed in much larger concentrations and their concentration–time profiles can be determined spectroscopically; see Sections 2.1.4 and 2.5.2. From this, rate constants for the overall formation and removal of these radicals can be found.

6.8.1 Kinetic features of photochemistry

- The term 'primary process' describes both the production of the excited species and its subsequent decomposition into atoms and free radicals. 'Secondary processes' describe all subsequent reactions of the radicals. The overall reaction can be chain or non-chain.

- The 'law of photochemical equivalence' states that one molecule absorbs one quantum, which allows a calculation of the rate of formation of radicals from the intensity of the radiation absorbed. This can be extremely useful in the kinetic analysis as it gives a rate constant for the first step in the sequence. For example

  $$CH_3COCH_3 + h\nu \rightarrow CH_3COCH_3^* \rightarrow 2\,CH_3^\bullet + CO$$

  produces excited CH_3COCH_3 molecules as the primary process, and the rate of absorption of photons will give

  $$\text{rate of formation of CO} = I_{abs} \tag{6.64}$$

  $$\text{rate of formation of } CH_3^\bullet = 2\,I_{abs} \tag{6.65}$$

  $$\text{rate of removal of } CH_3COCH_3 = I_{abs} \tag{6.66}$$

- The quantum yield is given by the number of reactant molecules transformed per quantum absorbed. The quantum yield of the primary process is unity, unless any collision processes, emission or fluorescence compete.

If the photochemical reaction were a single elementary step, then the quantum yield in terms of the reactants would also be unity. However, the quantum yield may differ from unity for several reasons.

(a) The quantum yield is a small integral number for complex, but non-chain reactions where more than one reactant molecule reacts per quantum absorbed

$$HBr + h\nu \rightarrow H^\bullet + Br^\bullet$$

$$HBr + H^\bullet \rightarrow H_2 + Br^\bullet$$

$$Br^\bullet + Br^\bullet + M \rightarrow Br_2 + M$$

where M is a third body which removes excess energy from the newly formed Br_2 molecule; see Section 6.12.4.

In the primary process one quantum is absorbed by one molecule of HBr, as a result of which a second molecule of HBr can react. For one quantum absorbed, effectively two molecules of HBr react, giving a quantum yield of two for the reactant.

(b) The quantum yield can be less than unity if

   (i) the radicals or intermediates formed in the primary process can recombine, so that the net removal of reactant will be less than the removal of reactant in the primary step, or

   (ii) the excited molecule can be deactivated before it can react.

(c) The quantum yield can be large, and often very large, and this is specific to, and diagnostic of, chain reactions. For every act of initiation, the primary process, the propagation steps ensure that large numbers of reactants can react. The mechanism for the reaction of CO and Cl_2 to give $COCl_2$ is

$$Cl_2 + h\nu \rightarrow 2\,Cl^\bullet$$
$$Cl^\bullet + CO \rightarrow COCl^\bullet$$
$$COCl^\bullet + Cl_2 \rightarrow COCl_2 + Cl^\bullet$$
$$COCl^\bullet + Cl^\bullet \rightarrow \text{inert species}$$

The $Cl^\bullet$ produced in the primary process is involved in the propagation steps which result in many molecules of Cl_2 and CO being transformed, giving a very large quantum yield.

The quantum yield is found by measuring the amount of reactant removed, or product formed, as a result of absorption of a given amount of radiation. The stoichiometry of the reactions involved may be such that the quantum yields for reactants and products differ. For instance, in the decomposition of HBr quoted the quantum yield in terms of HBr is two, and in terms of H_2 and Br_2 is unity.

6.8.2 The reaction of H_2 with I_2

For many years this reaction was quoted as the classic example of an elementary bimolecular reaction, and used as a test of collision and transition state theories, but it has now been shown to be a complex non-chain reaction.

Photochemical experiments show iodine atoms to be involved, and molecular beams subsequently confirmed this by shooting beams of H_2 and I_2 molecules at each other and finding no reaction. In contrast, beams of H_2 molecules and $I^\bullet$ atoms resulted in reaction.

When H_2 and I_2 react *thermally*

$$\text{observed rate} = k_{obs}^{thermal}[H_2][I_2] \tag{6.67}$$

There are two mechanisms which can account for this:

(a) direct molecular combination: $H_2 + I_2 \xrightarrow{k_{molec}} 2\,HI$

(b) an atomic mechanism: $I_2 + M \rightleftharpoons 2\,I^\bullet + M$

$$H_2 + I^\bullet + I^\bullet \xrightarrow{k_{atomic}} 2\,HI$$

These two schemes are kinetically equivalent, as will be shown below.

If both mechanisms contribute to the thermal rate

$$\text{rate} = k_{molec}[H_2][I_2] + k_{atomic}[H_2][I^\bullet]^2 \tag{6.68}$$
$$= k_{obs}^{thermal}[H_2][I_2] \tag{6.69}$$

Two situations can arise:

(a) the iodine atoms are in equilibrium with molecular iodine,

(b) equilibrium is not set up, and only a stationary state concentration of iodine atoms is present.

Equilibrium is set up between $I^\bullet$ and molecular I_2

$$I_2 \rightleftharpoons 2\,I^\bullet$$

where

$$K = \left(\frac{[I^\bullet]^2}{[I_2]}\right)_{\text{equil}} \tag{6.70}$$

and

$$[I^\bullet]^2_{\text{eq}} = K[I_2]_{\text{equil}} \tag{6.71}$$

Substitution into Equation (6.68) gives

$$\text{rate} = (k_{\text{molec}} + k_{\text{atomic}}K)[H_2]_{\text{equil}}[I_2]_{\text{equil}} \tag{6.72}$$

and this must be equal to the thermal rate, giving for the equilibrium situation

$$k_{\text{obs}}^{\text{thermal}} = k_{\text{molec}} + k_{\text{atomic}}K \tag{6.73}$$

$$= \text{a constant} \tag{6.74}$$

No matter whether reaction goes totally by the molecular mechanism, $k_{\text{atomic}} = 0$, or totally by the atomic mechanism, $k_{\text{molec}} = 0$, or by both mechanisms simultaneously, the rate law is predicted to be strictly second order and there is no kinetic means of determining which mechanism holds. The two schemes are kinetically equivalent.

A stationary state only is set up

Equilibrium is *not* set up and Equation (6.70) does not hold. However, the rate law still holds and Equation (6.68) can be transformed into

$$\text{rate} = k_{\text{molec}}[H_2][I_2] + \frac{k_{\text{atomic}}[H_2][I_2][I^\bullet]^2_{\text{steady state}}}{[I_2]} \tag{6.75}$$

$$= \left\{k_{\text{molec}} + k_{\text{atomic}}\frac{[I^\bullet]^2_{\text{steady state}}}{[I_2]}\right\}[H_2][I_2] \tag{6.76}$$

now giving

$$k_{\text{obs}}^{\text{thermal}} = k_{\text{molec}} + k_{\text{atomic}}\left\{\frac{[\text{I}^\bullet]^2_{\text{steady state}}}{[\text{I}_2]}\right\} \tag{6.77}$$

$$= \text{a constant} \tag{6.78}$$

provided that a steady state is maintained throughout the experiment. If Equation (6.77) holds then $k_{\text{obs}}^{\text{thermal}}$ takes a numerical value which depends on the particular steady state concentration set up during the experiment. If it were possible to set up a series of different steady state concentrations and measure $k_{\text{obs}}^{\text{thermal}}$ for each, then the value would depend on the steady state concentration.

There is no way that the steady state concentration can be varied if the thermal reaction is studied at constant temperature. However, if the reaction is studied photochemically with a series of differing intensities of radiation, then a series of different steady state concentrations can be set up. The k_{obs} value for the photochemical reaction would then depend on the intensity. Equation (6.77) is then transformed into

$$k_{\text{obs}}^{\text{photochemical}} = k_{\text{molec}} + k_{\text{atomic}}\left\{\frac{[\text{I}^\bullet]^2_{\text{steady state}}}{[\text{I}_2]}\right\} \tag{6.79}$$

$$= \text{a constant} \tag{6.80}$$

where the constant now depends on the intensity of the radiation used. The constant would only be independent of the intensity if k_{atomic} were zero.

Such experiments have been carried out, and the photochemical $k_{\text{obs}}^{\text{photochemical}}$ has been shown to vary with intensity in such a manner that the equation

$$k_{\text{obs}}^{\text{photochemical}} = k_{\text{molec}} + k_{\text{atomic}}\left(\frac{[\text{I}^\bullet]^2_{\text{steady state}}}{[\text{I}_2]}\right) \tag{6.79}$$

fits the data *only if* k_{molec} *equals zero*, showing that only the atomic mechanism is operative under photochemical conditions.

Extension of the rate constant data to the thermal reaction shows that this too can be accounted for totally by the atomic mechanism.

6.9 Chain Reactions

A chain reaction involves at least two highly reactive intermediates, called chain carriers, which take part in a cycle of reactions called propagation. In the first propagation step one intermediate is removed and the second formed. This intermediate then goes on to regenerate the first intermediate in the second propagation step. At least one step removes reactant, and at least one forms product. The cycle continues until either the reactant is all used up, or one, or both, of the chain carriers are removed by conversion into an unreactive species in the chain-breaking steps of termination. Most of the reaction, therefore, occurs in propagation, which is thus the source of the major products and the means whereby most of the reactant is removed.

Termination steps are generally recombination reactions, though disproportionation can occur. Recombination reactions occur when two radicals combine to give one molecule, while disproportionation occurs when two radicals react to form two molecules, rather than recombining to form one:

$$C_2H_5^\bullet + C_2H_5^\bullet \rightarrow C_4H_{10} \qquad \text{recombination}$$
$$C_2H_5^\bullet + C_2H_5^\bullet \rightarrow C_2H_4 + C_2H_6 \qquad \text{disproportionation}$$

Both are termination reactions, and the net effect is removal of chain carriers. Identification of the minor products will give evidence as to the relative proportions of each. Their essential feature is removal of chain carriers to form inert species incapable of carrying on the chain. They occur infrequently compared with propagation, and so produce the minor products.

The initiation step(s) first produces a chain carrier enabling propagation to be set up. Since one step of initiation can set up a long cycle of propagation events, very little reactant is used up in initiation.

Other types of step can occur, such as inhibition, but these are not essential for the characterization of a reaction as a chain. Inhibition occurs when a chain carrier reacts with a product in the reverse of a propagation step, or with an added substance, and as a consequence the overall rate is reduced.

6.9.1 Characteristic experimental features of chain reactions

- Chains always involve highly reactive atoms or radicals, often present in very low steady state concentrations.
- The rate of reaction can be of simple order, either first, second or non-integral, but it can also be complex. A complex order always means a complex mechanism, but a simple order *need not* mean a simple mechanism.
- The shape of the rate–time plot is significant: a maximum in rate suggests a chain reaction. Build-up to a maximum corresponds to build-up to the steady state, after which the concentration of chain carriers remains constant and the overall rate, *a composite quantity*, gradually decreases as in a typical non-chain reaction. When there is a complex mechanism the overall rate constant is also composite and is made up of one or more rate constants relating to individual steps in the mechanism; e.g. in Chapter 3, Problem 3.18, the mechanism for the decomposition of C_3H_8 is a chain mechanism and the analysis gave

$$\text{rate of reaction} = k_2(k_1/2k_4)^{1/2}[C_3H_8]^{3/2} \qquad (6.81)$$

The reaction is observed to be 3/2 order, so the observed rate constant

$$k_{obs} = k_2(k_1/2\,k_4)^{1/2} \tag{6.82}$$

It is in this respect that both the rate and rate constant are composite quantities.

The build-up is often very fast, e.g. over in 10^{-4} s, and not observed experimentally under normal steady state conditions. Special fast reaction techniques are required to study the build-up, and analysis of the data requires non-steady state methods. If build-up continues after the steady state concentrations occur, the rate of reaction continues to increase and explosion can occur.

- The induction period is the time taken before *observable* reaction occurs. It is linked to the time for build-up to the steady state to occur, but need not be equal to it. It is not an absolute quantity since it is determined by the sensitivity of the apparatus to detect very small amounts of reaction.
- Chain reactions can be accelerated by addition of radicals and retarded by addition of substances known to mop up radicals, and this can alter the length of the induction period.
- Termination can occur in the gas phase or at the surface of the reaction vessel. Addition of inert gas can alter the kinetics by increasing the efficiency of radical recombination, while changing the shape, size and nature of the surface will affect the rates of any reactions occurring at the walls.
- A large quantum yield is the *definitive characteristic* of a chain reaction. This is found from the photochemical reaction and is always measured if possible.

6.9.2 Identification of a chain reaction

This can best be illustrated by arguing from a specific reaction.

Worked Problem 6.6

Question. A simplified mechanism for the gas phase decomposition of methoxymethane, CH_3OCH_3, is as follows:

$$CH_3OCH_3 \xrightarrow{k_1} CH_3O^\bullet + CH_3^\bullet$$

$$CH_3^\bullet + CH_3OCH_3 \xrightarrow{k_2} CH_4 + {}^\bullet CH_2OCH_3$$

$${}^\bullet CH_2OCH_3 \xrightarrow{k_3} HCHO + CH_3^\bullet$$

$$CH_3^\bullet + CH_3^\bullet \xrightarrow{k_4} C_2H_6$$

$${}^\bullet CH_2OCH_3 + {}^\bullet CH_2OCH_3 \xrightarrow{k_5} CH_3OCH_2CH_2OCH_3$$

$$CH_3^\bullet + {}^\bullet CH_2OCH_3 \xrightarrow{k_6} CH_3CH_2OCH_3$$

1. Identify the intermediates, stating which are chain carriers. Why is this a chain reaction?
2. Identify, giving reasons, the initiation, propagation and termination steps.
3. State, giving reasons, which products are major and which minor.

Answer.

1. The intermediates are $CH_3O^\bullet$, $CH_3^\bullet$ and $CH_3OCH_2^\bullet$. The best way to identify a chain reaction is to look for a *cycle of steps* where intermediates are regenerated, and identify them. $CH_3^\bullet$ and $CH_3OCH_2^\bullet$ are recycled in steps 2 and 3 and are therefore chain carriers, and steps 2 and 3 are propagation steps. The reaction is a chain.
2. $CH_3O^\bullet$ is not recycled and cannot be a chain carrier. But step 1 in which it is produced also produces $CH_3^\bullet$, which enables the chain to start. Step 1 is thus initiation. The two chain carriers are removed to molecules incapable of carrying on the chain in steps 4, 5 and 6: these are termination steps. As shown above steps 2 and 3 are the propagation steps.
3. Steps 2 and 3 occur many times before they are interrupted by termination, or by depletion of reactant. Major products are thus formed in these steps and are CH_4 and HCHO. Termination occurs infrequently compared with propagation and so C_2H_6, $CH_3OCH_2CH_2OCH_3$ and $CH_3CH_2OCH_3$ are minor products. Likewise, initiation also occurs infrequently compared to propagation, and so the intermediate $CH_3O^\bullet$ might be found in trace amounts compared to the steady state amounts of the chain carriers, if subsequent reactions remove it sufficiently quickly.

6.9.3 Deduction of a mechanism from experimental data

The procedures involved here are the reverse of those in Problem 6.6 above, but the reasoning is similar.

Worked Problem 6.7

Question. The gas phase decomposition of ethanal occurs according to

$$CH_3CHO(g) \rightarrow CH_4(g) + CO(g)$$

The major products are CH_4 and CO, with C_2H_6, $CH_3COCOCH_3$ and CH_3COCH_3 being minor products. The radicals $CH_3^\bullet$, $CH_3CO^\bullet$ and $CHO^\bullet$ are present. Deduce a mechanism which fits these facts.

Answer.

- Initiation produces radicals by homolytic splitting of a bond. A rule of thumb (Problem 6.2) suggests that C–C splits more readily than C–H, which in turn splits more readily than C–O or C=O, and a good first guess is

$$CH_3CHO \rightarrow CH_3^\bullet + CHO^\bullet$$

confirmed by the presence of the two radicals.

- Propagation produces the main products, CH_4 and CO, often by H abstraction reaction from reactant. There are two H which can be abstracted, and the presence of $CH_3CO^\bullet$ suggests that

$$CH_3^\bullet + CH_3CHO \rightarrow CH_4 + CH_3CO^\bullet$$

is more likely than

$$CH_3^\bullet + CH_3CHO \rightarrow CH_4 + {}^\bullet CH_2CHO$$

- The second propagation step *must regenerate* the first chain carrier, here $CH_3^\bullet$, and produce a major product, suggesting

$$CH_3CO^\bullet \rightarrow CH_3^\bullet + CO$$

- The minor products are produced by reactions of the chain carriers. Here the nature of the products indicates recombination reactions.

$$CH_3^\bullet + CH_3^\bullet \rightarrow C_2H_6$$
$$CH_3^\bullet + CH_3CO^\bullet \rightarrow CH_3COCH_3$$
$$CH_3CO^\bullet + CH_3CO^\bullet \rightarrow CH_3COCOCH_3$$

The last two minor products confirm that H abstraction occurs on the CHO group of CH_3CHO.

- This leaves the non-chain carrier $CHO^\bullet$, produced in initiation, unaccounted for. It might only be present in trace amounts, as initiation occurs very infrequently compared to propagation, and if so it will be removed in a subsequent reaction to produce a trace product. No evidence is given as to what this will be.

A possible mechanism which fits the experimental facts:

$$CH_3CHO \xrightarrow{k_1} CH_3^\bullet + CHO^\bullet$$

$$CH_3^\bullet + CH_3CHO \xrightarrow{k_2} CH_4 + CH_3CO^\bullet$$

$$CH_3CO^\bullet \xrightarrow{k_3} CH_3^\bullet + CO$$

$$CH_3^\bullet + CH_3^\bullet \xrightarrow{k_4} C_2H_6$$

$$CH_3^\bullet + CH_3CO^\bullet \xrightarrow{k_5} CH_3COCH_3$$

$$CH_3CO^\bullet + CH_3CO^\bullet \xrightarrow{k_6} CH_3COCOCH_3$$

6.9.4 The final stage: the steady state analysis

Once a mechanism has been proposed, a kinetic analysis is required to see whether the kinetics predicted by the mechanism fits the experimentally determined kinetic rate expression. The predicted rate expression is found generally from a steady state analysis. Most mechanisms have at least three termination steps, and for a first analysis only one step is included at a time.

There are various examples of steady state treatments given throughout this chapter and in Chapter 3.

6.10 Inorganic Chain Mechanisms

Inorganic chain mechanisms are of much lower prevalence than the typical organic reactions that make up the vast bulk of chain reactions. Of the inorganic reactions, the classic examples are the H_2/Br_2 and H_2/Cl_2 reactions, with the first being the most straightforward. As mentioned earlier, the H_2/I_2 reaction was always used as an example of an elementary bimolecular reaction, but was finally shown to have a complex, non-chain mechanism. In contrast, the H_2/Br_2 and H_2/Cl_2 reactions are chain reactions.

6.10.1 The H_2/Br_2 reaction

$$H_2(g) + Br_2(g) \rightarrow 2\,HBr(g)$$

The rate of the thermal reaction between 200 °C and 300 °C is

$$\text{rate}_{\text{thermal}} = \frac{k[H_2][Br_2]^{1/2}}{1 + \dfrac{[HBr]}{c[Br_2]}} \tag{6.83}$$

where k and c are constants. The form of the rate expression shows that the product HBr inhibits this reaction, but that this inhibition is reduced by addition of Br_2. The term in [HBr] appears in the denominator, an indication of inhibition, and this term is decreased by addition of Br_2.

The photochemical reaction has virtually the same algebraic form as the thermal reaction:

$$\text{rate}_{\text{photochemical}} = \frac{k'[H_2]I_{\text{abs}}^{1/2}/[M]^{1/2}}{1 + \dfrac{[HBr]}{c[Br_2]}} \tag{6.84}$$

except that k' is different from k, and $\frac{I_{\text{abs}}^{1/2}}{[M]^{1/2}}$ replaces $[Br_2]^{1/2}$, where I_{abs} is the intensity of light absorbed.

The product of reaction is HBr, and experiment shows the presence of $H^\bullet$ and $Br^\bullet$ atoms. The generally accepted mechanism is

$$\begin{array}{ll} Br_2 + M \xrightarrow{k_1} 2\,Br^\bullet + M & \text{initiation} \\ Br^\bullet + H_2 \xrightarrow{k_2} HBr + H^\bullet & \text{propagation} \\ H^\bullet + Br_2 \xrightarrow{k_3} HBr + Br^\bullet & \text{propagation} \\ H^\bullet + HBr \xrightarrow{k_{-2}} H_2 + Br^\bullet & \text{inhibition} \\ 2\,Br^\bullet + M \xrightarrow{k_{-1}} Br_2 + M & \text{termination} \end{array}$$

where M is a third body allowing energy transfer.

The photochemical reaction differs only in the first step:

$$Br_2 + h\nu \rightarrow 2\,Br^\bullet \qquad \text{photochemical initiation}$$

Inhibition is accounted for by step -2, in which product removes one of the chain carriers. HBr competes with Br_2 for the chain carrier $H^\bullet$, and this is the chemical process which accounts for the term $[HBr]/[Br_2]$ in the rate expression. There are two reversible steps in this mechanism: steps (1) and (-1), and steps (2) and (-2).

6.10.2 The steady state treatment for the H_2/Br_2 reaction

The chain carriers are $H^\bullet$ and $Br^\bullet$ and

$$\frac{d[H^\bullet]}{dt} = k_2[Br^\bullet][H_2] - k_3[H^\bullet][Br_2] - k_{-2}[H^\bullet][HBr] = 0 \tag{6.85}$$

$$\begin{aligned} \frac{d[Br^\bullet]}{dt} &= 2\,k_1[Br_2][M] - k_2[Br^\bullet][H_2] + k_3[H^\bullet][Br_2] \\ &\quad + k_{-2}[H^\bullet][HBr] - 2\,k_{-1}[Br^\bullet]^2[M] = 0 \end{aligned} \tag{6.86}$$

The factors of two appear because two $Br^\bullet$ are produced per reaction step, and two $Br^\bullet$ are removed per reaction step in the back reaction. Adding the steady state equations gives

$$k_1[Br_2] = k_{-1}[Br^\bullet]^2 \tag{6.87}$$

which is the algebraic statement of the steady state postulate that there is no build-up of radicals, and so the rate of initiation equals the rate of termination, and

$$[Br^\bullet] = \left(\frac{k_1}{k_{-1}}\right)^{1/2}[Br_2]^{1/2} \tag{6.88}$$

The rate of reaction can be given in terms of removal of either reactant, H_2 or Br_2, or a rate of formation of the product HBr. The rate in terms of H_2 involves two terms, whereas the rate in terms of Br_2 or HBr involves three terms. The rate in terms of H_2 is chosen here.

$$\frac{-d[H_2]}{dt} = k_2[Br^\bullet][H_2] - k_{-2}[H^\bullet][HBr] \tag{6.89}$$

The concentration of the second chain carrier, $H^\bullet$, is required and this has to be found by substituting for $[Br^\bullet]$ into either of the steady state equations and solving for $[H^\bullet]$. Equation (6.85) has the smallest number of terms and is chosen.

$$k_2[Br^\bullet][H_2] - k_3[H^\bullet][Br_2] - k_{-2}[H^\bullet][HBr] = 0 \tag{6.90}$$

$$[H^\bullet] = \frac{k_2[H_2][Br^\bullet]}{k_3[Br_2] + k_{-2}[HBr]} = \frac{k_2[H_2]\left(\frac{k_1}{k_{-1}}\right)^{1/2}[Br_2]^{1/2}}{k_3[Br_2] + k_{-2}[HBr]} \tag{6.91}$$

Note well here: it would *not* be correct to use the long chains approximation here, even though the reaction has long chains. This is because there is the inhibition step, which removes and creates each of the chain carriers, and so the rates of the two propagation steps are not equal, as is demonstrated by Equation (6.90). It will be shown later that if initial rates were used, where there is very little inhibition, then the long chains approximation becomes valid (Section 6.10.3).

Substituting into Equation (6.89) gives

$$\frac{-\mathrm{d}[\mathrm{H}_2]}{\mathrm{d}t} = k_2\left(\frac{k_1}{k_{-1}}\right)^{1/2}[\mathrm{Br}_2]^{1/2}[\mathrm{H}_2] - k_{-2}[\mathrm{HBr}]\left\{\frac{k_2[\mathrm{H}_2]\left(\frac{k_1}{k_{-1}}\right)^{1/2}[\mathrm{Br}_2]^{1/2}}{k_3[\mathrm{Br}_2] + k_{-2}[\mathrm{HBr}]}\right\} \tag{6.92}$$

$$= k_2\left(\frac{k_1}{k_{-1}}\right)^{1/2}[\mathrm{Br}_2]^{1/2}[\mathrm{H}_2]\left\{1 - \frac{k_{-2}[\mathrm{HBr}]}{k_3[\mathrm{Br}_2] + k_{-2}[\mathrm{HBr}]}\right\} \tag{6.93}$$

$$= k_2\left(\frac{k_1}{k_{-1}}\right)^{1/2}[\mathrm{Br}_2]^{1/2}[\mathrm{H}_2]\left\{\frac{k_3[\mathrm{Br}_2]}{k_3[\mathrm{Br}_2] + k_{-2}[\mathrm{HBr}]}\right\} \tag{6.94}$$

$$= k_2\left(\frac{k_1}{k_{-1}}\right)^{1/2}[\mathrm{Br}_2]^{1/2}[\mathrm{H}_2]\left\{1 + \frac{k_{-2}[\mathrm{HBr}]}{k_3[\mathrm{Br}_2]}\right\}^{-1} \tag{6.95}$$

$$= \frac{k_2\left(\frac{k_1}{k_{-1}}\right)^{1/2}[\mathrm{Br}_2]^{1/2}[\mathrm{H}_2]}{1 + \frac{k_{-2}[\mathrm{HBr}]}{k_3[\mathrm{Br}_2]}} \tag{6.96}$$

which, when compared with Equation (6.83) for the thermal rate, gives

$$k = k_2\left(\frac{k_1}{k_{-1}}\right)^{1/2} \quad \text{and} \quad c = \frac{k_3}{k_{-2}} \qquad (6.97)\text{ and }(6.98)$$

A similar analysis for the photochemical rate replaces the rate of thermal initiation, $2k_1[\mathrm{Br}_2][\mathrm{M}]$, by the rate of photochemical initiation $2I_{\mathrm{abs}}$, giving

$$\text{photochemical rate} = \frac{\mathrm{d}[\mathrm{H}_2]}{\mathrm{d}t} = \frac{k_2\left(\frac{1}{k_{-1}[\mathrm{M}]}\right)^{1/2} I_{\mathrm{abs}}^{1/2}[\mathrm{H}_2]}{1 + \frac{k_{-2}[\mathrm{HBr}]}{k_3[\mathrm{Br}_2]}} \tag{6.99}$$

6.10.3 Reaction without inhibition

Inhibition can be reduced to virtually zero if initial rates are used. Very little product, HBr, will have built up during the period of experimental measurements and the rate of the inhibition step will be virtually zero. In the steady state expression for $[\mathrm{H}^\bullet]$, Equation (6.85), the term in k_{-2} will drop out, leaving

$$\frac{\mathrm{d}[\mathrm{H}^\bullet]}{\mathrm{d}t} = k_2[\mathrm{Br}^\bullet][\mathrm{H}_2] - k_3[\mathrm{H}^\bullet][\mathrm{Br}_2] = 0 \tag{6.100}$$

and

$$k_2[\mathrm{Br}^\bullet][\mathrm{H}_2] = k_3[\mathrm{H}^\bullet][\mathrm{Br}_2] \tag{6.101}$$

which is the long chains approximation that the rates of the two propagation steps are equal.

The steady state expression for $\mathrm{Br}^\bullet$, Equation (6.86), then reduces to

$$\frac{\mathrm{d}[\mathrm{Br}^\bullet]}{\mathrm{d}t} = 2k_1[\mathrm{Br}_2][\mathrm{M}] - 2k_{-1}[\mathrm{Br}^\bullet]^2[\mathrm{M}] = 0 \tag{6.102}$$

$$[\mathrm{Br}^\bullet] = \left(\frac{k_1}{k_{-1}}\right)^{1/2}[\mathrm{Br}_2]^{1/2} \tag{6.103}$$

The rate of reaction now becomes

$$\frac{-\mathrm{d}[\mathrm{H}_2]}{\mathrm{d}t} = k_2[\mathrm{Br}^\bullet][\mathrm{H}_2] \tag{6.104}$$

$$= k_2\left(\frac{k_1}{k_{-1}}\right)^{1/2}[\mathrm{Br}_2]^{1/2}[\mathrm{H}_2] \tag{6.105}$$

a much simpler expression.

6.10.4 Determination of the individual rate constants

This is a reaction where, providing data for the photochemical reaction are included, all the individual rate constants can be found.

- The equilibrium constant, K_1, for the reaction

  $$\mathrm{Br}_2 \rightleftharpoons 2\,\mathrm{Br}^\bullet$$

  can be independently measured, and gives the ratio k_1/k_{-1}.

- The equilibrium constant, K_2, for the reaction

  $$\mathrm{Br}^\bullet + \mathrm{H}_2 \rightleftharpoons \mathrm{HBr} + \mathrm{H}^\bullet$$

  can also be independently measured, and gives the ratio k_2/k_{-2}.

- $k_{\mathrm{obs}}^{\mathrm{thermal}} = k_2(k_1/k_{-1})^{1/2}$ (6.106)

  Since $k_{\mathrm{obs}}^{\mathrm{thermal}}$ and $(k_1/k_{-1})^{1/2}$ are known, k_2 follows.

- $K_2 = k_2/k_{-2}$ (6.107)

Since K_2 and k_2 are known, k_{-2} follows.

- $c = k_3/k_{-2}$ (6.108)

  Since c is known from experiment, and k_{-2} from the above argument, then k_3 follows.

- The individual values of k_1 and k_{-1} remain to be found, and these can be easily found if the photochemical rate is known. From Equations (6.96) and (6.99) the ratio of the two rates can be shown to be

$$\frac{\text{photochemical rate}}{\text{thermal rate}} = \left(\frac{1}{k_1}\right)^{1/2} \frac{I_{\text{abs}}^{1/2}}{[\text{Br}_2]^{1/2}[\text{M}]^{1/2}} \tag{6.109}$$

All the quantities except k_1 are known experimentally, hence k_1 can be found. Since K_1 is also known k_{-1} can then be found.

For this reaction, a steady state treatment, coupled with kinetic and non-kinetic observations, is sufficient to determine fully all the individual rate constants describing the mechanism. This need not always be the case; see Problem 6.8 below.

Note. the argument involving the ratio of the photochemical rate to the thermal rate as a route to the determination of k_1 *only holds* because the *mechanisms* for both reactions are the *same*. This method would not work with the photolysis and thermal decomposition of propanone, see Worked Problem 6.2 and Further Problem 4. These two reactions have totally different mechanisms: the photolytic reaction has a non-chain mechanism whereas the thermal decomposition, which occurs at a much higher temperature, has a chain mechanism.

As mentioned, by using the thermal and photochemical reactions all the rate constants in the H_2/Br_2 reactions can be found. The following problem relates to a reaction where this is not the case.

6.11 Steady State Treatments and Possibility of Determination of All the Rate Constants

It is important to realize that a steady state treatment on a mechanism does not necessarily generate a rate expression in which all the individual rate constants appear. If, as in the H_2/Br_2 reaction above, all the rate constants do appear in the rate expression, then it may be possible to determine the magnitudes of all the rate constants from a steady state analysis. But if they do not all appear, then the steady state treatment can only allow determination of those rate constants which do appear in the rate expression, and alternative ways will have to be found to give an independent determination of the remaining rate constants.

The following problem illustrates this.

Worked Problem 6.8

Question. The decomposition of methoxymethane, CH_3OCH_3, has already been used to illustrate the characteristic features of a chain reaction:

$$CH_3OCH_3 \xrightarrow{k_1} CH_3O^\bullet + CH_3^\bullet$$

$$CH_3^\bullet + CH_3OCH_3 \xrightarrow{k_2} CH_4 + {}^\bullet CH_2OCH_3$$

$${}^\bullet CH_2OCH_3 \xrightarrow{k_3} HCHO + CH_3^\bullet$$

$$CH_3^\bullet + CH_3^\bullet \xrightarrow{k_4} C_2H_6$$

$${}^\bullet CH_2OCH_3 + {}^\bullet CH_2OCH_3 \xrightarrow{k_5} CH_3OCH_2CH_2OCH_3$$

$$CH_3^\bullet + {}^\bullet CH_2OCH_3 \xrightarrow{k_6} CH_3CH_2OCH_3$$

Using this mechanism and considering step 4 to be the dominant termination step, show that this will lead to 3/2 order kinetics.

Can all the individual rate constants be determined from the reaction in its steady state?

Answer.

$$+\frac{d[CH_3^\bullet]}{dt} = k_1[CH_3OCH_3] - k_2[CH_3^\bullet][CH_3OCH_3] + k_3[{}^\bullet CH_2OCH_3] - 2k_4[CH_3^\bullet]^2 = 0 \qquad (6.110)$$

Note the factor of two: for every act of termination two $CH_3^\bullet$ are removed, and the rate is expressed in terms of production of $CH_3^\bullet$.

$$+\frac{d[{}^\bullet CH_2OCH_3]}{dt} = k_2[CH_3^\bullet][CH_3OCH_3] - k_3[{}^\bullet CH_2OCH_3] = 0 \qquad (6.111)$$

Both are equations in two unknowns, and cannot be solved on their own. Adding (6.110) and (6.111) gives

$$k_1[CH_3OCH_3] - 2k_4[CH_3^\bullet]^2 = 0 \qquad (6.112)$$

This is the algebraic statement of the steady state assumption that there is no build-up of chain carriers during reaction, and that termination balances initiation.

$$\therefore \quad [CH_3^\bullet] = \left(\frac{k_1[CH_3OCH_3]}{2k_4}\right)^{1/2} \qquad (6.113)$$

Rate of reaction can be expressed in three ways.

(a)
$$+\frac{d[CH_4]}{dt} = k_2[CH_3^\bullet][CH_3OCH_3] \quad (6.114)$$
$$= k_2\left(\frac{k_1}{2\,k_4}\right)^{1/2}[CH_3OCH_3]^{3/2} \quad (6.115)$$

(b)
$$-\frac{d[CH_3OCH_3]}{dt} = k_1[CH_3OCH_3] + k_2[CH_3^\bullet][CH_3OCH_3] \quad (6.116)$$
$$= k_1[CH_3OCH_3] + k_2\left(\frac{k_1}{2\,k_4}\right)^{1/2}[CH_3OCH_3]^{3/2} \quad (6.117)$$

This can be simplified. The first term can be dropped, since rate of initiation $\ll$ rate of propagation—the *long chains approximation*—leaving the rate as

$$-\frac{d[CH_3OCH_3]}{dt} = k_2\left(\frac{k_1}{2\,k_4}\right)^{1/2}[CH_3OCH_3]^{3/2} \quad (6.115)$$

as in (a).

(c)
$$+\frac{d[HCHO]}{dt} = k_3[^\bullet CH_2OCH_3] \quad (6.118)$$

$[^\bullet CH_2OCH_3]$ must be found. Substitute for $[CH_3^\bullet]$ into Equation (6.111):

$$[^\bullet CH_2OCH_3] = \frac{k_2}{k_3}\left(\frac{k_1}{2\,k_4}\right)^{1/2}[CH_3OCH_3]^{3/2} \quad (6.119)$$

giving

$$\frac{d[HCHO]}{dt} = k_3[^\bullet CH_2OCH_3] \quad (6.120)$$
$$= k_2\left(\frac{k_1}{2\,k_4}\right)^{1/2}[CH_3OCH_3]^{3/2} \quad (6.115)$$

as before.

Normally the easiest algebra is chosen: here $[CH_3^\bullet]$ was found first from solution of the steady state equations, and so either (a) or (b) is chosen in preference.

A very important point to notice. The steady state analysis shows that not all the individual rate constants can be found from the steady state reaction. In this case the steady state expression only involves three of the rate constants: k_3 does not appear.

6.11.1 Important points to note

- This analysis has used the *long chains approximation*:

  rate of initiation $\ll$ rate of propagation.

  This must only be used when simplifying an expression for the *overall rate*. It *must not* be used in steady state equations because these contain the difference of two large terms, i.e. the rates of the two propagation steps, and this difference is of similar magnitude to the rates of initiation and termination.

- A few chain reactions occur where the number of cycles in propagation, the chain length, is small and the long chains approximation will no longer be applicable. Such reactions will not be considered here.

- The long chains approximation must never be used when one or other or both of the chain carriers are removed or formed in steps other than propagation, initiation and termination. For example, when inhibition is present and a product removes or forms one or other of the chain carriers, the long chains approximation is invalid; see the H_2/Br_2 reaction, Section 6.10.

- The overall rate of reaction is always given by the rate of one of the propagation steps. *On no account* must it be given in terms of production of a minor product, as these are formed in termination, which accounts for only a small fraction of the reaction.

- Many chain reactions may involve two or more steps in initiation, and more than two propagation steps.

- Some reactions are also retarded in their later stages by a product as it builds up. This is called *inhibition*.

6.12 Stylized Mechanisms: A Typical Rice–Herzfeld Mechanism

In any mechanistic study the proposed mechanism must fit the observed kinetics, and clues as to likely fits could help eliminate unnecessary trial and error analyses. In 1934 Rice and Herzfeld proposed a set of mechanisms from which systematic rules were inferred relating the observed order to the type of initiation and termination steps. These mechanisms are highly stylized, but even today they often form the basis for the interpretation of the thermal decomposition of organic molecules, though extra steps often have to be added. They demonstrate how only a small change in mechanism will alter the kinetic features in a *systematic way*.

The overall reaction considered is

$$X \rightarrow P_1 + P_2$$

However, minor products, P_3, P_4 and P_5 are also formed, along with radical intermediates $R^\bullet$, $R_1^\bullet$, and $R_2^\bullet$. A possible schematic mechanism is

$$\begin{array}{rll} X + (M) \xrightarrow{k_1} R^\bullet + R_1^\bullet + (M) & & \text{initiation} \\ R_1^\bullet + X \xrightarrow{k_2} R_2^\bullet + P_1 & & \text{propagation} \\ R_2^\bullet + (M) \xrightarrow{k_3} R_1^\bullet + P_2 + (M) & & \text{propagation} \\ R_1^\bullet + R_1^\bullet + (M) \xrightarrow{k_4} P_3 + (M) & & \text{termination} \\ R_2^\bullet + R_2^\bullet + (M) \xrightarrow{k_5} P_4 + (M) & & \text{termination} \\ R_1^\bullet + R_2^\bullet + (M) \xrightarrow{k_6} P_5 + (M) & & \text{termination} \end{array}$$

$R^\bullet$ is formed in initiation, but as it is not involved in propagation it is not a chain carrier and will be removed in some subsequent reaction not associated with the chain. M is a third body whose significance will be explained below (Section 6.12.4).

A steady state treatment shows the overall order to depend on the type of step in which the radical is formed, and also on which termination step is dominant.

$R_1^\bullet + R_1^\bullet$ recombination $\rightarrow$ 3/2 order: $R_1^\bullet$ is formed in initiation, used in the first propagation step and re-formed in the second.

$R_2^\bullet + R_2^\bullet$ recombination $\rightarrow$ 1/2 order: $R_2^\bullet$ is formed in the first propagation step and is used in the second.

$R_1^\bullet + R_2^\bullet$ recombination $\rightarrow$ first order.

This treatment can be extended to include various types of initiation and propagation steps, and a whole systematic classification built up whereby the overall order can be predicted for any set of initiation, propagation and termination steps.

Several points can be made about this mechanism:

- Initiation is unimolecular, showing first order kinetics at high pressures where the reaction step is rate determining, moving to the low pressure second order region where activation is rate determining (see Chapter 1 and Section 4.5). The reaction step referred to here is that in the physical mechanism of an elementary step:

$$\begin{array}{ll} A + A \rightarrow A^* + A & \text{collisional activation step} \\ A^* + A \rightarrow A + A & \text{collisional deactivation step} \\ A^* + bA \rightarrow \text{products} & \text{reaction step} \end{array}$$

where $b = 0$ when reaction is unimolecular and $b = 1$ when reaction is bimolecular.

This behaviour is reflected in the presence of the third body, M, in the Rice–Herzfeld mechanism. The third body produces suitably activated molecules by an energy transfer collision. These can then pass through the activated complex and react. M can be a reactant molecule or an added inert substance capable of efficient energy transfer. It can also remove energy by an energy transfer step.

Change-over from first to second order, and the pressure at which this occurs, depends mainly on *three factors*. Low temperatures, high pressures and increasing complexity all favour first order kinetics, while high temperatures, low pressures and small numbers of atoms in the reactant favour second order.

- In this particular mechanism the second propagation step is also unimolecular and the same considerations apply, and the third body, M, is included to cover all situations.
- Three termination steps are proposed. All may occur simultaneously, though often one will predominate. Which termination step is most likely depends on the relative concentrations of the radicals and the rate constants of the termination steps.
- Generalizations can be made about the nature of the radicals in termination, and whether or not a third body is required in these steps.

6.12.1 Dominant termination steps

The rate constants of the propagation steps determine the relative concentrations of the chain carriers. If these only appear in propagation or termination, then, *and only then*, will the rates of the two propagation steps be equal.

$$k_2[R_1^\bullet][X] = k_3[R_2^\bullet] \tag{6.121}$$

$$\therefore \quad [R_1^\bullet] = \frac{k_3[R_2^\bullet]}{k_2[X]} \tag{6.122}$$

- If $k_2[X] \gg k_3$ then, for a given [X], $[R_1^\bullet] =$ constant $[R_2^\bullet]$ where the constant is very small, and

  $$\therefore \quad [R_1^\bullet] \ll [R_2^\bullet]$$

  so that $R_2^\bullet + R_2^\bullet$ recombination is the most likely termination.

- If $k_2[X] \ll k_3$ then, for a given [X], $[R_1^\bullet] =$ constant $[R_2^\bullet]$ where the constant is now very large, and

  $$\therefore \quad [R_1^\bullet] \gg [R_2^\bullet]$$

  so that $R_1^\bullet + R_1^\bullet$ recombination is the most likely termination.

- If $k_2[X]$ is comparable to k_3, then $[R_1^\bullet]$ and $[R_2^\bullet]$ are comparable. Discrimination as to which termination step is most likely is impossible. In this situation the relative rate constants of the termination steps are the major influence.

6.12.2 Relative rate constants for termination steps

Transition state theory (Section 4.4) gives the rate constant

$$k = \frac{\kappa kT}{h} \exp\left(\frac{+\Delta S^{\neq *}}{R}\right) \exp\left(\frac{-\Delta H^{\neq *}}{RT}\right) \quad (4.41)$$

where $\Delta H^{\neq *}$, the change in enthalpy on forming the activated complex, is closely related to the activation energy. $\Delta S^{\neq *}$ is the entropy of activation.

Radical recombinations have zero, or small positive or negative activation energies, and so to a first approximation

$$k = \frac{\kappa kT}{h} \exp\left(\frac{+\Delta S^{\neq *}}{R}\right) \quad (6.123)$$

for these reactions.

Since $\kappa kT/h$ is a constant, then the rate constant for each termination step reflects the magnitude and sign of the entropy of activation.

$\Delta S^{\neq *}$ is the entropy of activation, i.e. the change in entropy for the process of reactants moving into the activated complex. In the gas phase it is governed by the change in the number and types of degrees of freedom on activation, these degrees of freedom being related to translations, rotations and vibrations. Calculations show that reactions between large complex radicals have large negative $\Delta S^{*\neq}$, so $\exp(+\Delta S^{*\neq}/R)$ is very small, resulting in a small rate constant. Conversely, for small, simple radicals, $\Delta S^{*\neq}$ is considerably less negative and their rate constants are much higher than for large radicals (see Sections 4.4.2 and 4.4.3 and Worked Problems 4.15 and 4.16).

6.12.3 Relative rates of the termination steps

In practice these are a function of both the chain carrier concentrations and the rate constants for termination. Specific predictions can only be made if all rate constants for propagation and termination are known. It is only when the chain carrier concentrations are comparable that it is possible to say that reactions between the smaller radicals will dominate.

Worked Problem 6.9

Question. The following is a schematic mechanism for the bromination of an organic compound RH.

$$\mathrm{Br_2 + M \xrightarrow{k_1} 2\,Br^\bullet + M}$$

$$\mathrm{Br^\bullet + RH \xrightarrow{k_2} R^\bullet + HBr}$$

$$\mathrm{R^\bullet + Br_2 \xrightarrow{k_3} RBr + Br^\bullet}$$

$$\mathrm{2\,Br^\bullet + M \xrightarrow{k_4} Br_2 + M}$$

Using the rough values

$$k_1 = 10^{-2}\,\mathrm{dm^3\,mol^{-1}\,s^{-1}} \qquad k_2 = 10^{6}\,\mathrm{dm^3\,mol^{-1}\,s^{-1}}$$

$$k_3 = 10^{9}\,\mathrm{dm^3\,mol^{-1}\,s^{-1}} \qquad k_4 = 10^{10}\,\mathrm{dm^6\,mol^{-2}\,s^{-1}}$$

estimate

(a) the steady state concentrations of the chain carriers when $[\mathrm{Br_2}]$ and [RH] are both 10^{-2} mol dm^{-3} and

(b) the steady state rate for a mixture when $[\mathrm{Br_2}]$ and [RH] are both 10^{-2} mol dm^{-3}.

(c) Two other conceivable alternative termination steps are

$$\mathrm{2\,R^\bullet \xrightarrow{k_5} R_2}$$

$$\mathrm{R^\bullet + Br^\bullet \xrightarrow{k_6} RBr}$$

having approximate values for the rate constants $k_5 \approx k_6 \approx 10^{10}$ dm^3 mol^{-1} s^{-1}. Estimate the rates of the three alternative termination steps. Assume M to be RH or $\mathrm{Br_2}$.

Using the criteria given in Sections 6.12.1–6.12.3, suggest which is likely to be dominant.

Answer.

(a) *Considering the concentrations of chain carriers*

$$\frac{\mathrm{d}[\mathrm{Br^\bullet}]}{\mathrm{d}t} = 2\,k_1[\mathrm{Br_2}][\mathrm{M}] - k_2[\mathrm{Br^\bullet}][\mathrm{RH}] + k_3[\mathrm{R^\bullet}][\mathrm{Br_2}] - 2\,k_4[\mathrm{Br^\bullet}]^2[\mathrm{M}] = 0 \qquad (6.124)$$

$$\frac{\mathrm{d}[\mathrm{R^\bullet}]}{\mathrm{d}t} = k_2[\mathrm{Br^\bullet}][\mathrm{RH}] - k_3[\mathrm{R^\bullet}][\mathrm{Br_2}] = 0 \qquad (6.125)$$

Add these two equations:

$$2\,k_1[\mathrm{Br_2}][\mathrm{M}] = 2\,k_4[\mathrm{Br^\bullet}]^2[\mathrm{M}] \tag{6.126}$$

$$[\mathrm{Br^\bullet}] = \left(\frac{k_1}{k_4}\right)^{1/2}[\mathrm{Br_2}]^{1/2} \tag{6.127}$$

$$[\mathrm{Br^\bullet}] = \left(\frac{10^{-2}}{10^{10}}\right)^{1/2}(10^{-2})^{1/2} = 10^{-7}\,\mathrm{mol\,dm^{-3}}$$

$$[\mathrm{R^\bullet}] = \left(\frac{k_2[\mathrm{Br^\bullet}][\mathrm{RH}]}{k_3[\mathrm{Br_2}]}\right) \tag{6.128}$$

$$[\mathrm{R^\bullet}] = \left(\frac{10^6 \times 10^{-7} \times 10^{-2}}{10^9 \times 10^{-2}}\right) = 10^{-10}\,\mathrm{mol\,dm^{-3}}$$

Hence $[\mathrm{Br^\bullet}] \gg [\mathrm{R^\bullet}]$ and on this basis
$2\,\mathrm{Br^\bullet} + \mathrm{M} \rightarrow \mathrm{Br_2} + \mathrm{M}$ is the most likely termination.

(b) *The steady state rate.* This can be expressed in terms of the rates of steps 2 or 3.

Taking the rate to be

$$-\frac{\mathrm{d[RH]}}{\mathrm{d}t} = +\frac{\mathrm{d[HBr]}}{\mathrm{d}t} = k_2[\mathrm{Br^\bullet}][\mathrm{RH}]$$
$$= 10^6\,\mathrm{dm^3\,mol^{-1}\,s^{-1}} \times 10^{-7}\,\mathrm{mol\,dm^{-3}} \times 10^{-2}\,\mathrm{mol\,dm^{-3}}$$
$$= 10^{-3}\,\mathrm{mol\,dm^{-3}\,s^{-1}} \tag{6.129}$$

or

$$-\frac{\mathrm{d[Br_2]}}{\mathrm{d}t} = +\frac{\mathrm{d[RBr]}}{\mathrm{d}t} = k_3[\mathrm{R^\bullet}][\mathrm{Br_2}]$$
$$= 10^9\,\mathrm{dm^3\,mol^{-1}\,s^{-1}} \times 10^{-10}\,\mathrm{mol\,dm^{-3}} \times 10^{-2}\,\mathrm{mol\,dm^{-3}}$$
$$= 10^{-3}\,\mathrm{mol\,dm^{-3}\,s^{-1}} \tag{6.130}$$

As it should be, the rate is the same whichever way it is expressed.

(c) *Considering the rates of the possible termination steps.*

(i) $2\,\mathrm{Br^\bullet} + \mathrm{M} \xrightarrow{k_4} \mathrm{Br_2} + \mathrm{M}$

$$\text{rate of termination} = 2\,k_4[\mathrm{Br^\bullet}]^2[\mathrm{M}]$$
$$= 2 \times 10^{10} \times (10^{-7})^2 \times 10^{-2}\,\mathrm{mol\,dm^{-3}\,s^{-1}}$$
$$= 2 \times 10^{-6}\,\mathrm{mol\,dm^{-3}\,s^{-1}} \tag{6.131}$$

(ii) $2\,R^\bullet \xrightarrow{k_5} R_2$

$$\begin{aligned}\text{rate of termination} &= 2\,k_5[R^\bullet]^2\\ &= 2\times 10^{10}\times(10^{-10})^2\ \text{mol dm}^{-3}\ \text{s}^{-1}\\ &= 2\times 10^{-10}\ \text{mol dm}^{-3}\ \text{s}^{-1}\end{aligned} \tag{6.132}$$

(iii) $R^\bullet + Br^\bullet \xrightarrow{k_6} RBr$

$$\begin{aligned}\text{rate of termination} &= k_6[R^\bullet][Br^\bullet]\\ &= 10^{10}\times 10^{-10}\times 10^{-7}\ \text{mol dm}^{-3}\ \text{s}^{-1}\\ &= 10^{-7}\ \text{mol dm}^{-3}\ \text{s}^{-1}\end{aligned} \tag{6.133}$$

$\therefore$ rate of step 4 > rate of step 6 > rate of step 5
and on this basis step 4 would be dominant.
For this mechanism, both criteria reinforce each other by predicting that step 4 is dominant.

Note also that the slowest step involves the most complex radicals, and the fastest step involves the simplest radicals. This would be expected from transition state theory predictions (Section 6.12.2).

6.12.4 Necessity for third bodies in termination

The function of the third body is to facilitate energy transfer by collisions. The reverse of any radical–radical recombination is a unimolecular decomposition. In a unimolecular decomposition the third body provides activation energy; in the reverse reaction it removes excess energy. When the unimolecular reaction is in the first order region, the recombination is in the second order region. When the unimolecular reaction is in the second order region, the recombination is third order with energy transfer being rate determining for both. A third body is thus more likely to be required at high temperatures, low pressures and for small radicals, corresponding to the requirements for a third body in the decomposition reaction (see Section 4.5.12).

Worked Problem 6.10

Question. Explain the conclusions stated above in 6.12.4.

Answer. When the unimolecular reaction is in the first order region the rate-determining step is reaction, rather than activation. The reverse reaction must also have reaction as rate determining and this corresponds to the second order region of the recombination. Recombination and decomposition will be in the high pressure region.

At low pressures the unimolecular decomposition will have activation rate determining, and will be second order with a third body required for energy transfer.

The recombination must also have energy transfer as rate determining, and for this removal of energy a third body is required, leading to third order kinetics (see Section 4.5.12).

These generalizations about the need for a third body can be explained physically in terms of the three types of recombination:

$R_1^\bullet$ and $R_2^\bullet$ are atoms,

$R_1^\bullet$ and $R_2^\bullet$ are large radicals,

$R_1^\bullet$ and $R_2^\bullet$ are small radicals.

- *$R_1^\bullet$ and $R_2^\bullet$ are atoms.* When atoms recombine, the energy resulting from the highly exothermic reaction initially resides in the newly formed bond which is thus highly excited. If this energy is not removed before the first vibration occurs, then the diatomic molecule will dissociate on vibration and effective recombination will not occur. This energy can only be removed by a collision, which must occur in a time less than the time for one vibration. A third body is required, and can be either the reactant or another molecule capable of efficient energy transfer.

  $A + A \rightarrow A_2^*$ an excited diatomic molecule

  $M + A_2^* \rightarrow A_2 + M$ an energy transfer collision

  Third bodies are thus required for atom–atom recombinations.

- *$R_1^\bullet$ and $R_2^\bullet$ are large radicals.* These form a high energy complex molecule with the exothermicity of the recombination residing in the newly formed bond. This energy has to be removed before the bond executes a vibration. However, for large polyatomic molecules with a large number of vibrational modes there is an alternative route for stabilizing the newly formed bond. Because of anharmonicity, the excess energy can be distributed very, very rapidly around the molecule via the normal modes of vibration. The molecule, *as a whole*, is still highly excited, but the bond is stabilized before it has time to dissociate. This can be achieved in a time short compared to that for a vibration. This increases the lifetime of the excited molecule and the normal process of energy transfer by collision can occur after many vibrations of the bond. Third bodies are thus not required for large radical recombinations.

- *$R_1^\bullet$ and $R_2^\bullet$ are small radicals.* If the radical is small then the molecule may not have enough normal modes into which the excess energy can be channelled, and a third body will be necessary. As the radical size increases many more normal modes become available, and the chances of the bond being stabilized rapidly increase, reducing the need for a third body. The minimum number of normal modes required for adequate stabilization of the newly formed bond depends also on the temperature and pressure. If the number of normal modes exceeds around ten then a third body probably is not needed, provided the pressure is high enough. When the number is between six and ten there is a high chance that a third body will be required, and again the pressure may well prove decisive.

Worked Problem 6.11

Question. Predict whether a third body would be required in the following recombinations. State whether any provisos should be made on pressure and temperature considerations.

(a) $H^\bullet + H^\bullet \rightarrow H_2$ (b) $H^\bullet + OH^\bullet \rightarrow H_2O$

(c) $CH_3^\bullet + Cl^\bullet \rightarrow CH_3Cl$ (d) $CH_3^\bullet + CH_3^\bullet \rightarrow C_2H_6$

(e) $CH_3^\bullet + C_2H_5^\bullet \rightarrow C_3H_8$ (f) $C_3H_7^\bullet + C_3H_7^\bullet \rightarrow C_6H_{14}$

Answer.

(a) Two atoms recombining: third body always required.

(b) Atom + small radical recombining; non-linear triatomic molecule formed; $3N - 6 = 3$ vibrations: third body normally required.

(c) As in (b), but here the non-linear product has five atoms, i.e. nine vibrations. The third body possibly might not be required unless the pressure is low and the temperature is high. The reverse decomposition would then be expected to be second order under these conditions and require a third body in the activation step. This has been observed.

(d) As in (c), but eight atoms in product, i.e. 18 vibrations with conclusions similar to (c). At very low pressures and high temperatures a third body is required for the activation step. The reverse decomposition has been observed to move from first order towards second as the pressure is lowered, and so under these conditions of pressure the recombination would be third order, i.e. a third body is required.

(e) The product of (e) has 11 atoms and 27 vibrations. Although this has a fairly large number of normal modes of vibration and a third body would normally not be required, this reaction might still need a third body at very low pressures and high temperatures.

(f) The product of (f) has 20 atoms and 54 vibrations, and a third body requirement is not expected.

The results for (d) and (e) illustrate that, although a third body is not required under normal experimental conditions and the normal modes can cope with stabilizing the newly formed bond, this will not necessarily hold as the pressure is lowered.

This means that there is a range of sizes of the recombining radicals where there may be pressure and temperature regions where even a relatively large number of normal modes of vibration may not be enough to stabilize the newly formed bond.

6.12.5 The steady state treatment for chain reactions, illustrating the use of the long chains approximation

A simplified mechanism for the oxidation of hydrocarbons can be given. Such reactions often involve an initiator, I. This is a molecule which by decomposing produces a radical, $X^\bullet$. This, in turn, reacts with O_2 to produce a peroxy radical, $XO_2^\bullet$, which can then set the chain process going by producing the radical, $R^\bullet$.

One possible mechanism is

$$\left.\begin{array}{l} I \xrightarrow{k_1} X^\bullet + R'^\bullet \\ X^\bullet + O_2 \xrightarrow{k_2} XO_2^\bullet \\ XO_2^\bullet + RH \xrightarrow{k_3} XOOH + R^\bullet \end{array}\right\} \text{three stage initiation}$$

$$\left.\begin{array}{l} R^\bullet + O_2 \xrightarrow{k_4} RO_2^\bullet \\ RO_2^\bullet + RH \xrightarrow{k_5} ROOH + R^\bullet \end{array}\right\} \text{propagation}$$

$$RO_2^\bullet + RO_2^\bullet \xrightarrow{k_6} \text{inert molecule} + O_2 \quad \text{termination}$$

This is a highly simplified mechanism, but it does contain the essentials of the main chain process. This mechanism involves a three stage initiation process, something not uncommon in chain mechanisms. Only one of the possible propagation chains is given, and only one termination step is quoted.

The chain carriers are $R^\bullet$ and $RO_2^\bullet$, and the major product is ROOH. There are two radical intermediates, $X^\bullet$ and $XO_2^\bullet$, in this reaction which are not chain carriers, but are necessary for the production of the chain carriers. They must, therefore, be included in the steady state treatment. However, $R'^\bullet$, though a radical, is not involved in any chain and so does not come into the steady state treatment.

The steady state treatment is

$$\frac{d[X^\bullet]}{dt} = k_1[I] - k_2[X^\bullet][O_2] = 0 \tag{6.134}$$

$$\frac{d[XO_2^\bullet]}{dt} = k_2[X^\bullet][O_2] - k_3[XO_2^\bullet][RH] = 0 \tag{6.135}$$

$$\frac{d[R^\bullet]}{dt} = k_3[XO_2^\bullet][RH] - k_4[R^\bullet][O_2] + k_5[RO_2^\bullet][RH] = 0 \tag{6.136}$$

$$\frac{d[RO_2^\bullet]}{dt} = k_4[R^\bullet][O_2] - k_5[RO_2^\bullet][RH] - 2k_6[RO_2^\bullet]^2 = 0 \tag{6.137}$$

The standard procedure is to add all the steady state equations. This gives

$$k_1[I] = 2k_6[RO_2^\bullet]^2 \tag{6.138}$$

$$\therefore \quad [RO_2^\bullet] = \left(\frac{k_1}{2k_6}\right)^{1/2}[I]^{1/2} \tag{6.139}$$

Note. Formulating steady state equations for all the intermediates, chain and non-chain, allows elimination of the terms in $[X^\bullet]$ and $[XO_2^\bullet]$. If Equations (6.134) and (6.135) had not been set up and only the equations for the chain carriers had been considered, then adding these two equations would have given

$$k_3[XO_2^\bullet][RH] = 2\,k_6[RO_2^\bullet] \tag{6.140}$$

where the unknown term $[XO_2^\bullet]$ appears.

The rate of removal could be expressed either as a rate of removal of reactant, RH, or reactant, O_2, or formation of product ROOH.

$$-\frac{d[RH]}{dt} = k_3[XO_2^\bullet][RH] + k_5[RO_2^\bullet][RH] \tag{6.141}$$

$$-\frac{d[O_2]}{dt} = k_2[X^\bullet][O_2] + k_4[R^\bullet][O_2] - k_6[RO_2^\bullet]^2 \tag{6.142}$$

$$+\frac{d[ROOH]}{dt} = k_5[RO_2^\bullet][RH] \tag{6.143}$$

The simplest equation involving *known* quantities should be used. Since an expression has already been found for $[RO_2^\bullet]$, Equation (6.143) gives

$$+\frac{d[ROOH]}{dt} = k_5\left(\frac{k_1}{2\,k_6}\right)^{1/2}[I]^{1/2}[RH] \tag{6.144}$$

Had the reaction rate in terms of the reactants been used, this would have required finding an expression for either $[X^\bullet]$ or $[XO_2^\bullet]$ and $[RO_2^\bullet]$.

- Equation (6.134) gives $[X^\bullet]$

$$[X^\bullet] = \frac{k_1[I]}{k_2[O_2]}$$

- Adding Equations (6.134) and (6.135) eliminates $[X^\bullet]$, giving

$$k_1[I] = k_3[XO_2^\bullet][RH] \tag{6.145}$$

$$\therefore \quad [XO_2^\bullet] = \frac{k_1[I]}{k_3[RH]} \tag{6.146}$$

- Substitution into either of the steady state equations (6.136) or (6.137) is necessary to enable $[R^\bullet]$ to be found.

Choose (6.136) rather than the quadratic (6.137)

$$k_3\frac{k_1[\mathrm{I}][\mathrm{RH}]}{k_3[\mathrm{RH}]} - k_4[\mathrm{R}^\bullet][\mathrm{O}_2] + k_5\left(\frac{k_1}{2\,k_6}\right)^{1/2}[\mathrm{I}]^{1/2}[\mathrm{RH}] = 0 \qquad (6.147)$$

$$[\mathrm{R}^\bullet] = \frac{k_1[\mathrm{I}]}{k_4[\mathrm{O}_2]} + \frac{k_5}{k_4}\left(\frac{k_1}{2\,k_6}\right)^{1/2}\frac{[\mathrm{I}]^{1/2}[\mathrm{RH}]}{[\mathrm{O}_2]} \qquad (6.148)$$

The rate of reaction can now be expressed in terms of removal of RH or O_2. Taking each in turn gives the following.

(a)

$$-\frac{\mathrm{d}[\mathrm{RH}]}{\mathrm{d}t} = k_3[\mathrm{XO}_2^\bullet][\mathrm{RH}] + k_5[\mathrm{RO}_2^\bullet][\mathrm{RH}] \qquad (6.141)$$

$$= k_3\frac{k_1[\mathrm{I}][\mathrm{RH}]}{k_3[\mathrm{RH}]} + k_5\left(\frac{k_1}{2\,k_6}\right)^{1/2}[\mathrm{I}]^{1/2}[\mathrm{RH}] \qquad (6.149)$$

The term in k_1 relates to a rate of initiation. The long chains approximation assumes that rate of initiation $\ll$ rate of propagation, and so this term can drop out.

$$-\frac{\mathrm{d}[\mathrm{RH}]}{\mathrm{d}t} = k_5\left(\frac{k_1}{2\,k_6}\right)^{1/2}[\mathrm{I}]^{1/2}[\mathrm{RH}] \qquad (6.150)$$

(b)

$$-\frac{\mathrm{d}[\mathrm{O}_2]}{\mathrm{d}t} = k_2[\mathrm{X}^\bullet][\mathrm{O}_2] + k_4[\mathrm{R}^\bullet][\mathrm{O}_2] - k_6[\mathrm{RO}_2^\bullet]^2 \qquad (6.142)$$

This equation can be simplified immediately. The term in k_6 relates to a rate of termination. The long chains approximation assumes that rate of termination $\ll$ rate of propagation, and so this term can drop out.

$$-\frac{\mathrm{d}[\mathrm{O}_2]}{\mathrm{d}t} = k_2\frac{k_1[\mathrm{I}]}{k_2[\mathrm{O}_2]}[\mathrm{O}_2] + k_4\frac{k_1[\mathrm{I}]}{k_4[\mathrm{O}_2]}[\mathrm{O}_2] + k_4\frac{k_5}{k_4}\left(\frac{k_1}{2\,k_6}\right)^{1/2}\frac{[\mathrm{I}]^{1/2}[\mathrm{RH}]}{[\mathrm{O}_2]}[\mathrm{O}_2] \qquad (6.151)$$

The first two terms reduce to $2\,k_1[\mathrm{I}]$, which refers to a rate of initiation which by the long chains approximation can now drop out, leaving

$$-\frac{\mathrm{d}[\mathrm{O}_2]}{\mathrm{d}t} = k_5\left(\frac{k_1}{2\,k_6}\right)^{1/2}[\mathrm{I}]^{1/2}[\mathrm{RH}] \qquad (6.152)$$

All three expressions, Equations (6.144), (6.150) and (6.152), are the same, as they should be. If the long chains approximation had not been used, the terms involving the rates of termination and initiation would not have dropped out, and a more complex expression would have resulted.

6.12.6 Further problems on steady states and the Rice–Herzfeld mechanism

This section ends with two problems, which between them cover a lot of the aspects of chain reactions and use of one of the Rice–Herzfeld rules. The second problem explains how it is that many chain reactions have observed overall activation energies which are considerably less than the bond dissociation energy involved in the first step of the sequence.

Worked Problem 6.12

Question. The pyrolysis of ethanal

$$CH_3CHO(g) \rightarrow CH_4(g) + CO(g)$$

has been shown previously, Problem 6.7, to have the following simplified mechanism:

$$CH_3CHO \xrightarrow{k_1} CH_3^\bullet + CHO^\bullet$$

$$CH_3^\bullet + CH_3CHO \xrightarrow{k_2} CH_4 + CH_3CO^\bullet$$

$$CH_3CO^\bullet \xrightarrow{k_3} CH_3^\bullet + CO$$

$$CH_3^\bullet + CH_3^\bullet \xrightarrow{k_4} C_2H_6$$

$$CH_3^\bullet + CH_3CO^\bullet \xrightarrow{k_5} CH_3COCH_3$$

$$CH_3CO^\bullet + CH_3CO^\bullet \xrightarrow{k_6} CH_3COCOCH_3$$

1. Under certain conditions the reaction appears to have 3/2 order kinetics. Use the Rice–Herzfeld rules to identify the dominant termination step.
2. Using the mechanism which has this as the only termination step, verify that it does lead to 3/2 order kinetics.
3. Identify the unimolecular steps in this mechanism and indicate what conditions would lead to a requirement for a third body for these steps.
4. Is it likely that a third body would be required for this dominant termination step? Explain your reasoning.
5. Using the Rice–Herzfeld rules, predict the overall kinetics, if each of the other termination steps were dominant in turn.
6. Demonstrate the correctness of the predictions by carrying out a steady state analysis for the mechanism in which *unlike* radical recombination is the dominant termination.
7. Explain how individual rate constants could be found for the first and 3/2 order mechanisms.

Answer.

1. The Rice–Herzfeld rules indicate that 3/2 order occurs for termination via like radicals produced in initiation, which then undergo reaction of the type $R^\bullet$ + reactant $\rightarrow \cdots$. Hence step 4 is the dominant termination.

$$CH_3^\bullet + CH_3^\bullet \xrightarrow{k_4} C_2H_6$$

2. Applying the steady state treatment gives

$$+\frac{d[CH_3^\bullet]}{dt} = k_1[CH_3CHO] - k_2[CH_3^\bullet][CH_3CHO] + k_3[CH_3CO^\bullet] - 2\,k_4[CH_3^\bullet]^2 = 0 \tag{6.153}$$

Note the factor of two: for each termination step two $CH_3^\bullet$ are removed and the rate is expressed in terms of production of $CH_3^\bullet$.

$$+\frac{d[CH_3CO^\bullet]}{dt} = k_2[CH_3^\bullet][CH_3CHO] - k_3[CH_3CO^\bullet] = 0 \tag{6.154}$$

Add (6.153) + (6.154)

$$k_1[CH_3CHO] = 2\,k_4[CH_3^\bullet]^2 \tag{6.155}$$

Note this is the algebraic statement of the steady state assumption that there is no build-up of chain carriers during reaction, and so *the total rate of production must be balanced by the total rate of removal.*

$$\therefore \quad [CH_3^\bullet] = \left(\frac{k_1}{2\,k_4}\right)^{1/2}[CH_3CHO]^{1/2} \tag{6.156}$$

The mechanistic rate must be expressed in terms of the overall removal of reactants or the total production of the *major* products. For removal of reactant, CH_3CHO:

$$-\frac{d[CH_3CHO]}{dt} = k_1[CH_3CHO] + k_2[CH_3^\bullet][CH_3CHO] \tag{6.157}$$

This approximates to

$$-\frac{d[CH_3CHO]}{dt} = k_2[CH_3^\bullet][CH_3CHO] \tag{6.158}$$

This is a consequence of the long chains approximation, which states that initiation occurs only very infrequently compared to propagation.

Expressing the rate in terms of the major products:

$$+\frac{d[CH_4]}{dt} = k_2[CH_3^\bullet][CH_3CHO] \tag{6.159}$$

$$+\frac{d[CO]}{dt} = k_3[CH_3CO^\bullet] \tag{6.160}$$

Since $[CH_3^\bullet]$ has already been found, choose either (6.157) or (6.159).

$$-\frac{d[CH_3CHO]}{dt} \approx +\frac{d[CH_4]}{dt} = k_2\left(\frac{k_1}{2k_4}\right)^{1/2}[CH_3CHO]^{3/2} \tag{6.161}$$

which agrees with experiment and confirms recombination of $CH_3^\bullet$ as the dominant termination.

3. The initiation and second propagation steps are unimolecular.

$$CH_3CHO \xrightarrow{k_1} CH_3^\bullet + CHO^\bullet$$

$$CH_3CO^\bullet \xrightarrow{k_3} CH_3^\bullet + CO$$

Since these are complex species, with 15 and 12 vibrational degrees of freedom respectively, there will normally be no need for a third body. Reaction will be first order in both cases with the rate-determining step being reaction. It is only when activation is rate determining that there will be a need for a third body, in which case the rate of reaction would then be second order for both, and this would only occur at very low pressures and very high temperatures.

4. The recombination product is C_2H_6. This has 18 normal modes of vibration over which the exothermicity of the reaction can be spread, thereby stabilizing the C–C bond. A third body would not be required, unless at very low pressures and very high temperatures.

5. The alternative termination steps are

$$CH_3^\bullet + CH_3CO^\bullet \xrightarrow{k_5} CH_3COCH_3$$

$$CH_3CO^\bullet + CH_3CO^\bullet \xrightarrow{k_6} CH_3COCOCH_3$$

Step 5 is recombination of *unlike* radicals and would be predicted to show first order kinetics.

Step 6 is recombination of *like* radicals, where the *radical is formed in the first propagation step* and then undergoes an unimolecular reaction, and would be predicted to show 1/2 order kinetics.

6. The steady state equations for step 5 being the dominant termination are

$$+\frac{d[CH_3^\bullet]}{dt} = k_1[CH_3CHO] - k_2[CH_3^\bullet][CH_3CHO] + k_3[CH_3CO^\bullet] - k_5[CH_3CO^\bullet][CH_3^\bullet] = 0 \quad (6.162)$$

$$+\frac{d[CH_3CO^\bullet]}{dt} = k_2[CH_3^\bullet][CH_3CHO] - k_3[CH_3CO^\bullet] - k_5[CH_3CO^\bullet][CH_3^\bullet] = 0 \quad (6.163)$$

Add (6.162) + (6.163)

$$k_1[CH_3CHO] = 2k_5[CH_3CO^\bullet][CH_3^\bullet] \quad (6.164)$$

This is the steady state approximation.

$$\therefore \quad [CH_3^\bullet] = \frac{k_1[CH_3CHO]}{2k_5[CH_3CO^\bullet]} \quad (6.165)$$

This is an equation in two unknowns and cannot be solved. Another equation is needed.

As shown in Problems 6.3 and 6.8, there are two ways to tackle this:

(a) express either $[CH_3^\bullet]$ or $[CH_3CO^\bullet]$ in terms of the other, substitute in either steady state equation and solve the resulting quadratic equation; or

(b) use the long chains approximation, in the form that the rates of propagation are equal.

Method (b) is algebraically much simpler. However, the mechanism must be checked to see whether the approximation is legitimate. It is; the chain carriers are not involved in any steps other than initiation, propagation and termination, and so

$$k_2[CH_3^\bullet][CH_3CHO] = k_3[CH_3CO^\bullet] \quad (6.166)$$

This is the long chains approximation.

$$\therefore \quad [CH_3^\bullet] = \frac{k_3[CH_3CO^\bullet]}{k_2[CH_3CHO]} \quad (6.167)$$

$$\therefore \quad \frac{k_1[CH_3CHO]}{2k_5[CH_3CO^\bullet]} = \frac{k_3[CH_3CO^\bullet]}{k_2[CH_3CHO]} \quad (6.168)$$

$$\therefore \quad [CH_3CO^\bullet] = \left(\frac{k_1k_2}{2k_3k_5}\right)^{1/2}[CH_3CHO] \quad (6.169)$$

Since $[CH_3CO^\bullet]$ has been found, it is simpler to express the overall rate in terms of production of CO, giving

$$\frac{d[CO]}{dt} = k_3[CH_3CO^\bullet] \tag{6.160}$$

$$= k_3\left(\frac{k_1k_2}{2k_3k_5}\right)^{1/2}[CH_3CHO] \tag{6.170}$$

This is first order as predicted.

7. (a) The steady state expression for the 3/2 order mechanism is

$$-\frac{d[CH_3CHO]}{dt} \approx +\frac{d[CH_4]}{dt} = k_2\left(\frac{k_1}{2k_4}\right)^{1/2}[CH_3CHO]^{3/2} \tag{6.161}$$

This rate expression only contains three of the four individual rate constants of the mechanism. k_3 does not appear and thus cannot be determined from the steady state analysis.

- If the reaction can be photochemically initiated *and* occurs with the same mechanism, then the ratio of the photochemical and thermal rates should enable k_1 to be found.
- k_2 is the rate constant for the H abstraction reaction, and k_4 is the rate constant for the termination reaction. The termination reaction is a radical recombination reaction, which can be studied independently. It could also be calculated approximately as a collision number from collision theory or from transition state theory. Since it is a recombination reaction the activation energy can be taken as approximately zero.
- Since k_{obs} is known, and if k_1 and k_4 can be found as above, then k_2 can be found. Alternatively k_2 can be assumed to be equal to the rate constant given in tabulated data for this H abstraction reaction found from other reactions.
- As noted k_3 cannot be found from the steady state data.

(b) The steady state rate for the first order mechanism is

$$\frac{d[CO]}{dt} = k_3\left(\frac{k_1k_2}{2k_3k_5}\right)^{1/2}[CH_3CHO] \tag{6.170}$$

All four individual rate constants for this mechanism can, in principle, be found from the steady state analysis.

- Again k_1 can be found from the photochemical and thermal reactions, *provided* they have the same mechanism.
- k_5 is a rate constant for the recombination reaction and can be found in a similar manner to k_4 in the 3/2 order reaction.
- k_2 is the rate constant for the H abstraction reaction, and k_3 is a rate constant for a bond splitting process. The overall rate constant can be written as $(k_1k_2k_3/2k_5)^{1/2}$. If k_1 and k_5 can be found as above, this leaves only the product k_2k_3 able to be found. Tabulated data for either of the two relevant reactions may well be necessary for the other to be found.

Indeed tabulated data may often be necessary to determine other rate constants appearing in the steady state expression.

In early investigations of chain reactions it was sometimes argued that the high value of E_A for the initial breaking of a bond in the initiation step would preclude this step and the subsequent ones, especially as the observed overall activation energy was often considerably less than the bond dissociation energy. However, once it was realized that the overall observed rate constant was a combination of individual rate constants this objection was removed. The following problem illustrates this.

Worked Problem 6.13

Question. In Problem 6.12 where the mechanism for the pyrolysis of ethanal is

$$CH_3CHO \xrightarrow{k_1} CH_3^\bullet + CHO^\bullet$$

$$CH_3^\bullet + CH_3CHO \xrightarrow{k_2} CH_4 + CH_3CO^\bullet$$

$$CH_3CO^\bullet \xrightarrow{k_3} CH_3^\bullet + CO$$

$$CH_3^\bullet + CH_3^\bullet \xrightarrow{k_4} C_2H_6$$

the mechanistic rate was given by the equation

$$-\frac{d[CH_3CHO]}{dt} \approx +\frac{d[CH_4]}{dt} = k_2\left(\frac{k_1}{2k_4}\right)^{1/2}[CH_3CHO]^{3/2} \tag{6.161}$$

1. Find an expression for k_{obs}, E_{obs} and A_{obs} in terms of the individual reactions in this mechanism.
2. Using this result, explain why it is that chain reactions can occur despite the fact that the activation energy for the initiation step is often considerably higher than the overall observed activation energy.

Answer.

1.
$$\text{Observed rate} = k_{obs}[CH_3CHO]^{3/2} \quad (6.171)$$

$$\therefore \quad k_{obs} = k_2\left(\frac{k_1}{2\,k_4}\right)^{1/2} \quad (6.172)$$

$$k_{obs} = A_{obs}\exp\left(\frac{-E_{obs}}{RT}\right) \quad (6.173)$$

$$\log_e k_{obs} = \log_e\left\{k_2\left(\left(\frac{k_1}{2\,k_4}\right)^{1/2}\right)\right\} \quad (6.174)$$

$$= \log_e k_2 + \tfrac{1}{2}\log_e\left(\frac{k_1}{2\,k_4}\right) \quad (6.175)$$

$$= \log_e k_2 + \tfrac{1}{2}\log_e k_1 - \tfrac{1}{2}\log_e k_4 - \tfrac{1}{2}\log_e 2 \quad (6.176)$$

Since $\log_e k = \log_e A - E/RT$ (6.177)

$$\log_e A_{obs} - \frac{E_{obs}}{RT} = \log_e A_2 - \frac{E_2}{RT} + \tfrac{1}{2}\log_e A_1 - \frac{E_1}{2RT} - \tfrac{1}{2}\log_e A_4 + \frac{E_4}{2\,RT} - \tfrac{1}{2}\log_e 2 \quad (6.178)$$

$$\log_e A_{obs} = \log_e A_2 + \tfrac{1}{2}\log_e A_1 - \tfrac{1}{2}\log_e A_4 - \tfrac{1}{2}\log_e 2 \quad (6.179)$$

$$\therefore \quad A_{obs} = A_2\left(\frac{A_1}{2\,A_4}\right)^{1/2} \quad (6.180)$$

$$E_{obs} = E_2 + \frac{E_1}{2} - \frac{E_4}{2} \quad (6.181)$$

$$= E_2 + \left(\frac{E_1 - E_4}{2}\right) \quad (6.182)$$

2. Radical recombinations occur with zero or small negative/positive activation energies, i.e. here $E_4 \approx 0$.

$$\therefore \quad E_{obs} \approx E_2 + \frac{E_1}{2} \quad (6.183)$$

The overall activation energy is a sum of terms, but the larger activation energy, E_1, appears as $E_1/2$ and, provided E_2 is not too large, the sum can be less than E_1. E_2 relates to a H abstraction by a radical and is likely to be relatively small.

6.13 Special Features of the Termination Reactions: Termination at the Surface

Termination is generally a recombination of atoms or radicals in the gas phase, with a third body required to remove excess energy when recombination is between atoms or small radicals.

Termination can also occur on the surface of the vessel. Diffusion to the surface is generally rate determining. Other molecules will hinder diffusion and surface termination will depend on the pressure: high pressures will hinder diffusion and cut down on surface termination; low pressures favour it.

Surface termination can be demonstrated conclusively by experiment. Alteration of the size, shape and nature of the surface affect the rate, and systematic variation of these factors enables the contribution made by the surface termination to be assessed.

At low pressures surface termination is predominant, and altering the size, shape and nature of the surface has a major effect. Addition of inert gas cuts down diffusion to the surface, decreasing the rate of surface termination. As the pressure increases diffusion to the surface decreases, and gas phase termination becomes increasingly important until it is predominant at high pressures.

Gas phase terminations generally have very low, zero or negative activation energies, and their rate constants are governed by their pre-exponential factors. In contrast, surface terminations have a much wider range of activation energies.

Standard mechanisms for chain reactions generally miss out the surface termination steps, but these should be included. Such terminations are written as first order in radical since diffusion to the surface or adsorption on the surface are rate determining, rather than the second order bimolecular step of recombination of the two radicals adsorbed on the surface. A complete mechanism will also include the need for a third body in any unimolecular initiation or propagation steps, and in any gas phase termination steps.

6.13.1 A general mechanism based on the Rice–Herzfeld mechanism used previously

$$\begin{array}{ll}
\mathrm{X} + (\mathrm{M}) \xrightarrow{k_1} \mathrm{R}^\bullet + \mathrm{R}_1^\bullet + (\mathrm{M}) & \text{initiation} \\
\mathrm{R}_1^\bullet + \mathrm{X} \xrightarrow{k_2} \mathrm{R}_2^\bullet + \mathrm{P}_1 & \text{propagation} \\
\mathrm{R}_2^\bullet + (\mathrm{M}) \xrightarrow{k_3} \mathrm{R}_1^\bullet + \mathrm{P}_2 + (\mathrm{M}) & \text{propagation} \\
\mathrm{R}_1^\bullet + \mathrm{R}_1^\bullet + (\mathrm{M}) \xrightarrow{k_4} \text{inert} + (\mathrm{M}) & \text{gas phase termination} \\
\mathrm{R}_2^\bullet + \mathrm{R}_2^\bullet + (\mathrm{M}) \xrightarrow{k_5} \text{inert} + (\mathrm{M}) & \text{gas phase termination} \\
\mathrm{R}_1^\bullet + \mathrm{R}_2^\bullet + (\mathrm{M}) \xrightarrow{k_6} \text{inert} + (\mathrm{M}) & \text{gas phase termination} \\
\mathrm{R}_1^\bullet + \text{surface} \xrightarrow{k_7} \text{inert} + \text{surface} & \text{surface termination} \\
\mathrm{R}_2^\bullet + \text{surface} \xrightarrow{k_8} \text{inert} + \text{surface} & \text{surface termination}
\end{array}$$

Steps 7 and 8 are rate determining for surface termination, being diffusion of each type of radical to the surface or adsorption of each radical on to the surface, whichever is the slower process. The recombinations of adsorbed $R_1^\bullet$ and $R_2^\bullet$ by like–like and like–unlike radical recombinations are the fast steps in the surface termination process. Consequently surface termination consists of the two steps, 7 and 8, in contrast to the gas phase termination, which consists of the three steps, 4, 5 and 6.

Steady state procedures will be carried out under stylized schemes: ignoring the need for third bodies, assuming one recombination step to be dominant, and taking gas and surface terminations separately. This simplified analysis is sufficient to demonstrate that the order can change if gas or surface phase termination is dominant, or if both are significant. In reality, actual reactions are more complex.

Worked Problem 6.14

Using the above mechanism and a steady state treatment, but

- ignoring the effect of third bodies in initiation, propagation and termination and
- assuming $R_1^\bullet + R_1^\bullet$ is the dominant termination,

find the rate expressions for each of (a), (b) and (c) and compare them:

(a) taking gas phase termination to be dominant,

(b) taking surface termination to be dominant,

(c) taking both to be significant.

(a) *Gas phase termination dominant.*

$$\frac{d[R_1^\bullet]}{dt} = k_1[X] - k_2[R_1^\bullet][X] + k_3[R_2^\bullet] - 2k_4[R_1^\bullet]^2 = 0 \qquad (6.184)$$

$$\frac{d[R_2^\bullet]}{dt} = k_2[R_1^\bullet][X] - k_3[R_2^\bullet] = 0 \qquad (6.185)$$

Adding the two steady state equations gives

$$k_1[X] = 2k_4[R_1^\bullet]^2 \qquad (6.186)$$

$$[R_1^\bullet] = \left(\frac{k_1}{2k_4}\right)^{1/2}[X]^{1/2} \qquad (6.187)$$

$$\text{rate of reaction} = k_2[R_1^\bullet][X] \qquad (6.188)$$

$$= k_2\left(\frac{k_1}{2k_4}\right)^{1/2}[X]^{3/2} \qquad (6.189)$$

with

$$k_{obs} = k_2 \left(\frac{k_1}{2\,k_4} \right)^{1/2} \tag{6.190}$$

Reaction under these conditions is 3/2 order as is expected for a reaction where the dominant termination is recombination of like radicals produced in initiation.

(b) *Surface termination dominant.*

$$\frac{d[R_1^\bullet]}{dt} = k_1[X] - k_2[R_1^\bullet][X] + k_3[R_2^\bullet] - k_7[R_1^\bullet] = 0 \tag{6.191}$$

$$\frac{d[R_2^\bullet]}{dt} = k_2[R_1^\bullet][X] - k_3[R_2^\bullet] = 0 \tag{6.192}$$

Adding the two steady state equations gives

$$k_1[X] = k_7[R_1^\bullet] \tag{6.193}$$

$$[R_1^\bullet] = \left(\frac{k_1}{k_7} \right)[X] \tag{6.194}$$

$$\text{rate of reaction} = k_2[R_1^\bullet][X] \tag{6.188}$$

$$= k_2 \left(\frac{k_1}{k_7} \right)[X]^2 \tag{6.195}$$

with

$$k_{obs} = k_2 \frac{k_1}{k_7} \tag{6.196}$$

The reaction is now second order in contrast to 3/2 order when gas phase termination is dominant.

(c) *Both surface and gas termination significant.*

$$\frac{d[R_1^\bullet]}{dt} = k_1[X] - k_2[R_1^\bullet][X] + k_3[R_2^\bullet] - 2\,k_4[R_1^\bullet]^2 - k_7[R_1^\bullet] = 0 \tag{6.197}$$

$$\frac{d[R_2^\bullet]}{dt} = k_2[R_1^\bullet][X] - k_3[R_2^\bullet] = 0 \tag{6.198}$$

Adding the two steady state equations gives

$$k_1[X] = 2\,k_4[R_1^\bullet]^2 + k_7[R_1^\bullet] \tag{6.199}$$

This algebraic steady statement reads as

$$\textit{total}\text{ rate of initiation} = \textit{total}\text{ rate of termination}$$

and this is now a quadratic equation whose solution is required to get an explicit expression for $[R_1^\bullet]$ to insert into

$$\text{rate of reaction} = k_2[R_1^\bullet][X] \tag{6.188}$$

This will result in a complex expression for the rate of reaction. However, since the limits have been shown to be

gas phase termination dominant: 3/2 order,

surface termination dominant: second order

qualitatively it can be said that, when both contribute, the order will lie somewhere between 3/2 and 2.

The conclusion drawn from Worked Problem 6.14 is that changing the type of termination step from gas to surface alters the kinetics. This is because the order with respect to the radical differs between the second order recombination of the gas phase termination and surface termination where diffusion to the surface or adsorption on the surface is rate determining and first order. If, however, the rate-determining step in surface termination were bimolecular recombination on the surface, the order would not change between gas and surface termination. This is because both recombinations would now have the same order, i.e. $2k_4[R_1^\bullet]^2$ and $2k_7[R_1^\bullet]^2$, with the total rate of termination if both contributed being $2(k_1 + k_7)[R_1^\bullet]^2$.

Analysis for the recombination of

$$R_2^\bullet + R_2^\bullet$$
$$R_1^\bullet + R_2^\bullet$$

gives similar conclusions.

A more complex analysis would be needed if third bodies were included. Because unimolecular steps can show a pressure dependence, and because termination between atoms and small radicals needs a third body, often leading to a change in overall order with pressure, it is essential to establish first whether surface effects are significant. This is easily determined by altering the size, shape and nature of the vessel.

6.14 Explosions

Explosions occur when the rate of reaction increases dramatically over a short period of time, and the reactant is removed at an ever-increasing rate. The explosion is over when all the reactant is consumed. There are several types of explosion.

6.14.1 Autocatalysis and autocatalytic explosions

For most straight chains, termination balances initiation, and the radical concentrations remain in the steady state. However, there are some straight chains where the termination becomes negligible and the radical concentration can build up to large values well past the steady state region. The reaction becomes autocatalytic as the reaction rate increases dramatically with the build-up in radical concentration, and if the increase is very rapid an explosion can occur.

6.14.2 Thermal explosions

Under normal conditions, reactions are carried out at constant temperature where there is thermal equilibrium between the reaction vessel and thermostat. However, if the reaction is highly exothermic, the heat liberated cannot be dissipated quickly to the surroundings, the reaction will self-heat, leading to a rapid increase in rate which, if great enough, can lead to an explosion.

6.14.3 Branched chain explosions

In straight chain propagation one radical is destroyed while another is produced, so that the net change in the number of radicals is zero.

In contrast, in a branched chain, as one radical is destroyed more than one radical is produced, so that there is a net increase in the number of radicals as a result of propagation.

$$R_1^\bullet + X \rightarrow P + R_2^\bullet \quad \text{straight chain}$$

$$R_1^\bullet + X \rightarrow P + \alpha R_2^\bullet \quad \text{branched chain}$$

For branching to occur $\alpha > 1$, and can often be 2 or 3, though rarely greater. There are occasions (see Section 6.15 on degenerate branching) where α is just slightly greater than unity, and under these circumstances a different situation will occur.

Straight and branched chain propagation can occur simultaneously, or one can dominate. If a steady state concentration of radicals is set up and maintained throughout, reaction will occur at a finite rate, and termination will be able to cope with production of radicals from initiation, and straight and branched chain propagation. However, when branching occurs it allows the possibility of a continuing build-up of radicals if termination cannot cope with the branching, and if the build-up is sufficiently extensive explosions can occur.

Worked Problem 6.15

Question. In a branched chain reaction with $\alpha = 2$, there is a consequent build-up of radicals with each cycle of branched chain propagation:

$$R^\bullet + X \rightarrow 2R^\bullet$$

(a) Show that, if none of the radicals is destroyed, the number of radicals produced $= \alpha^n$ where n is the number of cycles.

(b) Construct a table showing that there has been a dramatic increase in the number of radicals after 30 cycles. Choose values of $n = 1, 2, 3, 4, 5, 10, 20$ and 30.

(c) If the rate constant for branching is $6 \times 10^6\ \text{mol}^{-1}\ \text{dm}^3\ \text{s}^{-1}$, and [reactant] $= 1 \times 10^{-2}\ \text{mol dm}^{-3}$, calculate the time taken for 30 cycles to occur.

Answer.

(a) Since $\alpha = 2$, then one radical produces two radicals, these two radicals will produce a total of four radicals, which in turn produce a total of eight, and these a total of 16. The number of radicals produced in each cycle is 2^n, or, in general, α^n.

(b) This gives the following table.

Number of cycles	Number of radicals
1	2
2	4
3	8
4	16
5	32
10	1024
20	1.05×10^6
30	1.07×10^9

(c) At $t = 0$, $\quad [\text{R}^\bullet] = c_0$

At time t, when 30 cycles have occurred, from the table

$$[\text{R}^\bullet] = 1.07 \times 10^9\ c_0$$

$$\text{rate of branching} = k[\text{R}^\bullet][\text{X}] \tag{6.200}$$

where k is a second order rate constant.

At the early stages of reaction, X will be initially in excess, and the branching reaction can be approximated to a pseudo-first order reaction with

$$\text{rate of branching} = k'[\text{R}^\bullet] \tag{6.201}$$

where k' is a pseudo-first order rate constant, with

$$k' = k[\text{X}] = 6 \times 10^6\ \text{mol}^{-1}\ \text{dm}^3\ \text{s}^{-1} \times 1 \times 10^{-2}\ \text{mol dm}^{-3} = 6 \times 10^4\ \text{s}^{-1}$$

$$\therefore \quad \log_e c_t = \log_e c_0 + k't \tag{6.202}$$

where the sign of $k't$ is positive since there is a *build-up* of radicals.

$$\therefore \quad \log_e (1.07 \times 10^9\, c_0) = \log_e c_0 + 6 \times 10^4\, t$$
$$6 \times 10^4\, t = \log_e 1.07 \times 10^9$$
$$\therefore \quad t = 3.5 \times 10^{-4}\ \text{s}$$

One of the remarkable features of these branched chain explosions is the astounding speed at which the system moves from a zero rate, through the steady state, through the build-up and thence to explosion. This induction period is short and can have a duration ranging down from seconds to milliseconds, or even less.

Note.

- In the steady state

$$[R^\bullet] = [R^\bullet]_{\text{steady state}} \tag{6.203}$$

and the radical concentration is *both small and constant*, and the steady state rate

$$-\frac{d[X]}{dt} = k_p[X][R^\bullet] \tag{6.204}$$

with the rate remaining finite throughout.

- In the explosive situation

$$[R^\bullet] \gg [R^\bullet]_{\text{steady state}} \tag{6.205}$$

and the radical concentration *builds up very, very rapidly*, so that in the explosive state

$$[R^\bullet] \to \infty$$

and the explosive rate

$$-\frac{d[X]}{dt} \to \infty$$

where the rate very rapidly builds up to an explosion.

6.14.4 A highly schematic and simplified mechanism for a branched chain reaction

This mechanism includes all the features necessary to describe a branched chain reaction. For simplicity, the initiation and second propagation steps are taken to be in their first order region, and the dominant termination is the recombination of like radicals.

$$\mathrm{X} \xrightarrow{k_1} \mathrm{R}^\bullet + \mathrm{R}_1^\bullet \qquad \text{initiation}$$

$$\mathrm{R}_1^\bullet + \mathrm{X} \xrightarrow{k_2} \mathrm{R}_2^\bullet + \mathrm{P}_1 \qquad \text{straight chain propagation}$$

$$\mathrm{R}_2^\bullet \xrightarrow{k_3} \mathrm{R}_1^\bullet + \mathrm{P}_2 \qquad \text{straight chain propagation}$$

$$\mathrm{R}_1^\bullet + \mathrm{Y} \xrightarrow{k_b} \mathrm{P}_3 + \alpha\,\mathrm{R}_2^\bullet \qquad \text{branched chain propagation}$$

$$\mathrm{R}_1^\bullet + \mathrm{R}_1^\bullet \xrightarrow{k_{tg}} \text{inert} \qquad \text{gas phase termination}$$

$$\mathrm{R}_1^\bullet + \text{surface} \xrightarrow{k_{ts}} \text{inert} \qquad \text{surface termination}$$

Here the value of α is unspecified, except that $\alpha > 1$, otherwise reaction would be simply straight chain.

6.14.5 Kinetic criteria for non-explosive and explosive reaction

- The *non-explosive* region where reaction is in the steady state:

$$\frac{\mathrm{d}[\mathrm{R}_1^\bullet]}{\mathrm{d}t} = k_1[\mathrm{X}] - k_2[\mathrm{R}_1^\bullet][\mathrm{X}] + k_3[\mathrm{R}_2^\bullet] - k_b[\mathrm{R}_1^\bullet][\mathrm{Y}] - 2\,k_{tg}[\mathrm{R}_1^\bullet]^2 - k_{ts}[\mathrm{R}_1^\bullet] = 0 \tag{6.206}$$

$$\frac{\mathrm{d}[\mathrm{R}_2^\bullet]}{\mathrm{d}t} = k_2[\mathrm{R}_1^\bullet][\mathrm{X}] - k_3[\mathrm{R}_2^\bullet] + \alpha\,k_b[\mathrm{R}_1^\bullet][\mathrm{Y}] = 0 \tag{6.207}$$

Adding gives

$$k_1[\mathrm{X}] + (\alpha - 1)k_b[\mathrm{R}_1^\bullet][\mathrm{Y}] = 2\,k_{tg}[\mathrm{R}_1^\bullet]^2 + k_{ts}[\mathrm{R}_1^\bullet] \tag{6.208}$$

which is the algebraic statement describing steady state conditions where

total rate of production of radicals = *total* rate of removal of radicals

and termination can cope with branching.

The total concentration of radicals present is

$$[\mathrm{R}^\bullet] = [\mathrm{R}_1^\bullet] + [\mathrm{R}_2^\bullet] \tag{6.209}$$

and under steady state conditions the change in concentration of radicals with time is zero,

$$\frac{\mathrm{d}[\mathrm{R}^\bullet]_{\text{steady state}}}{\mathrm{d}t} = \frac{\mathrm{d}([\mathrm{R}_1^\bullet] + [\mathrm{R}_2^\bullet])_{\text{steady state}}}{\mathrm{d}t} = 0 \tag{6.210}$$

The rate of reaction is

$$-\frac{d[X]}{dt} = k_1[X] + k_2[R_1^\bullet][X] \approx k_2[R_1^\bullet][X] \quad (6.211)$$

where $[R_1^\bullet]$ can be found from the solution of the quadratic equation (6.208).

- *The explosive reaction*. Since $\alpha > 1$, then it becomes possible for the total rate of production of radicals to exceed the total rate of removal, and a build-up of radicals to occur. Under these conditions termination can no longer cope with branching. Since the build-up is very rapid, the overall rate of reaction increases dramatically and explosion will occur. $d[R^\bullet]/dt$ will be large and positive if

$$k_1[X] + (\alpha - 1)k_b[R_1^\bullet][Y] \gg 2\,k_{tg}[R_1^\bullet]^2 + k_{ts}[R_1^\bullet] \quad (6.212)$$

This is possible if k_b or $(\alpha - 1)k_b[Y]$ are large; see Figure 6.1.

The change-over between the steady state and the explosion occurs when

$2\,k_{tg}[R_1^\bullet]^2 + k_{ts}[R_1^\bullet]$ *just becomes less* than $k_1[X] + (\alpha - 1)k_b[R_1^\bullet][Y]$.

There is obviously a range of conditions between the steady state, Equation (6.210), the point beyond which build-up from the steady state can start, and the start of the

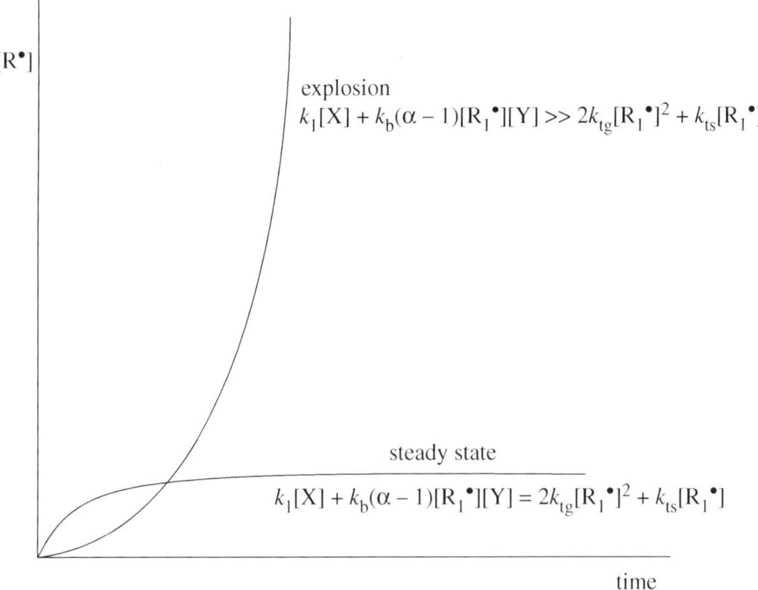

Figure 6.1 Diagram to show conditions leading to explosion or to steady state

explosive stage, Equation (6.212). This generally shows up as a very sharp change in the character of the reaction, though in fact, it is a *progressive* though very, very fast change. As with the build-up to the steady state, build-up from the steady state to the explosive state can only be studied using very fast techniques.

6.14.6 A typical branched chain reaction showing explosion limits

The reaction between hydrogen and oxygen is a branched chain reaction which shows explosive regions. It has been very extensively studied, and appears to have a very complex mechanism.

All mixtures explode above 600 °C, but in the range 450–600 °C there are certain pressures and temperatures at which reaction is explosive, while at others reaction occurs at a slow rate. These are related to the relative importance of gas phase and surface termination just balancing branching.

A simplified mechanism illustrating the essential features of the H_2/O_2 reaction

$$2\,H_2(g) + O_2(g) \rightarrow 2\,H_2O(g)$$

The main radicals present are $H^\bullet$, $OH^\bullet$, $O^\bullet$ and $HO_2^\bullet$, and the basic mechanism proposed is

$H_2 + O_2 \rightarrow 2\,OH^\bullet$ initiation

$OH^\bullet + H_2 \rightarrow H_2O + H^\bullet$ straight propagation

$H^\bullet + O_2 \rightarrow OH^\bullet + O^\bullet$ branching propagation

$O^\bullet + H_2 \rightarrow OH^\bullet + O^\bullet$ branching propagation

$M + H^\bullet + O_2 \rightarrow HO_2^\bullet + M$ gas phase termination since $HO_2^\bullet$ is relatively stable, even though it is a radical

$HO_2^\bullet + \text{surface} \rightarrow \frac{1}{2}H_2O + \frac{3}{4}O_2$ surface termination

$H^\bullet + \text{surface} \rightarrow \text{inert species}$

$O^\bullet + \text{surface} \rightarrow \text{inert species}$

$OH^\bullet + \text{surface} \rightarrow \text{inert species}$

Under certain conditions there is competition between

$$HO_2^\bullet + \text{surface} \rightarrow \frac{1}{2}H_2O + \frac{3}{4}O_2 \qquad \text{surface termination}$$

and the reactions

$$HO_2^\bullet + H_2 \rightarrow H_2O_2 + H^\bullet \quad \text{straight chain propagation}$$

$$HO_2^\bullet + H_2 \rightarrow H_2O + OH^\bullet \quad \text{straight chain propagation}$$

and with the consequent initiation reaction

$$H_2O_2 \rightarrow 2\,OH^\bullet \quad \text{initiation}$$

as well as many other radical reactions.

The pressure limits for the H_2/O_2 are discussed in the following section.

6.14.7 The dependence of rate on pressure and temperature

The pressure limits for the H_2/O_2 reaction are given in Figure 6.2. The three explosion limits define the transition between steady state regions of reaction and explosive

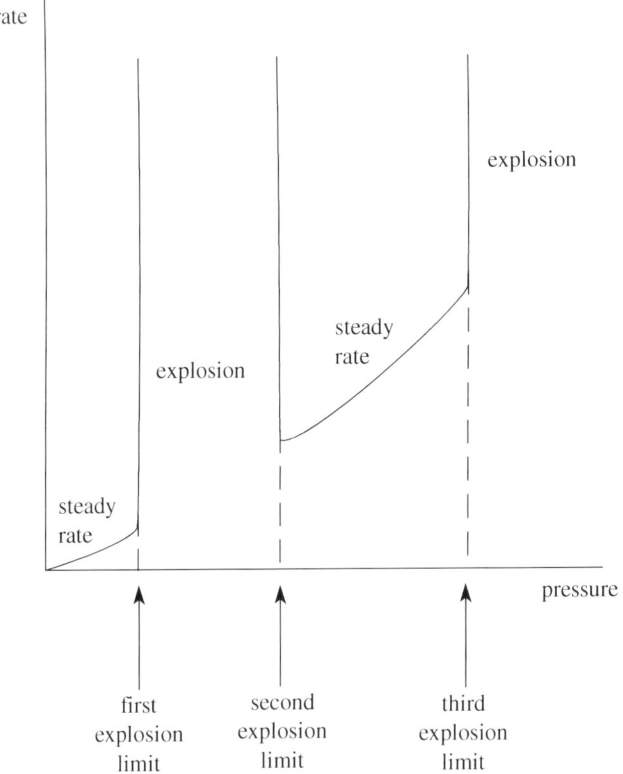

Figure 6.2 Dependence of rate on pressure for a reaction showing explosion limits

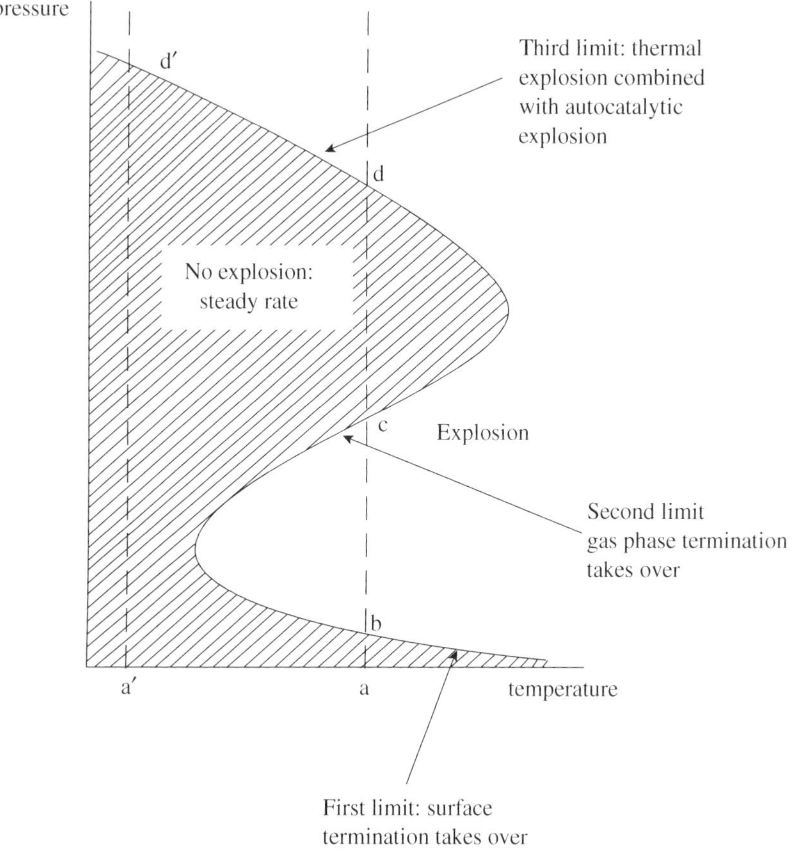

Figure 6.3 Temperature/pressure diagram for a reaction showing explosion limits

regions. Figure 6.3 shows the same information, but also includes the effect of temperature on the nature of this reaction. This diagram illustrates the regions of temperature and pressure in which the reaction is non-explosive and explosive. There are three explosion limits given. Figure 6.3 can be discussed in terms of what happens as the pressure changes along the lines a, b, c, d, and a′, d′.

- *The lower, or first explosion limit occurring at low pressures.* Below the pressure limit, i.e. along a, b, the reaction is in the steady state where branching is balanced by surface termination. As the pressure increases, termination at the surface becomes less effective because diffusion to the surface is progressively inhibited, but steady reaction is still maintained until the lower limit is *just* reached. At this pressure, surface termination *just* fails to balance branching, and the concentration of radicals can then build up very rapidly to a very, very large value from the steady state concentration. The reaction rate increases dramatically and explosion can occur. It does so with the appearance of a very sharp boundary, which in actuality is a progressive, albeit very rapid, change, from non-explosive to explosive conditions. Since the dominant termination is at the surface, the lower limit is altered by changing the size, shape and nature of the vessel. Addition of inert gas decreases diffusion of the radicals to the surface, decreasing the rate of termination and thereby shifting the lower limit. This can cause reaction in the steady state to become explosive.

As the pressure is increased above the lower limit, explosive conditions can be maintained since branching is the dominant reaction between the two limits. As the pressure is increased, termination in the gas phase becomes more and more effective, and surface termination less important.

- *The upper, or second, explosion limit occurs at higher pressures.* As the upper limit is approached from lower pressures, i.e. along b, c, gas phase termination now *just fails* to balance, or *just* balances branching, the radical concentrations can be reduced back to their steady state values and the reaction becomes non-explosive. Again the boundary is crossed in a very short time and has the appearance of a sharp boundary. Above the second limit, i.e. along c, d, gas phase termination balances branching and is dominant. Altering the size, shape and nature of the vessel has no great effect on the rate of termination, and this limit is independent of these factors. However, addition of inert gas will increase the chance of gas phase termination and will cause the limit to move to lower pressures.

- *The third pressure limit, when it exists, occurs at even higher pressures.* At the third limit, i.e. at pressures greater than at d, the reaction moves back into an explosive region which is in part thermal, and in part the result of some dramatic change in radical concentration. This results from another set of reactions coming into play so that gas phase termination can no longer cope with this new production of radicals.

- Along the line a', d', the reaction is always in the steady state with some termination reaction balancing production of radicals. Explosion cannot occur. At pressures greater than d' explosion always occurs.

6.15 Degenerate Branching or Cool Flames

Some reactions, especially oxidations of hydrocarbons, are characterized by regions of temperature and pressure in which 'mild' explosions occur. These 'mild' explosions are accompanied by 'cool' flames, which manifest themselves by the appearance of luminescence and by sudden changes in temperature and pressure, during which there is an audible 'click' from the reaction mixture. Figure 6.4 shows the increase in pressure found during a series of 'mild' explosions.

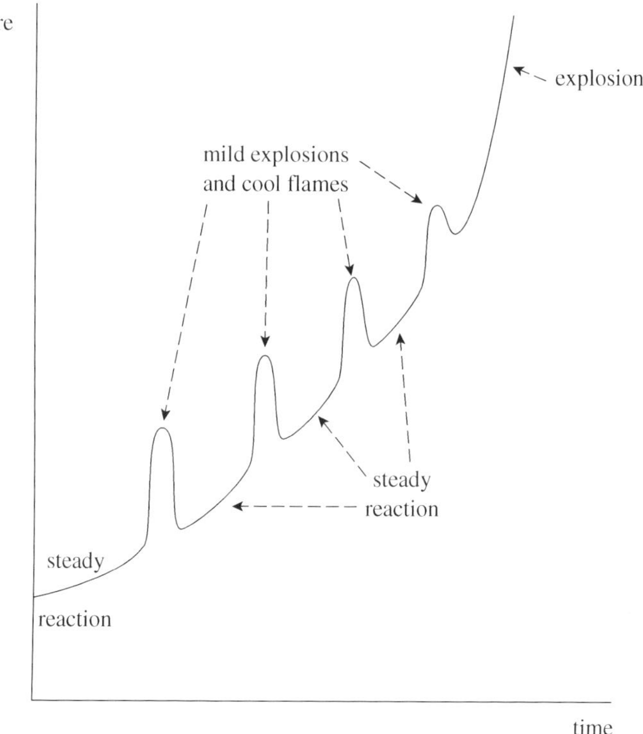

Figure 6.4 Diagram illustrating the occurrence of cool flames

The flames are usually blue in colour with peroxides and aldehydes (alkanals) present, and with the aldehydes showing a peak in concentration during the 'cool' flame episode. The main emitter in the 'cool' flame is electronically excited HCHO.

An observable induction period is often found, during which only traces of aldehydes are present. Addition of aldehydes can reduce or eliminate this induction period.

During a 'cool' flame episode the rate increases to a maximum and then falls off, often to be followed by another 'mild' explosion and 'cool' flame episode (Figure 6.4). During these episodes the rate is very high, but not so high as to be described as an explosion such as those already discussed. Also, during a 'cool' flame episode the temperature increase is between 10 and 100 K, in contrast to the genuine explosion, where it can be around 2000 K. During the period of increasing temperature, the reaction rate shows a negative temperature dependence, in contrast to the normal positive dependence, where the rate increases with increase in temperature.

If a 'cool' flame 'mild' explosion episode is followed by a second, then not all the reactant has been consumed in the first episode. What remains will be used up in the second, third and so forth episodes. This is in contrast to a genuine explosion, where all the reactant is rapidly removed.

Sometimes, a 'cool' flame may be followed by a genuine explosion. Here the 'cool' flame behaviour is lost after the first increase in rate, which does not fall off, but moves straight into the genuine explosion, believed to be thermal in nature.

A series of 'mild' explosions is more general at low pressures and low temperatures, in contrast to the high temperatures and pressures at which a 'mild' explosion leads into a genuine explosion. Typical temperature limits for hydrocarbons are approximately 280–410 °C, with pressure limits around 100–150 mmHg.

'Cool' flame behaviour is the result of highly limited degenerate branching, where the branching coefficient, α, is small, e.g. is of the order of 1.1, compared with being of the order of two or three in normal branching.

Most of the reaction occurs via straight chain propagation with only a small amount going via the branching propagation, in contrast to the earlier type of explosion, where most of the reaction goes via branching propagation.

Since branching is infrequent compared to straight chain propagation, there is only a very slow build-up of radicals beyond the steady state concentration. Because branching is so infrequent the build-up to large radical concentrations takes a long time, of the order of minutes, in contrast to the very rapid build-up into explosion, such as 10^{-3} s, for the normal branching reaction. During this period, termination is not able to cope with the infrequent branching of the 'cool' flame reaction. Each 'cool' flame often lasts for a time of the order of seconds.

Hydrocarbon combustions result in many varied products, and these depend on the temperature, in particular when the temperature is above or below the high temperature limit for 'cool' flame behaviour. At low temperatures, alcohols, aldehydes (alkanals) and ketones (alkanones) predominate, whereas at high temperatures the products are mainly saturated and unsaturated hydrocarbons of lower molecular weight than the reactant, along with hydrogen peroxide. The organic peroxides increase in concentration prior to the 'cool' flame episode and decrease as the 'cool' flame passes, though at this stage the hydrogen peroxide increases. Aldehydes show a peak in concentration during the 'cool' flame.

The effect of temperature is unusual as 'cool' flame combustion reactions show a region of temperatures and pressures in which the rate shows anomalous behaviour: the rate decreases as the temperature increases – the negative temperature coefficient effect.

Hydrocarbon combustions are highly complex with very many reactions participating. Nonetheless, a simplified mechanism can be written containing all the essential features to explain 'cool' flame behaviour.

6.15.1 A schematic mechanism for hydrocarbon combustion

$RH + O_2 \xrightarrow{k_1} R^\bullet + HO_2^\bullet$ initiation

$R^\bullet + O_2 \xrightarrow{k_2} RO_2^\bullet$ straight chain propagation

$RO_2^\bullet \xrightarrow{k_{-2}} R^\bullet + O_2$ straight chain propagation (reverse)

$$RO_2^\bullet + RH \xrightarrow{k_3} ROOH + R^\bullet \quad \text{straight chain propagation}$$

$$ROOH \xrightarrow{k_4} RO^\bullet + OH^\bullet \quad \text{degenerate branching}$$

$$RO^\bullet + RH \xrightarrow{k_5} ROH + R^\bullet \quad \text{propagation}$$

$$OH^\bullet + RH \xrightarrow{k_6} H_2O + R^\bullet \quad \text{propagation}$$

$$R^\bullet + R^\bullet \xrightarrow{k_{tg}} \text{inert} \quad \text{gas phase termination}$$

$$R^\bullet + \text{surface} \xrightarrow{k_{ts}} \quad \text{surface termination}$$

There are other very important reactions which must be considered. With increasing temperature there is competition between branching resulting from production of ROOH from $RO_2^\bullet$, step 3, and production of alkenes via $R^\bullet$:

$$R^\bullet + O_2 \xrightarrow{k_2} RO_2^\bullet \rightarrow \text{branching via ROOH}$$

$$R^\bullet + O_2 \rightarrow \text{alkene} + HO_2^\bullet$$

At low temperatures $HO_2^\bullet$ is relatively inert and can be ignored, but at high temperatures it forms hydrogen peroxide by H abstraction from reactant.

$$HO_2^\bullet + RH \rightarrow H_2O_2 + R^\bullet$$

At even higher temperatures H_2O_2 can decompose to produce a new branching sequence:

$$H_2O_2 \rightarrow 2\,OH^\bullet$$

$$RH + OH^\bullet \rightarrow R^\bullet + H_2O$$

which has a net overall production of radicals.

Since a wide variety of products are formed in combustion reactions, many more reactions should be included in the sequence. However, the nine reactions quoted in the simple mechanism are all that is needed to explain 'cool' flame behaviour. This is discussed in the following worked problem.

Worked Problem 6.16

Question.

(a) In the simplified mechanism, the first two steps of the straight chain propagation make up a reversible reaction. Write down the possible fates of the $RO_2^\bullet$ radical, and state what are the overall effects of these fates.

(b) Figure 6.5 gives the essential features of the energetics of this reaction. Work out the effect of temperature on the rates of the forward and back steps, and on the position of equilibrium.

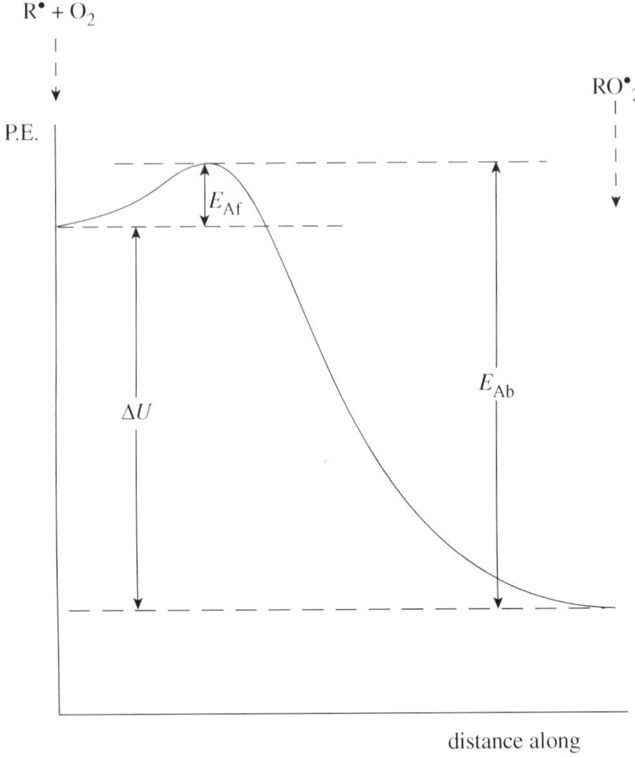

Figure 6.5 PE profile for the reaction $R^{\bullet} + O_2 \rightarrow RO_2^{\bullet}$

(c) ROOH is a relatively stable species. What does this imply about the likelihood of straight chain and branching chain propagation?

(d) What can be said about the relative importance of formation of radicals and their removal in the steady state and the 'cool' flame regions?

Answer.

(a) $RO_2^{\bullet}$ can either revert to reactants $R^{\bullet}$ and O_2, or it can react with RH to form the relatively stable ROOH as the second step of the straight chain sequence. ROOH can also set up the three-step sequence of reactions 4, 5 and 6, which make up the degenerate branching reaction.

(b) The forward reaction forming the peroxy radical, $RO_2^{\bullet}$, has a very low activation energy and so is only slightly dependent on temperature. It is also highly exothermic. The back reaction, in which a bond is broken, has a high activation energy and so is strongly temperature dependent. It is also highly endothermic. In consequence, the reversible reaction lies well to the right at low temperatures, favouring $RO_2^{\bullet}$ and its subsequent reaction. On the other hand, high temperatures will favour the back reaction leading to the breakdown of the $RO_2^{\bullet}$ back to $R^{\bullet}$ and O_2, decreasing the chances of ROOH being formed. Hence alternate increases and decreases of temperature can cause this equilibrium to oscillate.

(c) Since ROOH is relatively inert, very many cycles of the straight chain propagation sequence can occur before ROOH forms $RO^{\bullet}$ and $OH^{\bullet}$, the net effect of which is that two radicals produce four. However, this branching is relatively infrequent compared with straight chain propagation, which dominates, and this gives an overall α of close to unity. This places these combustion reactions in the class of degenerate branching in contrast to explosive reactions such as H_2/O_2 where most of the reaction goes via branching.

(d) In the steady state

total rate of production of radicals via initiation and branching propagation = *total rate of removal* of radicals via gas phase and surface termination

In the 'cool' flame episode the total rate of termination can no longer cope with the branching, the production of radicals increases and reaction moves into a 'cool' flame episode.

6.15.2 Chemical interpretation of 'cool' flame behaviour

When there are several episodes of 'cool' flames the reaction is oscillatory. The crucial feature is the reversible reaction

$$R^{\bullet} + O_2 \underset{k_{-2}}{\overset{k_2}{\rightleftharpoons}} RO_2^{\bullet}$$

At low temperatures, reaction lies well to the right favouring $RO_2^{\bullet}$, increasing the chance of ROOH being formed, leading to the branching reaction which increases the rate of oxidation. These subsequent reactions, 4, 5 and 6, are highly exothermic and lead to an increase in temperature due to self-heating.

As the temperature rises, the reversible reaction moves increasingly to the left favouring $R^{\bullet}$ over $RO_2^{\bullet}$. This increases the chance of termination rather than branching, leading to a decrease in the overall rate, and a decrease in the heat released by the reaction. The exothermic reactions 5 and 6 become less likely and self-heating decreases. This whole process occurs in a matter of seconds. The aldehydes, and particularly the emitting HCHO, build up during the period of increasing temperature and fade away during the period of decreasing temperature. Hence luminescence builds up and dies away during a 'cool' flame episode. If not all the reactant is used up the cycle can restart once the temperature has decreased sufficiently for the reversible reaction to move to the right, favouring $RO_2^\bullet$ and greater production of ROOH. When the rate of branching again becomes too high for termination to cope the system will move into a second 'cool' flame episode.

Worked Problem 6.17

Question. Using Figure 6.6, describing 'cool' flame behaviour, discuss what will happen to a mixture of given initial composition of RH and O_2 in the temperature regions defined by the lines a, b, c, d and m, n, p, q, r, s.

Answer.

- *Different temperature conditions along a, b, c, d at constant pressure and composition.* At temperatures between a and b reaction occurs non-explosively and is in the steady state, proceeding basically by straight chain propagation. The rate increases with increasing temperature in this region.

  For mixtures between b and c luminescence would appear in the vessel, the temperature and pressure would show a sudden jump and there would be an audible click. This is the 'cool' flame region, where degenerate branching becomes dominant. If not all the reactant is used up in the first 'cool' flame

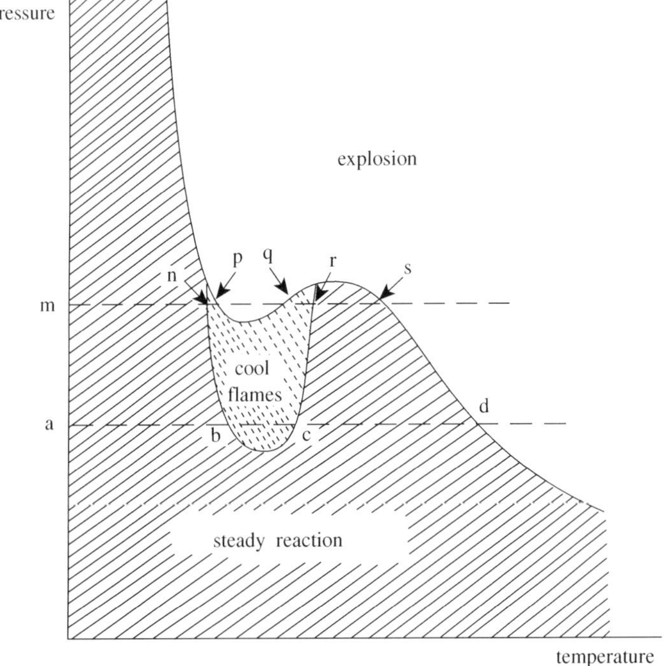

Figure 6.6 Temperature/pressure diagram for a reaction showing cool flames

episode there will be a succession of 'cool' flames. When the overall rate is studied, the rate increases with temperature as conditions are altered along b, c, but there comes a stage at which it *must start to decrease* with increase in temperature to enable the rate to decrease to the steady state rate which occurs at temperatures greater than c. This is the anomalous negative temperature coefficient region. For temperatures between c and d reaction is back to the steady state. At temperatures somewhat beyond c the rate increases with temperature. At temperatures greater than d reaction moves into a genuine explosion, thought to be partly thermal and partly due to other reactions becoming important; see Section 6.15.1.

- *Different temperatures along m, n, p, q, r, s for constant pressure and composition.* Between m and n reaction is in the steady state similar to along a, b. At temperatures between n and p the 'cool' flame region is reached with its accompanying characteristic features. Here the rate increases with increasing initial temperatures since, at temperatures greater than p but below q, explosion can occur. However, because the pressure is higher than along a, b, c, d, the sudden increase in pressure due to a 'cool' flame is sufficient to take the reaction into explosion. Degenerate branching is important in the 'cool' flame region, but thermal explosion can easily take over.

When the initial temperature lies between p and q all mixtures will explode.

For initial temperatures between q and r a 'cool' flame region exists, but, in contrast to region n, p, the overall rate must decrease with increasing temperatures since at r reaction enters the steady state region. Between q and r reaction has a negative temperature coefficient.

For initial temperatures between r and s reaction is again in the steady state with straight chain propagation dominant, but for temperatures greater than s explosion occurs.

Further reading

Laidler K.J., *Chemical Kinetics*, 3rd edn, Harper and Row, New York, 1987.

Laidler K.J. and Meiser J.H., *Physical Chemistry*, 3rd edn, Houghton Mifflin, New York, 1999.

Alberty R.A. and Silbey R.J., *Physical Chemistry*, 2nd edn, Wiley, New York, 1996.

Logan S.R., *Fundamentals of Chemical Kinetics*, Addison Wesley, Reading, MA, 1996.

*Nicholas J., *Chemical Kinetics*, Harper and Row, New York, 1976.

*Wilkinson F., *Chemical Kinetics and Reaction Mechanisms*, Van Nostrand Reinhold, New York, 1980.

Robson Wright M., *Fundamental Chemical Kinetics*, Horwood, Chichester, 1999.

*Out of print, but should be in university libraries.

Further problems

1. The bromination of propane C_3H_8 proceeds via a free radical chain reaction. The dominant major product is HBr, with 2-bromopropane produced in substantial amounts. 1-bromopropane, though produced in much smaller amounts, can still be classified as a major product. 2:3-dimethylbutane is a minor product, and hexane and 2-methylpentane are trace products.

   Deduce a mechanism that will account for these facts.

2. The decomposition of propanal, CH_3CH_2CHO, is a free radical chain process. The major products are C_2H_6 and CO. There are three minor products, butane, pentan-3-one and hexan-3:4-dione. Of these pentan-3-one is the most abundant. Pentanal is produced in trace amounts while H_2 is found as a very minor trace product. The reaction is found to be first order at high pressures moving to second order at low pressures. Deduce a mechanism to fit these facts.

   What methods could be used for the detection of the products and intermediates?

3. Using the steady state treatment deduce the rate expressions for the following mechanisms:

   (a)
$$C_2H_4I_2 \xrightarrow{k_1} C_2H_4I^\bullet + I^\bullet$$
$$I^\bullet + C_2H_4I_2 \xrightarrow{k_2} I_2 + C_2H_4I^\bullet$$
$$C_2H_4I^\bullet \xrightarrow{k_3} C_2H_4 + I^\bullet$$
$$I^\bullet + I^\bullet + M \xrightarrow{k_4} I_2 + M$$

   (b)
$$C_4H_9I \xrightarrow{k_1} C_4H_9^\bullet + I^\bullet$$
$$C_4H_9^\bullet + C_4H_9I \xrightarrow{k_2} C_4H_{10} + C_4H_8I^\bullet$$
$$C_4H_8I^\bullet \xrightarrow{k_3} C_4H_8 + I^\bullet$$
$$I^\bullet + C_4H_9I \xrightarrow{k_4} C_4H_9^\bullet + I_2$$
$$2I^\bullet + M \xrightarrow{k_5} I_2 + M$$

4. The pyrolysis of propanone is a chain reaction with possible mechanism

$$CH_3COCH_3 \xrightarrow{k_1} CH_3CO^\bullet + CH_3^\bullet$$
$$CH_3CO^\bullet \xrightarrow{k_2} CH_3^\bullet + CO$$
$$CH_3^\bullet + CH_3COCH_3 \xrightarrow{k_3} CH_4 + {}^\bullet CH_2COCH_3$$
$${}^\bullet CH_2COCH_3 \xrightarrow{k_4} CH_3^\bullet + CH_2CO$$
$$CH_3^\bullet + {}^\bullet CH_2COCH_3 \xrightarrow{k_5} CH_3CH_2COCH_3$$

Using this mechanism explain what is meant by the terms initiation, propagation and termination, and identify the major and minor products. Find the order of the reaction and an expression for the observed rate constant.

If step 5 is replaced by

$$CH_3^{\bullet} + CH_3^{\bullet} \xrightarrow{k_6} C_2H_6$$

predict what the order would become.

Likewise, predict the order if the major termination step became

$$^{\bullet}CH_2COCH_3 + {}^{\bullet}CH_2COCH_3 \xrightarrow{k_7} CH_3COCH_2CH_2COCH_3$$

5. The following is a possible mechanism for the decomposition of propane:

$$C_3H_8 \xrightarrow{k_1} CH_3^{\bullet} + C_2H_5^{\bullet}$$

$$CH_3^{\bullet} + C_3H_8 \xrightarrow{k_2} CH_4 + C_3H_7^{\bullet}$$

$$C_2H_5^{\bullet} + C_3H_8 \xrightarrow{k_3} C_2H_6 + C_3H_7^{\bullet}$$

$$C_3H_7^{\bullet} \xrightarrow{k_4} C_2H_4 + CH_3^{\bullet}$$

$$C_3H_7^{\bullet} + C_3H_7^{\bullet} \xrightarrow{k_5} C_3H_8 + C_3H_6$$

$$C_3H_7^{\bullet} + C_3H_7^{\bullet} \xrightarrow{k_6} C_6H_{14}$$

Deduce the rate expression for this reaction, and obtain an expression for the observed rate constant.

6. The following is a schematic mechanism for the bromination of an organic compound RH:

$$Br_2 + M \xrightarrow{k_1} 2\,Br^{\bullet} + M$$

$$Br^{\bullet} + RH \xrightarrow{k_2} R^{\bullet} + HBr$$

$$R^{\bullet} + Br_2 \xrightarrow{k_3} RBr + Br^{\bullet}$$

$$2\,Br^{\bullet} + M \xrightarrow{k_4} Br_2 + M$$

Derive the steady state rate for the production of HBr for this mechanism.

If the inhibition step

$$R^{\bullet} + HBr \xrightarrow{k_{-2}} Br^{\bullet} + RH$$

were included how would this affect the steady state rate expression?

Derive this rate expression.

7. The following is a possible simplified mechanism for the reaction of chlorine with ethene C_2H_4.

$$Cl_2 + M \xrightarrow{k_1} 2\,Cl^{\bullet} + M$$
$$Cl^{\bullet} + C_2H_4 \xrightarrow{k_2} C_2H_4Cl^{\bullet}$$
$$C_2H_4Cl^{\bullet} + Cl_2 \xrightarrow{k_3} C_2H_4Cl_2 + Cl^{\bullet}$$
$$Cl^{\bullet} + Cl^{\bullet} + M \xrightarrow{k_{-1}} Cl_2 + M$$
$$C_2H_4Cl^{\bullet} + C_2H_4Cl^{\bullet} \xrightarrow{k_4} C_4H_8Cl_2$$
$$Cl^{\bullet} + C_2H_4Cl^{\bullet} \xrightarrow{k_5} C_2H_4Cl_2$$

Why does M appear in steps 1 and -1 and not in steps 4 and 5?

Would a third body be expected to be required for the following termination reactions?

$$OH^{\bullet} + H^{\bullet} \rightarrow H_2O$$
$$CH_3^{\bullet} + CH_3^{\bullet} \rightarrow C_2H_6$$
$$H^{\bullet} + I^{\bullet} \rightarrow HI$$
$$CH_3^{\bullet} + (CH_3)_3C^{\bullet} \rightarrow (CH_3)_3CCH_3$$

Discuss the effect of temperature and pressure on your conclusion.

8. The following is a problem which combines knowledge about photochemical reactions and chain reactions.

In a photochemical reaction of overall stoichiometry

$$Cl_2(g) + \text{cyclo-}C_6F_{10}(g) \rightarrow C_6F_{10}Cl_2(g)$$

there was a quantum yield greater than 100. The rate was proportional to $[Cl_2]$ and to the square root of I_{abs}, but was independent of $[C_6F_{10}]$. No product other than $C_6F_{10}Cl_2$ could be detected. Devise a mechanism which fits the facts.

7 Reactions in Solution

Reaction is *fundamentally* the same in solution and in the gas phase, in so far as both involve a series of changing configurations of all species involved in the reaction step. But for solution reactions, these configurations are influenced by the solvent; and reactant/solvent and solvent/solvent interactions must be considered, with the solvent becoming an integral part of reaction and the reaction entity.

Aims

By the end of this chapter you should be able to

- know the ways in which the solvent can affect reactions in solution
- appreciate the differences between reactions in the gas phase and solution
- understand the ways in which charges or dipoles in the reactants can affect the rate and the rate constant
- adapt transition state theory from gas phase to solution
- discuss the primary salt effect and make corrections for non-ideality
- discuss the effect of the solvent on the rate constant in terms of charges of the reactants and the relative permittivity of the solvent and
- follow an outline of the effect of charge, solvent, reactant complexity and change in solvation pattern on the reaction rate.

7.1 The Solvent and its Effect on Reactions in Solution

There are three properties that are important here: the permittivity of the solvent, the viscosity of the solvent and the polarizability of the molecules of the solvent.

An Introduction to Chemical Kinetics. Margaret Robson Wright
 ISBNs: 0-470-09058-8 (hbk) 0-470-09059-6 (pbk)

The solvent is the medium, or dielectric, which modifies the electrostatic interactions between charges. It reduces the field strength due to the charges, it reduces the forces acting on the charges and it reduces the potential energy of interaction between the charges. The factor by which it reduces these quantities is the relative permittivity, ε_r. The larger the value of ε_r, the smaller will be the forces of interaction between the charges.

The viscosity, η, of a solvent restricts movement through it. The higher the viscosity, the slower will particles move around.

The polarizability, α, is a measure of the extent to which the electronic distribution over a molecule can be distorted by the electric field of charged particles, or dipolar molecules.

The solvent influences reaction in several ways.

- It can affect the mechanism;
- it can affect the magnitude of the rate constant through the relative permittivity, viscosity and polarizability and other electronic properties of the solvent;
- it can affect the species involved;
- it can affect any ion pairing, complexing, hydrogen bonding and other association phenomena.

Differences also arise.

- In contrast to gas phase reactions, reactions involving ions, polar molecules and charged transition states occur readily in solution.
- Typical molecular and free radical reactions of the gas phase also occur in solution, but are much less frequent, in contrast to the predominance of this type of reaction in the gas phase.
- In the gas phase, intermediates are often found in steady state concentrations. In solution, many of the intermediates are at equilibrium concentrations.
- Steady state concentrations in the gas phase cannot be studied *independently* of reaction, but equilibrium concentrations in the solution phase can be found readily. This can make the kinetic analysis for solution reactions easier than for the gas phase.
- The solvent often has a profound effect on the interpretation of rate constants, E_A or $\Delta H^{\neq *}$, A or $\Delta S^{\neq *}$ and $\Delta V^{\neq *}$.
- Gas phase reactions have predominantly molecular-type structures for the activated complex, whereas solution reactions often have predominantly charged or charge-separated structures.

- The effect of the solvent on the activated complex is often crucial:

  (a) the permittivity dictates the structure and the charge distribution in the activated complex;

  (b) the change in the structure of the solvent on forming the activated complex must be explicitly considered;

  (c) the shape and size of the activated complex is often to a large extent dictated by the solvent.

- In both phases, reactants must first diffuse together. In the gas phase the rate of diffusion is so fast that it has no effect on the rate of reaction. In solution, either diffusion or reaction can be rate determining. Normally reaction is the slower process, but in some reactions, such as very fast ionic reactions, diffusion can be rate determining.

- Molecules in a gas move in straight lines between collisions, but in solution Brownian motion occurs, with molecules following vastly more irregular paths. This can have an effect on some reactions.

- Theoretical calculations are less fundamental and rigorous for solution reactions. This is a consequence of the difficulty of calculating partition functions in solution. The main focus for solution reactions has been on the thermodynamic formulation of transition state theory.

7.2 Collision Theory for Reactions in Solution

Gas phase collision formulae (Section 4.1.2) can be used for many reactions in solution, i.e. for those involving uncharged species forming activated complexes with little or no charge separation. The theory of such reactions will not be considered further, and it will be assumed that the gas phase treatment is adequate.

However, modifications are required for reactions involving ions and charged or charge-separated activated complexes (Problem 4.5(ii) and Tables 7.1 to 7.3).

Experimental results show that

- reactions between like-charged ions have A factors considerably below the gas phase Z_{AB}, a consequence of repulsive forces between the ions giving a decreased rate of collision;

- reactions between unlike-charged ions have A factors considerably above Z_{AB}, a result of attractive forces between the ions increasing the collision rate;

- rate constants depend on the relative permittivity of the solvent.

Table 7.1 *A* factors for ion-ion reactions in aqueous solution

Reactants	$\log_{10} A$ (A/mol^{-1} dm^3 s^{-1})
$CH_2ClCOO^- + OH^-$	10.3
$CH_2ICOO^- + CN^-$	9.3
$ClO^- + ClO_2^-$	9.0
$CH_2BrCOO^- + S_2O_3^{2-}$	9.1
$N^+(CH_3)_3(CH_2)_2OCOCH_3 + H_3O^+$	8.6
$CH_2ICOO^- + SCN^-$	8.8
$Co(NH_3)_5Br^{2+} + Hg^{2+}$	8.1
$C_2H_5OCOCH_2CH_2SO_3^{2-} + OH^-$	6.3
$S_2O_3^{2-} + SO_3^{2-}$	6.3
$Co(NH_3)_5NO_2^{2+} + OH^-$	22.6
$Co(NH_3)_5Cl^{2+} + OH^-$	20.2
$Co(NH_3)_5NCS^{2+} + OH^-$	21.0
$Cr(H_2O)_6^{3+} + OH^-$	19.6
$Co(NH_3)_5F^{2+} + OH^-$	17.7
$NH_4^+ + OCN^-$	13.0

Table 7.2 *A* factors for ion–molecule reactions in aqueous solution

Reactants	$\log_{10} A$ (A/mol^{-1} dm^3 s^{-1})
$CH_2OHCH_2Cl + OH^-$	12.4
$CH_3I + S_2O_3^{2-}$	12.3
$CH_3I + OH^-$	12.1
$CH_2ICOOH + Cl^-$	11.9
$CH_3I + SCN^-$	11.2
$CH_2ICOOH + SCN^-$	10.6
$CH_3Br + I^-$	10.2
$CO_2 + OH^-$	10.2
$Co(NH_3)_5Cl^{2+} + H_2O$	9.3
$CH_3COOC_2H_5 + OH^-$	7.4
$CH_3COOC_4H_9 + OH^-$	6.9

These effects of charge and solvent are quantified by an exponential energy term related to the electrostatic interactions between the ions.

The *collision rate*

$$Z_{AB}^{soln} = (r_A + r_B)^2 n_A n_B \left(\frac{8\pi kT}{\mu}\right)^{1/2} \exp\left(\frac{-z_A z_B e^2}{4\pi\varepsilon_0\varepsilon_r rkT}\right) \tag{7.1}$$

Table 7.3 *A* factors for molecule–molecule reactions in a variety of solvents

Reactants	Solvent	$\log_{10} A$ (A/mol^{-1} dm^3 s^{-1})
$(C_2H_5)_3N + C_2H_5I$	C_6H_6	3.3
	$C_6H_5NO_2$	4.9
$(C_2H_5)_3N + C_6H_5CH_2Cl$	C_6H_6	3.6
	$C_6H_5NO_2$	4.1
	C_2H_5OH	7.0
$C_6H_5NH_2 + CH_3COC_6H_4Br$	C_6H_6	0.9
	$C_6H_5NO_2$	5.8
	C_2H_5OH	6.8
	$(CH_3)_2CO$	4.5
	$CHCl_3$	4.4
$(C_2H_5)_3N + C_2H_5Br$	100% $(CH_3)_2CO$	7.0
	80% $(CH_3)_2CO/C_6H_6$	7.1
	50% $(CH_3)_2CO/C_6H_6$	6.9
	20% $(CH_3)_2CO/C_6H_6$	6.7
	100% C_6H_6	4.1
$pNO_2C_6H_4CH_2Br + H_2O$	50% dioxane	6.9
	70% dioxane	5.4
	90% dioxane	3.4

The *rate constant*

$$k_{\text{soln}} = \left\{ (r_A + r_B)^2 \left(\frac{8\pi kT}{\mu} \right)^{1/2} \exp\left(\frac{-E'}{kT} \right) \right\} \exp\left(\frac{-z_A z_B e^2}{4\pi\varepsilon_0\varepsilon_r rkT} \right) \tag{7.2}$$

$$= \left\{ Z \exp\left(\frac{-E'}{kT} \right) \right\} \exp\left(\frac{-z_A z_B e^2}{4\pi\varepsilon_0\varepsilon_r rkT} \right) \tag{7.3}$$

- r is the distance between the charges, z_A and z_B, in the activated complex. However, r may be dependent on the solvent through solvation effects.
- ε_r is the relative permittivity of the solvent; ε_0 is a universal constant.
- E' is an activation energy, and Z the collision number for *molecules in the gas phase*, and neither has been modified to account for either electrostatic interactions, or the effect of the relative permittivity of the solvent.

Five conclusions follow.

- If z_A and z_B have *unlike* sign, the exponent in the exponential term involving them will be *positive*, and so the collision rate and the rate constant become *greater* than the gas phase values.

- If z_A and z_B have *like* sign, the exponent in the exponential is *negative*, and the solution phase quantities become *less* than the gas phase values.

- The effect of temperature on the collision rate is complex. In the gas phase the collision rate is proportional to $T^{1/2}$, but for solution this is modified by the exponential term in $1/(\varepsilon_r T)$. Since ε_r depends on T, it is not immediately clear how the exponential term modifies the $T^{1/2}$ term.

- Likewise, the effect of temperature on the solution rate constant is complex.

- The term $\exp(-z_A z_B e^2/4\pi\varepsilon_0\varepsilon_r rkT)$ modifies *both* the gas phase collision rate and the gas phase activation energy in a complex manner. Equation (7.3) *cannot* be written in a simple manner as

$$k_{\text{soln}} = (\text{a collision number}) \times (\text{exponential term in the observed activation energy}),$$

in contrast to the simple gas phase expression which can:

$$k = Z\exp(-E_A/RT). \tag{7.4}$$

For this reason, transition state theory (see Section 7.3) is a more direct and straightforward way to give a theoretical discussion of the effect of charge and solvent on rates.

7.2.1 The concepts of ideality and non-ideality

Because ionic interactions are significant in electrolyte solutions, reactions involving charges will occur under grossly non-ideal conditions. When the concentration of the solute in a solution tends to zero, i.e. $c \to 0$, the solution is regarded as *ideal*. Interactions present are solvent–solvent, solute–solvent and solute–solute, with only the solvent–solvent interactions being significant. The *non-ideal* solution corresponds to all finite concentrations of solute, other than infinite dilution, which describes the ideal solution. Non-ideality corresponds physically to all interactions that are over and above those present in the ideal solution. When the concentration of solute increases, solute–solute and solute–solvent interactions increase, and these cause modified solvent–solvent interactions, and all cause increasing non-ideality. The *major* factor giving rise to increasing non-ideality, as the solute concentration increases, is the solute–solute interactions, and these are particularly important when charges are involved. However, the solute–solvent interactions, i.e. solvation, or the modified solvent–solvent interactions, must not be forgotten.

Non-ideality is usually handled by the Debye–Hückel theory, and this can be easily done using transition state theory.

7.3 Transition State Theory for Reactions in Solution

The effect of the solvent on the rate constant is considered in terms of non-ideality, charge on reactants, relative permittivity and change in solvation pattern of the solvent. Because of the difficulty of assessing partition functions in solution, the thermodynamic formulation is used. A simplified version is given here.

The basic equation of transition state theory in this form is

$$k = \left(\frac{\kappa kT}{h}\right) K^{\neq *} \tag{4.39}$$

where $K^{\neq *}$ is a *concentration* equilibrium constant, with one term struck out, describing the formation of the activated complex (Sections 4.3.3, 4.3.4 and 4.4).

$$A + B \rightleftarrows (AB)^{\neq} \rightarrow \text{products}$$

and

$$K^{\neq} = \left(\frac{[AB]^{\neq}}{[A][B]}\right)_{\text{equil}} \tag{7.5}$$

is closely related to $K^{\neq *}$.

Remember the meanings of all the various k, κ and $K^{\neq *}$ in Equation (4.39).

- The k on the left-hand side of the equation is the theoretical rate constant.
- The k in the term $(\kappa kT/h)$ is the Boltzmann constant, and the κ is the transmission coefficient, generally taken to be unity.
- $K^{\neq *}$ is the equilibrium constant describing the formation of the activated complex but with one term corresponding to the free translation struck out.

7.3.1 Effect of non-ideality: the primary salt effect

Since $K^{\neq *}$ is a *concentration* equilibrium constant, it is a *non-ideal* equilibrium constant, and so k is also a *non-ideal* rate constant, which incorporates all the factors causing non-ideality. Since these factors will be different for different reactions, these rate constants and their derived kinetic parameters *should not* be used for comparisons of reactions, but must first be converted to ones where the effects of non-ideality have been taken care of, i.e. *ideal* values.

The Debye–Hückel theory gives a way of dealing with non-ideality in solutions of electrolytes. The ideal free energy can be calculated, and the difference between the actual free energy and this ideal value gives the non-ideal contribution resulting from all the ionic interactions. This difference, in turn, is considered to manifest itself as an activity coefficient, γ_i, for solute i, and is defined by

$$a_i = \gamma_i c_i \tag{7.6}$$

where a_i is the activity, and $\gamma_i \to 1$ when $c_i \to 0$, i.e. in the ideal solution.

Under non-ideal conditions

$$\mu_i = \mu_i^\theta + RT \log_e \gamma_i c_i \tag{7.7}$$

where μ_i, the chemical potential, refers to the solute species i.

The Debye–Hückel limiting law equation gives an expression for the activity coefficient of an ion as

$$\log_{10} \gamma_i = -A z_i^2 \sqrt{I} \tag{7.8}$$

where $A = 0.510\ \text{mol}^{-1/2}\ \text{dm}^{3/2}$ for water at 25 °C, z_i is the charge on the ion and I is the ionic strength defined as

$$I = \frac{1}{2} \sum_i (c_i z_i^2) \tag{7.9}$$

where *this sum* is over *all charged species* in the solution.

The *ideal* equilibrium constant for the formation of the activated complex is given in terms of activities:

$$K^{\neq}_{\text{ideal}} = \left(\frac{a_{\text{AB}^{\neq}}}{a_{\text{A}}\, a_{\text{B}}}\right) \tag{7.10}$$

This *ideal* equilibrium constant can also be written as

$$K^{\neq}_{\text{ideal}} = \left(\frac{a_{\text{AB}^{\neq}}}{a_{\text{A}}\, a_{\text{B}}}\right) = \left(\frac{[\text{AB}]^{\neq}}{[\text{A}][\text{B}]}\right)\left(\frac{\gamma_{\text{AB}^{\neq}}}{\gamma_{\text{A}}\gamma_{\text{B}}}\right) \tag{7.11}$$

$$= K^{\neq}_{\text{non-ideal}}\left(\frac{\gamma_{\text{AB}^{\neq}}}{\gamma_{\text{A}}\gamma_{\text{B}}}\right) \tag{7.12}$$

$$\therefore K^{\neq}_{\text{non-ideal}} = K^{\neq}_{\text{ideal}}\left(\frac{\gamma_{\text{A}}\gamma_{\text{B}}}{\gamma_{\text{AB}^{\neq}}}\right) \tag{7.13}$$

Each of $K^{\neq}_{\text{ideal}}$ and $K^{\neq}_{\text{non-ideal}}$ can be written as the product of the corresponding $K^{\neq *}$ and the factor relating to the internal translation in the activated complex. Since this factor appears on both sides of the equation it will cancel out, and so Equation (7.13) becomes

$$K^{\neq *}_{\text{non-ideal}} = K^{\neq *}_{\text{ideal}}\left(\frac{\gamma_{\text{A}}\gamma_{\text{B}}}{\gamma_{\text{AB}^{\neq}}}\right) \tag{7.14}$$

Since both k and $K^{\neq *}$ appearing in Equation (4.39) are *concentration* constants, and *not* constants in terms of activities,

$$k = \frac{\kappa kT}{h} K^{\neq *} \tag{4.39}$$

this equation can be rewritten in terms of a $k_{\text{non-ideal}}$ and a $K^{\neq *}_{\text{non-ideal}}$:

$$k_{\text{non-ideal}} = \frac{\kappa kT}{h} K^{\neq *}_{\text{non-ideal}} \tag{7.15}$$

which with Equation (7.14) gives

$$k_{\text{non-ideal}} = \frac{\kappa kT}{h} K^{\neq *}_{\text{ideal}} \frac{\gamma_A \gamma_B}{\gamma_{AB^{\neq}}} \tag{7.16}$$

$$= k_{\text{ideal}} \frac{\gamma_A \gamma_B}{\gamma_{AB^{\neq}}} \tag{7.17}$$

Note: the quotient of activity coefficients in the equation for the rate constant, i.e. Equation (7.17), is the reciprocal of the quotient of activity coefficients appearing in equations for the ideal equilibrium constants, Equations (7.11) and (7.12). This is because Equation (7.17) is in terms of the *non-ideal* rate constant.

Remember: *observed* experimental quantities inevitably are *non-ideal* quantities.

If all the activity coefficients are unity, the solution corresponds to ideality. However, for reactions in solution this is unlikely to happen, especially if there are ions or charge-separated species involved.

Remember: $K^{\neq *}_{\text{non-ideal}}$ is fundamentally a partition function quantity, and in the above treatment this has been converted to one involving activity coefficients.

In transition state theory the calculated rate constant is given by

$$k = \frac{\kappa kT}{h} \frac{Q^{\neq *}}{Q_A Q_B} \exp\left(\frac{-E_0}{RT}\right) \tag{4.20}$$

For gases, Q_A and Q_B are readily found from spectroscopic data, but $Q^{\neq *}$ for the activated complex can only be calculated if the potential energy surface is accurately known in the region of the critical configuration. It cannot be measured experimentally – the activated complex is merely a special critical configuration.

Likewise, $\gamma_{AB^{\neq}}$ also cannot be measured experimentally, although, like Q_A and Q_B, γ_A and γ_B can be measured, and at first sight the conversion given above may seem to give little improvement. However, for ion–ion and ion–molecule reactions, the Debye–Hückel theory, see Equation (7.8), can calculate the activity coefficient for any charged species and convert Equation (7.17) into a useful form. For other reactions the approach is only qualitative, but for them the effects of non-ideality are much smaller.

Taking *logarithms to base 10* Equation (7.17) gives

$$\log_{10} k_{\text{non-ideal}} = \log_{10} k_{\text{ideal}} + \log_{10} \gamma_A + \log_{10} \gamma_B - \log_{10} \gamma_{AB^{\neq}} \tag{7.18}$$

Note: logarithms are to base 10 because the Debye–Hückel equation is given in logarithms to base 10.

The Debye–Hückel equation for very dilute solutions can be used for each activity coefficient:

$$\log_{10} \gamma_i = -A\, z_i^2 \sqrt{I} \tag{7.8}$$

where A is the Debye–Hückel constant, which is dependent on temperature and takes the value 0.510 $\text{mol}^{-1/2}\ \text{dm}^{3/2}$ for water at 25 °C.

The activated complex has charge $z_A + z_B$. Taking Equations (7.18) and (7.8) together gives

$$\log_{10} k_{\text{non-ideal}} = \log_{10} k_{\text{ideal}} - A\, z_A^2 \sqrt{I} - A\, z_B^2 \sqrt{I} + A(z_A + z_B)^2 \sqrt{I} \tag{7.19}$$

$$= \log_{10} k_{\text{ideal}} - A\, z_A^2 \sqrt{I} - A\, z_B^2 \sqrt{I} + A\, z_A^2 \sqrt{I} + A\, z_B^2 \sqrt{I} + 2A\, z_A\, z_B \sqrt{I} \tag{7.20}$$

$$\log_{10} k_{\text{non-ideal}} = \log_{10} k_{\text{ideal}} + 2A\, z_A\, z_B \sqrt{I} \tag{7.21}$$

The treatment can be extended to higher ionic strengths, giving

$$\log_{10} k_{\text{non-ideal}} = \log_{10} k_{\text{ideal}} + 2A\, z_A\, z_B \frac{\sqrt{I}}{1 + \sqrt{I}} \tag{7.22}$$

This is called the *primary salt effect*.

Reactions involving charged reactants will occur under non-ideal conditions, and will show a dependence on ionic strength (Equations (7.21) and (7.22). This means that it is necessary to be able to work out the ionic strength from the experimental conditions. The following problem gives practice with this.

Worked Problem 7.1

Question. The reaction

$$Fe^{2+}(aq) + [Co(C_2O_4)_3]^{3-}(aq) \rightarrow Fe^{3+}(aq) + 3\,C_2O_4^{2-}(aq) + Co^{2+}(aq)$$

can be carried out at various ionic strengths, either without added salts, or with them. The Fe^{2+} is introduced as $Fe(NO_3)_2$, and the $Co(C_2O_4)_3^{3-}$ as $K_3Co(C_2O_4)_3$. Calculate the ionic strength for the following reaction conditions.

- $[Fe(NO_3)_2] = 2.00 \times 10^{-2}$ mol dm^{-3}, $[K_3Co(C_2O_4)_3] = 3.00 \times 10^{-2}$ mol dm^{-3}.
- $[Fe(NO_3)_2] = 2.00 \times 10^{-2}$ mol dm^{-3}, $[K_3Co(C_2O_4)_3] = 3.00 \times 10^{-2}$ mol dm^{-3} and $[KNO_3] = 5.00 \times 10^{-2}$ mol dm^{-3}.

Answer.

- $K_3Co(C_2O_4)_3$ gives $[K^+] = 9.00 \times 10^{-2}$ mol dm^{-3} and $[Co(C_2O_4)_3^{3-}] = 3.00 \times 10^{-2}$ mol dm^{-3}.
  $Fe(NO_3)_2$ gives $[NO_3^-] = 4.00 \times 10^{-2}$ mol dm^{-3} and $[Fe^{2+}] = 2.00 \times 10^{-2}$ mol dm^{-3}.

Charges on the ions are the following:

K^+ is +1, Fe^{2+} is +2, NO_3^- is −1, $Co(C_2O_4)_3^{3-}$ is −3.

$$I = 0.5 \sum_i (c_i z_i^2) \qquad (7.9)$$

= 0.5 (contribution from K^+ + contribution from Fe^{2+} + contribution from NO_3^- + contribution from $Co(C_2O_4)_3^{3-}$), giving

$$0.5 \times \{9.00 \times 10^{-2} \times (+1)^2 + 2.00 \times 10^{-2} \times (+2)^2 + 4.00 \times 10^{-2} \times (-1)^2 + 3.00 \times 10^{-2} \times (-3)^2\} \text{ mol dm}^{-3}$$

$$= 0.5 \times \{9.00 \times 10^{-2} + 8.00 \times 10^{-2} + 4.00 \times 10^{-2} + 27.0 \times 10^{-2}\} \text{ mol dm}^{-3}$$

$$= 24 \times 10^{-2} \text{ mol dm}^{-3}.$$

- $K_3Co(C_2O_4)_3$ gives $[K^+] = 9.00 \times 10^{-2}$ mol dm^{-3} and $[Co(C_2O_4)_3^{3-}] = 3.00 \times 10^{-2}$ mol dm^{-3}.

  $Fe(NO_3)_2$ gives $[NO_3^-] = 4.00 \times 10^{-2}$ mol dm^{-3} and $[Fe^{2+}] = 2.00 \times 10^{-2}$ mol dm^{-3}.

  KNO_3 gives $[K^+] = 5.00 \times 10^{-2}$ mol dm^{-3} and $[NO_3^-] = 5.00 \times 10^{-2}$ mol dm^{-3}.

  Total $[K^+] = 14.00 \times 10^{-2}$ dm^{-3} and total $[NO_3^-] = 9.00 \times 10^{-2}$ mol dm^{-3}.

$$I = 0.5 \times \{14.00 \times 10^{-2} \times (+1)^2 + 2.00 \times 10^{-2} \times (+2)^2 + 9.00 \times 10^{-2} \times (-1)^2 + 3.00 \times 10^{-2} \times (-3)^2\} \text{ mol dm}^{-3}$$

$$= 0.5 \times \{14.00 \times 10^{-2} + 8.00 \times 10^{-2} + 9.00 \times 10^{-2} + 27.00 \times 10^{-2}\} \text{ mol dm}^{-3}$$

$$= 29.0 \times 10^{-2} \text{ mol dm}^{-3}.$$

This treatment allows us to predict the effect of ionic strength on the rate constants for reactions in solution.

Worked Problem 7.2

Question. Predict the effect of increase of ionic strength on k for the following reactions:

- $CH_2ClCOO^- + OH^-$
- $Co(NH_3)_5Cl^{2+} + OH^-$
- $NH_4^+ + OCN^-$
- $S_2O_3^{2-} + SO_3^{2-}$
- $Fe^{2+} + Co(C_2O_4)_3^{3-}$

- $Co(NH_3)_3Br^{2+} + Hg^{2+}$
- $Fe(CN)_6^{4-} + S_2O_8^{2-}$
- $CH_3I + OH^-$
- $CH_3I + H_2O$

Answer.

Reaction	$z_A\, z_B$	k
• $CH_2ClCOO^- + OH^-$	+1	increases
• $Co(NH_3)_5Cl^{2+} + OH^-$	−2	decreases
• $NH_4^+ + OCN^-$	−1	decreases
• $S_2O_3^{2-} + SO_3^{2-}$	+4	increases
• $Fe^{2+} + Co(C_2O_4)_3^{3-}$	−6	decreases
• $Co(NH_3)_3Br^{2+} + Hg^{2+}$	+4	increases
• $Fe(CN)_6^{4-} + S_2O_8^{2-}$	+8	increases
• $CH_3I + OH^-$	0	no effect
• $CH_3I + H_2O$	0	no effect

Testing these equations requires measurements to be made at various ionic strengths. If the equations are valid then, for very dilute solutions, a plot of $\log_{10} k_{obs}$ versus $\sqrt{I}$ should be linear with intercept $\log_{10} k_{ideal}$ and slope $= 2A\, z_A\, z_B$, while at higher ionic strengths a plot of $\log_{10} k_{obs}$ versus $\sqrt{I}/(1+\sqrt{I})$ should be linear with the same intercept and slope (Figures 7.1 and 7.2). An extension to even higher ionic strengths uses an empirical B' factor corresponding to the empirical B factor used in the Debye–Hückel theory

$$\log_{10} k_{non\text{-}ideal} = \log_{10} k_{ideal} + 2A\, z_A\, z_B \frac{\sqrt{I}}{1+\sqrt{I}} + B'I \tag{7.23}$$

and a plot of $\log_{10} k_{non\text{-}ideal} - \log_{10} k_{ideal} - 2A\, z_A\, z_B \sqrt{I}/(1+\sqrt{I})$ versus I should be linear with zero intercept and slope B' (Figure 7.3).

It is absolutely vital that reactions in solution involving ions or charge-separated species are carried out at a variety of ionic strengths, and the rate constants extrapolated to zero ionic strength. The ionic strength can be varied by adding an appropriate amount of an electrolyte known to be fully dissociated in solution.

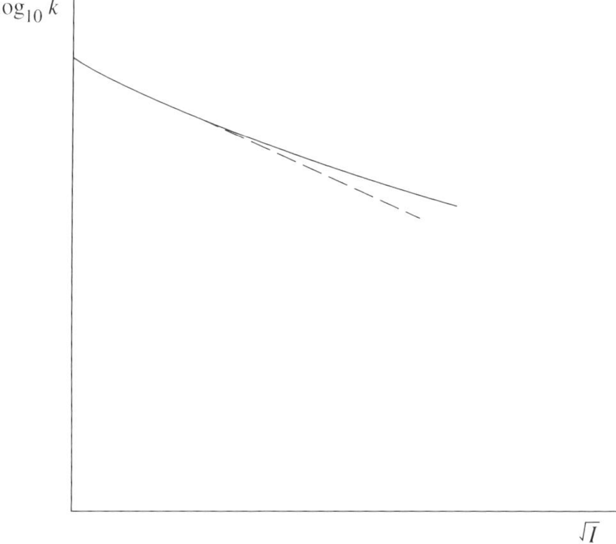

Figure 7.1 The primary salt effect: plot of $log_{10} k$ versus $\sqrt{I}$

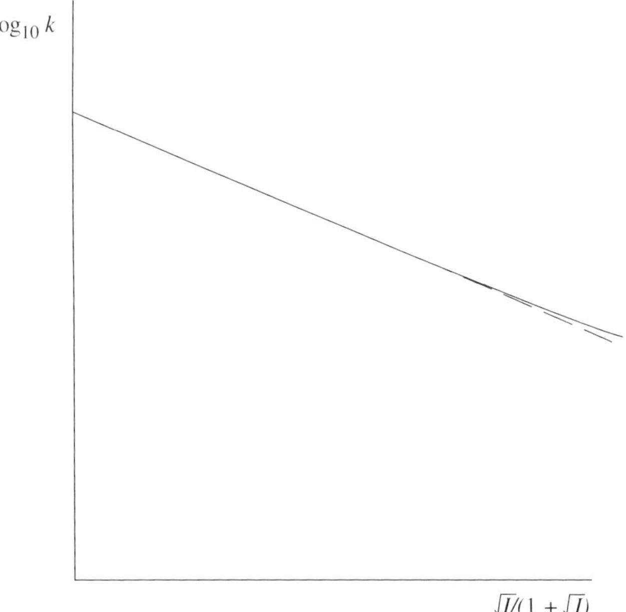

Figure 7.2 The primary salt effect: plot of $\log_{10} k$ versus $\sqrt{I}/(1 + \sqrt{I})$

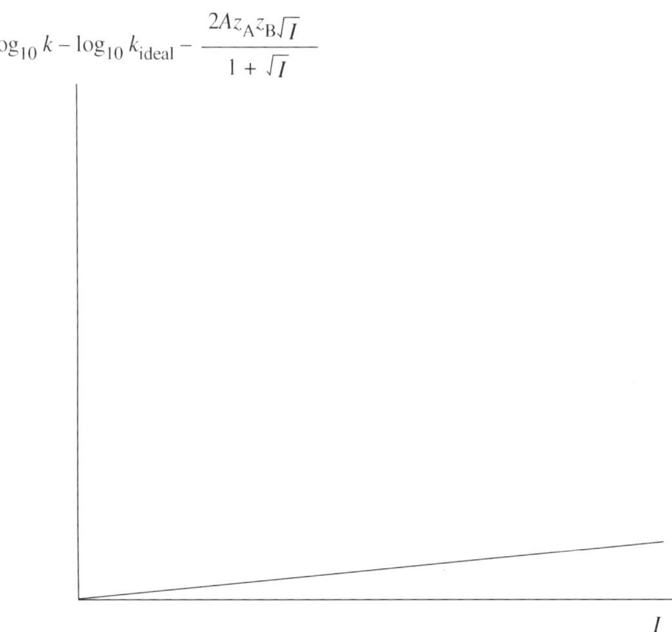

Figure 7.3 The primary salt effect: plot of $\log_{10} k - \log_{10} k_{ideal} - 2Az_Az_B\sqrt{I}/(1+\sqrt{I})$ versus I

Alternatively, reactions are studied at a constant ionic strength, often 0.100 mol dm^{-3}, with the ionic strength being made up to this value by adding an appropriate amount of e.g. $KClO_4$.

Pay attention here. The following two questions can be asked, and their answers are important.

- Why are some reactions studied at constant ionic strength and then compared?
- Which is the superior procedure: constant I, or extrapolation to zero I?

Answers.

- If the reactions are studied at constant ionic strength, then the comparison focuses on the chemistry of the reactions, without having to consider the effects of non-ideality. However, it may well be that there are specific effects which will manifest themselves in a different manner for each reaction, even though the ionic strength is the same, and these will not be picked up in this type of experiment.
- Extrapolation is superior because it corresponds to the ideal solution. Any specific effects pertinent to a given reaction will be taken care of automatically.

Worked Problem 7.3

Question. Use the following data relating to a reaction of the type $PtLX^{+}(aq) + Y^{-}(aq)$ to test the primary salt effect.

- Reaction at low ionic strengths

$\frac{10^2 I}{\text{mol dm}^{-3}}$	$\frac{\sqrt{I}}{\text{mol}^{1/2}\,\text{dm}^{-3/2}}$	$\frac{\sqrt{I}}{1+\sqrt{I}}$	$\frac{k}{\text{mol}^{-1}\,\text{dm}^3\text{min}^{-1}}$	$\log_{10} k$
0.160	0.040	0.0385	0.210	−0.678
0.423	0.065	0.061	0.194	−0.712
0.810	0.090	0.083	0.184	−0.735
1.21	0.110	0.099	0.180	−0.745
2.13	0.146	0.127	0.164	−0.785

- Reaction at higher ionic strengths

$\frac{10^2 I}{\text{mol dm}^{-3}}$	$\frac{\sqrt{I}}{\text{mol}^{1/2}\,\text{dm}^{-3/2}}$	$\frac{\sqrt{I}}{1+\sqrt{I}}$	$\frac{k}{\text{mol}^{-1}\,\text{dm}^3\,\text{min}^{-1}}$	$\log_{10} k$
3.76	0.194	0.162	0.152	−0.818
6.25	0.250	0.200	0.143	−0.845
9.73	0.312	0.238	0.127	−0.896
11.5	0.339	0.253	0.125	−0.903
18.9	0.435	0.303	0.112	−0.951

Answer. Low ionic strengths, see Figure 7.4.

- The plot of $\log_{10} k_{obs}$ versus $\sqrt{I}$ is linear with slope = $-0.97\ \text{mol}^{-1/2}\,\text{dm}^{3/2}$, and intercept $\log_{10} k_0 = -0.644$, giving the ideal $k = k_0 = 0.227\ \text{mol}^{-1}\,\text{dm}^3\,\text{min}^{-1}$.
- The predicted slope = $2A\,z_A\,z_B = 2 \times 0.51 \times (+1) \times (-1) = -1.02\ \text{mol}^{-1/2}$ $\text{dm}^{3/2}$, which is in agreement with the observed value.
- Since the plot is linear, the Debye–Hückel limiting law for low ionic strengths is adequate.

High ionic strengths, see Figures 7.5 and 7.6.

- The plot of $\log_{10} k_{obs}$ versus $\sqrt{I}$ is now non-linear, and the limiting law is no longer adequate for dealing with non-ideality. A plot of $\log_{10} k_{obs}$ versus $\sqrt{I}/(1+\sqrt{I})$ is necessary. This is linear, showing that the denominator can account for the increasing non-ideality at these higher ionic strengths.
- Slope = $-1.03\ \text{mol}^{-1/2}\ \text{dm}^{3/2}$, and intercept = -0.644, giving $k_0 = 0.227$ $\text{mol}^{-1}\ \text{dm}^3\ \text{min}^{-1}$.

These are the same values as found previously, which is as expected; see Figures 7.4–7.6.

This demonstrates that it is absolutely imperative to take account of interionic interactions for reactions involving charges.

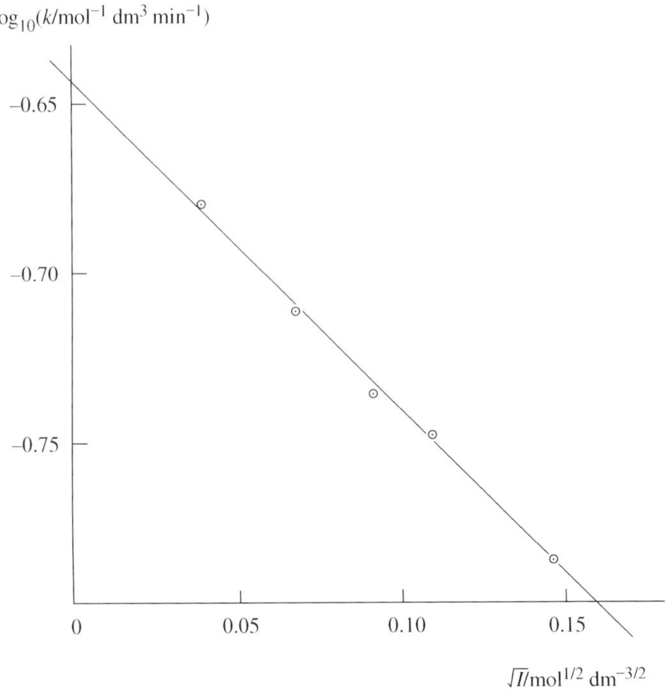

Figure 7.4 Graph of $\log_{10} k$ versus $\sqrt{I}$ for reaction $PtLX^+(aq) + Y^-(aq)$ at low ionic strengths

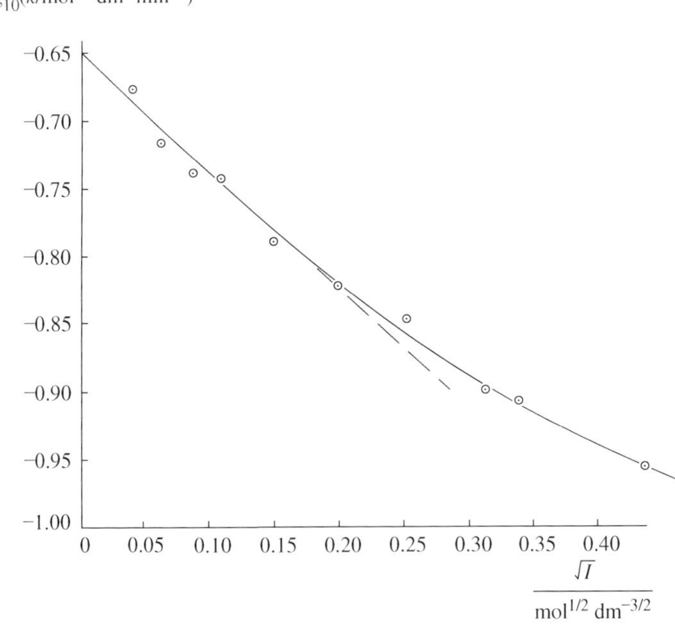

Figure 7.5 Graph of $\log_{10} k$ versus $\sqrt{I}$ for reaction $PtLX^+(aq) + Y^-(aq)$ at high ionic strengths

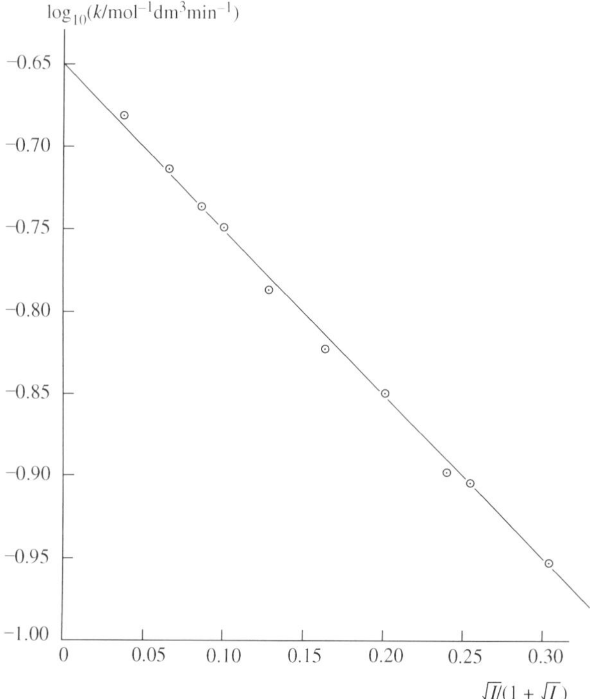

Figure 7.6 Graph of $\log_{10} k$ versus $\sqrt{I}/(1+\sqrt{I})$ for reaction $PtLX^{+}(aq) + Y^{-}(aq)$ over the whole range of ionic strengths

7.3.2 Dependence of $\Delta S^{\neq *}$ and $\Delta H^{\neq *}$ on ionic strength

If the observed k depends on ionic strength, then so will $\Delta S^{\neq *}$ and $\Delta H^{\neq *}$ (see Equation (4.41)) and extrapolation to zero ionic strength is therefore essential.

$$k_{\mathrm{obs}} = \frac{\kappa k T}{h} \exp\left(\frac{+\Delta S^{\neq *}}{R}\right) \exp\left(\frac{-\Delta H^{\neq *}}{RT}\right) \tag{7.24}$$

The theoretical treatment of the primary salt effect can be extended to $\Delta S^{\neq *}$ and $\Delta H^{\neq *}$, and magnitudes of the slope in plots of the observed values against $\sqrt{I}$ or $\sqrt{I}/(1+\sqrt{I})$ can be compared with the theoretical values.

However, unless a test for the theoretical predictions of the effect of ionic strength on $\Delta S^{\neq *}$ and $\Delta H^{\neq *}$ is required, it is simpler to use values of k_{obs} extrapolated to zero ionic strength. If these are found at various temperatures, then an Arrhenius plot will give the corresponding *ideal* values of $\Delta S^{\neq *}$ and $\Delta H^{\neq *}$, which are then suitable for comparison with those for other reactions.

Remember: the Arrhenius equation is (Section 3.23)

$$k = A \exp(-E_{\mathrm{A}}/RT) \tag{3.96}$$

$$\log_{\mathrm{e}} k = \log_{\mathrm{e}} A - E_{\mathrm{A}}/RT \tag{3.97}$$

and a plot of $\log_e k$ versus $1/T$ should be linear with slope $= -E_A/RT$ and intercept $= \log_e A$.

7.3.3 The effect of the solvent

The ideal k (corresponding to the extrapolated value at zero ionic strength) should *always be used* when the effect of solvent is considered, especially for ion–ion and ion–molecule reactions.

Equation (4.39) is the simplest starting point for the simplified treatment for non-ideality, but Equation (4.40) is the simplest starting point for the effect of the solvent on the observed rate constant, though it can also be used to derive the effect of non-ideality.

$$k = \frac{\kappa kT}{h} \exp\left(\frac{-\Delta G^{\neq *}}{kT}\right) \tag{7.25}$$

This is Equation (4.40) re-written with $\Delta G^{\neq *}$ expressed as a molecular instead of a molar quantity, and so k, the Boltzmann constant, is used instead of R, the gas constant.

For reactions in solution, the exponential term appropriate to the gas phase has to be modified to include the contribution accounting for the charges and the solvent. Calculations show that $\Delta G^{\neq *}$ has to be modified by a term involving ε_r, the relative permittivity of the solvent, and the charges on the ions. This term turns out to be the same as that appearing in the collision formula, Equation (7.3), i.e. $z_A z_B e^2/4\pi\varepsilon_0\varepsilon_r r_{\neq}$, so that

$$\Delta G^{\neq *}_{\text{ions in solution}} = \Delta G^{\neq *}_{\text{molecules in gas}} + \frac{z_A z_B e^2}{4\pi\varepsilon_0\varepsilon_r r_{\neq}} \tag{7.26}$$

The fact that the same expression modifies both collision theory and transition state theory should not cause surprise. This is because the modification is taking account of the same physical phenomenon in both theories, i.e. the interaction between the ions.

Equation (7.26) modifies the free energy of activation for a reaction between molecules in the gas phase by making allowance for the effects of the *charges and the solvent only.* A further modification allows for the effects of non-ideality, and ends up with the same dependence of $\Delta G^{\neq *}_{\text{ions in solution}}$ on ionic strength as results from the simpler approach given in Section 7.3.1.

Equation (7.26) must only be used to discuss reactions under ideal conditions and only applies to data extrapolated to zero ionic strength.

Since

$$k = \frac{\kappa kT}{h} \exp\left(\frac{-\Delta G^{\neq *}}{kT}\right) \tag{7.25}$$

then

$$\log_e k^{\text{ideal}}_{\text{ions in solution}} = log_e \frac{\kappa kT}{h} - \frac{\Delta G^{\neq *}_{\text{molecules in gas}}}{kT} - \frac{z_A\, z_B\, e^2}{4\,\pi\,\varepsilon_0\,\varepsilon_r\, r_{\neq}\, kT} \tag{7.27}$$

If all other variables are kept constant, then a decrease in ε_r will cause a decrease in k if $z_A\, z_B$ is positive, and will cause an increase if $z_A\, z_B$ is negative. However, this assumes that $r_{\neq}$ does not change as the permittivity changes. This is probably not justifiable.

The predicted $\log_e k_{\text{obs}} = \log_e k^{\text{ideal}}_{\text{ions in solution}}$ depends linearly on $1/\varepsilon_r$ for a given temperature, provided $r_{\neq}$ is the same in all solvents. If extrapolated experimental $\log_e k^{\text{ideal}}_{\text{obs}}$ values for a variety of solvents, and corresponding to zero ionic strength, i.e. an ideal solution, are plotted against $1/\varepsilon_r$, then predicted slope $= -z_A z_B\, e^2/4\,\pi\,\varepsilon_0\, r_{\neq}\, kT$. This can be compared with the observed slope.

This equation predicts that

- if the reacting ions are of like sign then the slope is negative;
- if the reacting ions are of unlike sign then the slope is positive;
- if $r_{\neq}$ varies with the solvent, then the value of the slope is unpredictable, and the experimental points may not even lie on a common straight line or curve.

However, it is *not* reasonable to assume that $r_{\neq}$ should be the same for a given reaction under all circumstances, for the following reasons.

(a) $r_{\neq}$ is the distance between the charges in the activated complex. As the relative permittivity decreases, the effect of associative phenomena increases, as does the possibility of clustering of solvent molecules and of charges, so the solvent has an effect on $r_{\neq}$.

(b) Temperature could also have an effect on $r_{\neq}$, e.g. by decreasing the effect of interactions between the charged species as the thermal energy becomes greater than or equal to the electrostatic energy of interaction.

- The experimental slope can be compared with the theoretical value. If it is assumed that the theory holds, then a comparison should give values of $r_{\neq}$, which can then be used to infer something about the structure and solvation pattern of the activated complex.
- The treatment assumes that the solvent is microscopically homogeneous. This need not happen with mixed solvents.
- This equation is approximate since it assumes that the solvent is a *structureless* continuum, and so cannot interact with the reactants and the activated complex. Electrolyte solution studies demonstrate conclusively that this is not the case. In

an electrolyte solution the effect of the solvent is to act as a medium that reduces the energy of interaction between charges, dipoles and quadrupoles by an amount given by the term $1/\varepsilon_r$. In deriving the equations the solvent is treated as a structureless continuum. In reality, the solvent molecules are polarizable, and generally have a dipole or a quadrupole with which the charges on the reactants can interact, and so the solvent is not a structureless continuum. It is through these interactions that the effect of the solvent *as a medium reducing the interaction energy between the charges* is made. However, the charge/solvent interactions *in this sense* are sufficiently small for the solvent structure to be almost unaltered, and for the solvent to be regarded as a structureless medium.

This may well be so in the bulk of the solution, but close to the ions or dipoles the structure of the solvent is considerably modified, so much so that the solvent molecules can be considered as moving with the charge or dipole. *In this sense* there is an *added* effect of the solvent, described as solvation, which is *over and above* the effect of the solvent acting as a medium reducing the interionic interactions, and which is described by Equation (7.26). The effect of the solvent in this sense is discussed under the heading of the effect of solvation on $\Delta H^{\neq *}$, $\Delta S^{\neq *}$ and $\Delta V^{\neq *}$ (Sections 7.5.5, 7.4.9 and 7.6.6 respectively).

These predicted expressions, Equations (7.26) and (7.27), are only approximate, and more sophisticated calculations are required. The following problem illustrates this approximate nature.

Worked Problem 7.4

Question. The $CH_2BrCOO^-/S_2O_3^{2-}$ reaction has been studied in solution over a range of relative permittivities. These were obtained using various mixtures of glycine, urea and sucrose in water. The following data are given for 25 °C, and have been extrapolated to zero ionic strength.

ε_r	135	134	100	92	88
$\dfrac{k}{\text{mol}^{-1}\text{dm}^3\text{s}^{-1}}$	0.0132	0.0138	0.00795	0.00640	0.00585
ε_r	83	78.5	76	73	70
$\dfrac{k}{\text{mol}^{-1}\text{dm}^3\text{s}^{-1}}$	0.00493	0.00412	0.00388	0.00381	0.00380

Use Equation (7.27) to interpret the data.

$$\left(\frac{e^2}{4\pi\varepsilon_0 kT} = 5.605 \times 10^{-8}\,\text{m}\right)$$

Answer.

$$\log_e k^{\text{ideal}}_{\text{ions in solution}} = \log_e \frac{\kappa kT}{h} - \frac{\Delta G^{\neq *}_{\text{molecules in gas}}}{kT} - \frac{z_A z_B e^2}{4\pi\varepsilon_0\varepsilon_r r_{\neq} kT} \qquad (7.27)$$

If the theory is adequate, then a plot of $\log_e k_{obs}$ versus $1/\varepsilon_r$ should be linear with slope $= -z_A z_B e^2/4\pi\varepsilon_0 r_{\neq} kT$.

$\frac{1}{\varepsilon_r}$	0.0074	0.0075	0.0100	0.0109	0.0114
$\log_e k_{obs}$	−4.33	−4.28	−4.83	−5.05	−5.14
$\frac{1}{\varepsilon_r}$	0.0120	0.0127	0.0132	0.0137	0.0143
$\log_e k_{obs}$	−5.31	−5.49	−5.55	−5.57	−5.57

Figure 7.7 shows this graph. This is linear within experimental error over most of the range of ε_r values, showing the treatment to be basically valid. It also suggests that the value of $r_{\neq}$ remains fairly constant over this range. Considerable deviations appear at low values of ε_r (high $1/\varepsilon_r$ values). This could be a consequence of the mixed nature of the solvent, resulting in a microscopically inhomogeneous mixture, and/or preferential solvation of the ions present by the water of the solvent. The very abrupt and dramatic change in slope is unlikely to be a result of a change in the value of $r_{\neq}$.

$$\text{slope} = -223 = -\frac{z_A z_B e^2}{4\pi\varepsilon_0 r_{\neq} kT} = -\frac{2 \times 5.605 \times 10^{-8}\,\text{m}}{r_{\neq}}$$

$$r_{\neq} = 503 \times 10^{-10}\,\text{m} = 5.03\,\text{pm}$$

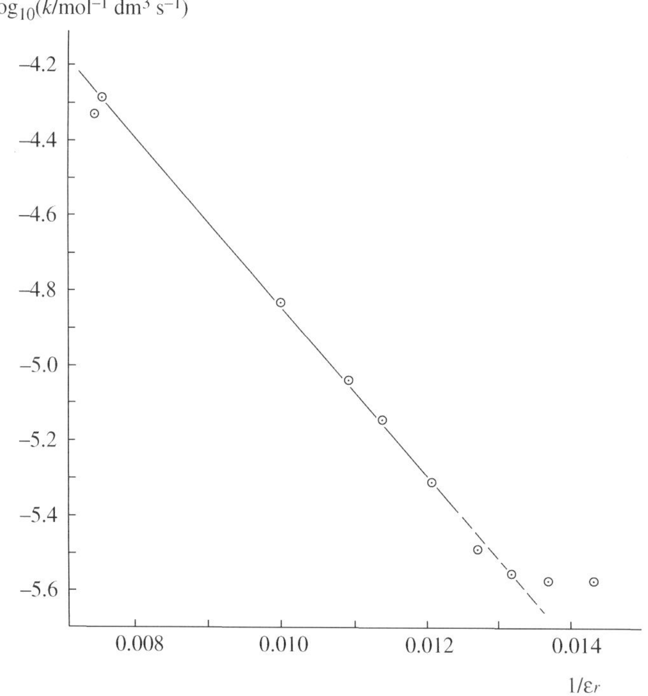

Figure 7.7 Graph of $\log_{10} k$ versus $1/\varepsilon_r$ for reaction of CH_2BrCOO^- with $S_2O_3^{2-}$

7.3.4 Extension to include the effect of non-ideality

If the effect of non-ideality due to long range interionic interactions is included, then it follows from Equation (7.25) that for any one solvent

$$\log_{10} k^{\text{non-ideal}}_{\text{ions in solution}} = \log_{10} k_0 + 2A\, z_A\, z_B \sqrt{I} \qquad (7.28)$$

which is the same as Equation (7.21), deduced by the more straightforward procedure.

Extensions to higher ionic strength give Equations (7.22) and (7.23).

7.3.5 Deviations from predicted behaviour

The following are the *major* sources of deviations from predicted behaviour.

- Ion association and other associative phenomena are not discussed. It is vital that such effects are considered, especially at higher ionic strengths, where significant association occurs. In kinetic studies, allowances for ion pairing are made by distinguishing between the *stoichiometric* ionic concentrations and the *actual* ones.

Note. The *stoichiometric* concentration is that based on the amount of the solute weighed out and the volume of the solution, i.e. it assumes that the solute in solution corresponds to that indicated by the formula.

The *actual* concentration takes account of the fact that the solute may alter character when in solution, e.g. the undissociated solute may be partially ionized in solution, or the free ions of an electrolyte may be partially removed to form ion pairs.

Kinetic results which apparently do not fit the above treatment of the primary salt effect do so when the observed rates are correlated with the *actual* ionic strengths rather than the *stoichiometric* values. The actual concentrations in the reaction solution are calculated using the known value of the equilibrium constant describing the ion pair. This is discussed in Problem 7.5.

Note. Ion pairing results when the electrostatic interactions between two oppositely charged ions become sufficiently large for the two ions to move around as one entity, the ion pair; e.g., in solutions of Na_2SO_4, the following occurs:

$$Na^+(aq) + SO_4^{2-}(aq) \rightleftarrows NaSO_4^-(aq)$$

The $Na^+(aq)$ is distributed between free Na^+ and the ion pair $NaSO_4^-$, so that

$$[Na^+]_{\text{total}} = [Na^+]_{\text{actual}} + [NaSO_4^-]_{\text{actual}} \qquad (7.29)$$

and the total concentration is equivalent to the stoichiometric value. Total and stoichiometric are in this sense interchangeable. The same argument can be applied to the SO_4^{2-}(aq).

$$[SO_4^{2-}]_{total} = [SO_4^{2-}]_{actual} + [NaSO_4^{-}]_{actual} \tag{7.30}$$

The same distinction can be made between the actual ionic strength, calculated from the actual concentrations, and the stoichiometric ionic strength.

There is also the possibility of complexing or chelation occurring between species in solution. This can be handled in the same way.

- Association phenomena increase and become much more important as the relative permittivity of the solvent decreases. Electrolytes which are completely, or nearly completely, dissociated in water are extensively associated in low relative permittivity solvents.

Since the potential energy of interaction between two ions is proportional to $1/\varepsilon_r$, then the lower the relative permittivity of a solvent, the greater will be the attraction between ions of opposite charge. A similar relation holds for the interaction between an ion and a dipole, and between two dipoles. So solvents of low relative permittivity encourage associative behaviour.

Quantitative allowance must be made for this if there is to be any hope of a meaningful interpretation of the results for reactions carried out in a variety of solvents with a range of relative permittivities.

- This treatment ignores solvation, but this is one of the most characteristic and important features of electrolyte solutions, and must be taken account of explicitly. When this is done, considerable insight can be given into the mechanistic details of reactions in solution.

Worked Problem 7.5

Question. The reaction:

$$(C_2H_5)_3N^+CH_2COOC_2H_5(aq) + OH^-(aq) \rightarrow (C_2H_5)_3N^+CH_2COO^-(aq) + C_2H_5OH(aq)$$

has $k_0 = 32.0\ \text{mol}^{-1}\ \text{dm}^3\ \text{s}^{-1}$ at 25 °C and zero ionic strength.

- *Set (a).* Experiments where the ionic strengths were made up to the quoted values with KCl gave the following results.

$\dfrac{\left(\frac{\sqrt{I}}{1+\sqrt{I}}\right)}{\text{mol}^{1/2}\text{dm}^{-3/2}}$	0.039	0.045	0.055	0.067
$\dfrac{k_{obs}}{\text{mol}^{-1}\text{dm}^3\text{s}^{-1}}$	29.2	28.7	28.2	27.3
$\log_{10} k_{obs}$	1.465	1.458	1.450	1.436

$\dfrac{\left(\frac{\sqrt{I}}{1+\sqrt{I}}\right)}{\text{mol}^{1/2}\text{dm}^{-3/2}}$	0.090	0.120	0.139	0.158
$\dfrac{k_{obs}}{\text{mol}^{-1}\text{dm}^3\text{s}^{-1}}$	25.8	24.1	23.0	22.0
$\log_{10} k_{obs}$	1.412	1.382	1.362	1.342

- *Set (b).* Experiments at high ionic strengths in the presence of various concentrations of $CaCl_2$ gave the following results.

$\dfrac{\left(\frac{\sqrt{I}}{1+\sqrt{I}}\right)}{\text{mol}^{1/2}\text{dm}^{-3/2}}$	0.090	0.120	0.139	0.158
$\dfrac{k_{obs}}{\text{mol}^{-1}\text{dm}^3\text{s}^{-1}}$	25.2	22.8	21.3	20.1
$\log_{10} k_{obs}$	1.401	1.358	1.328	1.303

- *Set (c).* Experiments at low ionic strengths in the presence of various concentrations of $CaCl_2$, gave the following results.

$\dfrac{\left(\frac{\sqrt{I}}{1+\sqrt{I}}\right)}{\text{mol}^{1/2}\text{dm}^{-3/2}}$	0.039	0.045	0.055	0.067
$\dfrac{k_{obs}}{\text{mol}^{-1}\text{dm}^3\text{s}^{-1}}$	29.2	28.6	27.9	26.8
$\log_{10} k_{obs}$	1.465	1.456	1.446	1.428

Interpret the data.

Answer.

- For *set (a)* data, for reactions in the absence of Ca^{2+}, a plot of $\log_{10} k_{obs}$ versus $\sqrt{I}/(1+\sqrt{I})$ is linear with *slope* $= -1.03$ mol$^{-1/2}$ dm$^{3/2}$, which agrees with the predicted slope of $2A\, z_A\, z_B = +2 \times 0.510 \times (+1) \times (-1)$ mol$^{-1/2}$ dm$^{3/2}$, i.e. -1.02 mol$^{-1/2}$ dm$^{3/2}$.

  Intercept $= log_{10}\, k_0 = 1.505$, giving the ideal $k_0 = 32.0$ mol^{-1} dm^3 s^{-1} (Figure 7.8(a)).

  This confirms the essential correctness of the theoretical treatment of the primary salt effect.

- For the second set of data, *set (b)*, for reactions in the presence of Ca^{2+}, but at *high ionic strengths*, the plot of $\log_{10}\, k_{obs}$ versus $\sqrt{I}/(1+\sqrt{I})$ is also linear, but there are several features that do not fit with the theoretical predictions (Figure 7.8(b)).

The slope of the graph is $= -1.44$ $\text{mol}^{-1/2}$ $\text{dm}^{3/2}$, and this does not fit the theoretical slope of -1.02 $\text{mol}^{-1/2}$ $\text{dm}^{3/2}$. The intercept on this basis would also not fit with the value obtained in the absence of $CaCl_2$.

Hence linearity of the plot of $\log_{10} k_{obs}$ versus $\sqrt{I}/(1+\sqrt{I})$ does not, in itself, necessarily give the correct interpretation of what is happening in this reaction.

- When the third set of data, *set (c)*, for reactions in the presence of Ca^{2+}, but for *low ionic strengths*, is plotted, the *apparent discrepancy* in the plot for the high ionic strength, added Ca^{2+}, data is immediately exposed. The full plot covering all data with added Ca^{2+} is shown to be a curve with the graph now having an intercept $\log_{10} k_0 = 1.505$, giving the ideal $k_0 = 32.0$ mol^{-1} dm^3 s^{-1}, which is that found for the 'no added' Ca^{2+} data (Figure 7.8(c)).

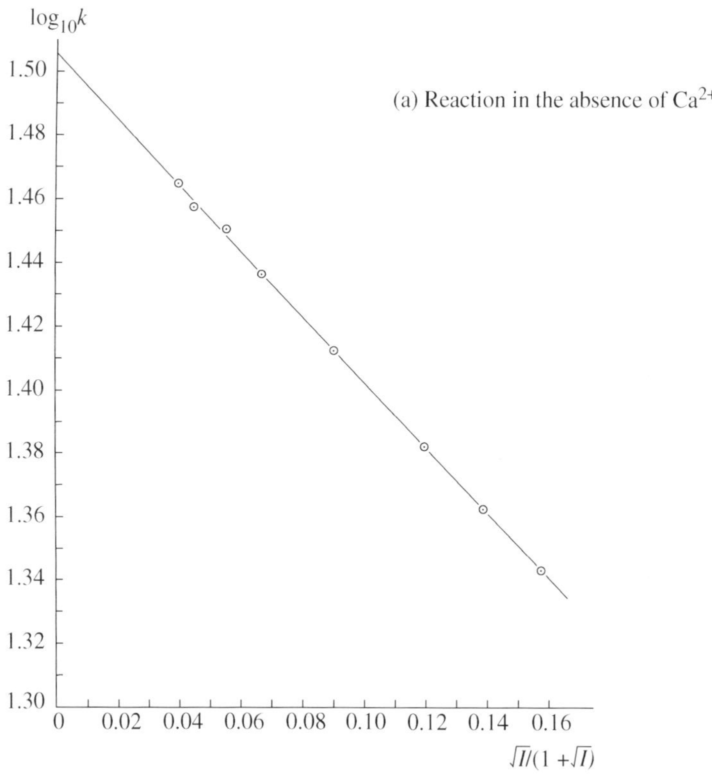

Figure 7.8 (a) Plot of $\log_{10} k$ versus $\sqrt{I}/(1+\sqrt{I})$ for reaction of $(C_2H_5)_3N^+CH_2COOCOOC_2H_5(aq)$ with $OH^-(aq)$ in the absence of Ca^{2+} at low ionic strengths. (b) Plot of $\log_{10} k$ versus $\sqrt{I}/(1+\sqrt{I})$ for reaction of $(C_2H_5)_3N^+CH_2COOC_2H_5(aq)$ with $OH^-(aq)$ in the presence of Ca^{2+} at high ionic strengths. (c) Plot of $\log_{10} k$ versus $\sqrt{I}/(1+\sqrt{I})$ for reaction of $(C_2H_5)_3N^+CH_2COOC_2H_5(aq)$ with $OH^-(aq)$ in the presence of Ca^{2+} for the full range of ionic strengths

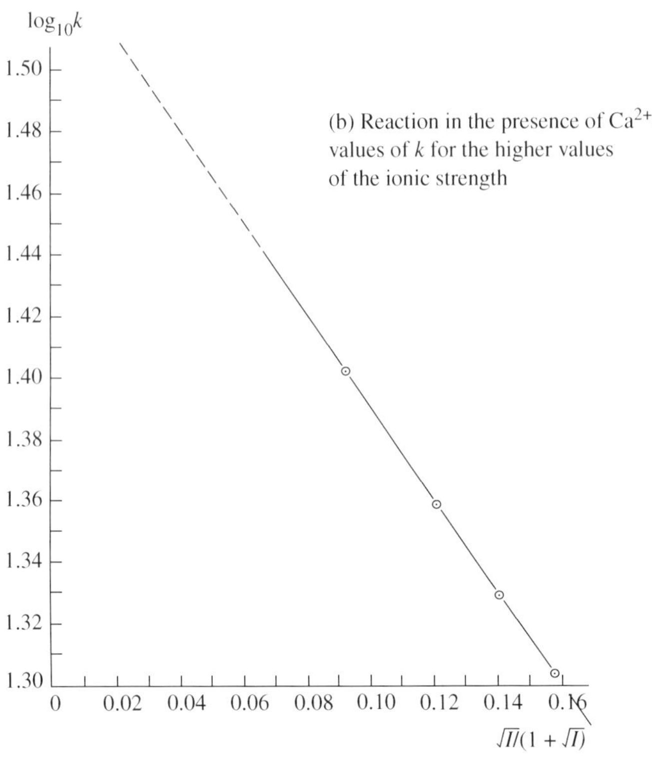

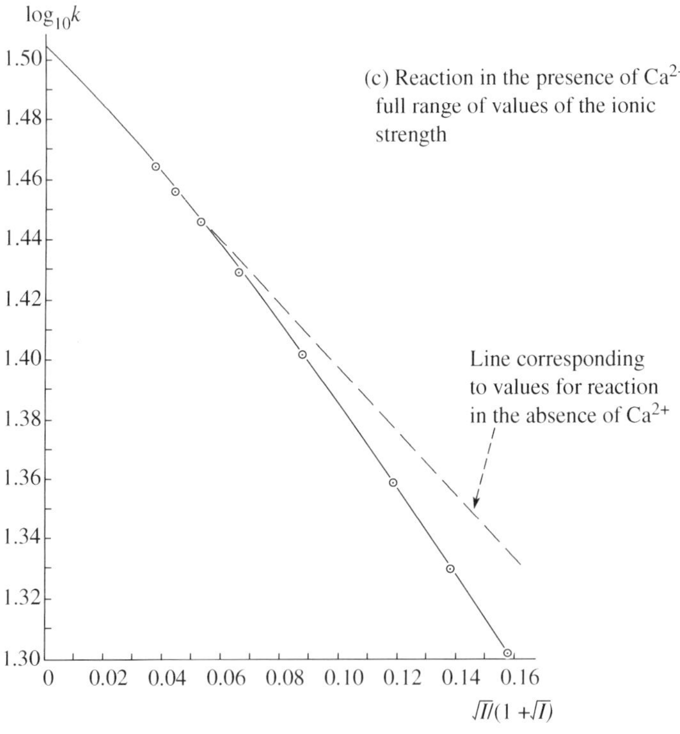

Figure 7.8 *(Continued)*

- A comparison of the graph for experiments in the absence of added $CaCl_2$ with that for experiments with added $CaCl_2$ is highly pertinent. For each value of $\sqrt{I}/(1+\sqrt{I})$, the points for the experiments with added $CaCl_2$ *lie below* those for experiments without, and, *even more importantly*, they lie progressively lower the greater is the concentration of added $CaCl_2$. This reaction has $OH^-(aq)$ as one reactant, and $OH^-(aq)$ is known to associate with Ca^{2+} ions.

$$Ca^{2+}(aq) + OH^-(aq) \rightleftharpoons CaOH^+(aq)$$

The higher the concentration of $CaCl_2$ added, the more ion pairs will be formed, and the more OH^- will be removed. This will reduce the rate of reaction for a given stoichiometric $[OH^-]$, with progressively greater reduction the more ion pairs are formed. This will show itself as a decrease in rate constant, with the decrease being proportional to the amount of OH^- removed in the formation of the ion pairs. It is possible to calculate the association constant for the ion pair from the decrease observed for each $[CaCl_2]$ added, and this was one of the original methods of finding this constant.

Problem 7.5 is particularly important as it illustrates conclusively that linearity of a plot of $\log_{10} k_{obs}$ versus $\sqrt{I}/(1+\sqrt{I})$ should not be taken to be the final word on the reaction. There are many reactions where a linear graph is obtained when the stoichiometric ionic strength is used. However, a further scrutiny of the results will reveal, as in the example quoted in Problem 7.5, that there are discrepancies which need to be explained. If this is done, considerable insight into the details of the reaction can be made. It will be of interest to compare this problem with Problem 8.3, where a similar hydrolysis is studied, but where the ion pair is found to be as reactive as OH^-.

7.4 $\Delta S^{\neq *}$ and Pre-exponential *A* Factors

Problem 4.14 showed that the observed *A* factor can be correlated with $\Delta S^{\neq *}$, and more importantly that '*p*' and $\exp(+\Delta S^{\neq *}/R)$ are related. Consideration of factors contributing to $\Delta S^{\neq *}$ gave considerable insight into the properties of the activated complex and the process of forming the activated complex for gas phase reactions. Such considerations have proved even more fruitful for reactions in solution.

Remember: when discussing microscopic quantities appearing in theoretical treatments, molecular units are generally used, e.g. k, the Boltzmann constant. In contrast, when discussing experimental quantities, molar units are generally used; e.g. R, the gas constant, is used rather than the Boltzmann constant; see Section 7.3, especially 7.3.1 and 7.3.3.

Worked Problem 7.6

Question. For the reaction

$$V^{2+}(aq) + [Co(NH_3)_5SO_4]^{+}(aq) \rightarrow V^{3+}(aq) + Co^{2+}(aq) + 5\,NH_3(aq) + SO_4^{2-}(aq)$$

the rate constant varies with temperature as follows:

$\frac{\text{temperature}}{^\circ\text{C}}$	15	25	35	45	50	55
$\frac{k}{\text{mol}^{-1}\text{dm}^3\text{s}^{-1}}$	12.1	25.9	50.3	95.8	125	154

Calculate values of $\Delta H^{\neq *}$ and $\Delta S^{\neq *}$.

Answer. The thermodynamic formulation of transition state theory (Section 4.4 and Equations (4.39) and (4.40)) is

$$k = \left(\frac{\kappa kT}{h}\right)K^{\neq *} = \left(\frac{\kappa kT}{h}\right)\exp\left(\frac{-\Delta G^{\neq *}}{RT}\right) \tag{4.39 and 4.40}$$

$$= \left(\frac{\kappa kT}{h}\right)\exp\left(\frac{-\Delta H^{\neq *}}{RT}\right)\exp\left(\frac{+\Delta S^{\neq *}}{R}\right) \tag{4.41}$$

$$\log_e \frac{k}{T} = \log_e\left(\frac{\kappa k}{h}\right) - \frac{\Delta H^{\neq *}}{RT} + \frac{\Delta S^{\neq *}}{R} \tag{4.42}$$

A plot of $\log_e k/T$ versus $1/T$ should have slope $= -\Delta H^{\neq *}/R$

$$\frac{\Delta S^{\neq *}}{R} = \log_e \frac{k}{T} - \log_e\left(\frac{\kappa k}{h}\right) + \frac{\Delta H^{\neq *}}{RT} \tag{7.31}$$

can then be calculated.

$\frac{\text{temperature}}{^\circ\text{C}}$	15	25	35	45	50	55
$\frac{k}{\text{mol}^{-1}\text{dm}^3\text{s}^{-1}}$	12.1	25.9	50.3	95.8	125	154
$\frac{T}{\text{K}}$	288	298	308	318	323	328
$\frac{10^3\text{K}}{T}$	3.470	3.356	3.245	3.143	3.095	3.047
$\frac{k}{T} \times \frac{K}{\text{mol}^{-1}\text{dm}^3\text{s}^{-1}}$	0.0420	0.0869	0.1632	0.3013	0.3870	0.4695
$\log_e \frac{k}{T}$	−3.17	−2.44	−1.81	−1.20	−0.95	−0.76

The graph of $\log_e(k/T)$ versus $1/T$ has slope $= -\Delta H^{\neq *}/R = -5.75 \times 10^3$ K, giving $\Delta H^{\neq *} = 5.75 \times 10^3\ \text{K} \times 8.3145\ \text{J mol}^{-1}\ \text{K}^{-1} = 47.8\ \text{kJ mol}^{-1}$ (Figure 7.9).

The intercept cannot be found by extrapolation (scale trouble, see comment below), and must be found by calculation.

$\dfrac{\kappa k}{h} = 2.084 \times 10^{10}\ \text{s}^{-1}\ \text{K}^{-1}$, assuming $\kappa = 1$

$$\frac{\Delta S^{\neq *}}{R} = \log_e \frac{k}{T} - \log_e\left(\frac{\kappa k}{h}\right) + \frac{\Delta H^{\neq *}}{RT} \qquad (7.31)$$

gives at 25 °C

$$\frac{\Delta S^{\neq *}}{R} = -2.44 - 23.76 + \frac{47.8 \times 10^3\ \text{J mol}^{-1}}{8.3145\ \text{J mol}^{-1}\ \text{K}^{-1} \times 298\ \text{K}}$$
$$= -2.44 - 27.36 + 19.29$$
$$\Delta S^{\neq *} = -6.91 \times 8.3145\ \text{J mol}^{-1}\ \text{K}^{-1} = -57.5\ \text{J mol}^{-1}\ \text{K}^{-1}$$

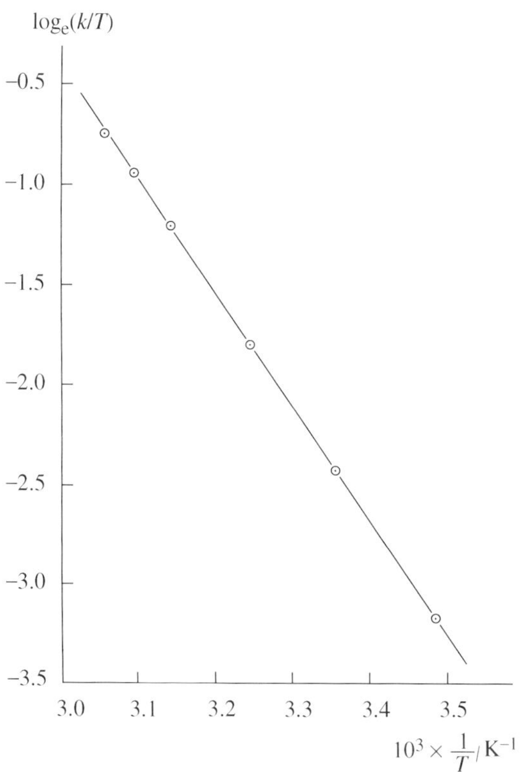

Figure 7.9 Plot of $\log_e(k/T)$ to determine $\Delta H^{\neq *}$

7.4.1 A typical problem in graphical analysis

Problem 7.6 illustrates a procedural problem often encountered by students.

$$k = \left(\frac{\kappa kT}{h}\right) K^{\neq *} = \left(\frac{\kappa kT}{h}\right) \exp\left(\frac{-\Delta G^{\neq *}}{RT}\right) \qquad (4.39) \text{ and } (4.40)$$

$$\frac{k}{T} = \left(\frac{\kappa k}{h}\right) \exp\left(\frac{-\Delta H^{\neq *}}{RT}\right) \exp\left(\frac{+\Delta S^{\neq *}}{R}\right) \qquad (7.32)$$

$$\log_e \frac{k}{T} = \log_e\left(\frac{\kappa k}{h}\right) - \frac{\Delta H^{\neq *}}{RT} + \frac{\Delta S^{\neq *}}{R} \qquad (4.42)$$

giving

$$\Delta S^{\neq *} = R \log_e \frac{k}{T} - R \log_e\left(\frac{\kappa k}{h}\right) + \frac{\Delta H^{\neq *}}{T} \qquad (7.33)$$

A plot of $\log_e(k/T)$ versus $1/T$ should have slope $= -\Delta H^{\neq *}/R$.

When the plot of $\log_e(k/T)$ versus $1/T$ is drawn, there is a very long extrapolation to zero for $1/T$. To carry out this extrapolation would require a ridiculous amount of graph paper, or a ridiculous bunching together of points on ordinary graph paper. Graphical extrapolation is impossible, and the value of the intercept must be found by calculation. This can happen in various aspects of physical chemistry, and very often occurs in plots vs $1/T$.

7.4.2 Effect of the molecularity of the step for which $\Delta S^{\neq *}$ is found

As in the gas phase, if the reaction step is bimolecular with two species forming an activated complex resembling a single species, there will be a decrease in entropy on activation. In solution this is called an *associative* reaction. If reaction is unimolecular and the activated complex resembles an incipient two (or more) species, then an increase in entropy would result. This is termed a *dissociative* reaction for solution reactions. A reaction somewhere in between these two extremes is termed *interchange*, and the entropy change is likely to be small.

7.4.3 Effect of complexity of structure

In the gas phase it was shown that as the complexity of the reacting molecules increased, $\Delta S^{\neq *}$ decreased; see Sections 4.3.5 and 4.4.2. This is a consequence of including the effect of the internal motions of rotation and vibration in reactants and activated complex. The change in the number of degrees of freedom is a major contribution to the entropy of activation; see Problems 4.12–4.15.

7.4.4 Effect of charges on reactions in solution

One of the most characteristic features of reactions in solution which involve ionic charges in reactants, or a charge distribution in the activated complex, is the quite decided dependence of the *A* factor on the charges involved. This has been hinted at previously in Section 7.1. Electrostatic interactions are important both between the reacting species themselves, and also especially with the solvent. This is not limited to reactions involving ionic species. Two neutral polar molecules can easily give a charge-separated activated complex which, though neutral overall, has an uneven distribution of charge over it.

7.4.5 Effect of charge and solvent on $\Delta S^{\neq *}$ for ion–ion reactions

The following *does not include* the effect of solvation and the change in the structure of the solvent on activation. This is discussed independently in Section 7.4.9.

For *ideal conditions*, corresponding to rate constants extrapolated to zero ionic strength,

$$\Delta G^{\neq *}_{\text{ions in solution}} = \Delta G^{\neq *}_{\text{molecules in gas}} + \frac{z_A z_B e^2}{4\pi\varepsilon_0\varepsilon_r r_{\neq}} \tag{7.26}$$

Since reactions should be compared under ideal conditions, equation (7.26) is used. The effect of temperature on $\Delta G^{\neq *}$ allows $\Delta S^{\neq *}$ to be found, since

$$\Delta S^{\neq *} = -\left(\frac{\partial \Delta \mathrm{G}^{\neq *}}{\partial \mathrm{T}}\right)_p \tag{7.34}$$

However, for reactions in solution it is necessary to consider also the fact that the relative permittivity ε_r decreases as temperature increases. Thus, for ions of like charge ($z_A z_B$ positive), the electrostatic term in Equation (7.26) *increases* as T increases. This gives a *positive* contribution to $(\partial \Delta G^{\neq *}/\partial T)_p$, and therefore a *negative* contribution to $\Delta S^{\neq *}$. Consequently, $\Delta S^{\neq *}$ is predicted to be less than the value it would have taken if there had been no electrostatic term.

From such arguments it can be shown that Equation (7.26) gives the following relations.

- For ions of like charge ($z_A z_B$ positive), $\Delta S^{\neq *}$ is predicted to be less than the value which it would have taken if there had been no electrostatic term.
- For ions of unlike charge ($z_A z_B$ negative), $\Delta S^{\neq *}$ is predicted to be greater than the value which it would have taken if there had been no electrostatic term.
- For ions of like sign $\Delta S^{\neq *}_{\text{ions in solution}} < \Delta S^{\neq *}_{\text{molecules in gas}}$.

- For ions of unlike sign $\Delta S^{\neq *}_{\text{ions in solution}} > \Delta S^{\neq *}_{\text{molecules in gas}}$.

Values of the A factor differ from each other because of differences in $\Delta S^{\neq *}$ (Worked Problem 4.14). Because A is proportional to $\exp(\Delta S^{\neq *}/R)$, a larger value of $\Delta S^{\neq *}$ means that there will be a larger value of A, and a smaller value of $\Delta S^{\neq *}$ means that there will be a smaller value of A. Thus

- for ions of like sign, the relation

$$\Delta S^{\neq *}_{\text{ions in solution}} < \Delta S^{\neq *}_{\text{molecules in gas}}$$

  requires that

$$A_{\text{ions in solution}} < A_{\text{molecules in gas}}$$

  and so

$$\log_{10} A_{\text{ions in solution}} < \log_{10} A_{\text{molecules in gas}}$$

- for ions of unlike sign, the relation

$$\Delta S^{\neq *}_{\text{ions in solution}} > \Delta S^{\neq *}_{\text{molecules in gas}}$$

  requires that

$$A_{\text{ions in solution}} > A_{\text{molecules in gas}}$$

  and so

$$\log_{10} A_{\text{ions in solution}} > \log_{10} A_{\text{molecules in gas}}$$

For both these situations a choice will have to be made as to the magnitude of $A_{\text{molecules in gas}}$, as this is the point of comparison. This is often taken to be around 3×10^{10} mol^{-1} dm^3 s^{-1}, $\log_{10} A = 10.5$.

Table 7.1 lists values of $\log_{10} A$ for some ionic reactions, and shows qualitative agreement of experiment with this electrostatically modified collision theory. At first sight this indicates the correctness of the electrostatic treatment.

However, a closer look shows that most of the reactions quoted are reactions of polyatomic ions, where it could be expected that the internal rotational and vibrational structures of the reactant ions and the activated complex will make a significant contribution. Values of gas phase $\Delta S^{\neq *}$ (Table 4.4) become increasingly more negative as the complexity of the reactants increases, with the corresponding calculated p factors likewise becoming increasingly smaller. This indicates that there cannot be a single point of comparison, and that there is considerable leeway in the value which can be chosen.

The conclusion is that, unless the internal structure is included in the calculations, or the point of comparison is chosen to take cognisance of this, the electrostatic theory may not be adequate.

7.4.6 Effect of charge and solvent on $\Delta S^{\neq *}$ for ion–molecule reactions

Here there is a discrete charge on one of the reactants and a dipole on the other. The electrostatic interactions arising from the charge and the solvent can be deduced similarly to those for ion–ion interactions.

$$\Delta G^{\neq *}_{\text{ion–molecule in solution}} = \Delta G^{\neq *}_{\text{molecules in gas}} + \frac{ze\mu \cos\theta}{4\pi\varepsilon_0\varepsilon_r r^2_{\neq}} \tag{7.35}$$

where μ is the dipole moment, and θ represents the angle of approach of the ion to the dipole.

This gives the relation

$$\Delta S^{\neq *}_{\text{ion-dipole in solution}} > \Delta S^{\neq *}_{\text{molecules in gas}}$$

from which it is predicted that $A_{\text{ion-dipole in solution}} >$ typical gas phase value, i.e. $>3 \times 10^{10}\,\text{mol}^{-1}\,\text{dm}^3\,\text{s}^{-1}$. This prediction is qualitatively the same as that for reaction between two ions of opposite charge. In physical terms, it arises because a positive ion would usually approach the negative end of a dipole, and a negative ion would usually approach the positive end of a dipole.

Detailed calculations suggest that p factors greater than unity and of the order of 1 to 10^3 are expected. This is not always observed; see Table 7.2, where some of the reactions have p factors in the predicted range and some have p factors less than unity.

The above treatment needs to be modified in some way. It can again be asked whether the typical value of $A \approx 3 \times 10^{10}\,\text{mol}^{-1}\,\text{dm}^3\,\text{s}^{-1}$ is the correct point of comparison. If the lower values suggested by transition state theory become the point of comparison, then the above treatment could be essentially correct.

Worked Problem 7.7

Question. Although there is a question as to whether $A \approx 3 \times 10^{10}\,\text{mol}^{-1}\,\text{dm}^3\,\text{s}^{-1}$ is the correct point of comparison for ion–molecule reactions, why is this ambiguity less likely to be a problem for ion–ion reactions?

Answer. The argument in Section 7.4.6 suggests that the discrepancy between predicted and observed values for ion–molecule reactions is a result of ignoring the change in the internal degrees of freedom. This is also relevant to ion–ion reactions when the ions are polyatomic species. But the charge effect for ion–ion reactions $\propto z_A z_B e^2/r_{\neq}$ is much larger than the charge-dipole effect for ion–molecule reactions $\propto ze\mu \cos\theta/r^2_{\neq}$. It is a matter of the relative magnitudes of these terms.

7.4.7 Effect of charge and solvent on $\Delta S^{\neq *}$ for molecule–molecule reactions

The theoretical treatment here is even more fraught with difficulties than that for ion–molecule reactions. Here, both the reactants and the activated complex are assumed to be dipolar. The theory predicts that there will be a spread in the values of $\Delta S^{\neq *}$ leading to p factors greater and less than unity, but these p factors are predicted to be close to unity. The observed p factors vary over a large range, Table 7.3, and are comparable to those for ion–ion-like charge reactions. The conclusion is that the electrostatic treatment may be totally inadequate, or that the effects of the internal structure are important.

7.4.8 Effects of changes in solvent on $\Delta S^{\neq *}$

It is much more difficult to predict the effect of change of solvent on $\Delta S^{\neq *}$ than it is to predict the effect on the ideal rate constant. This is because the analysis for the dependence of $\Delta S^{\neq *}$ on solvent includes the temperature dependence of the relative permittivity. This complexity in the analysis occurs for ion–ion and ion–molecule reactions, and, because of the difficulty in disentangling the effects, this will not be pursued further.

7.4.9 Changes in solvation pattern on activation, and the effect on *A* factors for reactions involving charges and charge-separated species in solution

A further complication in interpreting $\Delta S^{\neq *}$ values is the contribution made from solvation, and the alteration of the structure of the solvent on activation. This is dependent on the charges on the reactants and on the activated complex. The activated complex has a charge equal to the algebraic sum of the charges on the reactants.

Worked Problem 7.8

Question. Estimate the charges on the activated complexes formed during the following reactions.

Question	Answer
• $+3$ charge reacting with $+1$ charge	$+4$
• $+2$ charge reacting with -1 charge	$+1$
• -1 charge reacting with -2 charge	-3

- -3 charge reacting with a $+1$ charge — -2
- $+2$ charge reacting with a -2 charge — zero overall,

but take care, the activated complex is probably still charge separated.

- -1 charge reacting with an uncharged molecule — -1
- two uncharged molecules reacting with each other — zero overall,

but take care, the activated complex could be charge separated.

For reactions involving charges and charge-separated species the solvent will be orientated around the regions of charge, with the degree of orientation being greater the greater the charge. Neutral species can still orientate the solvent around the molecule by virtue of any asymmetric charge distribution, such as a dipole.

Worked Problem 7.9

Question. List the following species in order of ability to orientate solvent molecules.

$$OH^-,\ Ca^{2+},\ H_2O,\ H_3O^+,\ SO_4^{2-},\ NH_3^+CH_2COO^-,\ CCl_4,\ Al^{3+}$$

Answer.

$$Al^{3+} > Ca^{2+} \approx SO_4^{2-} > OH^- \approx H_3O^+ > NH_3^+CH_2COO^- \gg H_2O \gg CCl_4$$

Note.

- $NH_3{}^+CH_2COO^-$ is overall neutral, but charge separated;
- H_2O is also overall neutral, but has only a dipole;
- CCl_4 is also overall neutral, but has no dipole.

7.4.10 Reactions between ions in solution

Like charged ions — Values of *A*: 10^2–10^8 $dm^3\,mol^{-1}s^{-1}$
Unlike charged ions — Values of *A*: 10^{13}–10^{19} $dm^3\,mol^{-1}s^{-1}$

There is very little variation in the magnitude of *A* for a given reaction in different solvents.

When *reaction occurs between ions of like charge* the activated complex will have a larger charge than either reactant. All three will be solvated, but the activated complex can orientate the solvent considerably more than can the reactant ions. Orientation is equivalent to an increase in order for *solvent* molecules, which results in an increase in order for the solvent on activation, manifesting itself as a decrease in entropy on activation. Comparison with reactions for uncharged molecules in the gas phase predicts a decreased value for the entropy of activation, and hence low A factors.

When *reaction is between ions of unlike sign* the activated complex has a total net charge less than the reactants, and so the total degree of orientation of the solvent around the activated complex is now less than before activation. This leads to release of solvent molecules giving an increase in disorder of the *solvent*, leading to an increase in entropy on forming the activated complex. The $\Delta S^{\neq *}$ and the A factor are predicted to be larger than the values found for the reaction of molecules in the gas phase.

These predictions are confirmed by the typical values given in Table 7.1.

For both charge types it is not expected that there will be a great variation in $\Delta S^{\neq *}$ and A with change of solvent, and experiment bears this out. The *change* in orientation of solvent molecules on activation and the *change* of entropy are irrespective of the ease with which the permanent dipoles of the solvent can be oriented, even though there may be a wide spread in solvation of the reactant species with change of solvent. What is important is that there will be a similar spread in solvation of the activated complex with change of solvent, so there is little variation in the *change* in orientation on activation with change of solvent.

For reactions between ions of *like sign*, solvation effects give a decrease in entropy on activation. Consideration of the internal structure leads again to a decrease in entropy on activation. The two effects reinforce each other, and also are in the same direction as predicted by the electrostatic treatment as given in Section 7.4.5. This would indicate p factors of less than unity.

For reactions between ions of *unlike sign*, solvation effects give an increase in entropy on activation. Consideration of internal structure gives a decrease in $\Delta S^{\neq *}$ and the two effects could balance, or partially balance, each other. The electrostatic modifications would, however predict an increase in $\Delta S^{\neq *}$, indicating p factors greater than unity. Since the other two effects are in opposite directions to each other, the electrostatic modifications must dominate, with this being the major factor influencing the A factors.

7.4.11 Reaction between an ion and a molecule

Typical observed values of A are 10^{9}–10^{12} $mol^{-1}\ dm^{3}\ s^{-1}$, with very little variation in the magnitude of the A factor for a given reaction in different solvents.

The reactant molecule will be very much less solvated than the reactant ion, while the activated complex will be solvated to a similar extent to the initial state since the total charge remains the same on activation. The net change in orientation of the *solvent* will be small, giving a very small contribution to the entropy of activation and to the *A* factor from solvation effects. *A* factors will lie in the gas phase range of values. This is observed with a few exceptions (Table 7.2).

Since the total degree of orientation of the solvent is approximately the same for the initial state and activated complex, the change in orientation is small and close to zero, and approximately constant for a given reaction in a variety of solvents. Hence *A* factors are not expected to vary much with change of solvent.

Table 7.2 shows these predictions to be borne out by a large number of reactions, but there are exceptions where the *A* factors are much lower than expected. These can be explained if the internal degrees of freedom are considered. Effects of electrostatic interactions and of solvation are small, and could be in either direction, but the dominant effect is probably due to internal motions.

7.4.12 Reactions between uncharged polar molecules

Typical observed values for *A* are 10^2–10^7 $\text{mol}^{-1}\ \text{dm}^3\ \text{s}^{-1}$, and there is a large variation for a given reaction in different solvents.

Here the reactants are overall neutral, though dipolar, and the activated complex is also overall neutral. However, the charge distribution in the activated complex can vary from dipolar to an incipient charge distribution similar to an ion pair, with a range of different charge-separated intermediate structures possible. For instance, in the reaction between tertiary amines and halogenoalkanes the product is an ionic quaternary ammonium halide. For such reactions it is likely that the activated complex will have considerable charge separation, even to the extent of resembling an ion pair rather than a dipole:

$$\text{R}_3\text{N} + \text{R}'\text{I} \rightarrow [\text{R}_3\text{N}^{\delta+}\text{----R}'\text{----I}^{\delta-}]^{\neq} \rightarrow \text{R}_3\text{R}'\text{N}^+\text{I}^-$$

in contrast to reactions where molecular products are found, when the activated complex could be merely dipolar.

Existence of charge distributed over the activated complex would encourage solvation. The greater the degree of charge separation in the activated complex, the greater will be the solvation. This gives increased orientation of the *solvent* on activation and a decrease in entropy on activation. Hence low *A* factors are expected, with lower values correlating with increasing charge separation and solvation in the activated complex.

Since the reactants are molecules, the extent of solvation will be small and approximately the same for all solvents. But the extent of charge separation in the activated complex is very dependent on the solvent, by virtue of the magnitude of the relative permittivity, and a large variation in the extent of solvation of the activated complex is expected. The entropy of activation and the *A* factor are expected to vary with change of solvent.

Table 7.3 shows A factors that are much lower than theory would suggest and are more in keeping with those for ions of like charge reacting. It seems reasonable to assume that changes in the internal degrees of freedom are more important than are the charge and solvent factors. This is in keeping with the similar situation for ion–molecule reactions.

Worked Problem 7.10

Question. Assuming that the reaction step is associative, interpret the signs and magnitudes of the $\Delta S^{\neq *}$ values for the following reactions in terms of charge and complexity of structure:

Reaction	$\frac{\Delta S^{\neq *}}{\text{J mol}^{-1}\text{ K}^{-1}}$
$Co(NH_3)_5Cl^{2+}(aq) + H_2O(l)$	−32
$Co(NH_3)_5(NO_3)^{2+}(aq) + H_2O(l)$	+8
$Pd(CN)_4^{2-}(aq) + CN^-(aq)$	−178
$PtdienBr^+(aq) + I^-(aq)$	−104
$Co(NH_3)_5Br^{2+}(aq) + OH^-(aq)$	+92
$V^{2+}(aq) + Co(NH_3)_5SO_4^+(aq)$	−243
$CH_2BrCOO^-(aq) + S_2O_3^{2-}(aq)$	−71
$CH_3CONH_2(aq) + H_2O(l)$	−142
$PtdienBr^+(aq) + SCN^-(aq)$	−112

(dien is $NH_2CH_2CH_2NHCH_2CH_2NH_2$.)

Answer.

Reaction	$\frac{\Delta S^{\neq *}}{\text{J mol}^{-1}\text{ K}^{-1}}$	Type of reaction	Effect of charge	Effect of complexity	Effect of solvation	Comments
(a) $Co(NH_3)_5Cl^{2+}(aq) + H_2O(l)$	−32	ion + dipole	small, and +ve	−ve: perhaps biggest effect	small, +ve or −ve	seems to fit
(b) $Co(NH_3)_5(NO_3)^{2+}(aq) + H_2O(l)$	+8	ion + dipole	small, and +ve	−ve: perhaps biggest effect (similar to (a))	small, +ve or −ve	does not fit, but why the difference from (a)? −ve end of dipole simulates −ve charge under influence of 2+; (+2) + (−1). This effect would be +ve. But this would apply to (a) also?

Reaction	$\Delta S^{\neq *}$ / $J\,mol^{-1}\,K^{-1}$	Type of reaction	Effect of charge	Effect of complexity	Effect of solvation	Comments
(c) $Pd(CN)_4^{2-}(aq)$ + $CN^-(aq)$	−178	(−2) + (−1)	large, and −ve	−ve	−ve	fits; complexity enhances
(d) $PtdienBr^+(aq)$ + $I^-(aq)$	−104	(+1) + (−1)	+ve	large, and −ve	+ve	does not fit: complexity of dien compensates +ve effects?
(e) $Co(NH_3)_5Br^{2+}(aq)$ + $OH^-(aq)$	+92	(+2) + (−1)	large, and +ve	−ve	+ve	fits: complexity reduces value?
(f) $V^{2+}(aq)$ + $Co(NH_3)_5SO_4^+(aq)$	−243	(+2) + (+1)	large, and −ve	−ve	−ve	fits; compare with (c)
(g) $CH_2BrCOO^-(aq)$ + $S_2O_3^{2-}(aq)$	−71	(−1) + (−2)	large and −ve	−ve	−ve	fits; compare with (c) and (d)
(h) $CH_3CONH_2(aq)$ + $H_2O(l)$	−142	(0) + (0)	small, possibly −ve	small, −ve	small, possibly −ve	does not fit: mechanism may involve initial protonation
(i) $PtdienBr^+(aq)$ + $SCN^-(aq)$	−112	(+1) + (−1)	+ve	large, and −ve	+ve	does not fit: complexity of dien compensates +ve effects? compare with (d)

This question demonstrates the variety of effects contributing to the observed $\Delta S^{\neq *}$, and indicates how to attempt to rationalize them. However, this analysis assumes that the $\Delta S^{\neq *}$ values refer to a single associative step. A complete analysis *must* include a full mechanistic study before any definitive conclusions are reached.

7.5 $\Delta H^{\neq *}$ Values

Worked Problem 7.6 shows how $\Delta H^{\neq *}$ values can be found from the temperature dependence of the rate constant.

7.5.1 Effect of the molecularity of the step for which the $\Delta H^{\neq *}$ value is found

Again a distinction is possible between associative, dissociative and interchange mechanisms. In the dissociative mechanism at least one bond is broken. Dissociative reactions thus have higher $\Delta H^{\neq *}$ values than associative mechanisms. Interchange mechanisms, where synchronous bond breaking and bond forming occur, have values in between.

7.5.2 Effect of complexity of structure

$\Delta H^{\neq *}$ reflects the effect of bond breaking and bond making in the process of activation. Although the internal motions will contribute to this, they will not be expected to have a major effect.

7.5.3 Effect of charge and solvent on $\Delta H^{\neq *}$ for ion–ion and ion–molecule reactions

As in Sections 7.4.5 and 7.4.6, this does not include the effect of change in solvation pattern. Equation (4.40) is the best starting point and leads to Equation (7.27) as given in Section 7.3.3

$$\log_e \frac{k^{\text{ideal}}_{\text{ions in solution}}}{T} = \log_e \frac{\kappa k}{h} - \frac{\Delta G^{\neq *}_{\text{molecules in gas}}}{kT} - \frac{z_A z_B e^2}{4\pi \varepsilon_0 \varepsilon_r r_{\neq} kT} \tag{7.36}$$

Since, by analogy with Equation (4.33),

$$\Delta G^{\neq *} = \Delta H^{\neq *} - T\Delta S^{\neq *} \tag{7.37}$$

$$\log_e \frac{k^{\text{ideal}}_{\text{ions in solution}}}{T} = \log_e \frac{\kappa k}{h} + \frac{\Delta S^{\neq *}_{\text{molecules in gas}}}{R} - \frac{\Delta H^{\neq *}_{\text{molecules in gas}}}{RT} - \frac{z_A z_B e^2}{4\pi \varepsilon_0 \varepsilon_r r_{\neq} kT} \tag{7.38}$$

From this it can be shown that

for *ions of like sign*: $\frac{\Delta H^{\neq *}_{\text{ions in solution}}}{R} < \frac{\Delta H^{\neq *}_{\text{molecules in gas}}}{R}$

for *ions of unlike sign*: $\frac{\Delta H^{\neq *}_{\text{ions in solution}}}{R} > \frac{\Delta H^{\neq *}_{\text{molecules in gas}}}{R}$

An analogous treatment for reaction between an ion and a molecule suggests that for reaction of an

ion with a molecule: $\frac{\Delta H^{\neq *}_{\text{ion+molecule in solution}}}{R} > \frac{\Delta H^{\neq *}_{\text{molecules in gas}}}{R}$

As with the treatment for $\Delta S^{\neq *}_{\text{reactions in solution}}$ any comparisons with the gas phase values depends critically on what value we choose for $\Delta H^{\neq *}_{\text{molecules in gas}}$.

7.5.4 Effect of the solvent on $\Delta H^{\neq *}$ for ion–ion and ion–molecule reactions

The effect of change of solvent is made complex for both ion–ion and ion–molecule reactions because of the presence of the temperature dependence of the relative permittivity. This is similar to the corresponding situation for $\Delta S^{\neq *}_{\text{reactions in solution}}$ (Sections 7.4.5, 7.4.6 and 7.4.8).

7.5.5 Changes in *solvation pattern* on activation and the effect on $\Delta H^{\neq *}$

If reaction involves molecular reactants and a charge-separated activated complex, the activated complex is likely to be more solvated than the reactants. The activation energy is reduced by an amount similar to the ΔH of solvation, leading to an increase in rate relative to the gas phase reaction (Figure 7.10).

Reaction between two heavily solvated oppositely charged ions could produce a neutral, possibly charge-separated activated complex that is only lightly solvated,

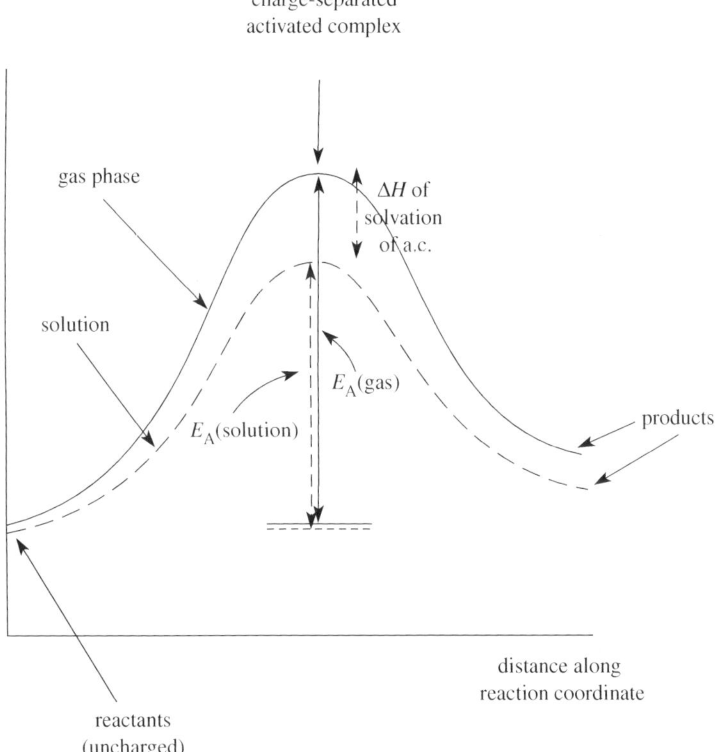

Figure 7.10 Potential energy profile showing the effect of the solvent for reaction between uncharged reactants → charge-separated activated complex

leading to an increased activation energy. This reinforces the effect of the electrostatic term. Likewise, if reaction is between two heavily solvated ions of like charge, the activated complex will have a higher charge than either of the reactants and be much more heavily solvated. This would lead to a decrease in the activation energy by an amount similar to the ΔH of solvation, but again this reinforces the effect of the electrostatic term.

These are extremes, and intermediate situations will also exist.

7.6 Change in Volume on Activation, $\Delta V^{\neq *}$

Such effects are confined to reactions in solution, and are of fundamental importance in mechanistic studies. As with entropy changes, there are several factors which can make a substantial contribution to the overall $\Delta V^{\neq *}$.

Worked Problem 7.11

Question. The following data give the dependence of rate constant on pressure for a second order rate constant at 25 °C:

$\frac{p}{\text{atm}}$	1.00	200	400	600	800
$\frac{10^4 k}{\text{mol}^{-1}\text{dm}^3\text{s}^{-1}}$	9.58	10.65	11.99	13.4	14.9

Find $\Delta V^{\neq *}$.

Answer.

$$k = \left(\frac{\kappa kT}{h}\right) \exp\left(\frac{-\Delta G^{\neq *}}{RT}\right) \tag{4.40}$$

$$\left(\frac{\partial \log_e k}{\partial p}\right)_T = -\frac{1}{RT}\left(\frac{\partial(\Delta G^{\neq *})}{\partial p}\right)_T = -\frac{\Delta V^{\neq *}}{RT} \tag{4.43}$$

The plot of $\log_e k$ versus p has a slope $= -\Delta V^{\neq *}/RT$

$\frac{p}{\text{atm}}$	1.00	200	400	600	800
$\frac{10^4 k}{\text{mol}^{-1}\text{dm}^3\text{s}^{-1}}$	9.58	10.65	11.99	13.4	14.9
$\log_e k$	−6.95	−6.85	−6.73	−6.62	−6.51

See Figure 7.11.

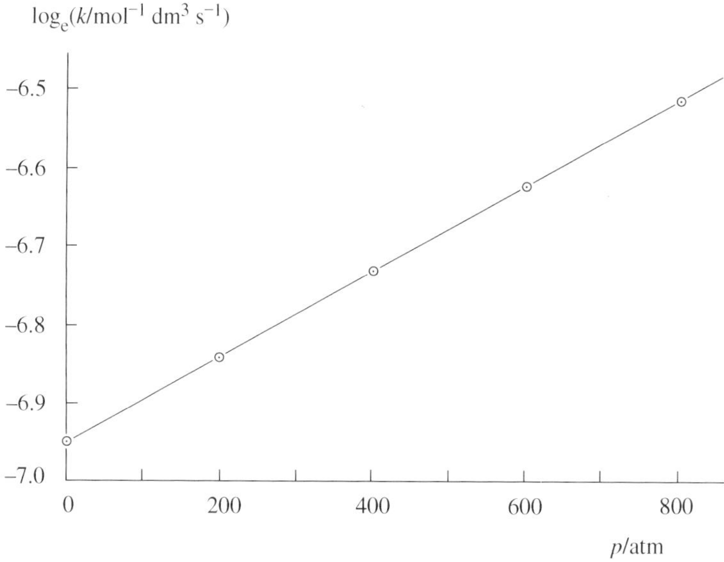

Figure 7.11 Graph of $\log_e k$ versus p to determine $\Delta V^{\neq *}$

$$\text{Slope of graph} = +5.5 \times 10^{-4}\,\text{atm}^{-1}$$

$$\begin{aligned}
\frac{\Delta V^{\neq *}}{RT} &= -5.5 \times 10^{-4}\,\text{atm}^{-1} \\
\Delta V^{\neq *} &= -5.5 \times 10^{-4}\,\text{atm}^{-1} \times RT \\
&= -\frac{5.5 \times 10^{-4}}{1.01325 \times 10^5}\text{N}^{-1}\,\text{m}^2 \times 8.3145\,\text{N m mol}^{-1}\,\text{K}^{-1} \times 298\,\text{K} \\
&= -1.34 \times 10^{-5}\,\text{m}^3\,\text{mol}^{-1} \\
&= -1.34 \times 10^{-5} \times 10^6\,\text{cm}^3\,\text{mol}^{-1} \\
&= -13.4\,\text{cm}^3\,\text{mol}^{-1}
\end{aligned}$$

One of the main difficulties in this problem is the matter of units!

Note. A positive pressure effect means that k increases as pressure increases, and $\Delta V^{\neq *}$ is negative.

A negative pressure effect means that k decreases as pressure increases, and $\Delta V^{\neq *}$ is positive.

7.6.1 Effect of the molecularity of the step for which $\Delta V^{\neq *}$ is found

In an associative step, two species form an activated complex which is smaller than the two reactants taken together. There is, thus, a decrease in volume on activation. In a dissociative step the reactant splits up. There is, thus, an increase in volume on activation. An interchange step gives intermediate changes, and the volume change is expected to be fairly small.

7.6.2 Effect of complexity of structure

Change in the internal motions of rotations and vibrations will make a very minor contribution. This contrasts with the much larger possible effect of a change in the overall volume, which reflects major changes in size between the overall size of the reactants and the activated complex.

7.6.3 Effect of charge on $\Delta V^{\neq *}$ for reactions between ions

Equation (4.40) is the starting point, with the corresponding expression for reaction between ions in solution being Equation (7.26).

$$\Delta G^{\neq *}_{\text{ions in solution}} = \Delta G^{\neq *}_{\text{molecules in gas}} + \frac{z_A z_B e^2}{4\pi\varepsilon_0\varepsilon_r r_{\neq}} \tag{7.26}$$

But ΔG^{θ} for reactions in solution depends on pressure, in contrast to gas phase reactions, for which ΔG^{θ} is independent of pressure.

One of the key equations of thermodynamics is

$$\mathrm{d}G = V\mathrm{d}p - S\mathrm{d}T \tag{7.39}$$

At constant T, $\mathrm{d}T = 0$, and $S\,\mathrm{d}T = 0$, giving

$$\mathrm{d}G = V\,\mathrm{d}p \tag{7.40}$$

$$\int_{p_1}^{p_2} \mathrm{d}G = \int_{p_1}^{p_2} V\,\mathrm{d}p \tag{7.41}$$

For an *ideal gas*

$$pV = nRT \tag{7.42}$$

giving

$$\int_{p_1}^{p_2} \mathrm{d}G = nRT\int_{p_1}^{p_2} \frac{\mathrm{d}p}{p} \tag{7.43}$$

from which

$$G_{p_2} - G_{p_1} = nRT\log_e\frac{p_2}{p_1} \tag{7.44}$$

and, if the standard state is 1 atm (more recently 1 bar $= 1.000\,00 \times 10^5$ N m^{-2}), then

$$\mu = \mu^{\theta} + RT \log_e p \tag{7.45}$$

where μ^{θ} is a constant dependent on T and *independent of p*. ΔG^{θ} is a combination of μ^{θ} values for products and reactants, and so is also *independent of pressure*.

For a *substance in solution*, the analogous equation for the ideal case is

$$\mu = \mu^{\theta} + RT \log_e c \tag{7.46}$$

where μ^{θ} is a constant dependent on T and *also on pressure*.

Correspondingly, ΔG^{θ} is a constant dependent on *both T and pressure*.

This means that the comparison has to be with *molecules in solution*, rather than molecules in the gas phase, as was appropriate for $\Delta S^{\neq *}$ and $\Delta H^{\neq *}$. The appropriate starting equation is thus not (7.26), but

$$\Delta G^{\neq *}_{\text{ions in solution}} = \Delta G^{\neq *}_{\text{molecules in solution}} + \frac{z_A z_B e^2}{4\pi \varepsilon_0 \varepsilon_r r_{\neq}} \tag{7.47}$$

Since for reactions in solution

$$\left(\frac{\partial \Delta G^{\theta}}{\partial p}\right)_T = +\Delta V^{\theta} \tag{4.36}$$

Equation (7.47) has to be differentiated with respect to pressure at constant temperature. This requires a knowledge of the pressure dependence of the relative permittivity at constant temperature.

The relative permittivity ε_r increases with increase in pressure. Thus for ions of like charge ($z_A z_B$ positive), the electrostatic term in Equation (7.47) *decreases* as p increases. This gives a *negative* contribution to $\left(\partial \Delta G^{\neq *}/\partial p\right)_T$, and therefore a *negative* contribution to $\Delta V^{\neq *}$. Consequently $\Delta V^{\neq *}$ is predicted to be less than the value it would have taken if there had been no electrostatic term.

For ions of unlike charge ($z_A z_B$ negative), $\Delta V^{\neq *}$ is predicted to be greater than the value which it would have taken if there had been no electrostatic term.

This can be summarized as

$$\text{for } \textit{ions of like sign}, \quad \Delta V^{\neq *}_{\text{ions in solution}} < \Delta V^{\neq *}_{\text{molecules in solution}}$$

$$\text{for } \textit{ions of unlike sign}, \quad \Delta V^{\neq *}_{\text{ions in solution}} > \Delta V^{\neq *}_{\text{molecules in solution}}$$

and so $\Delta V^{\neq *}_{\text{ions in solution}}$ for *ions of like sign* is likely to be less than $\Delta V^{\neq *}_{\text{ions in solution}}$ for *ions of unlike sign*.

7.6.4 Reactions between an ion and an uncharged molecule

In this case

$$\Delta G^{\neq *}_{\text{ion—molecule in solution}} = \Delta G^{\neq *}_{\text{molecules in solution}} + \frac{z e \mu \cos \theta}{4 \pi \varepsilon_0 \varepsilon_r r^2_{\neq}} \tag{7.48}$$

and the prediction is

$$\Delta V^{\neq *}_{\text{ion—polar molecule in solution}} > \Delta V^{\neq *}_{\text{molecules in solution}}$$

7.6.5 Effect of solvent on $\Delta V^{\neq *}$

Here the effect of the solvent on ion–ion reactions and ion–molecule reactions is complicated by the pressure dependence of ε_r for the solvent, and again this is not taken further.

7.6.6 Effect of change of *solvation pattern* on activation and its effect on $\Delta V^{\neq *}$

These effects arise from the same physical phenomenon as those contributing to $\Delta S^{\neq *}$, and consequently parallel them (Figure 7.12).

When an ion is solvated the solvent molecules are packed around the ion, and occupy a smaller volume than they would in pure solvent. For reaction between two ions of opposite charge, formation of the activated complex releases solvent molecules from their solvation sheaths, thereby increasing their volume compared

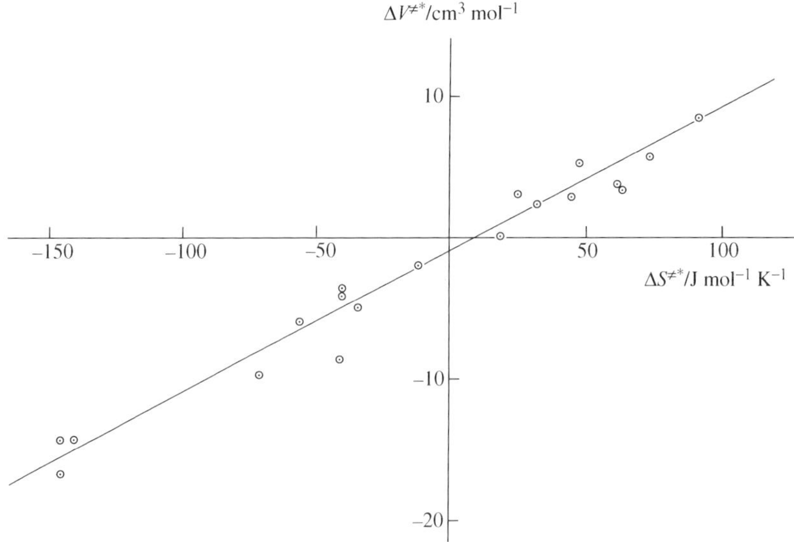

Figure 7.12 Correlation between $\Delta V^{\neq *}$ and $\Delta S^{\neq *}$ for a variety of reactions

to that when bound to the reactant ions. The *overall* volume increases, giving a negative pressure effect.

Remember the following.

- A plot of $\log_e k$ versus p has slope $= -\Delta V^{\neq *}/RT$.
- A positive pressure effect means that k increases as p increases, and $\Delta V^{\neq *}$ is negative.
- A negative pressure effect means that k decreases as p increases, and $\Delta V^{\neq *}$ is positive.

The opposite is found when like charged ions react. Overall, solvent molecules become more tightly packed on forming the activated complex and the *overall* volume decreases, giving a positive pressure effect.

For an ion–molecule reaction the effect will be small, while for reaction of neutral molecules there will be an overall decrease in volume if a charge-separated activated complex is formed. The greater the charge separation, the more negative will be the *overall* volume change, and the bigger the positive pressure effect.

Worked Problem 7.12

Question. *Complete* the following table by predicting the sign of $\Delta S^{\neq *}$ or $\Delta V^{\neq *}$ as appropriate. Also indicate the likely magnitude in terms of very large/small or large/small or close to zero.

Reaction	$\frac{\Delta S^{\neq *}}{\text{J mol}^{-1}\text{ K}^{-1}}$	$\frac{\Delta V^{\neq *}}{\text{cm}^3\text{ mol}^{-1}}$
$Co(NH_3)_5Cl^{2+}(aq) + H_2O(l)$	−32	
$Fe(H_2O)_6^{3+}(aq) + Cl^-(aq)$		+7.8
$Pd(CN)_4^{2-}(aq) + CN^-(aq)$	−178	
$PtdienBr^+(aq) + NO_2^-(aq)$	−56	−6.4
$Co(NH_3)_5Br^{2+}(aq) + OH^-(aq)$	+92	+8.5
$V^{2+}(aq) + Co(NH_3)_5SO_4^+(aq)$	−243	
$CH_2BrCOO^-(aq) + S_2O_3^{2-}(aq)$	−71	
$CH_3CONH_2(aq) + H_2O(l)$		−14.2

Answer.

Reaction	$\frac{\Delta S^{\neq *}}{\text{J mol}^{-1}\text{ K}^{-1}}$	$\frac{\Delta V^{\neq *}}{\text{cm}^3\text{ mol}^{-1}}$
$Co(NH_3)_5Cl^{2+}(aq) + H_2O(l)$	−32	small, and −ve
$Fe(H_2O)_6^{3+}(aq) + Cl^-(aq)$	small, and +ve	+7.8
$Pd(CN)_4^{2-}(aq) + CN^-(aq)$	−178	large, and −ve
$PtdienBr^+(aq) + NO_2^-(aq)$	−56	−6.4
$Co(NH_3)_5Br^{2+}(aq) + OH^-(aq)$	+92	+8.5
$V^{2+}(aq) + Co(NH_3)_5SO_4^+(aq)$	−243	large, and −ve
$CH_2BrCOO^-(aq) + S_2O_3^{2-}(aq)$	−71	moderate, and −ve
$CH_3CONH_2(aq) + H_2O(l)$	large, and negative	−14.2

7.7 Terms Contributing to Activation Parameters

7.7.1 $\Delta S^{\neq *}$

- *Is the step associative or dissociative or neither?*
  Associative, i.e. 2 → 1: expect a loss in entropy; $\Delta S^{\neq *}$ is −ve.
  Dissociative, i.e. 1 → 2: expect a gain in entropy; $\Delta S^{\neq *}$ is +ve.
  Interchange: expect a small change in entropy, i.e. $\Delta S^{\neq *}$ is close to zero.
  Remember to distinguish between the mechanism and the individual steps in it.

- *Change in internal modes for large species*
  Expect a loss in entropy. $\Delta S^{\neq *}$ is −ve.

- *Effect of charges: electrostatic term involving charge and solvent*

  1. Like charges: expect a loss of entropy. $\Delta S^{\neq *}$ is −ve.
  2. Unlike charges: expect a gain in entropy. $\Delta S^{\neq *}$ is +ve.
  3. Ion–molecule: expect a small increase in entropy. $\Delta S^{\neq *}$ is slightly +ve.
  4. Molecule–molecule with a charge-separated activated complex: expect a decrease. $\Delta S^{\neq *}$ is slightly −ve.
  5. Molecule–molecule without charge separation in the activated complex: expect a small increase or decrease. $\Delta S^{\neq *}$ is slightly +ve or −ve.

- *Change in solvation pattern on forming the activated complex*

  1. Like charges: expect a loss of entropy. $\Delta S^{\neq *}$ is −ve.
  2. Unlike charges: expect a gain in entropy. $\Delta S^{\neq *}$ is +ve.
  3. Ion–molecule: expect either a small loss or gain in entropy. $\Delta S^{\neq *}$ is slightly +ve or −ve.

  4 and 5. Molecule–molecule: expect a loss in entropy though it could be small. $\Delta S^{\neq *}$ is slightly −ve.

But, what is the basis of comparison; i.e., is the typical gas phase value to be used, and, if not, what value?

7.7.2 $\Delta V^{\neq *}$

- *Is the step associative or dissociative or neither?*
  Associative, i.e. 2 → 1: expect a decrease in volume; $\Delta V^{\neq *}$ is −ve.
  Dissociative, i.e 1 → 2: expect an increase in volume; $\Delta V^{\neq *}$ is +ve.
  Interchange: expect a small change in volume, i.e. $\Delta V^{\neq *}$ is close to zero.

- *Change in internal modes for large species*
  does not enter into the volume of activation. This makes the interpretation of volumes of activation more straightforward than that for entropies of activation.

- *Effect of charges: electrostatic term involving charge and solvent*

  1. Like charges: expect a decrease of volume; $\Delta V^{\neq *}$ is −ve.
  2. Unlike charges: expect an increase in volume; $\Delta V^{\neq *}$ is +ve.
  3. Ion–molecule: expect an increase in volume; $\Delta V^{\neq *}$ is +ve.
  4. Molecule–molecule with a charged-separated activated complex: expect a decrease in volume; $\Delta V^{\neq *}$ is −ve.
  5. Molecule–molecule without charge separation in the activated complex: expect a small decrease in volume; $\Delta V^{\neq *}$ is slightly −ve.

- *Change in solvation pattern on forming the activated complex*

  1. Like charges: expect a decrease in volume; $\Delta V^{\neq *}$ is −ve.
  2. Unlike charges: expect an increase in volume; $\Delta V^{\neq *}$ is +ve.
  3. Ion–molecule: expect either a small increase or decrease in volume; $\Delta V^{\neq *}$ is slightly +ve or −ve.
  4. Molecule–molecule: expect a decrease in volume, though it could be small; $\Delta V^{\neq *}$ is slightly −ve.

But, again, what value should be taken as the basis of comparison?

7.7.3 $\Delta H^{\neq *}$

- *Is the step associative or dissociative or neither?*
  Associative, i.e. 2 → 1: bond formation – expect a lower value of $\Delta H^{\neq *}$.
  Dissociative, i.e. 1 → 2: bond breaking – expect a higher value of $\Delta H^{\neq *}$.
  Interchange: expect $\Delta H^{\neq *}$ to take intermediate values.

  Change of internal modes for a large species
  Does not enter into the enthalpy of activation; this makes for a more straightforward interpretation of enthalpies of activation than entropies of activation.

- *Effect of charges: electrostatic term involving charge and solvent*

  1. Like charges: expect a smaller value for $\Delta H^{\neq *}$.

2. Unlike charges: expect a larger value for $\Delta H^{\neq *}$.
3. Ion–molecule: expect a slightly larger value for $\Delta H^{\neq *}$.
4. Molecule–molecule: expect a value similar to the standard point of comparison.

- *Changes in solvation pattern on activation.*

1. Like charges: expect a smaller value for $\Delta H^{\neq *}$.
2. Unlike charges: expect a larger value for $\Delta H^{\neq *}$.
3. Ion–molecule: expect no great change for $\Delta H^{\neq *}$.
4. Molecule–molecule: depends on the degree of charge separation; expect a smaller value for $\Delta H^{\neq *}$, but with no great change from the standard point of comparison.

But, again, what is the standard point of comparison?

Worked Problem 7.13

Question. Interpret the following data.

Reaction	$\frac{\Delta H^{\neq *}}{\text{kJ mol}^{-1}}$	$\frac{\Delta S^{\neq *}}{\text{J mol}^{-1}\text{ K}^{-1}}$	$\frac{\Delta V^{\neq *}}{\text{cm}^3\text{ mol}^{-1}}$
(a) $CH_3Br + H_2O$	100	−33	−14.5
(b) $PtdienBr^+ + N_3^-$	63	−71	−8.5
(c) $PtdienBr^+ + NO_2^-$	72	−56	−6.4
(d) $PtdienBr^+$ + pyridine	46	−136	−7.7
(e) $PtdienBr^+ + H_2O$	84	−63	−10
(f) $Pd(CN)_4^{2-} + CN^{*-}$	17	−178	—
(g) $Pt(CN)_4^{2-} + CN^{*-}$	26	−143	—
(h) $Au(CN)_4^- + CN^{*-}$	28	−100	—
(i) $CH_2BrCOO^- + S_2O_3^{2-}$	63.6	−71	−4.8
(j) $(C_2H_5)_3N + C_2H_5I$ in benzene	45.2	−243	−50.2

The data is given in sections. This should be helpful in the interpretation of the data. Assume that the data are for the reaction step as quoted.

- (a) $CH_3Br + H_2O$. This is a typical S_N2 reaction, and thus an interchange mechanism. This accounts for the relatively high $\Delta H^{\neq *}$. Interchange mechanism: expect a small, probably −ve $\Delta S^{\neq *}$ and −ve $\Delta V^{\neq *}$. Activated complex is charge separated and solvated: expect fairly large −ve contribution to both $\Delta S^{\neq *}$ and $\Delta V^{\neq *}$; observed.

- (b) $PtdienBr^+ + N_3^-$ and (c) $PtdienBr^+ + NO_2^-$. $\Delta H^{\neq *}$ suggests an interchange mechanism, but the values are low enough to correspond to interchange with associative characteristics. Expect a −ve $\Delta S^{\neq *}$ and −ve $\Delta V^{\neq *}$. But reaction is (+1) with (−1) charge, which should give a +ve contribution to both, because of the charge and solvation effects. The complexity of the dien would give a −ve contribution to $\Delta S^{\neq *}$. Overall a fairly small $\Delta S^{\neq *}$ and $\Delta V^{\neq *}$ is reasonable, as observed.

- (d) $PtdienBr^+$ + pyridine and (e) $PtdienBr^+ + H_2O$. $\Delta H^{\neq *}$ values again suggest an interchange mechanism with associative characteristics. The $\Delta S^{\neq *}$ and $\Delta V^{\neq *}$ values for (e), i.e. reaction with H_2O, are in line with those of reactions (b) and (c), even though this is an ion + dipole reaction, where the charge and solvation effects will make a less +ve contribution than for reactions (b) and (c).

  Reaction (d) i.e. with pyridine, has a surprisingly large −ve $\Delta S^{\neq *}$, but a $\Delta V^{\neq *}$ in keeping with those for reactions (b) and (c). Perhaps the high polarizability of the pyridine might lead it to be held relatively rigidly in the activated complex, causing a substantial −ve $\Delta S^{\neq *}$ contribution. This would be reinforced by a contribution from the relative complexity of pyridine compared with water. Neither factor would affect the $\Delta V^{\neq *}$.

- (f), (g) and (h), i.e. $Pd(CN)_4^{2-} + C^*N^-$, $Pt(CN)_4^{2-} + C^*N^-$ and $Au(CN)_4^- + C^*N^-$. All have small $\Delta H^{\neq *}$ values indicative of an associative mechanism. Reactions (f) and (g) are (−2) with (−1) charge reactions, and a large −ve contribution to $\Delta S^{\neq *}$ from the effect of charge and solvation would be expected. This would be reinforced by any −ve contribution from the complexity of the complexes.

  This can be contrasted with reaction (h), which has a less −ve $\Delta S^{\neq *}$ value, *but* this would be expected since the activated complex would have a −2 charge compared to −3 for reactions (f) and (g). This would lead to a less −ve contribution to $\Delta S^{\neq *}$ from the charge and solvation effects.

- (i) $CH_2BrCOO^- + S_2O_3^{2-}$. The $\Delta H^{\neq *}$ value suggests an interchange mechanism in keeping with the S_N2 characteristics of the substitution of $S_2O_3^{2-}$ for Br^-. The $\Delta S^{\neq *}$ and $\Delta V^{\neq *}$ values are relatively small and −ve, supporting the inference that this is an interchange rather an associative mechanism. Because of the charges, (−1) with (−2), there would be a fairly large −ve contribution to $\Delta S^{\neq *}$ and $\Delta V^{\neq *}$.

- (j) $(C_2H_5)_3N + C_2H_5I$ in benzene. This reaction produces a quaternary ammonium salt, which would be fully associated in benzene. It would, however, be charge separated. The very large −ve values for both $\Delta S^{\neq *}$ and $\Delta V^{\neq *}$ suggest that there is a considerable degree of charge separation in the activated complex, and that this is an associative mechanism. This would be reinforced by a contribution due to the complexity of both reactants. The fairly low value of $\Delta H^{\neq *}$ supports an associative mechanism.

Further reading

Laidler K.J., *Chemical Kinetics*, 3rd edn, Harper and Row, New York, 1987.
Laidler K.J. and Meiser J.H., *Physical Chemistry*, 3rd edn, Houghton Mifflin, New York, 1999.
Alberty R.A. and Silbey R.J., *Physical Chemistry*, 2nd edn, Wiley, New York, 1996.
Logan S.R., *Fundamentals of Chemical Kinetics*, Addison-Wesley, Reading, MA, 1996.
*Wilkinson F., *Chemical Kinetics and Reaction Mechanisms*, Van Nostrand Reinhold, New York, 1980.
Robson Wright M., *Fundamental Chemical Kinetics*, Horwood, Chichester, 1999.
*Laidler K.J., *Theories of Chemical Reaction Rates*, McGraw-Hill, New York, 1969.
*Glasstone S., Laidler K.J. and Eyring H., *Theory of Rate Processes*, McGraw-Hill, New York, 1941.
*Robson Wright M., *Nature of Electrolyte Solutions*, Macmillan, Basingstoke, 1988.
Robson Wright M., *Behaviour and Nature of Electrolyte Solutions*, Wiley, Chichester, in preparation.
Hay R.W., *Reaction Mechanisms of Metal Complexes*, Horwood, Chichester, 2000.
Cox B.G., *Modern Liquid Phase Kinetics*, Oxford University Press, Oxford, 1994.
Wilkins R.G., *Kinetics and Mechanisms of Transition Metal Complexes*, Verlagsgesselchaft, Cambridge, 1999.
Burgess J. and Tobe M., *Inorganic Reaction Mechanisms*, Longman, London, 2000.
*Out of print, but should be available in university libraries.

Further problems

1. When SO_4^{2-} is added to the reaction mixture for the acid-catalysed hydrolysis of acetylcholine, $(CH_3)_3N^+CH_2CH_2OCOCH_3$ (HE^+), with H_3O^+, added catalysis is observed. This can be interpreted as anion-catalysed hydrolysis by SO_4^{2-}, or as general acid catalysis by HSO_4^-. Predict the effect of ionic strength on the rate constants describing each mechanism, and comment on the result. Does this represent a means of distinguishing between the two mechanisms?

2. The following observations relate to an investigation of the reaction

$$CH_2ClCOO^-(aq) + OH^-(aq) \rightarrow CH_2OHCOO^-(aq) + Cl^-(aq)$$

at high pressures and 60 °C.

$\frac{10^{-5}p}{\mathrm{N\,m^{-2}}}$	0.98	2000	4000	6000
$\frac{10^{5}k}{\mathrm{mol^{-1}\,dm^{3}\,s^{-1}}}$	6.9	10.3	15.4	22.0

Obtain a value for the volume of activation.

3. The following are values of k for the reaction between pyridine and CH_3I in solution in propanone:

$\frac{\text{temperature}}{°\mathrm{C}}$	40	50	60	70
$\frac{10^{3}k}{\mathrm{mol^{-1}\,dm^{3}\,s^{-1}}}$	0.60	1.17	2.15	3.65

Find $\Delta H^{\neq *}$ and $\Delta S^{\neq *}$.

4. At an ionic strength of 0.100 mol dm^{-3} and a temperature of 25 °C, the observed rate constant for the reaction

$$CH_2BrCOO^-(aq) + S_2O_3^{2-}(aq) \rightarrow CH_2S_2O_3COO^{2-}(aq) + Br^-(aq)$$

is 1.07×10^{-2} mol^{-1} dm^3 s^{-1}, while that for a reaction of the type

$$\mathrm{PtdienX^+(aq) + Y^-(aq) \rightarrow PtdienY^+(aq) + X^-(aq)}$$

is 3.6×10^{-3} mol^{-1} dm^3 s^{-1}. Comment on these values.

Estimate the values of the rate constants for zero ionic strength, and comment on the relative values and compare with the values at $I = 0.100$ mol dm^{-3}.

5. Aqueous Ni^{2+} forms a 1:1 complex with the anion of murexide, mu$^-$,

$$\mathrm{Ni^{2+}(aq) + mu^-(aq) \underset{k_{-1}}{\overset{k_1}{\rightleftharpoons}} Ni\,mu^+(aq)}$$

At 25 °C observed values of the rate constant k_1 and the equilibrium constant K_1 depend on pressure as follows:

$\frac{p}{\mathrm{atm}}$	1	800	1500
$\frac{k_1}{\mathrm{mol^{-1}\,dm^{3}\,s^{-1}}}$	10 000	6900	4800
$\frac{K_1}{\mathrm{mol^{-1}\,dm^{3}}}$	1920	940	510

Determine the volume of activation $\Delta V^{\neq *}$ and standard ΔV^{θ} for the formation of the complex, and comment on the values obtained.

6. In a family of reactions

$$A + B \rightarrow C$$

in which cycloaddition occurs, some values of $\Delta V^{\neq *}$ and ΔV^{θ} are as follows:

A	B	Reaction medium	$\Delta V^{\neq *}$ / $cm^3\ mol^{-1}$	ΔV^{θ} / $cm^3\ mol^{-1}$
$CH_2{=}CH{-}C(CH_3){=}CH_2$	$CH_2{=}CHCOOCH_3$	BuBr (21 °C)	−30.8	−36.9
$CH_2{=}C(CH_3){-}C(CH_3){=}CH_3$	$CH_2{=}CHCOOCH_3$	BuBr (40 °C)	−30.2	−37.0
$CH_2{=}C(CH_3){-}C(CH_3){=}CH_3$	$CH_3OOC{-}C(H){=}C(H)COOCH_3$ (trans)	BuBr (40 °C)	−32.9	−37.2
$CH_2{=}C(CH_3){-}C(CH_3){=}CH_3$	HC=CH / CO CO \ O / (maleic anhydride ring)	BuBr (30 °C)	−41.3	−36.3
$CH_2{=}CH{-}C(CH_3){=}CH_2$	HC=CH / CO CO \ O / (maleic anhydride ring)	$ClCH_2CH_2Cl$ (35 °C)	−37	−35.5

Bu is $CH_3CH_2CH_2CH_2$.

Comment on these values, and give the values of $\Delta V^{\neq *}$ for the reverse reactions. Predict the trend in $\Delta S^{\neq *}$.

8 Examples of Reactions in Solution

This chapter applies many of the techniques discussed in previous chapters to the determination of rate constants for complex reactions in solution. It also focuses on a variety of types of reaction in solution.

Aims

By the end of this chapter you should be able to

- solve problems on reactions with two contributions to the overall rate
- solve problems on reactions with concurrent routes, where both reactants are in equilibrium
- recognize kinetically equivalent mechanisms
- handle metal ion and anion catalysis
- recognize and handle mechanisms involving pre-equilibria and
- handle steady state situations for reactions with an initial reversible step

8.1 Reactions Where More than One Reaction Contributes to the Rate of Removal of Reactant

This section will discuss reaction mechanisms of ever-increasing complexity.

An Introduction to Chemical Kinetics. Margaret Robson Wright
 ISBNs: 0-470-09058-8 (hbk) 0-470-09059-6 (pbk)

8.1.1 A simple case

Reactions in solution often have a contribution to the overall rate from reaction of the solvent with one or other of the reactants; e.g., hydrolyses of esters in aqueous solution often have a contribution from a 'water' rate.

A typical base hydrolysis may have two contributing reactions:

$$\text{ester(aq)} + \text{OH}^-\text{(aq)} \xrightarrow{k_1} \text{products}$$

$$\text{ester(aq)} + \text{H}_2\text{O(l)} \xrightarrow{k_2} \text{products}$$

where the products are the carboxylate ion and an alcohol.

Mechanistic rate:

$$-\frac{\text{d[ester]}}{\text{d}t} = k_1[\text{ester}][\text{OH}^-] + k_2[\text{ester}] \tag{8.1}$$

where the rate of the second reaction does not include $[H_2O]$, since this is in considerable excess and remains virtually constant throughout the reaction. k_2 is thus a pseudo-first order rate constant, with units time^{-1}.

Observed rate:

$$-\frac{\text{d[ester]}}{\text{d}t} = k_{\text{obs}}[\text{ester}][\text{OH}^-] \tag{8.2}$$

If the mechanism is correct, then

$$k_{\text{obs}}[\text{ester}][\text{OH}^-] = k_1[\text{ester}][\text{OH}^-] + k_2[\text{ester}] \tag{8.3}$$

giving

$$k_{\text{obs}} = k_1 + k_2/[\text{OH}^-] \tag{8.4}$$

Experimental rate constants gathered for experiments over a range of $[OH^-]$ should determine whether there is a 'water' rate or not, and allow k_1 and/or k_2 to be found.

If there is no water rate, then k_{obs} values will be independent of $[OH^-]$. If k_{obs} does depend on $[OH^-]$, then a graph of k_{obs} versus $1/[OH^-]$ should be linear, with slope $= k_2$ and intercept $= k_1$.

A very common problem in kinetics

$$k_{\text{obs}} = k_1 + k_2/[\text{OH}^-] \tag{8.4}$$

The method of analysis implies that k_{obs} is found at various $[OH^-]$. But there is a problem – OH^- is used up during each experiment. How can this be overcome?

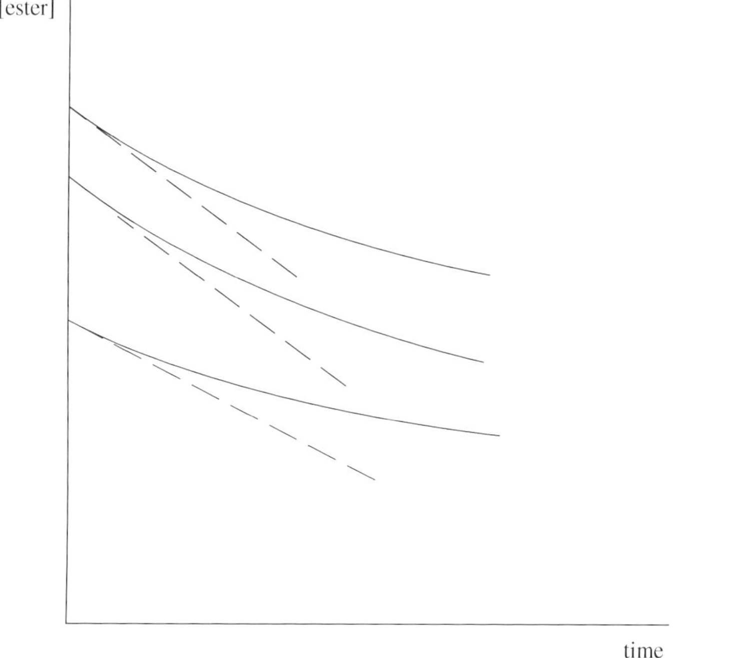

Figure 8.1 Determination of initial rates for various values of the initial concentrations: broken lines are tangents to the experimental curves at $t = 0$

This is actually a common problem in kinetics and is often overcome by the following.

- *Studying initial rates* (Figure 8.1). Under these conditions both the ester and OH^- concentrations remain effectively constant, and k_{obs} can be calculated from

$$k_{obs} = \frac{\text{rate}}{[\text{ester}][\text{OH}^-]} \tag{8.5}$$

- *Keeping [OH^-] constant throughout the experiment.* In this particular instance, because OH^- is used up, the reaction can be studied using a pH-stat, which adds OH^- continuously to keep the pH constant.

$$-\frac{\text{d}[\text{ester}]}{\text{d}t} = k_{obs}[\text{ester}][\text{OH}^-] \tag{8.6}$$

$$= k^{1\text{st}}[\text{ester}] \tag{8.7}$$

where

$$k^{1\text{st}} = k_{obs}[\text{OH}^-] \tag{8.8}$$

The volume of OH^- added gives a measure of how much ester has been hydrolysed at various times throughout the reaction, and enables the pseudo-first order rate constant k^{1st} to be found, from which the second order k_{obs} can be calculated.

The following problem illustrates the analysis given above.

Worked Problem 8.1

Question. The following kinetic data have been obtained for the reaction of pyridine with $Pt(dien)Cl^+$, where dien has the structure $NH_2(CH_2)_2NH(CH_2)_2NH_2$,

$$[Pt(dien)Cl]^+(aq) + pyridine(aq) \rightarrow [Pt(dien)pyridine]^{2+}(aq) + Cl^-(aq)$$

The concentration of the complex was 1.00×10^{-3} mol dm^{-3} in each experiment.

$\frac{[pyridine]}{mol\ dm^{-3}}$	0.100	0.200	0.300	0.400	0.500
$\frac{10^3 k_{obs}}{s^{-1}}$	1.75	2.70	3.65	4.60	5.55

- What can be said about the relative concentrations of pyridine and complex, and what does this imply for the conditions of the experiment? What does this imply about the observed rate constant and its units?
- Write down the steps involved in the reaction, and formulate an expression for the mechanistic rate.
- Formulate the expression for the observed rate.
- Calculate (a) the first order rate constant for the reaction of the complex with the solvent, water, (b) the pseudo-first order rate constant for reaction with pyridine and (c) the second order rate constant for reaction with pyridine.

Answer. In all of the experiments [pyridine] $\gg$ [complex]. Thus throughout each experiment [pyridine] is being held effectively constant, and each experiment is under pseudo-first order conditions, verified by the units of the observed rate constant, s^{-1}.

The two contributions to the overall reaction are

$$[Pt(dien)Cl]^+(aq) + pyridine(aq) \xrightarrow{k_1} [Pt(dien)\ pyridine]^{2+}(aq) + Cl^-(aq)$$

$$[Pt(dien)Cl]^+(aq) + H_2O(l) \xrightarrow{k_2} [Pt(dien)H_2O]^{2+}(aq) + Cl^-(aq)$$

Mechanistic rate:

$$-\frac{d[Pt(dien)Cl]^+}{dt} = k_1[Pt(dien)Cl^+][pyridine] + k_2[Pt(dien)Cl^+] \qquad (8.9)$$

Observed rate $= k_{obs}^{2nd}[Pt(dien)Cl^+][pyridine]$ (8.10)

$= k_{obs}^{1st}[Pt(dien)Cl^+]$ (8.11)

where k_{obs}^{2nd} is a second order rate constant, k_{obs}^{1st} is a pseudo-first order rate constant and

$$k_{obs}^{1st} = k_{obs}^{2nd}[pyridine] \quad (8.12)$$

If the mechanism is correct, then

$$k_{obs}^{1st}[Pt(dien)Cl^+] = k_1[Pt(dien)Cl^+][pyridine] + k_2[Pt(dien)Cl^+] \quad (8.13)$$

$$k_{obs}^{1st} = k_1[pyridine] + k_2 \quad (8.14)$$

A plot of k_{obs}^{1st} versus [pyridine] should be linear with slope $= k_1$ and intercept $= k_2$.
If there is no contribution from the reaction of the complex with water, then the graph will go through the origin, and k_2 will be zero.
From the data given, the graph of k_{obs}^{1st} versus [pyridine] is linear with a non-zero intercept (Figure 8.2), with

slope $= 9.5 \times 10^{-3}$ mol^{-1} dm^3 s^{-1} giving $k_1 = 9.5 \times 10^{-3}$ mol^{-1} dm^3 s^{-1}

intercept $= 0.80 \times 10^{-3}$ s^{-1} giving $k_2 = 8.0 \times 10^{-4}$ s^{-1}

8.1.2 A slightly more complex reaction where reaction occurs by two concurrent routes, and where both reactants are in equilibrium with each other

In the specific example chosen here the analysis is made easier because of *approximations* which can be made as a result of the magnitude of the equilibrium constant. This is not always the case (Section 8.2.2).

The base hydrolysis of aminoacid esters

This can be illustrated by the hydrolysis of a simple aminoacid ester. Glycine ethyl ester exists as

the unprotonated ester, E, $NH_2CH_2COOCH_2CH_3$

the protonated ester, HE^+, $NH_3^+CH_2COOCH_2CH_3$

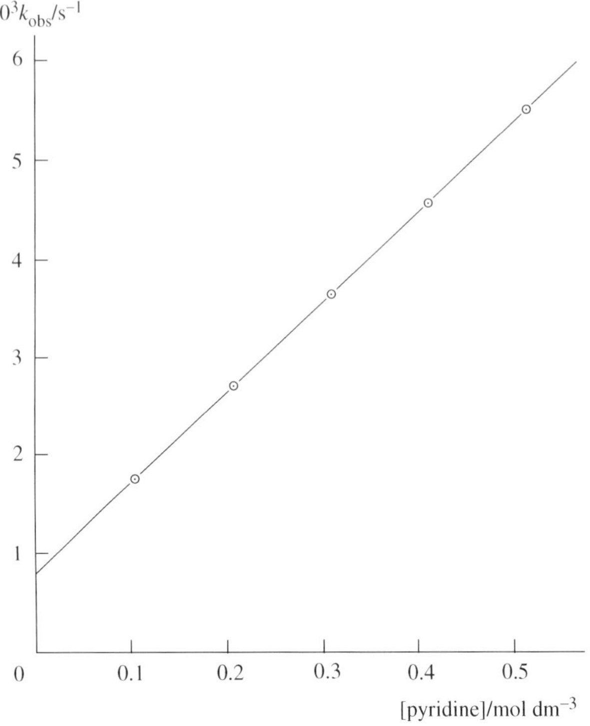

Figure 8.2 Graph of k_{obs} versus [pyridine] for reaction of $Pt(dien)Cl^+(aq)$ with pyridine(aq)

These species are in equilibrium with each other, with the position of equilibrium lying very much over to the unprotonated ester.

$$E(aq) + H_2O(l) \rightleftharpoons HE^+(aq) + OH^-(aq)$$

$$K_b = \left(\frac{[HE^+]_{actual}[OH^-]_{actual}}{[E]_{actual}} \right)_{equilibrium} = 5.01 \times 10^{-7} \text{ mol dm}^{-3} \tag{8.15}$$

Both of these species are hydrolysed by OH^- to form the corresponding carboxylate ion and C_2H_5OH.

The total concentration of ester initially present, $[E]_{total}$, is distributed between protonated ester actually present, $[HE^+]_{actual}$, and the actual remaining unprotonated ester, $[E]_{actual}$, so that

$$[E]_{total} = [E]_{actual} + [HE^+]_{actual} \tag{8.16}$$

This relation holds *at all times* throughout reaction, where $[E]_{total}$ is then the total concentration of ester left at time t, and $[HE^+]_{actual}$ and $[E]_{actual}$ are the actual concentrations of HE^+ and E left at each time t.

Reaction is first order in ester for experiments at constant pH, and further experiments at other constant pH values show reaction to be first order in OH^-.

The reaction can be followed very conveniently using a pH-stat, so that $[OH^-]$ is kept constant throughout reaction by continuous addition of OH^-. The volume of OH^- added at various times enables the total amount of ester hydrolysed at these times to be found.

The observed rate is defined in terms of the *total* amount of ester used up, without reference to the two forms of the ester. *But* the mechanism 'is interested' in the chemical steps involved in the reaction, and this must refer to the hydrolysis of both forms of the ester:

$$\mathrm{E} + \mathrm{OH}^- \xrightarrow{k_1} \text{products}$$

$$\mathrm{HE}^+ + \mathrm{OH}^- \xrightarrow{k_2} \text{products}$$

and both steps contribute to the predicted overall rate of removal of ester.

Mechanistic rate

$$-\frac{\mathrm{d}[\mathrm{E}]_{\text{total}}}{\mathrm{d}t} = k_1[\mathrm{E}]_{\text{actual}}[\mathrm{OH}^-]_{\text{actual}} + k_2[\mathrm{HE}^+]_{\text{actual}}[\mathrm{OH}^-]_{\text{actual}} \tag{8.17}$$

where $[\mathrm{E}]_{\text{total}}$, $[\mathrm{E}]_{\text{actual}}$ and $[\mathrm{HE}^+]_{\text{actual}}$ vary throughout the experiment, and will be the instantaneous values at the times at which the rate is measured (Figure 8.3).

Observed rate

This is measured in terms of the total ester used up at any given time:

$$-\frac{\mathrm{d}[\mathrm{E}]_{\text{total}}}{\mathrm{d}t} = k_{\text{obs}}[\mathrm{E}]_{\text{total}}[\mathrm{OH}^-]_{\text{actual}} \tag{8.18}$$

If the proposed mechanism is correct, then these rates must be equal.

$$k_{\text{obs}}[\mathrm{E}]_{\text{total}}[\mathrm{OH}^-]_{\text{actual}} = k_1[\mathrm{E}]_{\text{actual}}[\mathrm{OH}^-]_{\text{actual}} + k_2[\mathrm{HE}^+]_{\text{actual}}[\mathrm{OH}^-]_{\text{actual}} \tag{8.19}$$

$$k_{\text{obs}}[\mathrm{E}]_{\text{total}} = k_1[\mathrm{E}]_{\text{actual}} + k_2[\mathrm{HE}^+]_{\text{actual}} \tag{8.20}$$

$$k_{\text{obs}} = k_1\frac{[\mathrm{E}]_{\text{actual}}}{[\mathrm{E}]_{\text{total}}} + k_2\frac{[\mathrm{HE}^+]_{\text{actual}}}{[\mathrm{E}]_{\text{total}}} \tag{8.21}$$

All of the concentration terms are related to each other, so this equation can be simplified. Equation (8.15) rearranged gives

$$[\mathrm{HE}^+]_{\text{actual}} = \frac{K_b[\mathrm{E}]_{\text{actual}}}{[\mathrm{OH}^-]_{\text{actual}}} \tag{8.22}$$

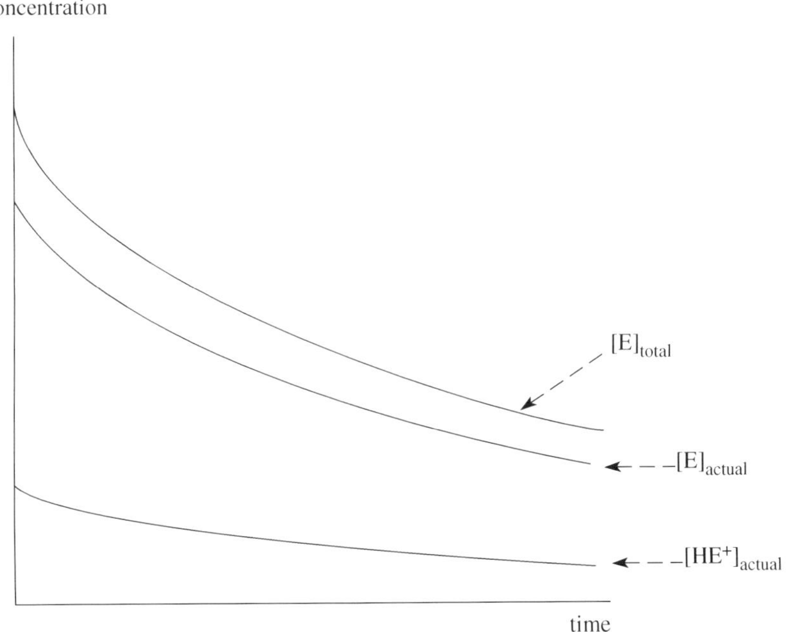

Figure 8.3 Graphs of concentration versus time for reaction of an amino-acid ester with OH^- showing total ester, $[E]_{total}$, protonated ester, $[HE^+]_{actual}$, and unprotonated ester, $[E]_{actual}$, with time

Substitution of this into Equation (8.20) gives

$$k_{obs}[E]_{total} = k_1[E]_{actual} + k_2 \frac{K_b[E]_{actual}}{[OH^-]_{actual}} \tag{8.23}$$

This can be simplified. Because $K_b = 5.01 \times 10^{-7}$ mol dm^{-3} (Equation (8.15)) and reaction is carried out in basic solutions, $[E]_{actual} \approx [E]_{total}$ and so

$$k_{obs}[E]_{total} = k_1[E]_{total} + k_2 \frac{K_b[E]_{total}}{[OH^-]_{actual}} \tag{8.24}$$

The validity of this argument can be shown by the following typical set of experimental conditions:

e.g. if pH = 10.00 then pOH = pK_w − pH = 4.00, and so $[OH^-] = 1.00 \times 10^{-4}$ mol dm^{-3}.

$$K_b = \frac{[HE^+]_{actual}[OH^-]_{actual}}{[E]_{actual}} \tag{8.15}$$

and

$$\frac{K_b}{[OH^-]_{actual}} = \frac{[HE^+]_{actual}}{[E]_{actual}} \tag{8.25}$$

$$\frac{K_b}{[OH^-]_{actual}} = \frac{5.01 \times 10^{-7}}{1.00 \times 10^{-4}} = 5.01 \times 10^{-3}, \text{ hence } [HE^+]_{actual} = 5.01 \times 10^{-3}[E]_{actual}$$

If $[E]_{total} = 2.00 \times 10^{-2}$ mol dm^{-3}, then

$$2.00 \times 10^{-2} = 5.01 \times 10^{-3}[E]_{actual} + [E]_{actual}$$

giving

$$[E]_{actual} = \frac{2.00 \times 10^{-2}}{1.00501} = 1.99 \times 10^{-2} \text{ mol dm}^{-3}$$

and

$$[HE^+]_{actual} = 1.00 \times 10^{-4} \text{ mol dm}^{-3}$$

and so throughout the experiment $[E]_{actual}$ at any stage of reaction is approximately equal to $[E]_{total}$ at these times, and so it follows that

$$k_{obs}[E]_{total} = k_1[E]_{total} + k_2 \frac{K_b[E]_{total}}{[OH^-]_{actual}} \tag{8.24}$$

with

$$k_{obs} = k_1 + k_2 \frac{K_b}{[OH^-]_{actual}} \tag{8.26}$$

This is the equation of a straight line, so that, if the mechanism is correct, a plot of k_{obs} versus $1/[OH^-]_{actual}$ should be linear with intercept $= k_1$ and slope $= k_2K_b$ (Figure 8.4).

Worked Problem 8.2

Question. The base hydrolysis of glycine ethyl ester, $NH_2CH_2COOCH_2CH_3$, can be followed at constant pH, as described in Section 8.1.2.

(a) The following data were collected at 25 °C. What inference can be drawn from this data?

pH	10.05	10.51	10.74	11.26
$\frac{k}{\text{mol}^{-1}\text{ dm}^3\text{ min}^{-1}}$	52	52	51	52

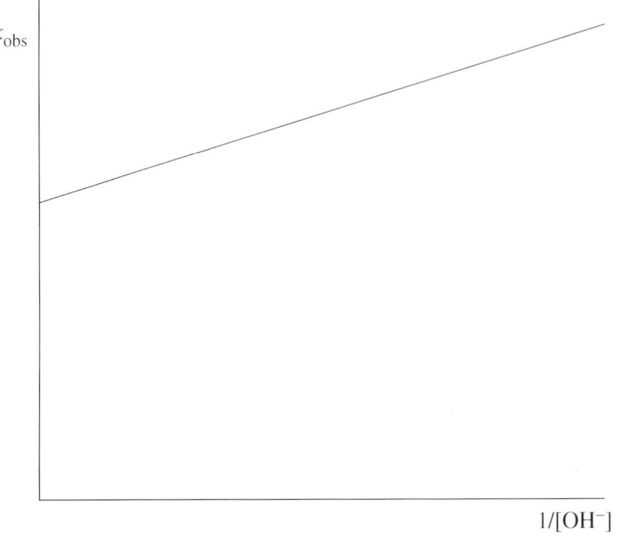

Figure 8.4 Schematic graph of k_{obs} versus $1/[OH^-]_{actual}$ for reaction of glycine ethyl ester with OH^-

(b) The following table gives data covering a wider range of pH. Calculate the rates of the reaction of OH^- with the protonated and unprotonated forms of the ester, for the conditions pH = 9.30, total [ester] = 2×10^{-2} mol dm^{-3}. $K_b = 5.01 \times 10^{-7}$ mol dm^{-3} at 25 °C.

pH	11.70	11.22	10.05	9.55	9.37
$\frac{k_{obs}}{\text{dm}^3\ \text{mol}^{-1}\ \text{min}^{-1}}$	43.3	45.0	50.2	64.0	74.3

(c) Comment on, and interpret the relative magnitudes of, the rates and the actual concentrations of protonated and unprotonated ester.

Answer. As shown in Equation (8.26)

$$k_{obs} = k_1 + k_2 \frac{K_b}{[OH^-]_{actual}} \tag{8.26}$$

(a) Looking at the first set of data, it is clear that the observed rate constant is independent of pH and thus of $[OH^-]$, and so the only reaction which appears to be occurring is the hydrolysis of the unprotonated ester by OH^-. This reaction is characterized by the rate constant, k_1, and the data suggests that $k_1 =$ 52 mol^{-1} dm^3 min^{-1}.

(b) When the second set of data is analysed, it becomes obvious that the apparent constancy of the observed rate constant in (a) is a consequence of the limited range of pH over which the first set of data has been collected. Fitting the second set of data to Equation (8.26) shows that the graph is not a horizontal straight line, but is one with a finite gradient

$pH[OH]_{actual}$	11.70	11.22	10.05	9.55	9.37
$\frac{10^5 \times [OH^-]_{actual}}{mol\ dm^{-3}}$	500	166	11.2	3.55	2.34
$\frac{10^{-3}}{[OH^-]_{actual}\ mol^{-1}\ dm^3}$	0.200	0.602	8.93	28.2	42.6
$\frac{k_{obs}}{dm^3\ mol^{-1}\ min^{-1}}$	43.3	45.0	50.2	64.0	74.3

A plot of k_{obs} versus $1/[OH^-]_{actual}$ is linear with slope $= k_2K_b = 7.1 \times 10^{-4}\ min^{-1}$, giving $k_2 = 7.1 \times 10^{-4}\ min^{-1}/5.01 \times 10^{-7} mol\ dm^{-3} = 1420\ dm^3\ mol^{-1}\ min^{-1}$ intercept $= k_1 = 44\ dm^3\ mol^{-1}\ min^{-1}$ (Figure 8.5).

(c) k_1 refers to reaction of an uncharged species with an ion, while k_2 refers to reaction of a cation with an anion. Because of the attractive forces between

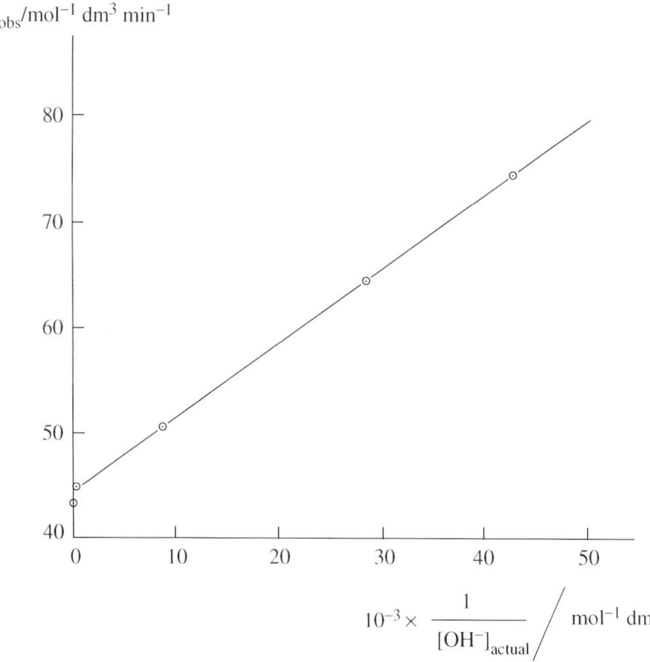

Figure 8.5 Graph of k_{obs} versus $1/[OH^-]$ for reaction of glycine ethyl ester with $[OH^-]_{actual}$

oppositely charged ions, the rate constant for such a reaction will be much higher than that for reaction between an ion and an uncharged molecule. The observed values are in keeping with this qualitative prediction.

At pH = 9.30, since pH + pOH = 14 then $[OH^-] = 2 \times 10^{-5}$ mol dm^{-3}; $[E]_{total} \approx [E]_{actual} = 2 \times 10^{-2}$ mol dm^{-3}

$$[HE^+]_{actual} = \frac{K_b[E]_{actual}}{[OH^-]_{actual}} = \frac{5.01 \times 10^{-7}\,\text{mol dm}^{-3} \times 2 \times 10^{-2}\,\text{mol dm}^{-3}}{2 \times 10^{-5}\,\text{mol dm}^{-3}}$$
$$= 5.01 \times 10^{-4}\,\text{mol dm}^{-3}$$

For the reaction $E + OH^- \xrightarrow{k_1}$

$$\begin{aligned}\text{rate} &= k_1[E][OH^-]_{actual} \\ &= 44\,\text{dm}^3\,\text{mol}^{-1}\,\text{min}^{-1} \times 2 \times 10^{-2}\,\text{mol dm}^{-3} \times 2 \times 10^{-5}\,\text{mol dm}^{-3} \\ &= 1.76 \times 10^{-5}\,\text{dm}^{-3}\,\text{mol min}^{-1} \end{aligned} \quad (8.27)$$

For the reaction $HE^+ + OH^- \xrightarrow{k_2}$

$$\begin{aligned}\text{rate} &= k_2[HE^+][OH^-]_{actual} \\ &= 1420\,\text{dm}^3\,\text{mol}^{-1}\,\text{min}^{-1} \times 5.01 \times 10^{-4}\,\text{mol dm}^{-3} \times 2 \times 10^{-5}\,\text{mol dm}^{-3} \\ &= 1.42 \times 10^{-5}\,\text{dm}^{-3}\,\text{mol min}^{-1} \end{aligned} \quad (8.28)$$

Although $[HE^+] \ll [E]$, the contributions to the overall rate from the protonated and unprotonated ester hydrolyses are approximately equal. This is because $k_2 \gg k_1$. A very small concentration combined with a large rate constant can still make a significant contribution to the rate.

An important point. The other important conclusion to be drawn from this problem is the limitations imposed on the interpretation by an insufficiently extensive range of the experimental conditions, *viz.* the limited range of pH over which the first set of data has been collected. Similar conclusions were drawn in the answer to the 'Further problems' of Chapter 4 Question 5, and Worked Problem 7.5.

In Problem 8.2, because the value of K_b is so low, the approximation $[E]_{total} \approx [E]_{actual}$ can be made, and this simplifies the analysis. If this is not the case, then an analysis analogous to that of Section 8.2.2 must be made.

8.1.3 Further disentangling of equilibria and rates, and the possibility of kinetically equivalent mechanisms

This can again be illustrated by a base hydrolysis of an ester, e.g. ethyl ethanoate, $CH_3COOC_2H_5$. If Ca^{2+} or Ba^{2+} salts are added to aqueous OH^- solution, the ion pairs $CaOH^+$ and $BaOH^+$ will be formed and will reduce the concentration of free OH^-. When the base hydrolysis is studied at constant ionic strength, but with added Ca^{2+} or Ba^{2+}, there is no change in rate from that found for hydrolyses at the same ionic strength, but without added cations.

The following questions can be asked.

- Why is this reaction studied at constant ionic strength?
- If addition of cations which reduce the actual concentration of free OH^- has no effect on the rate, what does this imply?

The answers to these questions are important.

- Reaction involves an ion and a molecule and could show a primary salt effect. The equilibrium constant describing the formation of the ion pairs also depends on ionic strength.
- $CaOH^+$ or $BaOH^+$ are formed. No change in the rate of reaction indicates $CaOH^+$ or $BaOH^+$ is as reactive as OH^-, suggesting a mechanism with two contributing steps.

$$CH_3COOC_2H_5 + OH^- \xrightarrow{k_{OH^-}} CH_3COO^- + C_2H_5OH$$
$$CH_3COOC_2H_5 + CaOH^+ \xrightarrow{k_{CaOH^+}} CH_3COO^- + C_2H_5OH + Ca^{2+}$$

Reaction of $CH_3COOC_2H_5$ with OH^- is specific base catalysis, while removal by $CaOH^+$ is general base catalysis.

Specific base catalysis: refers specifically to catalysis by OH^-.

General base catalysis: refers to catalysis by any species acting as a base by taking up a proton, H^+, i.e. acting as a Brønsted base.

$$CaOH^+(aq) + H_3^+O(aq) \rightarrow Ca^{2+}(aq) + 2H_2O(l)$$

But the mechanism with hydrolysis by $CaOH^+$ or $BaOH^+$ is kinetically indistinguishable from a metal ion catalysis mechanism:

$$CH_3COOC_2H_5 + OH^- \xrightarrow{k_{OH^-}} CH_3COO^- + C_2H_5OH$$
$$CH_3COOC_2H_5 + OH^- + Ca^{2+} \xrightarrow{k_{Ca^{2+}}} CH_3COO^- + C_2H_5OH + Ca^{2+}$$

Note. The constitution of the activated complex is the same in both cases: again a feature of kinetically equivalent mechanisms.

Worked Problem 8.3

Question. Show that the two mechanisms (given in Section 8.1.3) are kinetically equivalent and indicate how the respective rate constants could be found.

Answer.

- For the first mechanism involving the ion pair

$$\text{mechanistic rate} = k_{OH^-}[E][OH^-]_{actual} + k_{CaOH^+}[E][CaOH^+]_{actual} \tag{8.29}$$

$$\text{observed rate} = k_{obs}[E][OH^-]_{actual} \tag{8.30}$$

where E represents $CH_3COOC_2H_5$.

Note. The analysis is given in terms of $[OH^-]_{actual}$ rather than $[OH^-]_{total}$. This is because it is easy to follow the reaction at constant pH, using a pH-stat, where $[OH^-]_{actual}$ is measured. It is vital to make absolutely clear whether the rate is being measured in terms of a total concentration or an actual concentration. The decision is often made in terms of the experimental method.

If the mechanism is correct, then

$$k_{obs}[E][OH^-]_{actual} = k_{OH^-}[E][OH^-]_{actual} + k_{CaOH^+}[E][CaOH^+]_{actual} \tag{8.31}$$

$$k_{obs} = k_{OH^-} + k_{CaOH^+}\{[CaOH^+]_{actual}/[OH^-]_{actual}\} \tag{8.32}$$

The following equilibrium is set up in solution:

$$Ca^{2+}(aq) + OH^-(aq) \rightleftarrows CaOH^+(aq)$$

$$K_{association} = \left(\frac{[CaOH^+]_{actual}}{[Ca^{2+}]_{actual}[OH^-]_{actual}}\right)_{equil} \tag{8.33}$$

$$[CaOH^+]_{actual} = K_{association}[Ca^{2+}]_{actual}[OH^-]_{actual} \tag{8.34}$$

$$k_{obs} = k_{OH^-} + k_{CaOH^+}K_{association}[Ca^{2+}]_{actual} \tag{8.35}$$

A plot of k_{obs} versus $[Ca^{2+}]_{actual}$ will be linear with intercept $= k_{obs}$ and slope $= k_{CaOH^+}K_{association}$.

- For the second mechanism involving the free $Ca^{2+}(aq)$

$$\text{mechanistic rate} = k_{OH^-}[E][OH^-]_{actual} + k_{Ca^{2+}}[E][Ca^{2+}]_{actual}[OH^-]_{actual} \tag{8.36}$$

$$\text{observed rate} = k_{obs}[E][OH^-]_{actual} \tag{8.37}$$

and if this is the mechanism, then

$$k_{obs}[E][OH^-]_{actual} = k_{OH^-}[E][OH^-]_{actual} + k_{Ca^{2+}}[E][Ca^{2+}]_{actual}[OH^-]_{actual} \quad (8.38)$$

$$k_{obs} = k_{OH^-} + k_{Ca^{2+}}[Ca^{2+}]_{actual} \quad (8.39)$$

A plot of k_{obs} versus $[Ca^{2+}]_{actual}$ will be linear with intercept $= k_{OH^-}$ and slope $= k_{Ca^{2+}}$.

The two mechanisms are algebraically similar in that they both are of the form $y = mx + c$, the equation of a straight line, where y corresponds to k_{obs} and x to $[Ca^{2+}]_{actual}$.

These mechanisms are kinetically equivalent and cannot be distinguished. It could be argued, on plausibility grounds, that the second mechanism is less likely, because it involves the simultaneous coming together of three species instead of two.

Question. Why is it best that the reaction be carried out at constant pH using a pH-stat?

Answer. Constant pH means that $[CaOH^+]$ will remain constant throughout (see Equation (8.33)); otherwise, $[CaOH^+]$ would vary with pH and the subsequent analysis would be more complex.

Metal ion catalysis of this type is fairly common (Section 8.3). Likewise, anion catalysis can also occur.

8.1.4 Distinction between acid and base hydrolyses of esters

There is a fundamental difference between the acid and base hydrolyses of esters. In the former the H_3O^+ is a genuine catalyst, is incorporated in the activated complex and is regenerated after reaction. In base hydrolysis the OH^- is also incorporated in the activated complex, *but as a reactant*, and cannot be regenerated after reaction. In this respect the common habit of referring to base hydrolysis as base catalysed hydrolysis is misleading.

This can be contrasted with metal ion and anion catalyses, where the metal ion, or the anion, is incorporated into the activated complex, but is regenerated after the reaction. They are thus genuine catalysts for these hydrolyses, analogous to acid catalysis, and in direct contrast to the so-called base 'catalysed' hydrolysis. These comments are pertinent to the following problem.

Worked Problem 8.4

Question. The acid hydrolysis of a positively charged ester HE^+ (acetylcholine, $(CH_3)_3N^+CH_2CH_2OCOCH_3$) is first order in ester and first order in H_3O^+. It is catalysed by SO_4^{2-} ions. A possible mechanism is

$$HE^+ + H_3O^+ \xrightarrow{k_1} \text{products} + H_3O^+$$

$$HE^+ + H_3O^+ + SO_4^{2-} \xrightarrow{k_2} \text{products} + H_3O^+ + SO_4^{2-}$$

where the products are CH_3COOH and $(CH_3)_3N^+CH_2CH_2OH$. Since SO_4^{2-} is a catalyst, it is incorporated in the activated complex to be regenerated after reaction.

Show that the following data, collected at constant ionic strength, is consistent with this mechanism, and find k_1 and k_2. Comment on the relative value of the two contributions to the rate.

$\frac{10^5 k_{obs}}{\text{mol}^{-1}\,\text{dm}^3\,\text{s}^{-1}}$	3.10	3.34	3.82	4.46
$\frac{10^2 [SO_4^{2-}]_{actual}}{\text{mol dm}^{-3}}$	0.69	1.18	2.18	3.50

Answer. There are two contributions to the overall rate.
Mechanistic rate:

$$-\frac{d[HE^+]}{dt} = k_1[HE^+][H_3O^+]_{actual} + k_2[HE^+][H_3O^+]_{actual}[SO_4^{2-}]_{actual} \tag{8.40}$$

Observed rate:

$$-\frac{d[HE^+]}{dt} = k_{obs}[HE^+][H_3O^+]_{actual} \tag{8.41}$$

$$\therefore \quad k_{obs} = k_1 + k_2[SO_4^{2-}]_{actual} \tag{8.42}$$

Pay attention to the following comments. Since H_3O^+ is a catalyst and is not used up during the experiment, $[H_3O^+]$ is constant throughout. This means that the *observed* rate can be expressed in two ways:

- $$-\frac{d[HE^+]}{dt} = k^{\text{2nd order}}[HE^+][H_3O^+] \tag{8.43}$$

  where $k^{\text{2nd order}}$ is an observed second order rate constant with units conc.$^{-1}$ time^{-1}, or as

- $$-\frac{d[HE^+]}{dt} = k^{\text{1st order}}[HE^+] \tag{8.44}$$

where $k^{\text{1st order}}$ is a first order rate constant with units time^{-1} and

$$k^{\text{1st order}} = k^{\text{2nd order}}[H_3O^{3+}] \quad (8.45)$$

$$k^{\text{2nd order}} = \frac{k^{\text{1st order}}}{[H_3O^+]} \quad (8.46)$$

It is imperative when formulating the observed rate expression that it is consistent with the units of the quoted observed rate constant. In this case, the observed rate constant is given as a second order rate constant, and is thus consistent with the rate expression, Equation (8.43).

This situation arises in many cases of mechanisms, and it is worthwhile to get into the habit of checking the units of the observed rate constant and assessing the implications.

If the mechanism is correct then a plot of k_{obs} versus $[SO_4^{2-}]_{actual}$ should be linear with slope k_2, and intercept k_1.

This plot is linear (Figure 8.6), with intercept $= k_1 = 2.78 \times 10^{-5}$ mol^{-1} dm^3 s^{-1} and slope $= k_2 = 4.87 \times 10^{-4}$ mol^{-2} dm^6 s^{-1}.

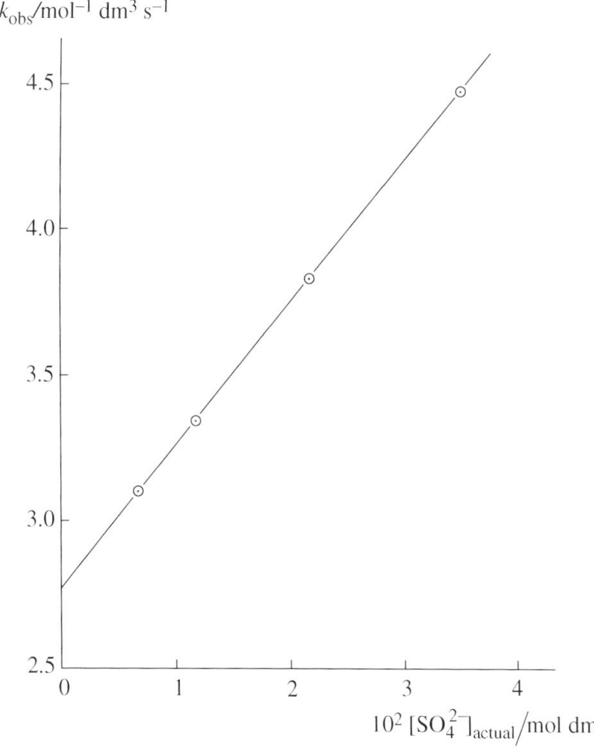

Figure 8.6 Graph of k_{obs} versus $[SO_4^{2-}]$, showing the catalytic effect of added sulphate in the acid hydrolysis of acetylcholine

Note. The units of k_2 are those for a third order rate constant, consistent with the contribution from the SO_4^{2-} catalysis being

$$k_2[HE^+][H_3O^+]_{actual}[SO_4^{2-}]_{actual}$$

The following questions relate to important points.

- Why should the rate expressions be written in terms of $[H_3O^+]_{actual}$ and $[SO_4^{2-}]_{actual}$ rather than in terms of the total concentrations?
- Comment on the relative values of the two contributions to the rate.

The answers are the following.

- The SO_4^{2-} is added to an acid solution, and is involved in the equilibrium

$$H_3O^+(aq) + SO_4^{2-}(aq) \rightleftharpoons HSO_4^-(aq) + H_2O(aq)$$

Significant amounts of SO_4^{2-} will be removed by protonation giving HSO_4^-, and this will also reduce the amount of H_3O^+, and so the actual concentrations will be significantly less than the total concentrations of these species. The actual concentrations can be calculated from the equilibrium quotient

$$\left(\frac{[HSO_4^-]_{actual}}{[H_3O^+]_{actual}[SO_4^{2-}]_{actual}}\right)_{equil} \quad (8.47)$$

with corrections for non-ideality.

- $k_2[SO_4^{2-}]$ ranges from 3.4×10^{-6} to 1.7×10^{-5} mol^{-1} dm^3 s^{-1}, which are values comparable to k_1 and not negligible compared with it. This is what would be expected for a catalytic reaction.

This is an anion-catalysed mechanism analogous to metal ion or cation-catalysed reactions. But the kinetically equivalent mechanism of general acid hydrolysis by HSO_4^-, analogous to general base hydrolysis, is also possible (Section 8.1.3 and Problem 8.3).

Since reaction is carried out in acid solution with added SO_4^{2-} there will be a substantial amount of HSO_4^- present in equilibrium with H_3O^+ and SO_4^{2-}.

$$SO_4^{2-}(aq) + H_3O^+(aq) \rightleftharpoons HSO_4^{2-}(aq) + H_2O(l)$$

$$K = \left(\frac{[HSO_4^-]_{actual}}{[H_3O^+]_{actual}[SO_4^{2-}]_{actual}}\right)_{equil} \quad (8.48)$$

with corrections for non-ideality.

A mechanism involving HSO_4^- can be written

$$HE^+ + H_3O^+ \xrightarrow{k_1} \text{products} + H_3O^+$$
$$HE^+ + HSO_4^- \xrightarrow{k_2} \text{products} + HSO_4^-$$

The *mechanistic rate*,

$$-\frac{d[HE^+]}{dt} = k_1[HE^+][H_3O^+]_{actual} + k_2[HE^+][HSO_4^-]_{actual} \tag{8.49}$$

The *observed rate* is again written as a second order process:

$$-\frac{d[HE^+]}{dt} = k_{obs}[HE^+][H_3O^+]_{actual} \tag{8.41}$$

$$\therefore \quad k_{obs} = k_1 + k_2\frac{[HSO_4^-]_{actual}}{[H_3O^+]_{actual}} \tag{8.50}$$

Substituting for $[HSO_4^-]_{actual}$ in terms of K and $[SO_4^{2-}]$ gives

$$k_{obs} = k_1 + k_2K[SO_4^{2-}]_{actual} \tag{8.51}$$

which is algebraically equivalent to the expression deduced by assuming catalysis is by SO_4^{2-}, and these mechanisms cannot be distinguished kinetically.

- $$k_{obs} = k_1 + k_2[SO_4^{2-}]_{actual} \tag{8.42}$$
  $$= k_1 + \text{constant}[SO_4^{2-}]_{actual} \tag{8.52}$$

- $$k_{obs} = k_1 + k_2K[SO_4^{2-}]_{actual} \tag{8.51}$$
  $$= k_1 + \text{constant}[SO_4^{2-}]_{actual} \tag{8.53}$$

The catalytic mechanisms proposed are

$$HE^+ + H_3O^+ + SO_4^{2-} \xrightarrow{k_2} \text{products} + H_3O^+ + SO_4^{2-}$$
$$HE^+ + HSO_4^- \xrightarrow{k_2} \text{products} + HSO_4^-$$

The second is more likely since it suggests that only two species have to come together to form the activated complex, whereas, in the first, three would have to come together: a less likely process. However, there are charges involved. The first mechanism is electrostatically more favourable, since it involves the reaction of *singly charged* cations with a *doubly charged* anion.

Whichever the mechanism, the activated complex will have the same constitution and charge.

8.2 More Complex Kinetic Situations Involving Reactants in Equilibrium with Each Other and Undergoing Reaction

There are a large number of inorganic and organic reactions which have these problems in the analysis. Three reactions will illustrate the ways to tackle the analysis.

8.2.1 A further look at the base hydrolysis of glycine ethyl ester as an illustration of possible problems

The simple treatment of Section 8.1.2 gave the *mechanistic rate* quoted as

$$-\frac{d[E]_{total}}{dt} = k_1[E]_{actual}[OH^-]_{actual} + k_2[HE^+]_{actual}[OH^-]_{actual} \tag{8.17}$$

The *observed rate* quoted was

$$-\frac{d[E]_{total}}{dt} = k_{obs}[E]_{total}[OH^-]_{actual} \tag{8.18}$$

giving

$$k_{obs}[E]_{total} = k_1[E]_{actual} + k_2[HE^+]_{actual} \tag{8.20}$$

$$k_{obs} = k_1\frac{[E]_{actual}}{[E]_{total}} + k_2\frac{[HE^+]_{actual}}{[E]_{total}} \tag{8.21}$$

Since K_b is small, it was possible to approximate $[E]_{actual} = [E]_{total}$ to give the simple relationship

$$k_{obs} = k_1 + k_2\frac{K_b}{[OH^-]_{actual}} \tag{8.26}$$

The approximation made for the glycine ethyl ester may not be justified in situations where there is a significant amount of all species involved in the equilibrium present, and a more rigorous analysis must be used.

In the present case, the way forward is to recognize that $[E]_{actual}/[E]_{total}$ is the fraction of the original ester that is unprotonated, $[HE^+]_{actual}/[E]_{total}$ is the fraction of the original ester that is protonated and

$$[E]_{total} = [HE^+]_{actual} + [E]_{actual} \tag{8.16}$$

Equation (8.16) can be rearranged to give $[E]_{actual}$ in terms of $[HE^+]_{actual}$ and $[E]_{total}$ and then with Equation (8.21) gives

$$k_{obs} = k_1\left(\frac{[E]_{total} - [HE^+]_{actual}}{[E]_{total}}\right) + k_2\left(\frac{[HE^+]_{actual}}{[E]_{total}}\right) \tag{8.54}$$

$$= k_1\left\{1 - \frac{[HE^+]_{actual}}{[E]_{total}} + \right\} + k_2\frac{[HE^+]_{actual}}{[E]_{total}} \tag{8.55}$$

$$k_{obs} = k_1 + (k_2 - k_1)\frac{[HE^+]_{actual}}{[E]_{total}} \tag{8.56}$$

The fraction protonated will depend on the pH; the higher the pH the smaller will be the amount of the ester which is protonated, and hence the smaller the ratio $[HE^+]_{actual}/[E]_{total}$. This quotient can be calculated for each pH, and a plot of the second order k_{obs} versus $[HE^+]_{actual}/[E]_{total}$ should be linear, with intercept k_1 and slope $(k_2 - k_1)$.

Worked Problem 8.5

Question. Show that an analysis of the data given in Problem 8.2 using this rigorous method gives the same results as previously. Values of the ratio $[HE^+]_{actual}/[E]_{total}$ have been calculated for each pH, using the known value for K_b.

$$K_b = 5.01 \times 10^{-7}\ \text{mol dm}^{-3} \text{ at } 25\ ^\circ\text{C}.$$

pH	11.70	11.22	10.05	9.55	9.37
$\frac{10^5 \times [OH^-]}{\text{mol dm}^{-3}}$	500	166	11.2	3.55	2.34
$\frac{10^4 \times [HE^+]_{actual}}{[E]_{total}}$	1.00	3.02	44.5	139	210
$\frac{k_{obs}}{\text{dm}^3\ \text{mol}^{-1}\ \text{min}^{-1}}$	43.3	45.0	50.2	64.0	74.3

Answer. The plot of k_{obs} versus $[HE^+]_{actual}/[E]_{total}$ is linear with intercept $= k_1 = 43\ \text{mol}^{-1}\ \text{dm}^3\ \text{min}^{-1}$, and slope $= k_2 - k_1 = 1480\ \text{mol}^{-1}\ \text{dm}^3\ \text{min}^{-1}$, giving $k_2 = 1523\ \text{mol}^{-1}\ \text{dm}^3\ \text{min}^{-1}$(Figure 8.7).

This agrees with the approximate treatment results, justifying the use of the approximation.

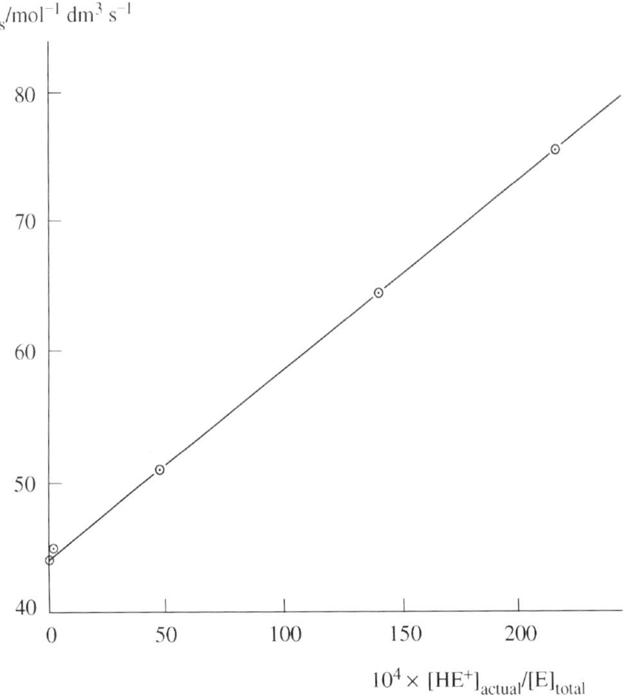

Figure 8.7 Graph of k_{obs} versus $[HE^+]_{actual}/[E]_{total}$ for the base hydrolysis of glycine ethyl ester

Another important point about analysis of experimental data and the drawing of graphs

Another way of getting a relationship between k_{obs} and quantities calculated from K_b and the pH will not result in a useful graph. This alternative procedure uses Equation (8.16) in terms of $[E]_{actual}$ and $[E]_{total}$ rather than $[HE^+]_{actual}$ and $[E]_{total}$.

$[E]_{actual}/[E]_{total}$ is the fraction of the original ester which is unprotonated, $[HE^+]_{actual}/[E]_{total}$ is the fraction of the original ester which is protonated and

$$[HE^+]_{actual} = [E]_{total} - [E]_{actual} \tag{8.57}$$

Substitution of this relation into Equation (8.21),

$$k_{obs} = k_1 \frac{[E]_{actual}}{[E]_{total}} + k_2 \frac{[HE^+]_{actual}}{[E]_{total}} \tag{8.21}$$

gives

$$k_{obs} = k_1 \frac{[E]_{actual}}{[E]_{total}} + k_2 \left(\frac{[E]_{total}}{[E]_{total}} - \frac{[E]_{actual}}{[E]_{total}} \right) \tag{8.58}$$

$$= k_2 + (k_1 - k_2) \frac{[E]_{actual}}{[E]_{total}} \tag{8.59}$$

where

$$\frac{[E]_{actual}}{[E]_{total}} = \frac{[OH^-]_{actual}}{[OH^-]_{actual} + K_b} \tag{8.60}$$

The expression for $[E]_{actual}/[E]_{total}$ given in Equation (8.60) follows from the following argument:

$$[HE^+]_{actual} = \frac{K_b[E]_{actual}}{[OH^-]_{actual}} \tag{8.22}$$

Substituting this into Equation (8.16) gives

$$[E]_{total} = \frac{K_b[E]_{actual}}{[OH^-]_{actual}} + [E]_{actual} = [E]_{actual}\left(\frac{K_b + [OH^-]_{actual}}{[OH^-]_{actual}}\right) \tag{8.61}$$

and so

$$\frac{[E]_{actual}}{[E]_{total}} = \frac{[OH^-]_{actual}}{[OH^-]_{actual} + K_b}$$

which is Equation (8.60).

Calculations show that this quotient does not vary much with pH, resulting in a 'bunched-up' graph from which no useful data can be found. This aspect of the graphical data should be compared with the similar conclusion mentioned in connection with determination of $\Delta S^{\neq *}$ and A factors using the Arrhenius equation, Worked Problems 3.22 and 7.6.

This fuller analysis is generally required, as will be shown in Sections 8.2.2 and 8.2.3 below.

8.2.2 Decarboxylations of β-keto-monocarboxylic acids

These involve both the undissociated acid, HA, and its anion, A^-, with mechanisms such as

$$RCOCH_2COOH \xrightarrow{k_1} RC(OH){=}CH_2 + CO_2$$

$$RCOCH_2COO^- \xrightarrow{k_2} RC(O^-){=}CH_2 + CO_2$$

where the β-carbonyl group has an electron withdrawing effect which assists the decarboxylation.

$$\mathrm{R{-}C({=}O){-}CH_2{-}C({=}O){-}O{-}H} \longrightarrow \mathrm{R{-}C(O{-}H){=}CH_2} + \mathrm{CO_2}$$

$$\mathrm{R{-}C({=}O){-}CH_2{-}C({=}O){-}O^-} \longrightarrow \mathrm{R{-}C(O^-){=}CH_2} + \mathrm{CO_2}$$

The *mechanistic rate* is

$$+\frac{\mathrm{d}[\mathrm{CO_2}]}{\mathrm{d}t} = -\frac{\mathrm{d}[\mathrm{HA}]_{\mathrm{total}}}{\mathrm{d}t} = k_1[\mathrm{HA}]_{\mathrm{actual}} + k_2[\mathrm{A^-}]_{\mathrm{actual}} \tag{8.62}$$

where

$$[\mathrm{HA}]_{\mathrm{total}} = [\mathrm{HA}]_{\mathrm{actual}} + [\mathrm{A^-}]_{\mathrm{actual}} \tag{8.63}$$

and

$$K_{\mathrm{a}} = \left(\frac{[\mathrm{H_3O^+}]_{\mathrm{actual}}[\mathrm{A^-}]_{\mathrm{actual}}}{[\mathrm{HA}]_{\mathrm{actual}}}\right)_{\mathrm{equil}} \tag{8.64}$$

The *observed rate* expression is given in terms of the total [HA]:

$$+\frac{\mathrm{d}[\mathrm{CO_2}]}{\mathrm{d}t} = -\frac{\mathrm{d}[\mathrm{HA}]_{\mathrm{total}}}{\mathrm{d}t} = k_{\mathrm{obs}}[\mathrm{HA}]_{\mathrm{total}} \tag{8.65}$$

$$k_{\mathrm{obs}} = k_1\frac{[\mathrm{HA}]_{\mathrm{actual}}}{[\mathrm{HA}]_{\mathrm{total}}} + k_2\frac{[\mathrm{A^-}]_{\mathrm{actual}}}{[\mathrm{HA}]_{\mathrm{total}}} \tag{8.66}$$

This is algebraically identical to that for the glycine ester hydrolysis, Equation (8.21). However, in this case the acid ionization constant is sufficiently large for significant amounts of both ionized and un-ionized acid to be present, and no approximations can be made, so that since

$$[\mathrm{HA}]_{\mathrm{actual}} = [\mathrm{HA}]_{\mathrm{total}} - [\mathrm{A^-}]_{\mathrm{actual}} \tag{8.67}$$

$$k_{\mathrm{obs}} = k_1\left\{\frac{[\mathrm{HA}]_{\mathrm{total}} - [\mathrm{A^-}]_{\mathrm{actual}}}{[\mathrm{HA}]_{\mathrm{total}}}\right\} + k_2\frac{[\mathrm{A^-}]_{\mathrm{actual}}}{[\mathrm{HA}]_{\mathrm{total}}} \tag{8.68}$$

$$= k_1\frac{[\mathrm{HA}]_{\mathrm{total}}}{[\mathrm{HA}]_{\mathrm{total}}} - k_1\frac{[\mathrm{A^-}]_{\mathrm{actual}}}{[\mathrm{HA}]_{\mathrm{total}}} + k_2\frac{[\mathrm{A^-}]_{\mathrm{actual}}}{[\mathrm{HA}]_{\mathrm{total}}} \tag{8.69}$$

$$= k_1 + (k_2 - k_1)\frac{[\mathrm{A^-}]_{\mathrm{actual}}}{[\mathrm{HA}]_{\mathrm{total}}} \tag{8.70}$$

and a plot of k_{obs} versus $[A^-]_{actual}/[HA]_{total}$ should be linear with slope $= k_2 - k_1$ and intercept k_1.

This requires that the decarboxylation is carried out over a range of pH values, for which the corresponding ratios $[A^-]_{actual}/[HA]_{total}$ can be calculated.

8.2.3 The decarboxylation of *β*-keto-dicarboxylic acids

Here the situation becomes one stage more complex. The acid can form both the anion and dianion and all three species can decarboxylate, e.g., for 2-oxobutanedioic acid (oxaloacetic acid)

$$\mathrm{HO{-}C({=}O){-}C({=}O){-}CH_2{-}C({=}O){-}O{-}H}$$

$$\mathrm{HO{-}C({=}O){-}C({=}O){-}CH_2{-}C({=}O){-}O^-}$$

$$\mathrm{^-O{-}C({=}O){-}C({=}O){-}CH_2{-}C({=}O){-}O^-}$$

$$H_2A + H_2O \rightleftharpoons HA^- + H_3O^+$$

$$HA^- + H_2O \rightleftharpoons A^{2-} + H_3O^+$$

$$H_2A \xrightarrow{k_1} \text{product} + CO_2$$

$$HA^- \xrightarrow{k_2} \text{product} + CO_2$$

$$A^{2-} \xrightarrow{k_3} \text{product} + CO_2$$

The *mechanistic rate* is

$$+\frac{d[CO_2]}{dt} = -\frac{d[H_2A]_{total}}{dt} = k_1[H_2A]_{actual} + k_2[HA^-]_{actual} + k_3[A^{2-}]_{actual} \tag{8.71}$$

and *the observed rate*

$$+\frac{d[CO_2]}{dt} = -\frac{d[H_2A]_{total}}{dt} = k_{obs}[H_2A]_{total} \tag{8.72}$$

$$k_{obs} = k_1\frac{[H_2A]_{actual}}{[H_2A]_{total}} + k_2\frac{[HA^-]_{actual}}{[H_2A]_{total}} + k_3\frac{[A^{2-}]_{actual}}{[H_2A]_{total}} \tag{8.73}$$

Each of the fractions will be pH dependent. At *very low* pH values, the fraction $[A^{2-}]_{actual}/[H_2A]_{total}$ will be small and, unless k_3 is sufficiently large, this term can be ignored. k_{obs} then reduces to a two-term expression from which k_1 and k_2 can be found as given for the β-keto-monocarboxylic acids. The value of k_3 can then be found from experiments at higher pH values where the fraction $[A^{2-}]_{actual}/[H_2A]_{total}$ makes a significant contribution.

For any given pH, and knowing the values of k_1 and k_2, the terms

$$k_1 \frac{[H_2A]_{actual}}{[H_2A]_{total}} + k_2 \frac{[HA^-]_{actual}}{[H_2A]_{total}}$$

can be calculated; k_{obs} can be measured and

$$k_{obs} - k_1 \frac{[H_2A]_{actual}}{[H_2A]_{total}} - k_2 \frac{[HA^-]_{actual}}{[H_2A]_{total}}$$

can be found. From Equation (8.73) this is equal to $k_3[A^{2-}]_{actual}/[H_2A]_{total}$. Since the ratio $[A^{2-}]_{actual}/[H_2A]_{total}$ can be calculated at the given pH, k_3 can be found.

This illustrates the complexities introduced into the kinetic analysis when reactant species are involved in equilibria, as well as in reaction. Most of the difficulties lie in the handling of the equilibria calculations, rather than in the kinetics.

All reactions of this type must be studied at constant ionic strength because of the considerable non-ideality of the solutions, and the equilibrium calculations should be given in terms of activities rather than concentrations.

Worked Problem 8.6

Question. The decarboxylation of 2-oxobutanedioic acid, p. 341 (oxaloacetic acid), H_2A, at constant pH has three contributing steps:

$$H_2A \xrightarrow{k_1} \text{product} + CO_2$$

$$HA^- \xrightarrow{k_2} \text{product} + CO_2$$

$$A^{2-} \xrightarrow{k_3} \text{product} + CO_2$$

and

$$-\frac{d[H_2A]_{total}}{dt} = k_1[H_2A]_{actual} + k_2[HA^-]_{actual} + k_3[A^{2-}]_{actual} \tag{8.71}$$

At low pH only a very small fraction of the acid is present as the dianion, and step 3 can be ignored with respect to steps 1 and 2. This is despite the similarities in the k values.

$$k_1 = 0.064 \times 10^{-3}\ \text{min}^{-1},\ k_2 = 3.4 \times 10^{-3}\ \text{min}^{-1},\ k_3 \approx 1 \times 10^{-3}\ \text{min}^{-1}$$

- Calculate the contribution to the reaction for each of the steps 1, 2 and 3 when the concentrations of the three species are

$$[H_2A]_{actual} = 2.80 \times 10^{-2}\ \text{mol dm}^{-3},\ [HA^-]_{actual} = 1.36 \times 10^{-3}\ \text{mol dm}^{-3},$$
$$[A^{2-}]_{actual} = 1.76 \times 10^{-6}\ \text{mol dm}^{-3}.$$

- If this experiment was carried out with the total acid concentration equal to $0.0294\ \text{mol dm}^{-3}$, find the overall observed rate constant.
- Why is it necessary to carry out the reaction at constant pH?

Answer.

(a)
$$-\frac{d[H_2A]_{total}}{dt} = k_1[H_2A]_{actual} + k_2[HA^-]_{actual} + k_3[A^{2-}]_{actual} \quad (8.71)$$
$$= 0.064 \times 10^{-3}\ \text{min}^{-1} \times 2.80 \times 10^{-2}\ \text{mol dm}^{-3}$$
$$+ 3.4 \times 10^{-3}\ \text{min}^{-1} \times 1.36 \times 10^{-3}\ \text{mol dm}^{-3}$$
$$+ 1 \times 10^{-3}\ \text{min}^{-1} \times 1.76 \times 10^{-6}\ \text{mol dm}^{-3}$$
$$= \{1.79 \times 10^{-6} + 4.62 \times 10^{-6}$$
$$+ 1.76 \times 10^{-9}\}\ \text{mol dm}^{-3}\ \text{min}^{-1}$$
$$= 6.41 \times 10^{-6}\ \text{mol dm}^{-3}\ \text{min}^{-1}$$

Hence, at this pH, the contribution to the overall rate from the monoanion is greater than that for the undissociated acid and the contribution from the dianion can be neglected.

Note. It is the product of rate constant and actual concentration which is important.

(b)
$$-\frac{d[H_2A]_{total}}{dt} = k_{obs}[H_2A]_{total} = 6.41 \times 10^{-6}\ \text{mol dm}^{-3}\ \text{min}^{-1}$$
$$[H_2A]_{total} = 0.0294\ \text{mol dm}^{-3}$$
$$k_{obs} = \frac{6.41 \times 10^{-6}\ \text{mol dm}^{-3}\ \text{min}^{-1}}{0.0294\ \text{mol dm}^{-3}} = 2.18 \times 10^{-4}\ \text{min}^{-1}$$

(c) It is vital to keep the pH constant because the concentrations of the reacting species are involved in acid–base equilibria, which are governed by the pH. This is something which must not be overlooked when reactants are involved in other equilibria or interlinked equilibria.

8.3 Metal Ion Catalysis

Many reactions of biological interest are catalysed by metal ions. Simple model systems for these complex reactions have been studied, e.g. metal-ion-catalysed decarboxylations of dicarboxylic acids, and metal-ion-catalysed hydrolyses.

In the decarboxylations, the catalytic activity is associated with the formation of a metal ion complex, formed with the carbonyl group and a carboxylic acid group which does *not* decarboxylate.

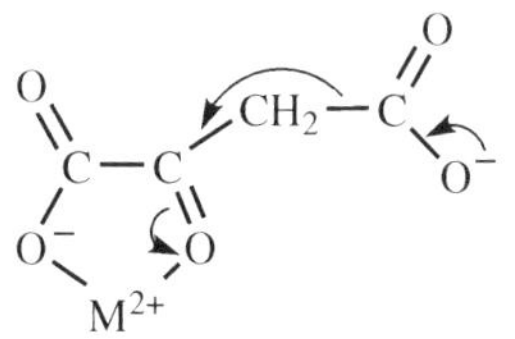

The formation of the complex creates a very strong electron-withdrawing group which helps the withdrawal of the electrons from the carbon–carbon bond which is breaking. Catalysis results because this electron-withdrawing effect is much stronger than the effect of the anion in the uncatalysed decarboxylations.

A similar situation arises when the metal-ion-catalysed base hydrolyses of aminoacids and peptides are studied. Such hydrolyses contrast with the base hydrolyses of simple esters, which are not catalysed by metal ions. Again, metal ion complexes with the aminoacid esters or the peptides are involved in the catalyses. The presence of the metal ion increases the electron-withdrawing power of the carbonyl group and helps attack by the hydroxide ion, leading to efficient hydrolysis.

Both the metal ion decarboxylations and the metal ion hydrolyses can be handled in similar fashion.

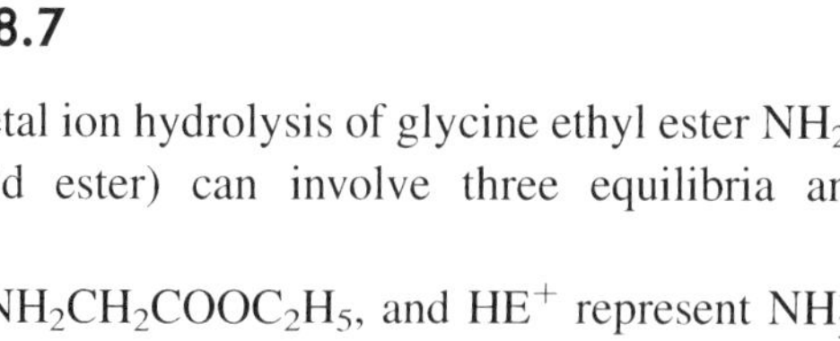

Worked Problem 8.7

Question. The metal ion hydrolysis of glycine ethyl ester $NH_2CH_2COOC_2H_5$ (the simplest amino-acid ester) can involve three equilibria and four concurrent reactions.

Let E represent $NH_2CH_2COOC_2H_5$, and HE^+ represent $NH_3^+CH_2COOC_2H_5$.

$$E + H_3O^+ \rightleftharpoons HE^+ + H_2O$$

$$K = \frac{[HE^+]_{actual}}{[E]_{actual}[H_3O^+]_{actual}} \tag{8.74}$$

$$E + M^{z+} \rightleftharpoons EM^{z+}$$

$$K_1 = \frac{[EM^{z+}]_{actual}}{[E]_{actual}[M^{z+}]_{actual}} \tag{8.75}$$

$$2E + M^{z+} \rightleftharpoons E_2M^{z+}$$

$$K_2 = \left(\frac{[E_2M^{z+}]_{actual}}{[E]^2_{actual}[M^{z+}]_{actual}}\right)_{equil} \tag{8.76}$$

$$E + OH^- \rightarrow \text{products}$$

$$HE^+ + OH^- \rightarrow \text{products}$$

$$E + OH^- + M^{z+} \rightarrow \text{products} + M^{z+}$$

$$2E + OH^- + M^{z+} \rightarrow \text{products} + M^{z+}$$

This scheme can be simplified by carrying out the reaction at sufficiently high pH values for the reaction of OH^- with the protonated ester to be ignored, and by having M^{z+} in such high excess that only the 1:1 chelate is formed. Under these circumstances, the reaction simplifies to

$$E + OH^- \xrightarrow{k_1} \text{products}$$

$$E + OH^- + M^{z+} \xrightarrow{k_{cat}} \text{products} + M^{z+}$$

Formulate the mechanistic rate expression and from this find an expression for the dependence of k_{obs} on $[M^{z+}]$ which will allow k_1 and k_{cat} to be found.

Answer. Mechanistic rate:

$$-\frac{d[E]_{actual}}{dt} = k_1[E]_{actual}[OH^-]_{actual} + k_{cat}[E]_{actual}[OH^-]_{actual}[M^{z+}]_{actual} \tag{8.77}$$

This can be simplified as follows.

- Reaction is carried out at pH values such that $[HE^+]_{actual}$ is very low, and so $[E]_{actual}$ can be approximated to $[E]_{total}$.
- Since the metal ion is present in large excess, $[M^{z+}]_{actual}$ can be approximated to $[M^{z+}]_{total}$.
- The reaction is carried out in a pH-stat, where the pH is kept constant by progressive addition of OH^- as reaction occurs. $[OH^-]_{actual}$ is thus kept constant throughout. The observed rate is thus in terms of $[OH^-]_{actual}$.

- These facts result in the *mechanistic rate* now being

$$-\frac{d[E]_{total}}{dt} = k_1[E]_{total}[OH^-]_{actual} + k_{cat}[E]_{total}[OH^-]_{actual}[M^{z+}]_{total} \quad (8.78)$$

The *observed rate* is

$$-\frac{d[E]_{total}}{dt} = k_{obs}[E]_{total}[OH^-]_{actual} \quad (8.79)$$

$$k_{obs} = k_1 + k_{cat}[M^{z+}]_{total} \quad (8.80)$$

If the mechanism is correct, then a plot of k_{obs} versus $[M^{z+}]_{total}$ will be linear with slope $= k_{cat}$ and intercept $= k_1$.

The value of k_1 from this plot should be equal to the value of k_1 for reaction in the absence of metal ions.

8.4 Other Common Mechanisms

Many inorganic mechanisms involve pre-equilibria (Section 3.22).

- $A \underset{k_{-1}}{\overset{k_1}{\rightleftarrows}} B \xrightarrow{k_2} C$
- $A + B \underset{k_{-1}}{\overset{k_1}{\rightleftarrows}} C \xrightarrow{k_2} D$
- $A + B \overset{k_1}{\rightleftarrows} C$

  $C + X \xrightarrow{k_2} D$

A mathematical analysis of such mechanisms shows that *simple* kinetics is only found when the equilibrium is *set up rapidly* and *maintained* throughout the reaction. In all other circumstances, complex kinetics would be expected, and computer analysis of the experimental data would be required. Fortunately, for most inorganic mechanisms the equilibrium *is* set up rapidly and maintained, and simple kinetics *is* observed.

The situations discussed in this book are for reactions where the pre-equilibrium is set up rapidly and maintained.

8.4.1 The simplest mechanism

$$A \underset{k_{-1}}{\overset{k_1}{\rightleftarrows}} B \xrightarrow{k_2} C$$

can be used to illustrate some of the features of such reaction types.

- If the pre-equilibrium is established rapidly, then the step

$$B \xrightarrow{k_2} C$$

will be rate determining, and reaction reduces effectively to B $\rightarrow$ C, so that

$$\text{mechanistic rate} = +\frac{d[C]}{dt} = k_2[B]_{\text{actual}} \tag{8.81}$$

- The kinetic requirements for this to be the case are
  k_1 *or* k_{-1} must be *large*
  and $k_1 + k_{-1} \gg k_2$
- When the setting up of the equilibrium is fast, and then maintained throughout the reaction, then *throughout the reaction*
- $$\frac{[B]_{\text{actual at time } t}}{[A]_{\text{actual at time } t}} = K \tag{8.82}$$

  $[B]_{\text{actual at time } t}$ is the equilibrium value of $[B]_{\text{actual}}$ at that time, and likewise for A.
- B is always being removed to form C, and so throughout reaction both A and B are being removed. $[A]_{\text{actual}}$ and $[B]_{\text{actual}}$ are, therefore, not constant throughout the reaction, but vary continually throughout reaction, even though their quotient $[B]_{\text{actual at time } t}/[A]_{\text{actual at time } t}$ always remains constant.

The analysis follows a similar pattern for all types of pre-equilibrium reaction, and this will be illustrated below with each step in the argument being listed.

8.4.2 Kinetic analysis of the simplest mechanism

The mechanism

$$A \underset{k_{-1}}{\overset{k_1}{\rightleftharpoons}} B \xrightarrow{k_2} C$$

has the following features.

- Reaction is found to be first order in A.
- The *total stoichiometric* amount of A present at any given time may be regarded as made up of the *actual amounts* of A and B present at that time, so that

$$[A]_{\text{total}} = [A]_{\text{actual}} + [B]_{\text{actual}} \tag{8.83}$$

  These actual concentrations which are present in the solution are related by the equilibrium constant:

$$K = \left(\frac{[B]_{\text{actual}}}{[A]_{\text{actual}}}\right)_{\text{equil}} \tag{8.84}$$

so that

$$[\mathrm{A}]_{\text{actual}} = [\mathrm{B}]_{\text{actual}}/K \tag{8.85}$$

and

$$[\mathrm{B}]_{\text{actual}} = K[\mathrm{A}]_{\text{actual}} \tag{8.86}$$

- From the mechanism

  increase in $[\mathrm{B}]_{\text{actual}}$ = decrease in $[\mathrm{A}]_{\text{actual}}$ – increase in [C]

$$+\frac{\mathrm{d}[\mathrm{B}]_{\text{actual}}}{\mathrm{d}t} = -\frac{\mathrm{d}[\mathrm{A}]_{\text{actual}}}{\mathrm{d}t} - \frac{\mathrm{d}[\mathrm{C}]}{\mathrm{d}t} \tag{8.87}$$

$$+\frac{\mathrm{d}[\mathrm{C}]}{\mathrm{d}t} = -\frac{\mathrm{d}[\mathrm{A}]_{\text{actual}}}{\mathrm{d}t} - \frac{\mathrm{d}[\mathrm{B}]_{\text{actual}}}{\mathrm{d}t} = -\frac{\mathrm{d}([\mathrm{A}]_{\text{actual}} + [\mathrm{B}]_{\text{actual}})}{\mathrm{d}t} = -\frac{\mathrm{d}[\mathrm{A}]_{\text{total}}}{\mathrm{d}t} \tag{8.88}$$

  The rate-determining step is $\mathrm{B} \xrightarrow{k_2} \mathrm{C}$

$$\text{mechanistic rate} = +\frac{\mathrm{d}[\mathrm{C}]}{\mathrm{d}t} = k_2[\mathrm{B}]_{\text{actual}} \tag{8.81}$$

$$\therefore \quad -\frac{\mathrm{d}([\mathrm{A}]_{\text{actual}} + [\mathrm{B}]_{\text{actual}})}{\mathrm{d}t} = -\frac{\mathrm{d}[\mathrm{A}]_{\text{total}}}{\mathrm{d}t} = k_2[\mathrm{B}]_{\text{actual}} \tag{8.89}$$

- It is necessary to find an expression for $[\mathrm{B}]_{\text{actual}}$ in terms of constants, e.g. K, or k_1 and k_2, and/or known quantities, e.g. $[\mathrm{A}]_{\text{total}}$.

  So substitute for $[\mathrm{A}]_{\text{actual}}$ in Equation (8.83) to get an expression for $[\mathrm{B}]_{\text{actual}}$ in terms of K and $[\mathrm{A}]_{\text{total}}$:

$$[\mathrm{A}]_{\text{total}} = \frac{[\mathrm{B}]_{\text{actual}}}{K} + [\mathrm{B}]_{\text{actual}} = [\mathrm{B}]_{\text{actual}}\left(\frac{1+K}{K}\right) \tag{8.90}$$

$$[\mathrm{B}]_{\text{actual}} = \left(\frac{K}{1+K}\right)[\mathrm{A}]_{\text{total}} \tag{8.91}$$

- The *mechanistic rate* can now be given as

$$\text{rate of reaction} = +\frac{\mathrm{d}[\mathrm{C}]}{\mathrm{d}t} = -\frac{\mathrm{d}[\mathrm{A}]_{\text{total}}}{\mathrm{d}t} = k_2[\mathrm{B}]_{\text{actual}} = k_2\left(\frac{K}{1+K}\right)[\mathrm{A}]_{\text{total}} \tag{8.92}$$

- It now remains to decide how the experimental rate is to be followed. It is important to pay attention to this step in the analysis.
  There are four possibilities:

  1. in terms of the product, [C]
  2. in terms of the total [A]

3. in terms of the actual A present, $[A]_{actual}$
4. in terms of the actual B present, $[B]_{actual}$.

The chemistry of the system will determine which is chosen, but usually the change in $[A]_{total}$ or [C] is monitored when

- *observed rate* $= -\frac{d[A]_{total}}{dt} = +\frac{d[C]}{dt} = k_{obs}[A]_{total}$ (8.93)

But note: in the examples in Problems 8.3 and 8.7, the rate is given in terms of an *actual* concentration of a reactant. In these examples the experimental technique actually measured the *actual* concentration rather than the *total* concentration. Care must be taken to distinguish between these two situations.

- If the mechanism is correct, then the observed and mechanistic rates must be equal and

$$k_{obs}[A]_{total} = k_2\left(\frac{K}{1+K}\right)[A]_{total} \quad (8.94)$$

$$k_{obs} = k_2\left(\frac{K}{1+K}\right) \quad (8.95)$$

Since the reaction is first order in A, a plot of $\log_e[A]_{total\ at\ time\ t}$ versus t is linear with slope $= -k_{obs} = -k_2K/(1+K)$, from which k_2 can be found, provided K is known independently. $K = k_1/k_{-1}$, and other experiments will be required to determine k_1 and k_{-1} individually.

There are two limiting cases for k_{obs}.

- If K is large, then most of the reactant present is in the form of B, and $1 + K \approx K$, so that $k_{obs} \approx k_2$.
- If K is small, then there is very little B present and $1 + K \approx 1$, so that $k_{obs} = k_2K$.

Worked Problem 8.8

Question. Formulate expressions for k_{obs} for the above mechanism for cases where the overall rate is found by following

- $[B]_{actual}$ with time
- $[A]_{actual}$ with time.

Answer.

- Following $[B]_{actual}$ with time:

observed rate of reaction $= -\frac{d[B]_{actual}}{dt} = k_{obs}[B]_{actual}$ (8.96)

$$\textit{mechanistic rate} \text{ of reaction} = +\frac{d[C]}{dt} = -\frac{d[A]_{actual}}{dt} - \frac{d[B]_{actual}}{dt}$$
$$= -\frac{d([A]_{actual} + [B]_{actual})}{dt} \quad (8.97)$$

From Equation (8.85): $[A]_{actual} = [B]_{actual}/K$

$$+\frac{d[C]}{dt} = -\frac{d([A]_{actual} + [B]_{actual})}{dt}$$
$$= -\frac{d\left\{\frac{[B]_{actual}}{K} + [B]_{actual}\right\}}{dt} = -\frac{d[B]_{actual}}{dt}\left\{\frac{1+K}{K}\right\} \quad (8.98)$$

$$-\frac{d[B]_{actual}}{dt} = \frac{K}{1+K}\frac{d[C]}{dt} \quad (8.99)$$

$$+\frac{d[C]}{dt} = k_2[B]_{actual} \quad (8.81)$$

$$-\frac{d[B]_{actual}}{dt} = \frac{k_2 K}{1+K}[B]_{actual} \quad (8.100)$$

If the mechanism is correct, Equations (8.96) and (8.100) are equal, giving:

$$k_{obs} = \frac{k_2 K}{1+K} \quad (8.101)$$

- Following $[A]_{actual}$ with time,

$$\textit{observed rate} = -\frac{d[A]_{actual}}{dt} = k_{obs}[A]_{actual} \quad (8.102)$$

$$\textit{mechanistic rate} = +\frac{d[C]}{dt} = -\frac{d([A]_{actual} + [B]_{actual})}{dt}$$
$$= -\frac{d([A]_{actual} + K[A]_{actual})}{dt} \quad (8.103)$$
$$= -\frac{d[A]_{actual}}{dt}\{1+K\} \quad (8.104)$$

$$-\frac{d[A]_{actual}}{dt} = \frac{1}{1+K}\frac{d[C]}{dt} \quad (8.105)$$

$$\frac{d[C]}{dt} = k_2[B]_{actual} \quad (8.81)$$

$$-\frac{d[A]_{actual}}{dt} = \frac{k_2}{1+K}[B]_{actual} \quad (8.106)$$

If the mechanism is correct, then the observed and mechanistic rates are equal:

$$k_{obs}[A]_{actual} = \frac{k_2}{1+K}[B]_{actual} \quad (8.107)$$

$$k_{obs} = \frac{k_2}{1+K}\frac{[B]_{actual}}{[A]_{actual}} = \frac{k_2 K}{1+K} \quad (8.108)$$

The analyses in Section 8.4.2 and in this problem are all self-consistent, as is shown by all resulting in the same expression for k_{obs}.

Note how absolutely crucial it is to specify exactly how the experimental rate is found, and how vital it is to distinguish between *actual* and *total* concentrations.

8.4.3 A slightly more complex scheme

$$A + B \underset{k_{-1}}{\overset{k_1}{\rightleftharpoons}} C \xrightarrow{k_2} D$$

This is a typical inorganic mechanism where there is a pre-equilibrium formation of an ion pair or complex, followed by reaction to product, e.g.

$$Cr^{3+}(aq) + edta^{4-}(aq) \underset{k_{-1}}{\overset{k_1}{\rightleftharpoons}} Cr(edta)^{-}(aq)$$

$$Cr(edta)^{-}(aq) \xrightarrow{k_2} \text{products}$$

The forward step of the reversible reaction is second order, and this poses the usual problems.

If the initial conditions correspond to

- $[A]_0 = [B]_0$, then the situation is similar to that discussed in Section 8.4.2, except that the forward step 1 is second order;
- $[A]_0 \neq [B]_0$, then complex kinetics are observed, and computer solutions are necessary.

However, if either A or B is held in excess then

- the forward step of formation of C is pseudo-first order.

Normally one of the reactants is held in excess, and the reaction is studied under pseudo-first order conditions, when

- the forward step of formation of C is pseudo-first order;
- the reaction is found to be first order in the reactant not in excess, and this is the reactant which is monitored throughout;
- the reaction is found to be of complex kinetics for the substance in excess, with the observed pseudo-first order rate constant depending on the concentration of this substance in a complex manner;
- this is resolved by carrying out experiments at various concentrations, *all of which are in excess.*

8.4.4 Standard procedure for determining the expression for k_{obs} for the given mechanism (Section 8.4.3)

This should be compared with the previous analysis (Section 8.4.2).

- In this reaction a decision has to be made as to which of the reactants will always be held in excess. The analysis is similar whichever is chosen. Let reactant in excess be B.
- The *total* amounts of A and B are distributed by the equilibrium into the *actual* amounts of A, B and C present at any given time.

  These are related by the equilibrium constant, K:

$$K = \left(\frac{[\mathrm{C}]_{\mathrm{actual}}}{[\mathrm{A}]_{\mathrm{actual}}[\mathrm{B}]_{\mathrm{actual}}} \right)_{\mathrm{equil}} \tag{8.109}$$

so that

$$[\mathrm{A}]_{\mathrm{actual}} = \frac{[\mathrm{C}]_{\mathrm{actual}}}{\mathrm{K}[\mathrm{B}]_{\mathrm{actual}}} \tag{8.110}$$

and

$$[\mathrm{C}]_{\mathrm{actual}} = K[\mathrm{A}]_{\mathrm{actual}}[\mathrm{B}]_{\mathrm{actual}} \tag{8.111}$$

- From the mechanism

increase in $[\mathrm{C}]_{\mathrm{actual}}$ = decrease in $[\mathrm{A}]_{\mathrm{actual}}$ – increase in [D]

$$+\frac{\mathrm{d}[\mathrm{C}]_{\mathrm{actual}}}{\mathrm{d}t} = -\frac{\mathrm{d}[\mathrm{A}]_{\mathrm{actual}}}{\mathrm{d}t} - \frac{\mathrm{d}[\mathrm{D}]}{\mathrm{d}t} \tag{8.112}$$

$$+\frac{\mathrm{d}[\mathrm{D}]}{\mathrm{d}t} = -\frac{\mathrm{d}[\mathrm{A}]_{\mathrm{actual}}}{\mathrm{d}t} - \frac{\mathrm{d}[\mathrm{C}]_{\mathrm{actual}}}{\mathrm{d}t} = -\frac{\mathrm{d}([\mathrm{A}]_{\mathrm{actual}} + [\mathrm{C}]_{\mathrm{actual}})}{\mathrm{d}t} = -\frac{\mathrm{d}[\mathrm{A}]_{\mathrm{total}}}{\mathrm{d}t} \tag{8.113}$$

Since the rate-determining step is $\mathrm{C} \xrightarrow{k_2} \mathrm{D}$

$$\textit{mechanistic rate} = +\frac{\mathrm{d}[\mathrm{D}]}{\mathrm{d}t} = k_2[\mathrm{C}]_{\mathrm{actual}} \tag{8.114}$$

$$\therefore \quad -\frac{\mathrm{d}[\mathrm{A}]_{\mathrm{total}}}{\mathrm{d}t} = k_2[\mathrm{C}]_{\mathrm{actual}} \tag{8.115}$$

A similar expression can be written for the changes involving B.

$$+\frac{\mathrm{d}[\mathrm{D}]}{\mathrm{d}t} = -\frac{\mathrm{d}[\mathrm{B}]_{\mathrm{actual}}}{\mathrm{d}t} - \frac{\mathrm{d}[\mathrm{C}]_{\mathrm{actual}}}{\mathrm{d}t} = -\frac{\mathrm{d}([\mathrm{B}]_{\mathrm{actual}} + [\mathrm{C}]_{\mathrm{actual}})}{\mathrm{d}t} = -\frac{\mathrm{d}[\mathrm{B}]_{\mathrm{total}}}{\mathrm{d}t}$$
$$= k_2[\mathrm{C}]_{\mathrm{actual}} \qquad (8.116)$$

- It is necessary to find an expression for $[\mathrm{C}]_{\mathrm{actual}}$ in terms of constants, e.g. K, and/or known quantities, e.g. $[\mathrm{A}]_{\mathrm{total}}$ or $[\mathrm{B}]_{\mathrm{total}}$. Carrying out the argument in terms of $[\mathrm{C}]_{\mathrm{actual}}$ and $[\mathrm{A}]_{\mathrm{total}}$ requires substitution of $[\mathrm{A}]_{\mathrm{actual}} = [\mathrm{C}]_{\mathrm{actual}}/K[\mathrm{B}]_{\mathrm{actual}}$ from Equation (8.110).

- $$[\mathrm{A}]_{\mathrm{total}} = [\mathrm{A}]_{\mathrm{actual}} + [\mathrm{C}]_{\mathrm{actual}} \qquad (8.117)$$

giving

$$[\mathrm{A}]_{\mathrm{total}} = \frac{[\mathrm{C}]_{\mathrm{actual}}}{K[\mathrm{B}]_{\mathrm{actual}}} + [\mathrm{C}]_{\mathrm{actual}} = [\mathrm{C}]_{\mathrm{actual}}\left\{\frac{1 + K[\mathrm{B}]_{\mathrm{actual}}}{K[\mathrm{B}]_{\mathrm{actual}}}\right\} \qquad (8.118)$$

$$\therefore\ [\mathrm{C}]_{\mathrm{actual}} = [\mathrm{A}]_{\mathrm{total}}\left\{\frac{K[\mathrm{B}]_{\mathrm{actual}}}{1 + K[\mathrm{B}]_{\mathrm{actual}}}\right\} \qquad (8.119)$$

- The *mechanistic rate* can now be given as

$$\text{rate of reaction} = +\frac{\mathrm{d}[\mathrm{D}]}{\mathrm{d}t} = -\frac{\mathrm{d}[\mathrm{A}]_{total}}{\mathrm{d}t} = k_2[\mathrm{C}]_{\mathrm{actual}} \qquad (8.120)$$

$$= k_2\left\{\frac{K[\mathrm{B}]_{\mathrm{actual}}}{1 + K[\mathrm{B}]_{\mathrm{actual}}}\right\}[\mathrm{A}]_{\mathrm{total}} \qquad (8.121)$$

This equation involves $[\mathrm{B}]_{\mathrm{actual}}$. Since B is being held in excess throughout all experiments, very little B will actually be removed by reaction, with $[\mathrm{B}]_{\mathrm{actual}} \approx [\mathrm{B}]_{\mathrm{total}}$, and so

$$\text{rate of reaction} = +\frac{\mathrm{d}[\mathrm{D}]}{\mathrm{d}t} = -\frac{\mathrm{d}[\mathrm{A}]_{\mathrm{total}}}{\mathrm{d}t} = k_2\left\{\frac{K[\mathrm{B}]_{\mathrm{total}}}{1 + K[\mathrm{B}]_{\mathrm{total}}}\right\}[\mathrm{A}]_{\mathrm{total}} \qquad (8.122)$$

- It now remains to decide how the experimental rate is to be followed. Again it is important to pay attention to this step in the analysis.

  There are four possibilities:

  (a) in terms of the product, [D]

  (b) in terms of the *total* [A]

  (c) in terms of the *actual* A present, $[\mathrm{A}]_{\mathrm{actual}}$

  (d) in terms of the *actual* C present, $[\mathrm{C}]_{\mathrm{actual}}$.

The simplest choice often is to monitor the concentration of the total [A] or the formation of product, D, with time:

$$\textit{observed rate} = -\frac{d[A]_{total}}{dt} = +\frac{d[D]}{dt} = k_{obs}[A]_{total} \tag{8.123}$$

where k_{obs} is the pseudo-first order rate constant for a given $[B]_{total}$, and

$$k_{obs} = k'_{obs}[B]_{excess} \tag{8.124}$$

for *that* experiment.

Note. The observed rate constant can be tabulated either as k_{obs} or as k'_{obs}. It is important to realize this, and to know which is being quoted. k_{obs} and k'_{obs} can be easily distinguished by the units:

- if they are time^{-1}, then a pseudo-first order constant is given, i.e. k_{obs};
- if they are conc.$^{-1}$ time^{-1}, then the second order k'_{obs} is quoted.

If the mechanism is correct, then the observed and mechanistic rates must be equal, and

$$k_{obs}[A]_{total} = k_2\left\{\frac{K[B]_{total}}{1 + K[B]_{total}}\right\}[A]_{total} \tag{8.125}$$

$$\therefore\ k_{obs} = \frac{k_2 K[B]_{total}}{1 + K[B]_{total}} \tag{8.126}$$

and k_{obs} depends on $[B]_{total}$ in a complex manner (Figure 8.8).

- There are two limiting cases for k_{obs}:

the excess of B is very large, called 'swamping conditions' or 'saturation', when

$$1 + K[B]_{total} \rightarrow K[B]_{total} \text{ and } k_{obs} \approx k_2 \tag{8.127}$$

and the reaction rate is independent of [B], i.e. is zero order in B;
the excess of B is only small, when

$$1 + K[B]_{total} \rightarrow 1 \text{ and } k_{obs} \approx k_2 K[B]_{total} \tag{8.128}$$

and the reaction rate is proportional to [B], i.e. is first order in B.

For all other values of the excess of B, the reaction shows complex kinetics (Figure 8.8).

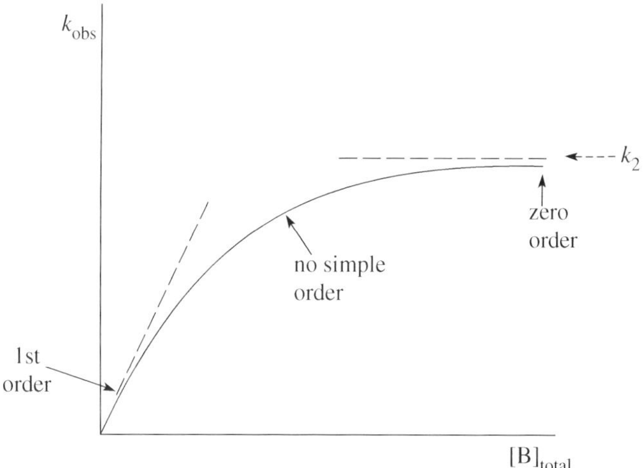

Figure 8.8 Graph of k_{obs} versus $[B]_{total}$ for a reaction involving a pre-equilibrium: $A + B \underset{k_{-1}}{\overset{k_1}{\rightleftharpoons}} C \xrightarrow{k_2} D$

From the graph in Figure 8.8, k_{obs} approaches the plateau region asymptotically, and extrapolation to the y-axis should give a value for k_2. This, however, is not an accurate method, and is crucially dependent on being able to observe the plateau region accurately: something which often is not possible. Another method is required.

If a denominator involves a sum or a difference, the relationship can always be simplified by taking the reciprocal. This is a standard mathematical trick for converting a non-linear equation to a linear one.

Doing so with Equation (8.126) gives

$$\frac{1}{k_{obs}} = \frac{1 + K[B]_{total}}{k_2 K[B]_{total}} = \frac{1}{k_2 K[B]_{total}} + \frac{1}{k_2} \tag{8.129}$$

This is the equation of a straight line and a plot of $1/k_{obs}$ versus $1/[B]_{total}$ should be linear with intercept $= 1/k_2$ and slope $= 1/k_2K$ (Figure 8.9) and so k_2 and K can be found independently.

For other pre-equilibria where the reaction step is second order, i.e.

$$A + B \underset{k_{-1}}{\overset{k_1}{\rightleftharpoons}} C$$

$$C + D \xrightarrow{k_2} \text{products}$$

the observed rate constant will be quoted either as a pseudo-second order rate constant with units conc.$^{-1}$ time^{-1}, or as a third order rate constant with units conc.$^{-2}$ time^{-1}.

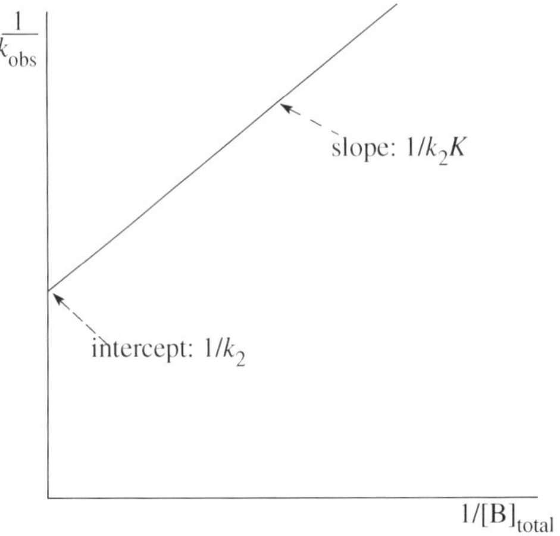

Figure 8.9 Graph of $1/k_{obs}$ versus $1/[B]_{total}$ for the pre-equilibrium reaction of Figure 8.8

Worked Problem 8.9

Question. The reaction $[Fe^{III}(CN)_6]^{3-}(aq) + [Co^{II}(edta)]^{2-}(aq) \rightarrow [Fe^{II}(CN)_6]^{4-}(aq) + [Co^{III}(edta)]^{-}(aq)$ is thought to proceed via the mechanism

$$[Fe(CN)_6]^{3-} + [Co(edta)]^{2-} \underset{k-1}{\overset{k_1}{\rightleftharpoons}} [Fe(CN)_6Co(edta)]^{5-}$$

$$[Fe(CN)_6Co(edta)]^{5-} \xrightarrow{k_2} [Fe(CN)_6]^{4-} + [Co(edta)]^{-}$$

If the equilibrium is set up *rapidly* and *maintained* throughout reaction, and if $[Co(edta)]^{2-}$ *is present in large excess*, show that

$$k_{obs} = \frac{k_2 K_1 [Co(edta)^{2-}]_{total}}{1 + K_1 [Co(edta)^{2-}]_{total}} \tag{8.130}$$

Show that the following data is consistent with the above mechanism, and find k_2 and K_1.

$\frac{10^3[Co(edta)^{2-}]}{mol\ dm^{-3}}$	2.56	3.98	5.52	7.03
$\frac{10^3 k_{obs}}{s^{-1}}$	3.64	4.15	4.44	4.63

Answer.

$$K_1 = \frac{[\mathrm{Fe(CN)_6Co(edta)^{5-}}]_{\mathrm{actual}}}{[\mathrm{Fe(CN)_6^{3-}}]_{\mathrm{actual}}[\mathrm{Co(edta)^{2-}}]_{\mathrm{actual}}} \tag{8.131}$$

$$[\mathrm{Fe(CN)_6^{3-}}]_{\mathrm{total}} = [\mathrm{Fe(CN)_6^{3-}}]_{\mathrm{actual}} + [\mathrm{Fe(CN)_6Co(edta)^{5-}}]_{\mathrm{actual}} \tag{8.132}$$

Step 2 is the rate-determining step:

$$\text{Mechanistic rate} = k_2[\mathrm{Fe(CN)_6Co(edta)^{5-}}]_{\mathrm{actual}} \tag{8.133}$$

An expression for $[\mathrm{Fe(CN)_6Co(edta)^{5-}}]_{\mathrm{actual}}$ is required in terms of constants and $[\mathrm{Fe(CN)_6^{3-}}]_{\mathrm{total}}$.

From Equation (8.131),

$$[\mathrm{Fe(CN)_6^{3-}}]_{\mathrm{actual}} = \frac{[\mathrm{Fe(CN)_6Co(edta)^{5-}}]_{\mathrm{actual}}}{K_1[\mathrm{Co(edta)^{2-}}]_{\mathrm{actual}}} \tag{8.134}$$

This is substituted into Equation (8.132) to give

$$[\mathrm{Fe(CN)_6^{3-}}]_{\mathrm{total}} = \frac{[\mathrm{Fe(CN)_6Co(edta)^{5-}}]_{\mathrm{actual}}}{K_1[\mathrm{Co(edta)^{2-}}]_{\mathrm{actual}}} + [\mathrm{Fe(CN)_6Co(edta)^{5-}}]_{\mathrm{actual}} \tag{8.135}$$

$$= [\mathrm{Fe(CN)_6Co(edta)^{5-}}]_{\mathrm{actual}}\left(\frac{1}{K_1[\mathrm{Co(edta)^{2-}}]_{\mathrm{total}}} + 1\right) \tag{8.136}$$

$$= [\mathrm{Fe(CN)_6Co(edta)^{5-}}]_{\mathrm{actual}}\left(\frac{1 + K_1[\mathrm{Co(edta)^{2-}}]_{\mathrm{actual}}}{K_1[\mathrm{Co(edta)^{2-}}]_{\mathrm{actual}}}\right) \tag{8.137}$$

$$[\mathrm{Fe(CN)_6Co(edta)^{5-}}]_{\mathrm{actual}} = \left(\frac{K_1[\mathrm{Co(edta)^{2-}}]_{\mathrm{actual}}[\mathrm{Fe(CN)_6^{3-}}]_{\mathrm{total}}}{1 + K_1[\mathrm{Co(edta)^{2-}}]_{\mathrm{actual}}}\right) \tag{8.138}$$

$$\textit{Mechanistic rate} = k_2[\mathrm{Fe(CN)_6Co(edta)^{5-}}]_{\mathrm{actual}} \tag{8.133}$$

$$= k_2\left(\frac{K_1[\mathrm{Co(edta)^{2-}}]_{\mathrm{actual}}[\mathrm{Fe(CN)_6^{3-}}]_{\mathrm{total}}}{1 + K_1[\mathrm{Co(edta)^{2-}}]_{\mathrm{actual}}}\right) \tag{8.139}$$

But $\mathrm{Co(edta)^{2-}}$ is present in large excess, and so $[\mathrm{Co(edta)^{2-}}]_{\mathrm{actual}} \approx [\mathrm{Co(edta)^{2-}}]_{\mathrm{total}}$

$$\therefore\ \text{rate} = k_2\left(\frac{K_1[\mathrm{Co(edta)^{2-}}]_{\mathrm{total}}[\mathrm{Fe(CN)_6^{3-}}]_{\mathrm{total}}}{1 + K_1[\mathrm{Co(edta)^{2-}}]_{\mathrm{total}}}\right) \tag{8.140}$$

The observed rate is measured in terms of the removal of the reactant *not* in excess.

$$\textit{Observed rate} = -\frac{d[Fe(CN)_6^{3-}]_{total}}{dt} = k[Fe(CN)_6^{3-}]_{total}[Co(edta)^{2-}]_{total} \quad (8.141)$$

$$= k_{obs}[Fe(CN)_6^{3-}]_{total} \quad (8.142)$$

where k_{obs} is a pseudo-first order rate constant, equal to $k[Co(edta)^{2-}]_{total}$.

Note. This is in agreement with the units of k_{obs} quoted in the question.
If the mechanism is correct, then

$$k_{obs}[Fe(CN)_6^{3-}]_{total} = k_2\left(\frac{K_1[Co(edta)^{2-}]_{total}[Fe(CN)_6^{3-}]_{total}}{1 + K_1[Co(edta)^{2-}]_{total}}\right) \quad (8.143)$$

$$k_{obs} = k_2\left(\frac{K_1[Co(edta)^{2-}]_{total}}{1 + K_1[Co(edta)^{2-}]_{total}}\right) \quad (8.144)$$

$$\frac{1}{k_{obs}} = \frac{1 + K_1[Co(edta)^{2-}]_{total}}{k_2K_1[Co(edta)^{2-}]_{total}} = \frac{1}{k_2K_1[Co(edta)^{2-}]_{total}} + \frac{1}{k_2} \quad (8.145)$$

A plot of $1/k_{obs}$ versus $1/[Co(edta)^{2-}]_{total}$ is given in Figure 8.10. This is linear with
intercept $= 185$ s $= 1/k_2$, giving $k_2 = 5.41 \times 10^{-3}$ s^{-1}
slope $= 0.230$ mol dm^{-3} s $= 1/k_2K_1$, giving $K_1 = 1/(5.41 \times 10^{-3} \times 0.221)$ mol^{-1} dm^3 $= 805$ mol^{-1} dm^3.

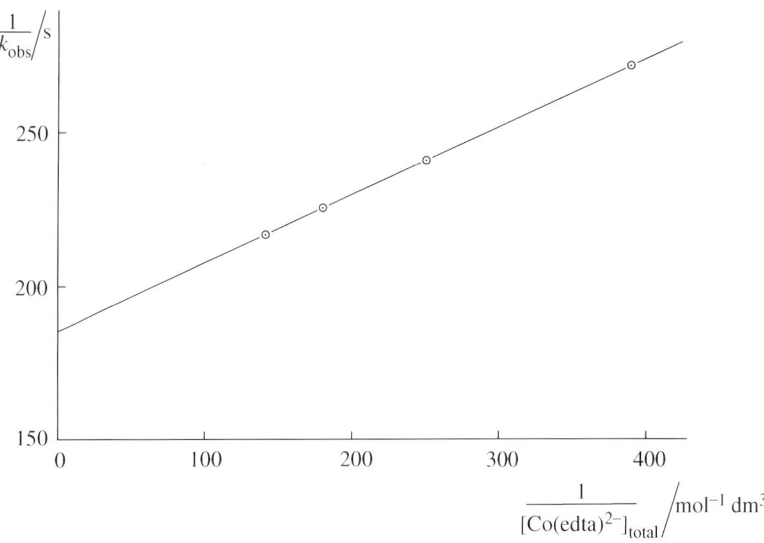

Figure 8.10 Graph of $1/k_{obs}$ versus $1/[CO(edta)^{2-}]$ for reaction of $Fe(CN)_6^{3-}$ with $Co(edta)^{2-}$

8.5 Steady States in Solution Reactions

Some reactions in solution can be analysed by the steady state treatment, and the principles discussed for the gas phase are also relevant to solution reactions.

- A proposed mechanism is written down;
- highly reactive intermediates are present;
- these intermediates are present in very, very low but almost constant concentrations;
- $+\mathrm{d}[\mathrm{I}]_{\mathrm{ss}}/\mathrm{d}t = 0$;
- steady state equations are set up for each intermediate;
- these are solved to give each $[\mathrm{I}]_{\mathrm{ss}}$;
- the mechanistic rate is then derived, and compared with the observed rate.

8.5.1 Types of reaction for which a steady state treatment could be relevant

Although free radical reactions are found less often in solution than in the gas phase, they do occur, and are generally handled by steady state methods. There are also organic and inorganic reactions that involve non-radical intermediates in *steady state concentrations*. These intermediates are often produced by an initial reversible reaction, or a set of reversible reactions. This can be compared with the pre-equilibria discussed in Section 8.4, where the intermediates are in *equilibrium concentrations*. The steady state treatment is also used extensively in acid–base catalysis and in enzyme kinetics.

Question. How could a distinction be made between the situation where a pre-equilibrium is set up and maintained throughout reaction, and one where a steady state is set up?

Answer. If there are measurable concentrations of intermediates, then there will not be a steady state. If the intermediates are detectable but are in very, very low concentrations, then a steady state is probably set up. This could also be a pre-equilibrium state with the equilibrium lying very far to the left, in which case the equilibrium concentration of intermediate probably approximates to a steady state. An analysis of the algebraic forms of the rate expressions for the steady state and pre-equilibrium situations may be needed to see whether they can be distinguished (Section 8.5.2 and Problem 8.11).

S_N1 reactions, such as hydrolyses and substitution by anions of tertiary halogenoalkanes, are good examples of reactions with a reversible first step. In these a carbonium ion is produced, which then reacts with water or anions. Chapter 6, Problem 6.5, illustrates the different rate expressions found after applying the steady state treatment, and after assuming a pre-equilibrium where the equilibrium lies very far to the left, i.e. where K is very small, and only a very, very small amount of R^+ is present. It also looks at the conditions under which the amount of R^+ present in a steady state could approximate to a pre-equilibrium. The discussion *did not include* the situation where K is *not small.*

8.5.2 A more detailed analysis of Worked Problem 6.5

For the reaction

$$\mathrm{RX} \underset{k_{-1}}{\overset{k_1}{\rightleftharpoons}} \mathrm{R}^+ + \mathrm{X}^-$$

$$\mathrm{R}^+ + \mathrm{Y}^- \xrightarrow{k_2} \mathrm{RY}$$

- *The steady state analysis*:

$$+\frac{\mathrm{d}[\mathrm{R}^+]}{\mathrm{d}t} = k_1[\mathrm{RX}] - k_{-1}[\mathrm{R}^+][\mathrm{X}^-] - k_2[R^+][\mathrm{Y}^-] = 0 \tag{8.146}$$

$$[\mathrm{R}^+] = \frac{k_1[\mathrm{RX}]}{k_{-1}[\mathrm{X}^-] + k_2[\mathrm{Y}^-]} \tag{8.147}$$

$$\textit{Steady state rate} = -\frac{\mathrm{d}[\mathrm{RX}]}{\mathrm{d}t} = k_2[\mathrm{R}^+][\mathrm{Y}^-] = \frac{k_2k_1[\mathrm{RX}][\mathrm{Y}^-]}{k_{-1}[\mathrm{X}^-] + k_2[\mathrm{Y}^-]} \tag{8.148}$$

- *The equilibrium analysis*:

$$[\mathrm{RX}]_{\mathrm{total}} = [\mathrm{RX}]_{\mathrm{actual}} + [\mathrm{R}^+]_{\mathrm{actual}} \tag{8.149}$$

$$K = \frac{[\mathrm{R}^+]_{\mathrm{actual}}[\mathrm{X}^-]_{\mathrm{actual}}}{[\mathrm{RX}]_{\mathrm{actual}}} \tag{8.150}$$

$$[\mathrm{RX}]_{\mathrm{actual}} = \frac{[\mathrm{R}^+]_{\mathrm{actual}}[\mathrm{X}^-]_{\mathrm{actual}}}{K} \tag{8.151}$$

$$[\mathrm{RX}]_{\mathrm{total}} = \frac{[\mathrm{R}^+]_{\mathrm{actual}}[\mathrm{X}^-]_{\mathrm{actual}}}{K} + [\mathrm{R}^+]_{\mathrm{actual}} \tag{8.152}$$

$$= [\mathrm{R}^+]_{\mathrm{actual}}\left(\frac{[\mathrm{X}^-]_{\mathrm{actual}} + K}{K}\right) \tag{8.153}$$

$$[\mathrm{R}^+]_{\mathrm{actual}} = \frac{[\mathrm{RX}]_{\mathrm{total}}K}{K + [\mathrm{X}^-]_{\mathrm{actual}}} \tag{8.154}$$

The pre-equilibrium rate of reaction for all values of K

$$\frac{-\mathrm{d}[\mathrm{RX}]_{\mathrm{total}}}{\mathrm{d}t} = k_2[\mathrm{R}^+]_{\mathrm{actual}}[\mathrm{Y}^-] = \frac{k_2K[\mathrm{RX}]_{\mathrm{total}}[\mathrm{Y}^-]}{K + [\mathrm{X}^-]_{\mathrm{actual}}} \tag{8.155}$$

The pre-equilibrium rate for all values of K can also be written as

$$\frac{-\mathrm{d}[\mathrm{RX}]_{\mathrm{total}}}{\mathrm{d}t} = k_2[\mathrm{R}^+]_{\mathrm{actual}}[\mathrm{Y}^-] = \frac{k_2k_1[\mathrm{RX}]_{\mathrm{total}}[\mathrm{Y}^-]}{k_1 + k_{-1}[\mathrm{X}^-]_{\mathrm{actual}}} \tag{8.156}$$

These two expressions for the steady state (Equation 8.148), and the equilibrium state (Equation 8.156), are kinetically distinct, and so the two mechanisms could be distinguished kinetically. However, there is a limiting case in which they cease to be distinguishable.

The steady state expression has two limiting forms.

- If $k_2[\mathrm{Y}^-] \ll k_{-1}[\mathrm{X}]$, the rate approximates to

$$\frac{k_2k_1[\mathrm{RX}][\mathrm{Y}^-]}{k_{-1}[\mathrm{X}^-]} = \frac{k_2K[\mathrm{RX}][\mathrm{Y}^-]}{[\mathrm{X}^-]} \tag{8.157}$$

and corresponds to removal of the carbonium ion by reaction being insignificant compared to removal by the back step of the reversible reaction.

- If $k_2[\mathrm{Y}^-] \gg k_{-1}[\mathrm{X}]$, the rate approximates to

$$k_1[\mathrm{RX}] \tag{8.158}$$

and corresponds to removal of R^+ in the back reaction being insignificant compared with removal by reaction.

The pre-equilibrium rate expression has two limiting forms.

- If K is small,

$$\text{pre-equilibrium rate} = \frac{k_2K[\mathrm{RX}]_{\mathrm{total}}[\mathrm{Y}^-]}{[\mathrm{X}^-]_{\mathrm{actual}}} \tag{8.159}$$

$$= \frac{k_2k_1[\mathrm{RX}]_{\mathrm{total}}[\mathrm{Y}^-]}{k_{-1}[\mathrm{X}^-]_{\mathrm{actual}}} \tag{8.160}$$

This corresponds to $k_1 \ll k_{-1}[\mathrm{X}^-]_{\mathrm{actual}}$, i.e. $K \ll [\mathrm{X}^-]_{\mathrm{actual}}$.

Since $K = ([R^+]_{actual}\ [X^-]_{actual}/[RX]_{actual})_{equil}$, if $K \ll [X^-]_{actual}$ then $[R^+]_{actual}/[RX]_{actual} \ll 1$. This implies that there is only a very small amount of R^+ present, and that $[RX]_{actual} \approx [RX]_{total}$.

This is exactly the same expression as the approximate steady state expression, Equation (8.157), corresponding to the condition, $k_2[Y^-] \ll k_{-1}[X^-]$, which, in turn, corresponds to most of the R^+ being removed by the back reaction, rather than by reaction with Y^-.

Under these conditions the steady state and the pre-equilibrium state are indistinguishable. However, the conditions under which they become indistinguishable are different in each, i.e.

for the pre-equilibrium, $k_1 \ll k_{-1}[X^-]_{actual}$;

for the steady state rate, $k_2[Y^-] \ll k_{-1}[X^-]_{actual}$

- If K is large, the pre-equilibrium rate reduces to

$$k_2[RX]_{total}[Y^-] \tag{8.161}$$

corresponding to $k_1 \gg k_{-1}[X^-]_{actual}$, i.e. $K \gg [X^-]_{actual}$, which corresponds to $[R^+]_{actual}/[RX]_{actual} \gg 1$, which implies a large amount of R^+ present.

Under these conditions, the steady state rate, Equation (8.158), and the pre-equilibrium rate are totally distinct and experiment should be able to distinguish between them, i.e.

for the pre-equilibrium, $k_1 \gg k_{-1}[X^-]_{actual}$, and rate $= k_2[RX]_{total}[Y^-]$

for the steady state, $k_2[Y^-] \gg k_{-1}[X^-]_{actual}$, and rate $= k_1[RX]$.

Worked Problem 8.10

Question. The previous section derived the steady state and the pre-equilibrium rate expressions for the reaction

$$RX \underset{k_{-1}}{\overset{k_1}{\rightleftharpoons}} R^+ + X^-$$

$$R^+ + Y^- \xrightarrow{k_2} RY$$

$$\text{steady state rate} = \frac{k_2k_1[RX][Y^-]}{k_{-1}[X^-] + k_2[Y^-]} \tag{8.148}$$

$$\text{pre-equilibrium rate} = \frac{k_2K[RX]_{total}[Y^-]}{[X^-]_{actual} + K} \tag{8.155}$$

If Y^- is OH^-, and the product is the alcohol,

- write an expression for the overall rate in terms of RX, ROH or X^-,

- what is the order with respect to RX?
- explain why a plot of $\log_e$ [RX] versus t would be non-linear, and why k_{obs} cannot be found using this simple integrated plot.

Answer. The amount of ROH formed is equal to the amount of RX removed. The amount of X^- formed equals the amount of RX removed, and this equals the initial amount of RX present minus the amount of RX present at time t.

- For the steady state

$$-\frac{d[RX]}{dt} = +\frac{d[ROH]}{dt} = +\frac{d([RX]_0 - [RX]_t)}{dt} = \frac{k_2 k_1 [RX][OH^-]}{k_{-1}[X^-] + k_2[OH^-]} \tag{8.162}$$

- For the pre-equilibrium

$$-\frac{d[RX]}{dt} = +\frac{d[ROH]}{dt} = +\frac{d([RX]_0 - [RX]_t)}{dt} = \frac{k_2 K [RX]_{total}[OH^-]}{[X^-]_{actual} + K} \tag{8.163}$$

- The reaction is predicted to be first order in RX. Nonetheless, the observed rate cannot be written as rate $= k_{obs}[RX]$ because, if the mechanism is correct, the rate at any time will also depend on $[X^-]$ and $[OH^-]$ in both cases. A plot of $\log_e$[RX] would therefore be non-linear, and the first order integrated rate expression in [RX] cannot be used.

It is possible to integrate the above rate expressions to get an equation which is linear, but such integrations are beyond the scope of this book.

Note. It is absolutely vital to check the predicted rate expression to see what the dependencies on the *concentrations of other substances present are, and to see whether these concentrations vary throughout the given experiment.* This should be checked with the pre-equilibria discussed in Section 8.4.

The necessity of checking whether the observed rate fits a steady state or equilibrium situation has been emphasized in Problem 8.10 above. The following problem gives further practice on this important point.

Worked Problem 8.11

Question. Discuss whether it is possible to determine kinetically whether the following reactions are in a pre-equilibrium or a steady state.

- $A \underset{k_{-1}}{\overset{k_1}{\rightleftharpoons}} B \overset{k_2}{\longrightarrow} C$
- $A + B \underset{k_{-1}}{\overset{k_1}{\rightleftharpoons}} C \overset{k_2}{\longrightarrow} D$

Answer.

- For $A \underset{k_{-1}}{\overset{k_1}{\rightleftharpoons}} \mathrm{B} \xrightarrow{k_2} \mathrm{C}$

  the pre-equilibrium rate was shown to be

$$-\frac{\mathrm{d}[\mathrm{A}]}{\mathrm{d}t} = k_2\left(\frac{K}{1+K}\right)[\mathrm{A}]_{\text{total}} \tag{8.92}$$

$$= \frac{k_2 k_1 [\mathrm{A}]}{k_{-1} + k_1} \tag{8.164}$$

  Applying the steady state treatment to B:

$$+\frac{\mathrm{d}[\mathrm{B}]}{\mathrm{d}t} = k_1[\mathrm{A}] - k_{-1}[\mathrm{B}] - k_2[\mathrm{B}] = 0 \tag{8.165}$$

$$[\mathrm{B}] = \frac{k_1[\mathrm{A}]}{k_{-1} + k_2} \tag{8.166}$$

$$-\frac{\mathrm{d}[\mathrm{A}]}{\mathrm{d}t} = k_2[\mathrm{B}] = \frac{k_2 k_1 [\mathrm{A}]}{k_{-1} + k_2} \tag{8.167}$$

  The steady state and pre-equilibrium rate expressions are kinetically equivalent, so it is impossible to distinguish between them experimentally, *but note* that the expressions are different, because the denominators are different.

- *In contrast*, for the reaction

$$\mathrm{A} + \mathrm{B} \underset{k_{-1}}{\overset{k_1}{\rightleftharpoons}} \mathrm{C} \xrightarrow{k_2} \mathrm{D}$$

  it is possible to distinguish between the steady state and pre-equilibrium situations.

  The pre-equilibrium rate of reaction in the presence of excess B is

$$-\frac{\mathrm{d}[\mathrm{A}]_{\text{total}}}{\mathrm{d}t} = k_2\left\{\frac{K[\mathrm{B}]_{\text{total}}}{1 + K[\mathrm{B}]_{\text{total}}}\right\}[\mathrm{A}]_{\text{total}} \tag{8.122}$$

$$= \frac{k_2 k_1 [\mathrm{B}]_{\text{total}}[\mathrm{A}]_{\text{total}}}{k_{-1} + k_1[\mathrm{B}]_{\text{total}}} \tag{8.168}$$

  The observed rate

$$-\frac{\mathrm{d}[\mathrm{A}]_{\text{total}}}{\mathrm{d}t} = k_{\text{obs}}[\mathrm{A}]_{\text{total}}[\mathrm{B}]_{\text{total}} \tag{8.169}$$

$$\therefore\ k_{\text{obs}} = \frac{k_2 k_1}{k_{-1} + k_1[\mathrm{B}]_{\text{total}}} \tag{8.170}$$

The steady state expression for the rate of reaction

$$+\frac{d[C]}{dt} = k_1[A][B] - k_{-1}[C] - k_2[C] = 0 \tag{8.171}$$

$$[C] = \frac{k_1[A][B]}{k_{-1} + k_2} \tag{8.172}$$

$$-\frac{d[A]}{dt} = k_2[C] = \frac{k_2 k_1[A][B]}{k_{-1} + k_2} = k_{obs}[A][B] \tag{8.173}$$

$$k_{obs} = \frac{k_2 k_1}{k_{-1} + k_2} \tag{8.174}$$

These two situations can easily be distinguished kinetically. The steady state rate has a denominator which is *independent* of [B], while the pre-equilibrium rate *depends* on [B].

It is essential to check whether a reaction is in a steady state or in a pre-equilibrium state. If this is to be done kinetically, it is vital to derive the mechanistic rate expressions and see whether they are kinetically equivalent or not. If they are, then recourse to non-kinetic methods becomes necessary.

In all of these reactions the species involved in the reversible reactions are progressively used up with time. However, there are reactions such as acid–base catalysis and enzyme catalysis where one of the species in the reversible reaction is the catalyst, and as such is regenerated so that the total catalyst concentration remains constant throughout the whole of the reaction. In such cases it is often essential to use initial rates or rates at various stages during reaction for analysis.

8.6 Enzyme Kinetics

Enzymes are proteins which act as catalysts in many reactions of biological and biochemical importance. Because of their efficiency, enzymes are effective at very low concentrations, of the order of 10^{-8} mol dm^{-3} to 10^{-10} mol dm^{-3}. The molecule whose reaction is being studied is called the substrate, and typical concentrations are 10^{-6} mol dm^{-3} or greater. This means that the substrate is always in large excess.

The simplest mechanistic scheme describing the action of an enzyme is

$$E + S \underset{k_{-1}}{\overset{k_1}{\rightleftharpoons}} ES \underset{k_{-2}}{\overset{k_2}{\rightleftharpoons}} E + \text{product}$$

Enzyme kinetics are usually studied as initial rates and step (-2) can then be ignored.

The rate of reaction is defined as

$$k_2[\mathrm{ES}]_{\mathrm{actual}} \tag{8.175}$$

and

$$[\mathrm{E}]_{\mathrm{total}} = [\mathrm{E}]_{\mathrm{actual}} + [\mathrm{ES}]_{\mathrm{actual}} \tag{8.176}$$

As in so many of the examples studied previously, the aim is to express $[\mathrm{ES}]_{\mathrm{actual}}$ in terms of $[\mathrm{E}]_{\mathrm{total}}$ and constants. In this case a steady state is assumed, and the constants will be the rate constants k_1, k_{-1} and k_2. Assuming a steady state,

$$+\frac{\mathrm{d[ES]}}{\mathrm{d}t} = k_1[\mathrm{E}]_{\mathrm{actual}}[\mathrm{S}] - k_{-1}[\mathrm{ES}]_{\mathrm{actual}} - k_2[\mathrm{ES}]_{\mathrm{actual}} = 0 \tag{8.177}$$

$$k_{-1}[\mathrm{ES}]_{\mathrm{actual}} + k_2[\mathrm{ES}]_{\mathrm{actual}} = k_1[\mathrm{E}]_{\mathrm{actual}}[\mathrm{S}] \tag{8.178}$$

$$[\mathrm{ES}]_{\mathrm{actual}} = \frac{k_1[\mathrm{E}]_{\mathrm{actual}}[\mathrm{S}]}{k_{-1} + k_2} \tag{8.179}$$

An expression for $[\mathrm{ES}]_{\mathrm{actual}}$ in terms of constants and $[\mathrm{E}]_{\mathrm{total}}$ is required, and this requires elimination of $[\mathrm{E}]_{\mathrm{actual}}$ from Equations (8.177)–(8.179)

Equation (8.177) can be rewritten using Equation (8.176) in the form

$$[\mathrm{E}]_{\mathrm{actual}} = [\mathrm{E}]_{\mathrm{total}} - [\mathrm{ES}]_{\mathrm{actual}} \tag{8.180}$$

$$k_{-1}[\mathrm{ES}]_{\mathrm{actual}} + k_2[\mathrm{ES}]_{\mathrm{actual}} = k_1[\mathrm{S}]\{[\mathrm{E}]_{\mathrm{total}} - [\mathrm{ES}]_{\mathrm{actual}}\} \tag{8.181}$$

$$= k_1[\mathrm{E}]_{\mathrm{total}}[\mathrm{S}] - k_1[\mathrm{ES}]_{\mathrm{actual}}[\mathrm{S}] \tag{8.182}$$

$$[\mathrm{ES}]_{\mathrm{actual}}\{k_1[\mathrm{S}] + k_{-1} + k_2\} = k_1[\mathrm{E}]_{\mathrm{total}}[\mathrm{S}] \tag{8.183}$$

$$[\mathrm{ES}]_{\mathrm{actual}} = \frac{k_1[\mathrm{E}]_{\mathrm{total}}[\mathrm{S}]}{k_1[\mathrm{S}] + k_{-1} + k_2} \tag{8.184}$$

giving

$$\mathrm{rate} = \frac{k_2 k_1[\mathrm{E}]_{\mathrm{total}}[\mathrm{S}]}{k_1[\mathrm{S}] + k_{-1} + k_2} = \frac{k_2[\mathrm{E}]_{\mathrm{total}}[\mathrm{S}]}{[\mathrm{S}] + (k_{-1} + k_2)/k_1} \tag{8.185}$$

This is equivalent to

$$\mathrm{rate} = \frac{\mathrm{const}[\mathrm{S}]}{[\mathrm{S}] + \mathrm{const.}'} \tag{8.186}$$

This expression shows complex kinetics with no simple order in S, and none of the standard integrated expressions is applicable; compare this with the similar situation given in Problem 8.10. Analysis using rate/concentration data is necessary, and, to eliminate effects of any contribution from the reverse of step 2, initial rates are used.

In practice, most enzyme reactions are considered to involve a steady state, with only a few being pre-equilibria. Other investigations have examined the transient conditions preceding the establishment of a steady state.

Worked Problem 8.12

Question. Show how it is possible to determine the rate constants k_1, k_{-1} and k_2 for the reaction

$$\mathrm{E} + \mathrm{S} \underset{k_{-1}}{\overset{k_1}{\rightleftharpoons}} \mathrm{ES} \xrightarrow{k_2} \mathrm{E} + \text{product}$$

Assume the reaction to be in the steady state.

Answer. The rate equation has been given as

$$\text{rate} = \frac{k_2 k_1 [\mathrm{E}]_{\text{total}}[\mathrm{S}]}{k_1[\mathrm{S}] + k_{-1} + k_2} = \frac{k_2 [\mathrm{E}]_{\text{total}}[\mathrm{S}]}{[\mathrm{S}] + (k_{-1} + k_2)/k_1} \qquad (8.185)$$

Carry out experiments at a known $[\mathrm{E}]_{\text{total}}$, and a variety of [S]. For each different experiment draw graphs of [S] versus time. Measure the initial rates for each (Figure 8.11). A plot of initial rate against [S] will be a curve with a plateau, reminiscent of Figure 8.8. As in Section 8.4.4, there is a denominator which

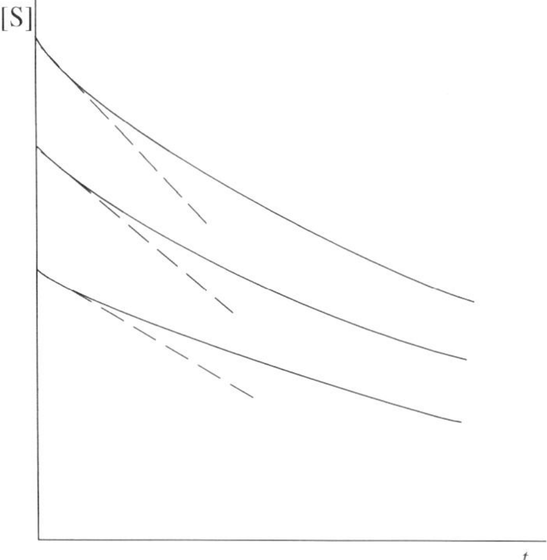

Figure 8.11 Determination of initial rates for an enzymic reaction: broken lines are tangents to the experimental curves at $t = 0$

involves a sum of terms, and so the reciprocal of both sides is taken.

$$\frac{1}{\text{rate}} = \frac{[\text{S}] + (k_{-1} + k_2)/k_1}{k_2[\text{S}][\text{E}]_{\text{total}}} = \frac{1}{k_2[\text{E}]_{\text{total}}} + \left(\frac{k_{-1} + k_2}{k_1 k_2[\text{E}]_{\text{total}}}\right)\frac{1}{[\text{S}]} \tag{8.187}$$

A plot of 1/rate versus 1/[S] should be linear with

$$\text{intercept} = 1/k_2[\text{E}]_{\text{total}}$$

Since $[\text{E}]_{\text{total}}$ is known, hence k_2.

$$\text{slope} = \left(\frac{k_{-1} + k_2}{k_1 k_2[\text{E}]_{\text{total}}}\right)$$

Since $k_2[\text{E}]_{\text{total}}$ is known from the intercept, $(k_{-1} + k_2)/k_1$ can be found. Unless k_1 or k_{-1} can be found independently, the other cannot be found.

In standard enzyme kinetics texts $k_2[\text{E}]_{\text{total}}$ is termed V_s and $(k_{-1} + k_2)/k_1$ is termed the Michaelis constant, K_{M}. The equation

$$\text{rate} = \frac{k_2 k_1[\text{E}]_{\text{total}}[\text{S}]}{k_1[\text{S}] + k_{-1} + k_2} = \frac{k_2[\text{E}]_{\text{total}}[\text{S}]}{[\text{S}] + (k_{-1} + k_2)/k_1} = \frac{V_s[\text{S}]}{K_{\text{M}} + [\text{S}]} \tag{8.188}$$

is called the Michaelis–Menten equation.

Further reading

Laidler K.J., *Chemical Kinetics*, 3rd edn, Harper and Row, New York, 1987.

Laidler K.J. and Meiser J.H., *Physical Chemistry*, 3rd edn, Houghton Mifflin, New York, 1999.

Alberty R.A. and Silbey R.J., *Physical Chemistry*, 2nd edn, Wiley, New York, 1996.

Logan S.R., *Fundmentals of Chemical Kinetics*, Addison Wesley, Reading, MA, 1996.

*Wilkinson F., *Chemical Kinetics and Reaction Mechanisms*, Van Nostrand Reinhold, New York, 1980.

*Robson Wright M., *Nature of Electrolyte Solutions*, Macmillan, Basingstoke, 1988.

Robson Wright M., *Behaviour and Nature of Electrolyte Solutions*, Wiley, Chichester, in preparation.

Hay R.W., *Reaction Mechanisms of Metal Complexes*, Horwood, Chichester, 2000.

Cox B.G., *Modern Liquid Phase Kinetics*, Oxford University Press, Oxford, 1994.

Wilkins R.G., *Kinetics and Mechanisms of Transition Metal Complexes*, John Wiley & Sons (Australia) Ltd. (in press).

Burgess J. and Tobe M., *Inorganic Reaction Mechanisms*, Longman, London, 2000.

*Out of print, but should be in university libraries.

Further problems

1. The following kinetic data were found for the reaction of a complex with NH_3 in aqueous solution at 25 °C, with the $[NH_3]$ being held in excess

$$A(aq) + NH_3(aq) \xrightarrow{k_1} \text{products}$$

The complex also reacts with the solvent, water

$$A(aq) + H_2O(l) \xrightarrow{k_2} \text{products}$$

Derive the mechanistic rate expression and thence find an expression for k_{obs}, the pseudo-first order rate constant.

Use the following data to find values for k_1 and k_2.

$\frac{[NH_3]}{\text{mol dm}^{-3}}$	0.050	0.100	0.150	0.200	0.250
$\frac{10^5 k_{obs}}{\text{s}^{-1}}$	6.0	7.6	9.7	11.3	13.1

2. The base hydrolysis of α-amino-acid esters is catalysed by transition metal ions, and the following reactions contribute to the mechanism:

$$E + H_2O \underset{k_{-1}}{\overset{k_1}{\rightleftharpoons}} HE^+ + OH^-$$

$$E + OH^- \xrightarrow{k_2} \text{products}$$

$$HE^+ + OH^- \xrightarrow{k_3} \text{products}$$

$$E + OH^- + M^{2+} \xrightarrow{k_4} \text{products} + M^{2+}$$

Deduce an expression for k_{obs} where k_{obs} has units mol^{-1} dm^3 s^{-1} (p$K_1 = 6.3$). Indicate how k_2, k_3 and k_4 could be found.

3. The base hydrolysis of an ester in constant ionic strength aqueous solutions of $TlNO_3$ shows a dependence of the observed rate constant on $[Tl^+]$. Explain this observation, and deduce the expression for k_{obs}. Why is it more sensible to follow this reaction at constant pH, rather than following the decrease in pH with time?

4. The reaction between the ammonium ion and the cyanate ion to form urea

$$NH_4^+(aq) + OCN^-(aq) \rightarrow CO(NH_2)_2(aq)$$

proceeds via the pre-equilibrium

$$NH_4^+(aq) + OCN^-(aq) \underset{k_{-1}}{\overset{k_1}{\rightleftharpoons}} NH_3(aq) + HNCO(aq)$$

followed by the step

$$NH_3(aq) + HNCO(aq) \xrightarrow{k_2} CO(NH_2)_2(aq)$$

Deduce the rate expression, and use the following data for 50 °C to find k_2.

$K_1 = 1.3 \times 10^{-5}$, and equal initial concentrations of NH_4^+ and OCN^- were used.

$\frac{10^3[NH_4^+]}{\text{mol dm}^{-3}}$	8.85	8.26	7.75	7.30	6.94	6.62
$\frac{\text{time}}{\text{min}}$	0	170	340	500	650	790

5. The reaction

$$[L_4Co(NHR_2)X]^{2+} + OH^-(aq) \rightarrow [L_4Co(NHR_2)OH]^{2+} + X^-(aq)$$

is believed to occur via an initial pre-equilibrium process

$$[L_4Co(NHR_2)X]^{2+} + OH^- \underset{k_{-1}}{\overset{k_1}{\rightleftharpoons}} [L_4Co(NR_2)X]^+ + H_2O$$

followed by

$$[L_4Co(NR_2)X]^+ \underset{\text{slow}}{\xrightarrow{k_2}} [L_4Co(NR_2)]^{2+} + X^-$$

$$[L_4Co(NR_2)]^{2+} + H_2O \underset{\text{fast}}{\xrightarrow{k_3}} [L_4Co(NHR_2)OH]^{2+}$$

Show that

- $+\dfrac{d[L_4Co(NHR_2)OH^{2+}]_{\text{actual}}}{dt} = -\dfrac{d[L_4Co(NHR_2)X^{2+}]_{\text{total}}}{dt}$
- $k_{\text{obs}} = \dfrac{k_2K_1[OH^-]_{\text{total}}}{1 + K_1[OH^-]_{\text{total}}}$

where k_{obs} has the units time^{-1}. Indicate how k_2 and K_1 could be found from experimental data.

6. The mechanism

$$\mathrm{M} + \mathrm{L} \underset{k_{-1}}{\overset{k_1}{\rightleftharpoons}} \mathrm{ML}$$

$$\mathrm{ML} + \mathrm{X} \xrightarrow{k_2} \text{products}$$

is a standard inorganic mechanism. If the pre-equilibrium is set up rapidly and is maintained throughout reaction, use the following data to show that at high ligand concentration the reaction becomes zero order in ligand, and calculate the values of k_2 and K_1.

$\frac{[\mathrm{L}]}{\mathrm{mol\ dm^{-3}}}$	0.008 56	0.0192	0.0301	0.0423	0.0654	0.0988
$\frac{10^4 k_{obs}}{\mathrm{mol^{-1}\ dm^3\ s^{-1}}}$	0.83	1.28	1.51	1.67	1.84	1.96

7. The mechanism for the ligand exchange reaction

$$\mathrm{ML_6} + \mathrm{L^*} \xrightarrow{k_{obs}} \mathrm{ML_5L^*} + \mathrm{L}$$

could be described by the following:

$$\mathrm{ML_6} \underset{k_{-1}}{\overset{k_1}{\rightleftharpoons}} \mathrm{ML_5} + \mathrm{L}$$

$$\mathrm{ML_5} + \mathrm{L^*} \xrightarrow{k_2} \mathrm{ML_5L^*}$$

(a) For experiments where $\mathrm{L^*}$ is held in excess, and assuming that the pre-equilibrium is set up rapidly and is maintained throughout the reaction, show that

$$k_{obs} = \frac{k_2 K_1}{K_1 + [\mathrm{L}]_{actual}} \qquad \text{where } K_1 = k_1/k_{-1}$$

(b) If L were also present in excess at the beginning of the reaction, how could K_1 and k_2 be found?

8. A possible mechanism for the addition of bromine to an alkene in aqueous solution is

$$\text{alkene(aq)} + \mathrm{Br_2(aq)} \underset{k_{-1}}{\overset{k_1}{\rightleftharpoons}} \text{adduct(aq)}$$

$$\text{adduct(aq)} + \mathrm{Br_2(aq)} \underset{\text{slow}}{\overset{k_2}{\longrightarrow}} \text{alkeneBr}^+\mathrm{Br_3^-}\text{(aq)}$$

$$\text{alkeneBr}^+\mathrm{Br_3^-}\text{(aq)} \underset{\text{fast}}{\overset{k_3}{\longrightarrow}} \text{products}$$

(a) Show that, provided the equilibrium is set up rapidly and is maintained throughout reaction, and Br_2 is in large excess,

$$k_{obs} = \frac{k_2 K_1}{1 + K_1[Br_2]_{total}}$$

(b) Show how, under differing conditions of excess Br_2, this reaction can have first order or second order kinetics with respect to bromine.

(c) How can k_2 and K_1 be found?

(d) If K_1 were very small, what effect would this have on the kinetics? Find an expression for k_{obs}. Would it be possible to find k_2 and K_1 using this expression?

9. The following data relate to the decarboxylation of 0.0300 mol dm^{-3} oxaloacetic acid in the presence of Zn^{2+} carried out at constant pH and at an ionic strength of 0.100 mol dm^{-3}. Under these conditions $[A^{2-}]_{actual} = 6 \times 10^{-5}$ mol dm^{-3}.

The equilibrium constant for the formation of the 1:1 chelate of the dianion, ZnA, $K = 1580$ mol^{-1} dm^3.

$\frac{[Zn^{2+}]_{total}}{\text{mol dm}^{-3}}$	0	0.0025	0.0050	0.0075	0.0100
$10^4\, k_{obs}/\text{s}^{-1}$	0.725	1.10	1.48	1.86	2.18

Under the conditions of the experiment the chelate formed is the 1:1 chelate of the metal ion with the dianion A^{2-}, i.e. ZnA, and the concentrations are such that most of the metal is unchelated so that $[Zn^{2+}]_{actual} \approx [Zn^{2+}]_{total}$. The 1:1 chelate is the species which decarboxylates. The total contribution from the uncatalysed decarboxylation can be given as $k_{uncat}[H_2A]_{total}$.

Formulate the rate expression and find the values of k_{uncat} and k_{cat}.

Answers to Problems

Chapter 2

1. • (a) Spectrophotometry: Br_2 is coloured. (b) Br_2-stat.
   • (a) esr; (b) spectroscopy.
   • Dilatometry.
   • (a) Spectrophotometry: I_2 is coloured. (b) Titration with thiosulphate.
   • Total pressure decrease, or spectroscopy on NO.
   • Flow.
   • Molecular beams.
   • (a) Conductance increase, or (b) $N_2(g)$ produced.
   • Relaxation: ultrasonics.
   • (a) Spectrophotometry; (b) titration with thiosulphate.
   • Laser spectroscopic methods.

2. Plot absorbances versus concentration to give a Beer's law plot; note that this goes through the origin. The slope gives $\varepsilon d = 85\ \text{mol}^{-1}\ \text{dm}^3$. Since $d = 1.00$ cm, $\varepsilon = 85\ \text{mol}^{-1}\ \text{dm}^3\ \text{cm}^{-1}$.

   Read off the concentrations corresponding to the quoted absorbances using the calibration graph. This gives the following table. The plot of concentration versus time is a curve.

An Introduction to Chemical Kinetics. Margaret Robson Wright
 ISBNs: 0-470-09058-8 (hbk) 0-470-09059-6 (pbk)

A	1.420	1.139	0.904	0.724	0.584
$\frac{\text{conc.}}{\text{mol dm}^{-3}}$	0.0167	0.0134	0.106	0.0085	0.0069
$\frac{\text{time}}{\text{s}}$	0	100	200	300	400

3. Curve (a) shows the intensity decreasing with time, hence reactant is being used up. Curve (b) shows intensities increasing with time, hence formation of product. Curve (c) shows intensities increasing with time, reaching a maximum and then decreasing, hence formation and removal of an intermediate.

4.

$\frac{\text{time}}{\text{s}}$	0	100	200	350	600	900
$\frac{[\text{A}]}{\text{mol dm}^{-3}}$	0.0200	0.0161	0.0137	0.0111	0.0083	0.0065
$\log_e[\text{A}]$	−3.91	−4.13	−4.29	−4.50	−4.79	−5.04
$\frac{\text{mol}^{-1}\,\text{dm}^3}{[\text{A}]}$	50	62	73	90	120	154

Plot graphs of [A] versus t, $\log_e[\text{A}]$ versus t and $1/[\text{A}]$ versus t.

The graphs of [A] versus t and $\log_e[\text{A}]$ versus t are both curves, but the graph of $1/[\text{A}]$ versus time is linear. Linearity always implies a simple relationship between the two quantities plotted, and is significant for interpretation of the data (see Chapter 3).

5.

$\frac{\text{distance}}{\text{cm}}$	2.0	4.0	6.0	8.0	10.0
A	0.575	0.457	0.346	0.232	0.114

Converting distances to times:

$v =$ distance/time $\therefore\ t = \frac{d}{v}$. Watch units: v is m s^{-1}, but d is cm.

Converting absorbances to concentrations:

Beer's law states $A = \varepsilon c d$

$\therefore\quad c = \frac{A}{\varepsilon d}\qquad \varepsilon = 1180$ mol^{-1} dm^3 cm^{-1}; pathlength is 1.00 cm.

distance/cm	2.0	4.0	6.0	8.0	10.0
time/ms	0.8	1.6	2.4	3.2	4.0
A	0.575	0.457	0.346	0.232	0.114
10^4conc./mol dm^{-3}	4.87	3.87	2.93	1.97	0.97

The data can be converted as in the table. The graph of concentration versus time is linear. As in questions 2 and 4, this is highly significant as to how the rate of the reaction depends on concentration (see Chapter 3).

6. The initial height of the solution in the dilatometer corresponds to the solution of ethylene oxide at zero time, i.e. $[C_2H_4O] = 0.100$ mol dm^{-3}, and $[C_2H_6O_2] = 0$.

The height of the solution at ∞ corresponds to $[C_2H_4O] = 0$, and $[C_2H_6O_2] = 0.100$ mol dm^{-3}.

These two points can be plotted on graph paper and joined by a straight line, thereby creating a calibration graph. From this $[C_2H_4O]_{remaining}$ at the quoted times can be interpolated, giving the table below from which the curved plot of $[C_2H_4O]_{remaining}$ versus time can be drawn.

$[C_2H_4O]$/moldm^{-3}	0.100	0.0932	0.0865	0.0805	0.0747	0.0557	0.0481	0.0414	0.0000
time/min	0	20	40	60	80	160	200	240	∞

7.

pH	12.81	12.50	12.40	12.30	12.20	12.10	12.00	11.90	11.80
$10^2[OH^-]$/moldm^3	6.5	3.2	2.5	2.0	1.6	1.3	1.0	0.79	0.63
time/s	0	35	45	57	70	80	90	100	114

The plot of pH versus time is linear, but the plot of $[OH^-]$ versus time is a curve. As in Worked Problem 2.4, this indicates a direct relation between $\log_e [OH^-]_{remaining}$ and time, and is significant for the interpretation of the dependence of rate on concentration (see Chapter 3).

$$\text{pH} = -\log_{10}[H_3O^+] \text{ and since } K_w = [OH^-][H_3O^+],\ [OH^-] = K_w/[H_3O^+].$$

$\therefore$ pOH = pK_w − pH, and so if pH versus time is linear, then so also will pOH versus time be linear.

8. This requires conversion of the total pressure to the partial pressure of monomer remaining.

   Let the initial partial pressure of the monomer, A, be p_A^0.
   At time t, let the partial pressure of monomer have dropped by y.
   $\therefore$ at time t, the partial pressure of monomer remaining is $p_A^0 - y$, and the partial pressure of dimer formed is $y/2$.
   The total pressure at time t, $p_{total} = p_A^0 - y + y/2 = p_A^0 - y/2$.

$$\therefore \quad y = 2(p_A^0 - p_{total})$$

$$\begin{aligned}\text{partial pressure of monomer remaining} &= p_A^0 + 2p_{total} - 2p_A^0 \\ &= 2p_{total} - p_A^0\end{aligned}$$

$\frac{p_{total}}{\text{mm Hg}}$	632	591	533	484	453	432	416
$\frac{2p_{total}}{\text{mm Hg}}$	1264	1182	1066	968	906	864	832
$\frac{2p_{total} - p_A^0}{\text{mm Hg}}$	632	550	434	336	274	232	200
$\frac{\text{time}}{\text{min}}$	0	10	30	60	90	120	150

9. In this experiment one highly conducting ion is replaced by a less conducting ion. Since the initial and final conductivities are known, this enables a straight line calibration to be prepared.

   At $t = 0$, $\kappa = 1.235 \times 10^{-3}$ ohm^{-1} cm^{-1}, and this corresponds to the reactant initially, when $[OH^-] = 5.00 \times 10^{-3}$ mol dm^{-3}.

   At $t = \infty$, $\kappa = 0.443 \times 10^{-3}$ ohm^{-1} cm^{-1}, and this corresponds to product, $[CH_3COO^-] = 5.00 \times 10^{-3}$ mol dm^{-3}.

   Using this the following table can be constructed.

$\frac{\text{time}}{\text{min}}$	0	10	20	30	40	∞
$\frac{10^3\kappa}{\text{ohm}^{-1}\ \text{cm}^{-1}}$	1.235	1.041	0.925	0.842	0.787	0.443
$\frac{10^3[OH^-]}{\text{mol dm}^{-3}}$	5.00	3.78	3.04	2.52	2.17	0

Chapter 3

1. $$A \rightarrow B + C;$$
$$\text{rate} = k[A]^n$$
$$\log_{10} \text{rate} = \log_{10} k + n \log_{10}[A]$$

(a) Data is rate/concentration data. Plot $\log_{10}$ rate versus $\log_{10}$ concentration. Graph should be linear with slope $= n$ and intercept $= \log_{10} k$.

$\log_{10}$ rate	−5.52	−4.42	−3.21	−2.32
$\log_{10}[A]$	−2.30	−1.75	−1.14	−0.70

$$\text{slope} = n = 2, \quad \therefore \quad \text{order} = 2$$
$$\text{intercept} = \log_{10} k = -0.92, \quad \therefore \quad k = 0.120\,\text{mol}^{-1}\,\text{dm}^3\,\text{min}^{-1}.$$

(b) The reaction has been shown to be second order. Data is now concentration/time data $\therefore$ use a second order integrated plot to confirm conclusion in (a).
$\dfrac{1}{[A]_t} = \dfrac{1}{[A]_o} + kt$; plot $1/[A]_t$ versus t should be linear with slope $= k$, if reaction is second order.

$\dfrac{\text{mol dm}^{-3}}{[A]}$	100	124	148	200
$\dfrac{\text{time}}{\text{min}}$	0	200	400	800

Graph is linear: confirms second order.
Slope $= k = 0.122\ \text{mol}^{-1}\ \text{dm}^3\ \text{min}^{-1}$.

2. Total pressure/time data given $\therefore$ use the integrated rate expression method. But first have to convert to partial pressure of reactant remaining.

Let initial partial pressure of C_2H_4O be p_0, and let y be the decrease in partial pressure by time t.

Partial pressure of C_2H_4O remaining at time t is $p_0 - y$, and the sum of the partial pressures of CH_4 and CO formed by time t is $y + y = 2y$.

Total pressure at time $t = p_{\text{total}} = p_0 - y + 2y = p_0 + y$, giving $y = p_{\text{total}} - p_0$
Partial pressure of C_2H_4O left at time $t = 2p_0 - p_{\text{total}}$.
To find the order, try first and second order integrated equations in turn.

- First order plot: $\log_e p(C_2H_4O)_t = \log_e p(C_2H_4O)_0 - kt$
plot of $\log_e p(C_2H_4O)_t$ versus t should be linear if first order, with slope $= -k$.

$\frac{\text{time}}{\text{min}}$	0	9	13	18	50	100
$\frac{p_{\text{total}}}{\text{mm Hg}}$	116.5	129.4	134.7	141.3	172.1	201.2
$\frac{(2p_0 - p_{\text{total}})}{\text{mm Hg}}$	116.5	103.6	98.3	91.7	60.9	31.8
$\log_e\left(\frac{2p_0 - p_{\text{total}}}{\text{mm Hg}}\right)$	4.76	4.64	4.59	4.52	4.11	3.46

- Second order plot: $1/p_t = 1/p_0 + kt$
  Plot of $1/p_t$ versus t should be linear if second order, with slope $= +k$.

$\frac{\text{time}}{\text{min}}$	0	9	13	18	50	100
$\frac{10^3}{(2p_0 - p_{\text{total}})}/\text{mm Hg}^{-1}$	8.58	9.65	10.1	10.9	16.4	31.4

The first order plot is linear, while the second order plot is curved. This shows the reaction to be first order.

$$\text{slope} = -0.013\ \text{min}^{-1}$$
$$\therefore \quad k = 0.013\ \text{min}^{-1}.$$

Data required: rate constants over a range of temperatures.

3. The aim is to determine the dependence of rate on concentration for each reactant in turn. Choose one experiment and compare it with each of the others in turn.

- Compare 1 with 2:
  [A] is the same in each experiment, as is [B],
  [C] increases by a factor of 1.5; rate increases by a factor of 1.5.
  $\therefore$ first order in C.

- Compare 1 with 3:
  [A] is the same in each experiment, as is [C],
  [B] increases by a factor of two; rate increases by a factor of four.
  $\therefore$ second order in B.

- Compare 1 with 4:
  [B] is the same in each experiment, as is [C],
  [A] doubles; rate remains the same.
  $\therefore$ zero order in A.

$$\text{rate} = k[\text{A}]^0[\text{B}]^2[\text{C}] = k[\text{B}]^2[\text{C}]$$

$$\text{from the first set of results:} \quad k = \frac{2.5 \times 10^{-5}\,\text{mol dm}^{-3}\,\text{s}^{-1}}{(0.005)^2 \times 0.010\,\text{mol}^3\,\text{dm}^{-9}}$$

$$= 100\,\text{mol}^{-2}\,\text{dm}^6\,\text{s}^{-1}$$

$$\text{Other results give an average} \quad k = 100\,\text{mol}^{-2}\,\text{dm}^6\,\text{s}^{-1}.$$

4. (a) Let the initial pressure of NO_2Cl be p_0, and let the decrease in partial pressure by time t be y. The partial pressure of NO_2Cl remaining is $p_0 - y$, and the partial pressures of NO_2 and Cl_2 formed are y and $y/2$ respectively.

The total pressure, p_{total}, at time t is $p_0 - y + y + y/2 = p_0 + y/2$

$\therefore \quad y = 2(p_{\text{total}} - p_0)$

$p(NO_2Cl)_{\text{remaining}} = p_0 - 2(p_{\text{total}} - p_0) = 3p_0 - 2p_{\text{total}}$.

If the reaction is first order, then a plot of $\log_e(3p_0 - 2p_{\text{total}})$ versus time should be linear with slope $= -k$.

$\frac{\text{time}}{\text{s}}$	0	300	900	1500	2000
$\frac{p_{\text{total}}}{\text{mm Hg}}$	19.9	21.7	24.4	26.3	27.3
$\frac{2p_{total}}{\text{mm Hg}}$	39.8	43.4	48.8	52.6	54.6
$\frac{3p_0}{\text{mm Hg}}$	59.7	59.7	59.7	59.7	59.7
$\frac{3p_0 - 2p_{total}}{\text{mm Hg}}$	19.9	16.3	10.9	7.1	5.1
$\log_e\left(\frac{3p_0 - 2p_{total}}{\text{mm Hg}}\right)$	2.99	2.79	2.39	1.96	1.63

A plot of $\log_e((3p_0 - 2p_{total})/\text{mm Hg})$ versus time is linear, with

$$\text{slope} = -6.77 \times 10^{-4}\,\text{s}^{-1}.$$

$$\therefore \quad k = 6.77 \times 10^{-4}\,\text{s}^{-1}.$$

At 60% reaction the partial pressure of NO_2Cl remaining is 40% of p_0.

$$\log_e p_t = \log_e p_0 - kt, \quad \text{so that}$$

$$\log_e(0.4p_0) - \log_e p_0 = -kt$$

$$\log_e 0.4 = -6.77 \times 10^{-4}\,\text{s}^{-1} \times t$$

$$t = -\frac{-0.916}{6.77 \times 10^{-4}\,\text{s}^{-1}} = 1350\,\text{s}.$$

(b) The data given is rate/concentration data, and simple inspection will not give the order. Use the systematic log rate/log concentration method.

$$\text{rate} = k\, p(\mathrm{NO_2Cl})^n \therefore \log \text{rate} = \log k + n \log p(\mathrm{NO_2Cl})$$

Plot of $\log_{10}$ rate versus $\log_{10} p(\mathrm{NO_2Cl})$ should be linear with slope $= n$ and intercept $= \log_{10} k$.

$\frac{10^3 \text{ rate}}{\text{mm Hg s}^{-1}}$	0.96	2.23	4.06	10.4
$\log_{10}$ rate	−3.018	−2.652	−2.391	−1.983
$\frac{p(\mathrm{NO_2Cl})}{\text{mm Hg}}$	1.45	3.39	6.17	15.9
$\log_{10} p(\mathrm{NO_2Cl})$	0.161	0.530	0.790	1.201

A plot of $\log_{10}$ rate versus $\log_{10} p(\mathrm{NO_2Cl})$ is linear with slope $= 1$, $\therefore$ reaction is first order. Intercept $= -3.18$, $\therefore k = 6.64 \times 10^{-4}\ \mathrm{s}^{-1}$. Data is consistent with part (a).

5. The total pressure has to be converted into partial pressure of reactant remaining at the various times.

Let the initial partial pressure of A be p_0, and let the decrease in partial pressure by time t be y.

Partial pressure of A at time $t = p_0 - y$, and partial pressure of A_2 formed $= (1/2)y$. Total pressure at time $t = p_{\text{total}} = p_0 - y + y/2 = p_0 - y/2 \therefore y = 2(p_0 - p_{\text{total}})$. Partial pressure of A remaining $= p_0 - y = p_0 - 2(p_0 - p_{\text{total}}) = 2p_{\text{total}} - p_0$.

Data is pressure/time data; the appropriate integrated rate equation should be used. Reaction is stated to be second order:

$$\frac{1}{(p_\mathrm{A})_t} = \frac{1}{(p_\mathrm{A})_0} + kt$$

A plot of $1/(p_\mathrm{A})_t$ versus t should be linear with slope $= +k$, and intercept $= 1/(p_\mathrm{A})_0$.

$\frac{p_{\text{total}}}{\text{mm Hg}}$	645	561	476	436	412	390
$\frac{2p_{\text{total}}}{\text{mm Hg}}$	1290	1122	952	872	824	780
$\frac{p_0}{\text{mm Hg}}$	645	645	645	645	645	645
$\frac{2p_{\text{total}} - p_0}{\text{mm Hg}}$	645	477	307	227	179	135
$\frac{10^3}{(2p_{\text{total}} - p_0)\text{mm Hg}}$	1.55	2.10	3.26	4.40	5.59	7.41
$\frac{\text{time}}{\text{s}}$	0	50	150	250	350	500

Plot is linear, confirming reaction to be second order with slope $= k = 1.17 \times 10^{-5} (\text{mm Hg})^{-1}\,\text{s}^{-1}$.
At the first half-life $p_A = 1/2\,p_0$

$$1/(p_A)_t = 1/(p_A)_0 + kt \text{ gives}$$
$$\frac{1}{(1/2)(p_A)_0} = \frac{1}{(p_A)_0} + k t_{1/2} \quad \therefore \quad \frac{1}{(p_A)_0} = k t_{1/2}$$

and

$$t_{1/2} = \frac{1}{k(p_A)_0} = \frac{1}{1.17 \times 10^{-5} (\text{mm Hg})^{-1} s^{-1} \times 645\,\text{mm Hg}} = 132\,\text{s}$$

The first half-life is 132 s.

At the second half-life, $p_A = (1/4)p_0$, and the time to reach this partial pressure $= 3/k(p_A)_0 = 397$ s.

The second half-life is the difference in time for the partial pressure of A to decrease to $(1/4)p_0$ from $(1/2)p_0 = (397 - 132)$ s $= 265$ s.

Second half-life > first half-life, confirming the second order behaviour.

6. Orders with respect to two reactants are required. Look to see whether there are any experiments where one substance is at the same pressure, while the pressure of the other reactant is altered, i.e. look for pseudo-order conditions.

   - In experiments 3 and 1, $p(N_2O)$ initially is the same while the initial $p(Cl_2)$ is altered.
     $p(Cl_2)$ increases by a factor of four; rate increases by a factor of two.
     rate $\propto (\text{pressure})^n$, $\therefore\ 2 = 4^n$ and $n = 1/2$. Order with respect to Cl_2 is 1/2.

   - In experiments 2 and 1, $p(Cl_2)$ initially is the same while the initial $p(N_2O)$ is altered.

     $p(N_2O)$ increases by a factor of two; rate also increases by a factor of two.

     Reaction is first order with respect to N_2O.

     Rate $= k\,p(N_2O)p(Cl_2)^{1/2}$ with overall order $= 3/2$.

7. Data given is concentration/time data, $\therefore$ use the integrated rate equation to confirm first order kinetics.

   The reaction is catalysed by acids, but during the experiment quoted [acid] remains constant, and reaction is being carried out under pseudo-order conditions.

   Toluene is both a reactant and the solvent, and so the toluene will always be in excess, and again reaction is carried out under pseudo-order conditions.

$$\text{rate} = k[C_6H_5CH_3]^m[\text{catalyst}]^n[Cl_2]^p$$

Since during any given experiment, $[C_6H_5CH_3]$ and [catalyst] are held constant, and the order with respect to Cl_2 is stated to be first, then

$$\text{rate} = k'[Cl_2]$$
$$\log_e[Cl_2]_t = \log_e[Cl_2]_0 - k't.$$

Data given is $[Cl_2]$ remaining at time t ∴ no need to carry out any stoichiometric conversion.
Plot of $\log_e[Cl_2]_t$ versus t should be linear with slope $= -k'$.

$\frac{[Cl_2]}{mol\,dm^{-3}}$	0.0775	0.0565	0.0410	0.0265	0.0175
$\frac{time}{min}$	0	3	6	10	14
$\log_e[Cl_2]$	−2.557	−2.874	−3.194	−3.631	−4.046

Plot is linear, confirming reaction to be first order, with slope $= -10.7 \times 10^{-2}$ min^{-1}, and $k' = 10.7 \times 10^{-2}$ min^{-1}.

Part (b) gives values of the pseudo-first order rate constant for a series of [catalyst]. For any given [catalyst],

rate $= k'[Cl_2]$ where k' is the pseudo-first order rate constant found in part (a), and $k' = k''[\text{catalyst}]^n$ where $k'' = k[C_6H_5CH_3]^m$

$$\log_{10} k' = \log_{10} k'' + n \log_{10}[\text{catalyst}]$$

∴ plot of $\log_{10} k'$ versus $\log_{10}$ [catalyst] should be linear with slope $= n$ and intercept $= \log_{10} k''$.
From part (a) $k' = 10.7 \times 10^{-2}$ min^{-1} when [catalyst] $= 0.117$ mol dm^{-3}.

$\log_{10}$[catalyst]	−1.301	−1.125	−1.000	−0.932	−0.815	−0.699
$\log_{10} k'$	−1.556	−1.276	−1.046	−0.971	−0.764	−0.569

Plot is linear with slope = 1.6, and it is likely that the order is 3/2, i.e. 1.5, with respect to catalyst.

Chapter 4

1. • Atom has three translations.

   Non-linear molecule has three translations, three rotations, and $(3N - 6)$ vibrations.

   Total for reactants: six translations, three rotations, $(3N - 6)$ vibrations.

   Non-linear activated complex has three translations, three rotations, one internal translation and $(3(N + 1) - 7)$ vibrations, i.e $(3N - 4)$ vibrations.

On activation: lose three translations; gain two vibrations and one internal translation.

- Linear molecule has three translations, two rotations and $(3N_A - 5)$ vibrations.

  Non-linear molecule has three translations, three rotations and $(3N_B - 6)$ vibrations.

  Total for reactants: six translations, five rotations and $(3N_A + 3N_B - 11)$ vibrations.

  Non-linear activated complex has three translations, three rotations, one internal translation and $(3N_A + 3N_B - 7)$ vibrations.

  On activation: lose three translations and two rotations; gain four vibrations and one internal translation.

- Two linear molecules have six translations, four rotations and $(3N_A + 3N_B - 10)$ vibrations.

  Linear activated complex has three translations, two rotations, one internal translation and $(3N_A + 3N_B - 6)$ vibrations.

  On activation: lose three translations and two rotations; gain four vibrations and one internal translation.

- Two linear molecules have six translations, four rotations and $(3N_A + 3N_B - 10)$ vibrations.

  Non-linear activated complex has three translations, three rotations, one internal translation and $(3N_A + 3N_B - 7)$ vibrations.

  On activation: lose three translations and one rotation; gain three vibrations and one internal translation.

- Non-linear molecule has three translations, three rotations and $(3N - 6)$ vibrations.

  Non-linear activated complex has three translations, three rotations, one internal translation and $(3N - 7)$ vibrations.

  On activation: lose one vibration; gain one internal translation.

2. The temperature dependence of k_{obs} will give values to $\Delta H^{\neq *}$ and $\Delta S^{\neq *}$.

$$k = \frac{\kappa k T}{h} K^{\neq *} = \frac{\kappa k T}{h} \exp\left(\frac{-\Delta H^{\neq *}}{RT}\right) \exp\left(\frac{+\Delta S^{\neq *}}{R}\right)$$

$$\log_e \frac{k}{T} = \log_e \frac{\kappa k}{h} - \frac{\Delta H^{\neq *}}{RT} + \frac{\Delta S^{\neq *}}{R}$$

A plot of $\log_e(k/T)$ versus $1/T$ should be linear with slope $= -\Delta H^{\neq *}/R$. The intercept should be $\log_e(\kappa k/h) + \Delta S^{\neq *}/R$. But, because this would involve a ridiculous extrapolation, $\Delta S^{\neq *}$ will have to be calculated.

ΔG^{θ} for reactions in solution depends on pressure, in contrast to gas phase reactions, for which ΔG^{θ} is independent of pressure.

One of the key equations of thermodynamics is

$$dG = V\,dp - S\,dT$$

At constant T, $dT = 0$ and $S\,dT = 0$, giving

$$dG = V\,dp$$

$$\int_{p_1}^{p_2} dG = \int_{p_1}^{p_2} V\,dp$$

For an ideal gas $pV = nRT$ giving

$$\int_{p_1}^{p_2} dG = nRT \int_{p_1}^{p_2} \frac{dp}{p}$$

from which

$$G_{p_2} - G_{p_1} = nRT \log_e \frac{p_2}{p_1}$$

and, if the standard state is 1 atm (more recently 1 bar), then $\mu = \mu^\theta + RT \log_e p$, where μ^θ is a constant dependent on T and *independent of p*.

$$1\ \text{bar} = 1.000\cdots \times 10^5\ \text{N m}^{-2}(\text{Pa})$$
$$1\ \text{atm} = 1.01325 \times 10^5\ \text{N m}^{-2}(\text{Pa})$$

ΔG^θ is a combination of μ^θ values for products and reactants, and so is also *independent of pressure*.

For a substance in solution, the analogous equation for the ideal case is $\mu = \mu^\theta + RT \log_e c$, where μ^θ is a constant dependent on T and *also on pressure*. Correspondingly, ΔG^θ is a constant dependent on *both T and pressure*.

3. $$Br^\bullet + H_2 \rightarrow HBr + H^\bullet$$

$Br^\bullet$: atom with three translations

H_2: two atoms, linear molecule with three translations, two rotations and one vibration.

Activated complex: three atoms, linear, three translations, two rotations, one internal translation and three vibrations.

On activation: lose three translations, and gain two vibrations.

$$\Delta S^{\neq *} = -(3 \times 50)\ \text{J mol}^{-1}\ \text{K}^{-1} + (2 \times 10)\ \text{J mol}^{-1}\ \text{K}^{-1}$$
$$= -130\ \text{J mol}^{-1}\ \text{K}^{-1}$$

$$CH_3^\bullet + H_2 \rightarrow CH_4 + H^\bullet$$

$CH_3^\bullet$: four atoms, non-linear molecule with three translations, three rotations and six vibrations.

H_2: two atoms, linear molecule with three translations, two rotations and one vibration.

Total for reactants: six translations, five rotations and seven vibrations.

Activated complex: six atoms, non-linear, three translations, three rotations, one internal translation and 11 vibrations.

On activation: lose three translations, lose two rotations and gain four vibrations.

$$\Delta S^{\neq *} = -(3 \times 50)\,\mathrm{J\,mol^{-1}\,K^{-1}} - (2 \times 40)\,\mathrm{J\,mol^{-1}\,K^{-1}} + (4 \times 10)\,\mathrm{J\,mol^{-1}\,K^{-1}}$$
$$= -190\,\mathrm{J\,mol^{-1}K^{-1}}.$$

$$F_2 + ClO_2^{\bullet} \rightarrow FClO_2 + F^{\bullet}$$

F_2: two atoms, linear molecule with three translations, two rotations and one vibration.

ClO_2: three atoms, non-linear molecule with three translations, three rotations and three vibrations.

Total for reactants: six translations, five rotations and four vibrations.

Activated complex: five atoms, non-linear, three translations, three rotations, one internal translation and eight vibrations.

On activation: lose three translations, lose two rotations and gain four vibrations.

$$\Delta S^{\neq *} = -(3 \times 50)\,\mathrm{J\,mol^{-1}\,K^{-1}} - (2 \times 40)\,\mathrm{J\,mol^{-1}\,K^{-1}} + (4 \times 10)\,\mathrm{J\,mol^{-1}\,K^{-1}}$$
$$= -190\,\mathrm{J\,mol^{-1}\,K^{-1}}.$$

$$C_4H_6 + C_2H_4 \rightarrow \text{cyclo-}C_6H_{10}$$

C_4H_6: 10 atoms, non-linear molecule with three translations, three rotations and 24 vibrations.

C_2H_4: six atoms, non-linear molecule with three translations, three rotations and 12 vibrations.

Total for reactants: six translations, six rotations and 36 vibrations.

Activated complex: 16 atoms, non-linear, three translations, three rotations, one internal translation and 41 vibrations.

On activation: lose three translations, lose three rotations and gain five vibrations.

$$\Delta S^{\neq *} = -(3 \times 50)\,\mathrm{J\,mol^{-1}\,K^{-1}} - (3 \times 40)\,\mathrm{J\,mol^{-1}\,K^{-1}} + (5 \times 10)\,\mathrm{J\,mol^{-1}\,K^{-1}}$$
$$= -220\,\mathrm{J\,mol^{-1}\,K^{-1}}.$$

Note the following.

- As the complexity of the reactants increases, the entropy loss becomes greater and consequently the *A* factor will decrease.

- The reactions

$$CH_3^{\bullet} + H_2 \rightarrow CH_4 + H^{\bullet}$$
$$F_2 + ClO_2^{\bullet} \rightarrow FClO_2 + F^{\bullet}$$

are chemically totally different, but because the changes in the numbers of rotational and vibrational modes are the same, the approximate entropy change is the same.

- These are only rough guides to the true entropy change on activation. Detailed calculations using spectroscopic data for the reactants and a calculated potential energy surface for the activated complex will yield accurate partition functions for the translational, rotational and vibrational terms involved. Since the quantities contributing to the partition functions for each molecule will be different, then accurate calculations will be able to differentiate between such reactions as

$$CH_3^{\bullet} + H_2 \rightarrow CH_4 + H^{\bullet}$$
$$F_2 + ClO_2 \rightarrow FClO_2 + F^{\bullet}$$

4. C_4H_6: relative molar mass $= 4 \times 12.011 + 6 \times 1.008 = 54.092$
C_3H_4O: relative molar mass $= 3 \times 12.011 + 4 \times 1.008 + 15.999 = 56.064$

$$m_{C_4H_6} = \frac{54.092}{6.0221 \times 10^{23}}\,\text{g} = 8.9822 \times 10^{-23}\,\text{g} = 8.9822 \times 10^{-26}\,\text{kg}$$

$$m_{C_3H_4O} = \frac{56.064}{6.0221 \times 10^{23}}\,\text{g} = 9.3097 \times 10^{-23}\,\text{g} = 9.3097 \times 10^{-26}\,\text{kg}$$

$$\frac{1}{\mu} = \frac{1}{m_{C_4H_6}} + \frac{1}{m_{C_3H_4O}}$$

$$\therefore \quad \mu = \frac{m_{C_4H_6} \times m_{C_3H_4O}}{m_{C_4H_6} + m_{C_4H_4O}} = \frac{8.9822 \times 10^{-26} \times 9.3097 \times 10^{-26}}{(8.9822 + 9.3097) \times 10^{-26}}\,\text{kg}$$

$$= 4.5715 \times 10^{-26}\,\text{kg}$$

$$\left(\frac{8\,k\,T}{\pi\,\mu}\right)^{1/2} = \left(\frac{8 \times 1.38066 \times 10^{-23}\,\text{J}\,\text{K}^{-1} \times 500\,\text{K}}{\pi \times 4.5715 \times 10^{-26}\,\text{kg}}\right)^{1/2} = 620\,\text{J}^{1/2}\,\text{kg}^{-1/2}$$

$$= 620\,\text{m}\,\text{s}^{-1}$$

$$\pi\,d^2 = 5.03 \times 10^{-19}\,\text{m}^2$$

$$Z_{(\text{molecular})} = \sigma\,v = 5.03 \times 10^{-19}\,\text{m}^2 \times 620\,\text{m}\,\text{s}^{-1} = 3.12 \times 10^{-16}\,\text{m}^3\,\text{s}^{-1}$$

$$Z_{(\text{molar})} = 3.12 \times 10^{-16} \times 6.022 \times 10^{23}\,\text{m}^3\,\text{mol}^{-1}\,\text{s}^{-1}$$

$$= 1.88 \times 10^{8}\,\text{m}^3\,\text{mol}^{-1}\,\text{s}^{-1}$$

$$= 1.88 \times 10^{11}\,\text{mol}^{-1}\,\text{dm}^3\,\text{s}^{-1}$$

If the p factor is taken to be unity, then the predicted A factor for collision theory $=$ $p\,Z_{(\text{molar})} = 1.88 \times 10^{11}\ \text{mol}^{-1}\ \text{dm}^3\ \text{s}^{-1}$.

Entropies of activation.

C_4H_6: 10 atoms, non-linear ∴ three translations, three rotations and 24 vibrations

C_3H_4O: eight atoms, non-linear ∴ three translations, three rotations and 18 vibrations

activated complex: 18 atoms ∴ three translations, three rotations, one internal translation and 47 vibrations

∴ on forming the activated complex

lose three translations and three rotations,

gain five vibrations.

$$\Delta S^{\neq *} \approx -(3 \times 40) - (3 \times 25) + (5 \times 10)\,\mathrm{J\,mol^{-1}\,K^{-1}}$$
$$= -145\,\mathrm{J\,mol^{-1}\,K^{-1}}$$

p corresponds to $\exp(\Delta S^{\neq *}/R) = \exp(-145/8.3145) = 2.7 \times 10^{-8}$

$$A = \mathrm{e}^2 \times \frac{\kappa k T}{h} \exp\left(\frac{\Delta S^{\neq *}}{R}\right) = \mathrm{e}^2 \times \frac{1 \times 1.38066 \times 10^{-23}\,\mathrm{J\,K^{-1}} \times 500\,\mathrm{K}}{6.6261 \times 10^{-34}\,\mathrm{J\,s}} \times 2.7 \times 10^{-8}$$
$$= 2.1 \times 10^6\,\mathrm{mol^{-1}\,dm^3\,s^{-1}}$$

(The factor e^2 takes account of the difference between the Arrhenius equation given in terms of the experimental activation energy, and that in terms of the enthalpy of activation.)

collision theory A factor $= 1.88 \times 10^{11}\,\mathrm{mol^{-1}\,dm^3\,s^{-1}}$

approximate transition state theory A factor $= 2.1 \times 10^6\,\mathrm{mol^{-1}\,dm^3\,s^{-1}}$

experimental A factor $= 1.6 \times 10^6\,\mathrm{mol^{-1}\,dm^3\,s^{-1}}$

There is a large discrepancy between the experimental and the collision theory values, demonstrating the inadequacy of collision theory. There is excellent agreement between the experimental and transition state theory values, though, in part, the very close agreement is merely a reflection of the particular typical magnitudes taken for the entropy contributions. However, even taking upper and lower limits to these contributions still gives magnitudes for the A factor which are much closer to the experimental A factor than the collision theory value.

5. The Lindemann mechanism gives

$$k_{\mathrm{obs}}^{\mathrm{1st}} = \frac{k_1 k_2[\mathrm{A}]}{k_{-1}[\mathrm{A}] + k_2}$$
$$\frac{1}{k_{\mathrm{obs}}^{\mathrm{1st}}} = \frac{k_{-1}}{k_1 k_2} + \frac{1}{k_1[\mathrm{A}]}$$

If k_1, k_{-1} and k_2 are all constants, then the graph of $1/k_{\mathrm{obs}}^{\mathrm{1st}}$ versus $1/[\mathrm{A}]$ should be linear with slope $= 1/k_1$ and intercept $= k_{-1}/k_1k_2$.

Using the first set of data

$\dfrac{10^{-3}\dfrac{1}{[\mathrm{A}]}}{\mathrm{mol^{-1}\,dm^3}}$	100	67	50	25
$\dfrac{1}{k_{\mathrm{obs}}^{\mathrm{1st}}/\mathrm{s}^{-1}}$	3.00	2.56	2.35	2.02

The graph of $1/k_{obs}^{1st}$ versus $1/[A]$ is linear, and the data fits the simple Lindemann mechanism, giving

$$\text{slope} = \frac{1}{k_1} = 1.30 \times 10^{-5}\,\text{mol dm}^{-3}\,\text{s}$$

$$\therefore \quad k_1 = 7.7 \times 10^4\,\text{mol}^{-1}\,\text{dm}^3\,\text{s}^{-1}$$

$$\text{intercept} = \frac{k_{-1}}{k_1\,k_2} = 1.70\,\text{s}$$

$$\therefore \quad k_2 = \frac{5.0 \times 10^{10}\text{mol}^{-1}\,\text{dm}^3\text{s}^{-1}}{1.70\,\text{s} \times 7.7 \times 10^4\text{mol}^{-1}\,\text{dm}^3\text{s}^{-1}}$$

$$= 3.8 \times 10^5\,\text{s}^{-1}$$

the mean life-time, τ, of the activated molecule $= 1/k_2 = 2.6 \times 10^{-6}$ s.

Inclusion of the remaining three values drastically alters the intepretation. A grossly non-linear plot results, clearly illustrating the inadequacy of the simple Lindemann scheme.

$\dfrac{10^{-3}\,\dfrac{1}{[A]}}{\text{mol}^{-1}\,\text{dm}^3}$	10	5	2.5
$\dfrac{1}{k_{obs}/\text{s}^{-1}}$	1.60	1.28	1.05

This problem also illustrates the importance of covering a sufficient range of values of the variables in order to test a theory adequately. The same point arises in Worked Problem 7.5 of Chapter 7 where extension of the data totally alters the interpretation.

6. $$Br^\bullet + H_2 \rightarrow HBr + H^\bullet$$

$Br^\bullet$: atom with three translations

H_2: two atoms, linear molecule with three translations, two rotations and one vibration.

Activated complex: three atoms, linear, three translations, two rotations, one internal translation and three vibrations.

Quotient of partition functions: $\dfrac{f_{trans}^3 f_{rot}^2 f_{vib}^3}{f_{trans}^3 f_{trans}^3 f_{rot}^2 f_{vib}}$, i.e. $\dfrac{f_{vib}^2}{f_{trans}^3}$

Temperature dependence: $T^2/T^{3/2}$ i.e. $T^{1/2}$ at high temperatures and $1/T^{3/2}$ i.e. $T^{-3/2}$ at low temperatures.

$$CH_3^\bullet + H_2 \rightarrow CH_4 + H^\bullet$$

$CH_3^\bullet$: four atoms, non-linear molecule with three translations, three rotations and six vibrations.

H_2: two atoms, linear molecule with three translations, two rotations and one vibration.

Activated complex: six atoms, non-linear, three translations, three rotations, one internal translation and 11 vibrations.

Quotient of partition functions: $\dfrac{f^3_{\text{trans}} f^3_{\text{rot}} f^{11}_{\text{vib}}}{f^3_{\text{trans}} f^3_{\text{rot}} f^6_{\text{vib}} f^3_{\text{trans}} f^2_{\text{rot}} f_{\text{vib}}}$ i.e. $\dfrac{f^4_{\text{vib}}}{f^3_{\text{trans}} f^2_{\text{rot}}}$

Temperature dependence: $\dfrac{T^4}{T^{3/2}T}$ i.e. $T^{3/2}$ at high temperatures, and $\dfrac{1}{T^{3/2}T}$ i.e. $T^{-5/2}$ at low temperatures.

$$F_2 + ClO_2^{\bullet} \rightarrow FClO_2 + F^{\bullet}$$

F_2: two atoms, linear molecule with three translations, two rotations and one vibration.

ClO_2: three atoms, non-linear molecule with three translations, three rotations and three vibrations.

Activated complex: five atoms, non-linear, three translations, three rotations, one internal translation and eight vibrations.

Quotient of partition functions: $\dfrac{f^3_{\text{trans}} f^3_{\text{rot}} f^8_{\text{vib}}}{f^3_{\text{trans}} f^2_{\text{rot}} f_{\text{vib}} f^3_{\text{trans}} f^3_{\text{rot}} f^3_{\text{vib}}}$ i.e. $\dfrac{f^4_{\text{vib}}}{f^3_{\text{trans}} f^2_{\text{rot}}}$

Temperature dependence: $\dfrac{T^4}{T^{3/2}T}$ i.e. $T^{3/2}$ at high temperatures, and $\dfrac{1}{T^{3/2}T}$ i.e. $T^{-5/2}$ at low temperatures.

$$C_4H_6 + C_2H_4 \rightarrow cyclo\text{-}C_6H_{10}$$

C_4H_6: 10 atoms, non-linear molecule with three translations, three rotations and 24 vibrations.

C_2H_4: six atoms, non-linear molecule with three translations, three rotations and 12 vibrations.

Activated complex: 16 atoms, non-linear, three translations, three rotations, one internal translation and 41 vibrations.

Quotient of partition functions: $\dfrac{f^3_{\text{trans}} f^3_{\text{rot}} f^{41}_{\text{vib}}}{f^3_{\text{trans}} f^3_{\text{rot}} f^{24}_{\text{vib}} f^3_{\text{trans}} f^3_{\text{rot}} f^{12}_{\text{vib}}}$ i.e. $\dfrac{f^5_{\text{vib}}}{f^3_{\text{trans}} f^3_{\text{rot}}}$

Temperature dependence: $\dfrac{T^5}{T^{3/2}T^{3/2}}$ i.e. T^2 at high temperatures, and $\dfrac{1}{T^{3/2}T^{3/2}}$ i.e. T^{-3} at low temperatures.

7. • cyclopropane → propene

   The $\Delta S^{\neq *}$ and ΔS^{θ} values are similar: reaction is $1 \rightarrow 1$ and so the entropy change is not associated with a change in the number of molecules. The activated complex resembles the product and so bond breaking and making is well developed in the activated complex. On isomerization there is release of the strain in the three-membered ring.

- cyclobutane → $2\,C_2H_4$
  $\Delta S^{\neq *} \ll \Delta S^{\theta}$: reaction is $1 \rightarrow 2$, and so an increase in entropy will be expected. Since $\Delta S^{\neq *} \ll \Delta S^{\theta}$, the activated complex closely resembles the reactant, and the C–C bond breaking is not at all well developed. There is little strain to be released in the four-membered ring.

- CH_2—CH_2 (ring closed by O) ⟶ CH_3CHO
  The two values are virtually equal and so the activated complex very closely resembles the products, and all bond breaking and electronic rearrangements are virtually complete. Again, there is the driving force of release of strain in the three-membered ring.

- $CH_3NC \rightarrow CH_3CN$
  The two values are similar and very close to zero. Reaction is $1 \rightarrow 1$ and so there will be no entropy change as a consequence of a change in number of molecules. Because of the very small change in the standard change in entropy for the overall reaction very little can be said, despite the similarity in the values.

- cyclo-$C_4F_8 \rightarrow 2\,C_2F_4$
  $\Delta S^{\neq *} \ll \Delta S^{\theta}$: reaction is $1 \rightarrow 2$, and so an increase in entropy will be expected. Since $\Delta S^{\neq *} \ll \Delta S^{\theta}$, then the activated complex closely resembles the reactant, and the C–C bond breaking is not at all well developed. There is little strain to be released in the four-membered ring.

  This reaction closely resembles cyclobutane → $2\,C_2H_4$ both chemically and physically.

  For all reactions, $\Delta S^{\neq *}$ is positive, implying that the activated complex is looser than the reactant (though the value for $CH_3NC \rightarrow CH_3CN$ is only just positive).

- $\Delta S^{\theta} = \Delta S_f^{\neq *} - \Delta S_b^{\neq *}$, and so the following table can be drawn up for the reverse reactions.

reaction	$\Delta S_b^{\neq *}/\mathrm{J\,mol^{-1}\,K^{-1}}$	$\Delta S_b^{\theta}/\mathrm{J\,mol^{-1}\,K^{-1}}$
propene → cyclopropane	+9.5	−29.5
$2C_2H_4$ → cyclobutane	−104.9	−146.9
CH_3CHO ⟶ CH_2—CH_2 (ring closed by O)	−0.8	−21.8
$CH_3CN \rightarrow CH_3NC$	+7.3	+3.3
$2C_2F_4$ → cyclo-C_4F_8	−123.9	−172.9

The two association reactions are completely different from the others. Here $2 \rightarrow 1$, and a large decrease in entropy would be expected. For both, $\Delta S^{\neq *}$ is strongly negative, confirming this expectation. However, $\Delta S^{\neq *}$ is considerably less negative than ΔS^{θ} implying an activated complex which is less tightly bound than the cyclic product.

For the other reactions $\Delta S^{\neq *}$ is small and much more positive than for the two associations, as would be expected for $1 \rightarrow 1$ reactions compared to $2 \rightarrow 1$. All are more positive than the overall ΔS^{θ}, which is as expected for an activated complex that is looser than the products.

Chapter 5

1. • Since the M—N distance in the activated complex is only slightly greater than the internuclear distance in the reactant molecule, then the reaction entity has only moved very slightly along the reactant valley. This means that the activated complex lies in the entrance valley, giving an early barrier. This conclusion is verified by the large P—M distance in the activated complex, which indicates that P is still far from MN in the activated complex.

The forward reaction is exothermic, and so the back reaction must be endothermic, as is shown on the potential energy profile.

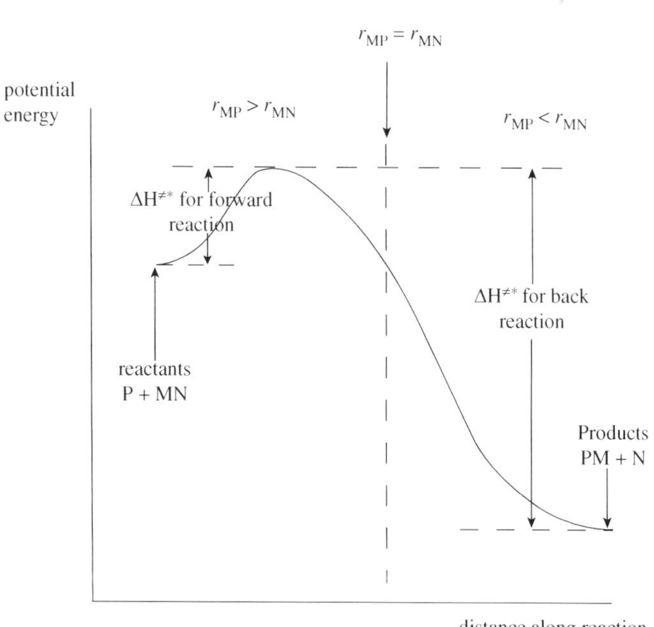

- Since the activated complex lies in the entrance valley, there is a straight run up the valley to the critical configuration. Translational energy will favour reaction, with vibrational energy playing a minor role. Since the critical configuration is reached before the bond M—N has extended much, the bond P—M will be formed as M and N separate and the bond breaks. The recoil energy will reside in the forming P—M bond, and will set up vibrations in this bond, which will be vibrationally excited. The products will thus have a high vibrational energy compared with the translational energy.

  Since the critical configuration is reached when P is far from M, then the reaction will have a large cross section. The velocity of the products will be low, and the products will initially be in excited vibrational levels. The molecular beam contour diagram will show predominantly forward scattering typical of a stripping mechanism.

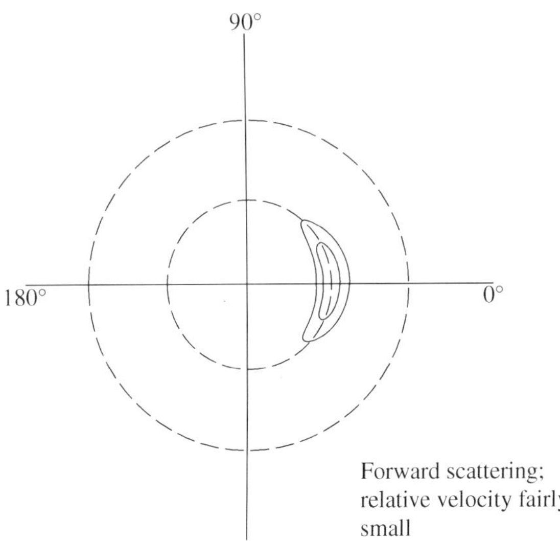

- The back reaction will have the exact reverse characteristics. The activated complex will lie in the exit valley, and reaction will be enhanced by high vibrational energy. There will be high translational energy in the products, the cross section will be small, and the molecular beam contour diagram will show predominantly backward scattering, typical of a rebound mechanism.

2. (a) This diagram shows predominantly backward scattering, with very high translational energies and low vibrational energy in the products. The cross section is small. This indicates an activated complex in the exit valley, and suggests a rebound mechanism.

(b) This diagram suggests that the reaction occurs via a 'cutting the corner' trajectory where the usual minimum energy path is not followed.

3. As the vibrational ladder is climbed, i.e. v the vibrational quantum number increases, the vibrational energy increases. The question thus states that reaction becomes faster as the vibrational energy increases.
This means that vibrational energy favours reaction, from which it can be inferred that the activated complex lies in the exit valley, and that there is a late barrier. This is a rebound mechanism where the reactants have to get very close together before reaction can occur.

4. (a) If the cross section is large, then reaction can occur when the reactants are far apart. This is a stripping mechanism where the activated complex lies in the entrance valley, there is an early barrier and products will separate with predominantly vibrational energy. The reaction will be favoured by high translational energies. In a stripping mechanism, the Ca will pull off the O from the NO_2 at a distance, and continue on with only a small deflection from the original trajectory. This implies that there must be strong forces of interaction between the reactant molecules for this to be possible.

(b) In contrast to the above reaction, a small cross section implies that reaction can only occur at small distances apart. This is typical of a rebound mechanism, where the activated complex occurs in the exit valley, where there is a late barrier and products will separate with predominantly high translational energy. The reaction will be favoured by high vibrational energies. In a rebound mechanism the $CH_3^{\bullet}$ will only react with the Br_2 when they are almost in contact, and on rebound it will fly off in a backwards direction, so that there is backwards scattering of products on the molecular beam contour diagram. Since reaction can only occur when the reactants are close together, the forces of interaction must be relatively weak.

(c) In this reaction the molecular contour diagram shows both forward and backward scattering. This is typical of a reaction with a collision complex where many rotations, and perhaps several vibrations occur, before the products separate. The directions in which the product molecules separate in this situation will be random, and a fairly symmetrical scattering pattern will result.

(d) If the products are predominantly in the ground vibrational state this implies that the energy of the products is predominantly translational, characteristic of a rebound mechanism with a late barrier where the activated complex is in the exit valley. The other characteristics will be as in (b) above.

(e) The cross section is related to the rate constant for the reaction. The larger the cross section the greater will be the rate constant. The data given states that the rate constant for reaction is greater if the reactant is vibrationally excited. This is a characteristic of a rebound mechanism where the activated complex is in the exit valley and the potential energy profile has a late barrier. For the other characteristics of a rebound mechanism, see (b) and (d) above.

5. (a) The radii are 80 pm and 190 pm, so that when the reactant molecules are just touching the centres of the molecules are 270 pm apart. This would correspond to a cross section $= \pi d^2$, where $d = 270$ pm.
Cross section corresponding to just touching $= \pi \times (270)^2\,\text{pm}^2 = 0.23\,\text{nm}^2$.
Observed cross section $= 0.20\,\text{nm}^2$.
Since these two cross sections have similar magnitudes, this would correspond to a rebound mechanism where the two molecules have to come very close together before reaction will occur. The activated complex thus lies in the exit valley and the potential energy profile has a late barrier.

(b) If the mechanism is indeed a rebound mechanism then vibrational energy in the reactants will favour reaction, and the products will have predominantly translational energy. This conclusion can be confirmed by the information given.

$$\text{translational energy} = 42.5\,\text{kJ}\,\text{mol}^{-1} = \frac{42.5 \times 10^3}{6.022 \times 10^{23}}\,\text{J} = 7.06 \times 10^{-20}\,\text{J}$$

$$\text{vibrational energy} = (v + 1/2)h\nu_0$$

The product molecule is predominantly in the ground vibrational state, i.e. $v = 0$.

$$\begin{aligned} \text{vibrational energy} &= 1/2 \times 6.626 \times 10^{-34}\,\text{J}\,\text{s} \times 3 \times 10^{13}\,\text{s}^{-1} \\ &= 0.994 \times 10^{-20}\,\text{J} \end{aligned}$$

$$\therefore \quad \frac{\text{translational energy}}{\text{vibrational energy}} = \frac{7.06 \times 10^{-20}\,\text{J}}{0.994 \times 10^{-20}\,\text{J}} = 7.1$$

Hence the predominant energy in the products is translational, which confirms the conclusions from the cross sections.

Reaction is a rebound mechanism, with the critical configuration in the exit valley.

(c) In a rebound mechanism the reactants have to collide head on, and in doing so the products rebound backwards in a direction opposite to the original line of approach of the colliding molecules. The molecular beam contour diagram will show backward scattering.

(d) The potential energy contour diagram and profile are shown in next page.

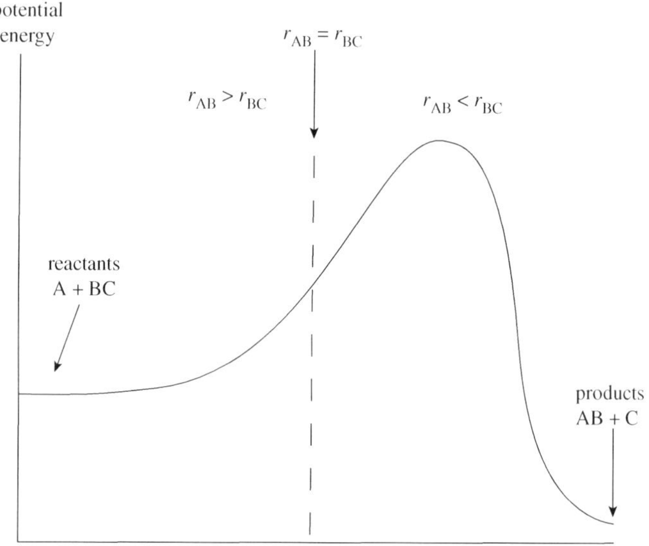
potential
energy
$r_{AB} = r_{BC}$
$r_{AB} > r_{BC}$
$r_{AB} < r_{BC}$
reactants
A + BC
products
AB + C
distance along reaction
coordinate

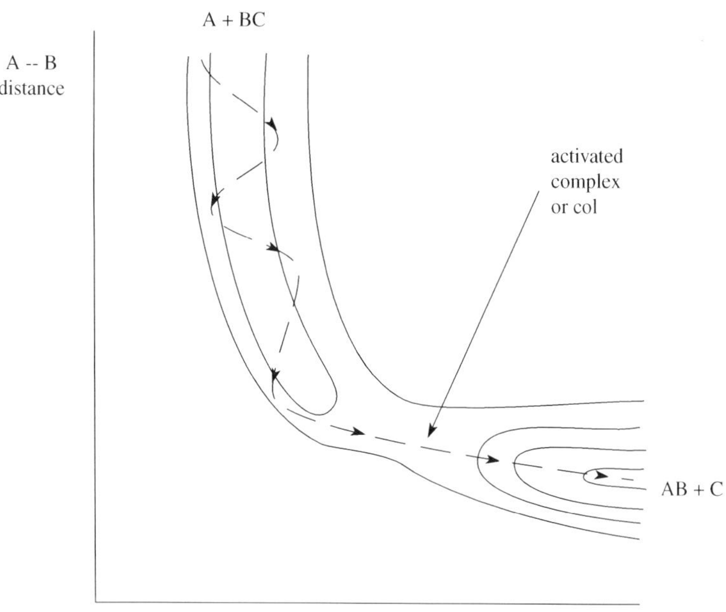
A + BC
A -- B
distance
activated
complex
or col
AB + C
B -- C distance

Chapter 6

1. Major products found in propagation: HBr and $CH_3CHBrCH_3$ with $BrCH_2CH_2CH_3$.

Minor products produced in termination: $CH_3CH(CH_3)CH(CH_3)CH_3$ with $CH_3(CH_2)_4CH_3$ and $(CH_3)_2CH(CH_2)_2CH_3$ as trace products.

The reaction is a bromination reaction, and the initiation step is likely to be production of $Br^•$:

$$Br_2 + M \rightarrow 2\,Br^• + M$$

The first propagation step is likely to be a H abstraction from the other reactant by $Br^•$. There are two possibilities:

$$CH_3CH_2CH_3 + Br^• \rightarrow HBr + CH_3CH_2CH_2^•$$
$$CH_3CH_2CH_3 + Br^• \rightarrow HBr + CH_3CH^•CH_3$$

This produces the major product HBr and a second radical which must undergo reaction to regenerate $Br^•$ and form the other major product, $CH_3CHBrCH_3$, along with a similar reaction to produce the less substantial major product $BrCH_2CH_2CH_3$. These must be produced by reaction of the two alkyl radicals produced in the two possible first steps of the propagation cycle.

The 2-bromopropane must come from the $CH_3CH^•CH_3$ radical, which must thus be the dominant chain carrier, while the lesser amount product, 1-bromo-propane, will come from the radical $CH_3CH_2CH_2^•$

$$CH_3CH^•CH_3 + Br_2 \rightarrow CH_3CHBrCH_3 + Br^• \quad \text{dominant propagation}$$
$$CH_3CH_2CH_2^• + Br_2 \rightarrow BrCH_2CH_2CH_3 + Br^• \quad \text{minor propagation}$$

The termination steps will be recombination of the chain carriers:

$$Br^• + Br^• + M \rightarrow Br_2 + M$$

The Br_2 is not picked up as a minor product since it is a reactant. The dominant minor product will be produced by recombination of the major intermediate,

$$CH_3CH^•CH_3 + CH_3CH^•CH_3 \rightarrow CH_3CH(CH_3)CH(CH_3)CH_3$$

with the trace minor products being produced from the minor intermediate,

$$CH_3CH_2CH_2^• + CH_3CH_2CH_2^• \rightarrow CH_3CH_2CH_2CH_2CH_2CH_3$$
$$CH_3CH^•CH_3 + CH_3CH_2CH_2^• \rightarrow (CH_3)_2CHCH_2CH_2CH_3$$

Proposed mechanism:

$$Br_2 + M \rightarrow 2\,Br^• + M \qquad \text{initiation}$$

$$CH_3CH_2CH_3 + Br^• \rightarrow HBr + CH_3CH^•CH_3 \qquad \text{dominant propagation}$$

$CH_3CH^{\bullet}CH_3 + Br_2 \rightarrow CH_3CHBrCH_3 + Br^{\bullet}$ dominant propagation

$CH_3CH_2CH_3 + Br^{\bullet} \rightarrow HBr + CH_3CH_2CH_2^{\bullet}$ minor propagation

$CH_3CH_2CH_2^{\bullet} + Br_2 \rightarrow BrCH_2CH_2CH_3 + Br^{\bullet}$ minor propagation

$Br^{\bullet} + Br^{\bullet} + M \rightarrow Br_2 + M$ termination

$CH_3CH^{\bullet}CH_3 + CH_3CH^{\bullet}CH_3 \rightarrow CH_3CH(CH_3)CH(CH_3)CH_3$ termination

$CH_3CH_2CH_2^{\bullet} + CH_3CH_2CH_2^{\bullet} \rightarrow CH_3CH_2CH_2CH_2CH_2CH_3$ trace termination

$CH_3CH^{\bullet}CH_3 + CH_3CH_2CH_2^{\bullet} \rightarrow (CH_3)_2CHCH_2CH_2CH_3$ trace termination

2. (a) The major products, C_2H_6 and CO, must be produced in propagation.

   (b) The three minor products are produced in termination; pentan-3-one is the most abundant and is formed in the dominant termination step.

   (c) Pentan-3-one, butane and hexan-3:4-dione are formed by recombination of chain carriers. The structures of these minor products could give clues about the nature of the chain carriers.

   (d) Pentanal is a trace product and H_2 a very trace product. These could be produced from the non-chain radical formed in the initiation step, or could be produced in a chain carrier reaction which occurs very infrequently.

   (e) k_{obs} depends on pressure. This suggests the presence of at least one unimolecular step.

   (f) Reaction is first order at high pressures, which suggests that the dominant termination step is recombination of unlike radicals. The dominant minor product is pentan-3-one, $CH_3CH_2COCH_2CH_3$. If the chain carriers are unlike, one possibility suggested by the structure of the product is $CH_3CH_2CO^{\bullet}$ and $CH_3CH_2^{\bullet}$.

   (g) The structures of the other two minor products suggest the same radicals: $CH_3CH_2CH_2CH_3$ suggests recombination of $CH_3CH_2^{\bullet}$ with $CH_3CH_2^{\bullet}$. $CH_3CH_2COCOCH_2CH_3$ suggests recombination of $CH_3CH_2CO^{\bullet}$ with $CH_3CH_2CO^{\bullet}$.

   (h) Initiation is often, though not always, a C—C split. This is more likely than a C—H or C—O split.
   $CH_3CH_2CHO \rightarrow CH_3CH_2^{\bullet} + CHO^{\bullet}$, which would fit in with the deductions above, leaving $CHO^{\bullet}$ as the non-chain carrier and possible source of the trace products.

   (i) The first propagation step is often, though not always, a H abstraction on the reactant by the chain carrier produced in initiation.

$$CH_3CH_2^{\bullet} + CH_3CH_2CHO \rightarrow ?$$

The query is where the H abstraction occurs. There are three possibilities:

$$CH_3CH_2^{\bullet} + CH_3CH_2CHO \rightarrow C_2H_6 + {}^{\bullet}CH_2CH_2CHO$$

$$CH_3CH_2^{\bullet} + CH_3CH_2CHO \rightarrow C_2H_6 + CH_3CH^{\bullet}CHO$$

$$CH_3CH_2^{\bullet} + CH_3CH_2CHO \rightarrow C_2H_6 + CH_3CH_2CO^{\bullet}$$

The dominant termination being recombination of unlike radicals and production of $CH_3CH_2COCOCH_2CH_3$ suggests that the dominant first propagation step is

$$CH_3CH_2^{\bullet} + CH_3CH_2CHO \rightarrow C_2H_6 + CH_3CH_2CO^{\bullet}$$

(j) The second propagation step must produce the second major product, CO, and regenerate $CH_3CH_2^{\bullet}$, suggesting

$$CH_3CH_2CO^{\bullet} \rightarrow CH_3CH_2^{\bullet} + CO$$

though the other two possibilities are

$$^{\bullet}CH_2CH_2CHO \rightarrow CH_3CH_2^{\bullet} + CO$$

$$CH_3CH^{\bullet}CHO \rightarrow CH_3CH_2^{\bullet} + CO$$

(k) The termination reactions are determined by the chain carriers argued as above, i.e. $CH_3CH_2CO^{\bullet}$, $CH_3CH_2^{\bullet}$.

$$CH_3CH_2^{\bullet} + CH_3CH_2CO^{\bullet} \rightarrow CH_3CH_2COCH_2CH_3$$
$$CH_3CH_2^{\bullet} + CH_3CH_2^{\bullet} \rightarrow CH_3CH_2CH_2CH_3$$
$$CH_3CH_2CO^{\bullet} + CH_3CH_2CO^{\bullet} \rightarrow CH_3CH_2COCOCH_2CH_3$$

(l) There remains the trace pentanal and very trace H_2 to account for.

(i) Pentanal: the radicals that are around are

$$CH_3CH_2^{\bullet}, CH_3CH_2CO^{\bullet}, {}^{\bullet}CH_2CH_2CHO, CH_3CH^{\bullet}CHO, CHO^{\bullet}$$

Pentanal could be produced by recombination of $CH_3CH_2^{\bullet}$ and $^{\bullet}CH_2CH_2CHO$

$$CH_3CH_2^{\bullet} + {}^{\bullet}CH_2CH_2CHO \rightarrow CH_3CH_2CH_2CH_2CHO$$

(ii) The very trace H_2 could come from the non-chain carrier produced in initiation:

$$CHO^{\bullet} + M \rightarrow H^{\bullet} + CO + M \quad \text{a known reaction}$$
$$H^{\bullet} + CH_3CH_2CHO \rightarrow H_2 + CH_3CH_2CO^{\bullet}$$
$$\rightarrow H_2 + {}^{\bullet}CH_2CH_3CHO$$
$$\rightarrow H_2 + CH_3CH^{\bullet}CHO$$

All observations are accounted for and a possible mechanism could be

$$CH_3CH_2CHO \rightarrow CH_3CH_2^\bullet + CHO^\bullet \qquad \text{initiation}$$
$$CH_3CH_2^\bullet + CH_3CH_2CHO \rightarrow C_2H_6 + CH_3CH_2CO^\bullet \qquad \text{major propagation}$$
$$CH_3CH_2CO^\bullet \rightarrow CH_3CH_2^\bullet + CO \qquad \text{major propagation}$$
$$CH_3CH_2^\bullet + CH_3CH_2CO^\bullet \rightarrow CH_3CH_2COCH_2CH_3 \qquad \text{dominant termination}$$
$$CH_3CH_2^\bullet + CH_3CH_2^\bullet \rightarrow CH_3CH_2CH_2CH_3 \qquad \text{minor termination}$$
$$CH_3CH_2CO^\bullet + CH_3CH_2CO^\bullet \rightarrow CH_3CH_2COCOCH_2CH_3 \qquad \text{minor termination}$$
$$CH_3CH_2^\bullet + CH_3CH_2CHO \rightarrow C_2H_6 + {}^\bullet CH_2CH_2CHO \qquad \text{minor propagation}$$
$${}^\bullet CH_2CH_2CHO \rightarrow CH_3CH_2^\bullet + CO \qquad \text{minor propagation}$$
$$CH_3CH_2^\bullet + CH_3CH_2CHO \rightarrow C_2H_6 + CH_3CH^\bullet CHO \qquad \text{minor propagation}$$
$$CH_3CH^\bullet CHO \rightarrow CH_3CH_2^\bullet + CO \qquad \text{minor propagation}$$
$$CH_3CH_2^\bullet + {}^\bullet CH_2CH_2CHO \rightarrow CH_3CH_2CH_2CH_2CHO \qquad \text{trace termination}$$
$$H^\bullet + CH_3CH_2CHO \rightarrow H_2 + CH_3CH_2CO^\bullet \qquad \text{very trace propagation}$$
$$\rightarrow H_2 + {}^\bullet CH_2CH_3CHO \qquad \text{very trace propagation}$$
$$\rightarrow H_2 + CH_3CH^\bullet CHO \qquad \text{very trace propagation}$$

(m) Methods of detection:
products, chromatography, spectroscopy, mass spectrometry
radicals, esr, spectroscopy, mass spectrometry

3. (a) Chain carriers are $I^\bullet$, $C_2H_4I^\bullet$

$$\frac{d[I^\bullet]}{dt} = k_1[C_2H_4I_2] - k_2[I^\bullet][C_2H_4I_2] + k_3[C_2H_4I^\bullet] - 2k_4[I^\bullet]^2[M] = 0 \qquad (I)$$

$$\frac{d[C_2H_4I^\bullet]}{dt} = k_1[C_2H_4I_2] + k_2[I^\bullet][C_2H_4I_2] - k_3[C_2H_4I^\bullet] = 0 \qquad (II)$$

Add (I) and (II):

$$2k_1[C_2H_4I_2] = 2k_4[I^\bullet]^2[M]$$

$$[I^\bullet] = \left(\frac{k_1}{k_4}\right)^{1/2}\frac{[C_2H_4I_2]^{1/2}}{[M]^{1/2}} \qquad (III)$$

Choose to express the rate in terms of $[I^\bullet]$

$$-\frac{d[C_2H_4I_2]}{dt} = k_1[C_2H_4I_2] + k_2[C_2H_4I_2][I^\bullet]$$

$$\approx k_2[C_2H_4I_2][I^\bullet]$$

(i.e. by using the long chains approximation rate of initiation $\ll$ rate of propagation)

$$= k_2 \left(\frac{k_1}{k_4}\right)^{1/2} \frac{[C_2H_4I_2]^{3/2}}{[M]^{1/2}}$$

or

$$+\frac{d[I_2]}{dt} = k_2[C_2H_4I_2][I^\bullet] + k_4[I^\bullet]^2[M]$$

$$\approx k_2[C_2H_4I_2][I^\bullet]$$

(i.e. by using the long chains approximation rate of termination $\ll$ rate of propagation)

$$= k_2 \left(\frac{k_1}{k_4}\right)^{1/2} \frac{[C_2H_4I_2]^{3/2}}{[M]^{1/2}}$$

It would not be sensible to express the overall rate in terms of

$$+\frac{d[C_2H_4]}{dt} = k_3[C_2H_4I^\bullet]$$

as this requires a knowledge of the as yet unknown $[C_2H_4I^\bullet]$, which would have to be found by substituting into one or other of the steady state equations and solving for $[C_2H_4I^\bullet]$, or by using the long chains approximation in the form that the rates of the two propagation steps are equal.

$$k_2[I^\bullet][C_2H_4I_2] = k_3[C_2H_4I^\bullet] \quad \therefore \quad [I^\bullet] = \frac{k_3[C_2H_4I^\bullet]}{k_2[C_2H_4I_2]} \qquad \text{(IV)}$$

But

$$[I^\bullet] = \left(\frac{k_1}{k_4}\right)^{1/2} \frac{[C_2H_4I_2]^{1/2}}{[M]^{1/2}} \qquad \text{(III)}$$

Solving the simultaneous equations (III) and (IV) gives

$$[C_2H_4I^\bullet] = \frac{k_2}{k_3}\left(\frac{k_1}{k_4}\right)^{1/2} \frac{[C_2H_4I_2]^{3/2}}{[M]^{1/2}}$$

$$+\frac{d[C_2H_4]}{dt} = k_3[C_2H_4I^\bullet] = k_2\left(\frac{k_1}{k_4}\right)^{1/2} \frac{[C_2H_4I_2]^{3/2}}{[M]^{1/2}}$$

as previously.

(b) Chain carriers: $C_4H_9^\bullet$, $C_4H_8I^\bullet$, $I^\bullet$
Note: this is a three-stage propagation reaction, but can still be handled in the standard manner.

$$+\frac{d[C_4H_9^\bullet]}{dt} = k_1[C_4H_9I] - k_2[C_4H_9^\bullet][C_4H_9I] + k_4[I^\bullet][C_4H_9I] = 0 \quad (I)$$

$$+\frac{d[I^\bullet]}{dt} = k_1[C_4H_9I] + k_3[C_4H_8I^\bullet] - k_4[I^\bullet][C_4H_9I] - 2k_5[I^\bullet]^2[M] = 0 \quad (II)$$

$$+\frac{d[C_4H_8I^\bullet]}{dt} = k_2[C_4H_9^\bullet][C_4H_9I] - k_3[C_4H_8I^\bullet] = 0 \quad (III)$$

Add (I), (II) and (III). There are three steady state equations because there are three chain carriers.

$$2k_1[C_4H_9I] = 2k_5[I^\bullet]^2[M] \quad (IV)$$

$$[I^\bullet] = \left(\frac{k_1}{k_5}\right)^{1/2}\frac{[C_4H_9I]^{1/2}}{[M]^{1/2}}$$

The overall rate can be expressed in terms of

removal of C_4H_9I; this requires $[C_4H_9^\bullet]$ and $[I^\bullet]$;
formation of I_2; this requires $[I^\bullet]$;
production of C_4H_{10}; this requires $[C_4H_9^\bullet]$;
production of C_4H_8; this requires $[C_4H_8I^\bullet]$.

The simplest procedure is to choose the rate which requires $[I^\bullet]$, since it is already known.

$$+\frac{d[I_2]}{dt} = k_4[I^\bullet][C_4H_9I] + k_5[I^\bullet]^2[M]$$
$$\approx k_4[I^\bullet][C_4H_9I]$$

(by the long chains approximation rate of termination $\ll$ rate of propagation)

$$= k_4\left(\frac{k_1}{k_5}\right)^{1/2}\frac{[C_4H_9I]^{3/2}}{[M]^{1/2}}$$

(i) could substitute for $[I^\bullet]$ into (I), and find $[C_4H_9^\bullet]$, thence $+d[C_4H_{10}]/dt$
(ii) could substitute for $[C_4H_9^\bullet]$ into (III), and find $[C_4H_8I^\bullet]$, thence $+d[C_4H_8]/dt$
(iii) having found both $[I^\bullet]$ and $[C_4H_9^\bullet]$, thence $-d[C_4H_9I]/dt$. This would be a three term rate expression involving k_1, k_2 and k_4. The long chains approximation rate of initiation $\ll$ rate of propagation would eliminate the term in k_1.

4. The essential first step here is to recognize the chain carriers and the propagation steps. There are three radicals, $CH_3^\bullet$, $CH_3CO^\bullet$ and $^\bullet CH_2COCH_3$. Two of these, $CH_3^\bullet$ and $^\bullet CH_2COCH_3$, are recycled in steps 3 and 4, which are therefore the propagation steps and produce the major products, CH_4 and CH_2CO. These radicals are removed from the chain in step 5, which is thus a termination step, producing the minor product, $CH_3CH_2COCH_3$. $CH_3^\bullet$ and $CH_3CO^\bullet$ are first produced in step 1, which is thus an initiation step. $CH_3CO^\bullet$ breaks down to give $CH_3^\bullet$, which is a chain carrier, and so this is also initiation. This mechanism thus involves a two-stage initiation. The breakdown of $CH_3CO^\bullet$ also gives CO. Since $CH_3CO^\bullet$ is not recycled, the CO will only be present in trace amounts.

Because $CH_3CO^\bullet$ generates a chain carrier it must be included in the steady state analysis.

$$+\frac{d[CH_3^\bullet]}{dt} = k_1[CH_3COCH_3] + k_2[CH_3CO^\bullet] - k_3[CH_3^\bullet][CH_3COCH_3] + k_4[^\bullet CH_2COCH_3] - k_5[CH_3^\bullet][^\bullet CH_2COCH_3] = 0 \qquad \text{(I)}$$

$$+\frac{d[CH_3CO^\bullet]}{dt} = k_1[CH_3COCH_3] - k_2[CH_3CO^\bullet] = 0 \qquad \text{(II)}$$

$$+\frac{d[^\bullet CH_2COCH_3]}{dt} = k_3[CH_3^\bullet][CH_3COCH_3] - k_4[^\bullet CH_2COCH_3] - k_5[CH_3^\bullet][^\bullet CH_2COCH_3] = 0 \qquad \text{(III)}$$

Adding (I), (II) and (III):

$$2k_1[CH_3COCH_3] = 2k_5[CH_3^\bullet][^\bullet CH_2COCH_3] \qquad \text{(IV)}$$

This is the algebraic form of the steady state assumption, *viz.*, the total rate of initiation is balanced by the total rate of termination.

Equation (IV) contains two unknowns, and so cannot be solved. A second independent equation in these two unknowns is required.

Since the chain carriers are not involved in any steps other than initiation, propagation and termination, the long chains assumption can be used in the form of 'the rates of the two propagation steps are equal'.

$$k_3[CH_3^\bullet][CH_3COCH_3] = k_4[^\bullet CH_2COCH_3] \qquad \text{(V)}$$

$$\text{(IV) gives } [CH_3^\bullet] = \frac{k_1[CH_3COCH_3]}{k_5[^\bullet CH_2COCH_3]}$$

$$\text{(V) gives } [CH_3^\bullet] = \frac{k_4[^\bullet CH_2COCH_3]}{k_3[CH_3COCH_3]}$$

$$\therefore \quad [^\bullet CH_2COCH_3] = \left(\frac{k_1k_3}{k_4k_5}\right)^{1/2}[CH_3COCH_3] \qquad \text{(VI)}$$

Express the rate of reaction in terms of a rate expression for removal of reactant or formation of a major product. It is sensible to chose one that involves $[^{\bullet}CH_2COCH_3]$:

$$+\frac{d[CH_2CO]}{dt} = k_4[^{\bullet}CH_2COCH_3]$$

$$= \left(\frac{k_1\,k_3\,k_4}{k_5}\right)^{1/2}[CH_3COCH_3]$$

Reaction is first order in reactant and $k_{obs} = (k_1\,k_3\,k_4/k_5)^{1/2}$.
Note: the rate of reaction in terms of removal of reactant or production of the other major product CH_4 requires $[CH_3^{\bullet}]$.

The Rice–Herzfeld rules state the following.

Recombination of like radicals used up in the first propagation step gives overall 3/2 order in reactant.

Recombination of like radicals produced in the first propagation step gives overall 1/2 order in reactant.

5. With this mechanism the main difficulty lies in figuring out what is going on.

Initiation produces $CH_3^{\bullet}$ and $C_2H_5^{\bullet}$. $CH_3^{\bullet}$ is undoubtedly a chain carrier: it is used up in step 2, and in doing so produces $C_3H_7^{\bullet}$, and it is regenerated in step 4, which uses up $C_3H_7^{\bullet}$. Therefore, $C_3H_7^{\bullet}$ is also a chain carrier.

$C_2H_5^{\bullet}$ is first produced in initiation, and it then is used up in step 3, which produces $C_3H_7^{\bullet}$. $C_3H_7^{\bullet}$ then forms $CH_3^{\bullet}$, which participates in the $CH_3^{\bullet}/C_3H_7^{\bullet}$ chain. But $C_2H_5^{\bullet}$ is *not* regenerated and is, therefore, not a chain carrier. But, because it produces a chain carrier, it must be included in the steady state equations. Had it only been produced in initiation and never appeared again, it could have been ignored. It would then have eventually produced a very trace amount of a minor product by some reaction not associated with the chain. But because it produces a chain carrier it cannot be ignored, and a steady state equation must be produced for it. However, it must *not* be regarded as a chain carrier. Chain carriers are $CH_3^{\bullet}$ and $C_3H_7^{\bullet}$.

$$+\frac{d[CH_3^{\bullet}]}{dt} = k_1[C_3H_8] - k_2[CH_3^{\bullet}][C_3H_8] + k_4[C_3H_7^{\bullet}] = 0 \qquad \text{(I)}$$

$$+\frac{d[C_2H_5^{\bullet}]}{dt} = k_1[C_3H_8] - k_3[C_2H_5^{\bullet}][C_3H_8] = 0 \qquad \text{(II)}$$

$$\frac{d[C_3H_7^{\bullet}]}{dt} = k_2[CH_3^{\bullet}][C_3H_8] + k_3[C_2H_5^{\bullet}][C_3H_8] - k_4[C_3H_7^{\bullet}] - 2k_5[C_3H_7^{\bullet}]^2 - 2k_6[C_3H_7^{\bullet}]^2 = 0 \qquad \text{(III)}$$

Add (I), (II) and (III):

$$2k_1[C_3H_8] = 2(k_5 + k_6)[C_3H_7^\bullet]^2 = 0$$

$$\therefore \quad [C_3H_7^\bullet] = \left(\frac{k_1}{k_5 + k_6}\right)^{1/2}[C_3H_8]^{1/2}$$

The overall rate of reaction can be expressed in terms of *removal of reactant* C_3H_8: this requires $[CH_3^\bullet]$ and $[C_2H_5^\bullet]$.

$[C_2H_5^\bullet]$ appears in step 3 and this term will drop out from the overall rate because it is not a propagation step, and although it produces $C_3H_7^\bullet$ this step is not essential for the continuance of the chain. Rate of propagation $\gg$ rate of step 3. This will leave a two-term equation, which can be approximated by the long chains approximation, since rate of initiation $\ll$ rate of propagation.

$$-\frac{d[C_3H_8]}{dt} = k_1[C_3H_8] + k_2[CH_3^\bullet][C_3H_8] + k_3[C_2H_5^\bullet][C_3H_8]$$

With the above approximations, this reduces to

$$-\frac{d[C_3H_8]}{dt} = k_2[CH_3^\bullet][C_3H_8]$$

$[CH_3^\bullet]$ has to be found by substituting $[C_3H_7^\bullet] = (k_1/(k_5 + k_6))^{1/2}[C_3H_8]^{1/2}$ into equation (I)

$$+\frac{d[CH_3^\bullet]}{dt} = k_1[C_3H_8] - k_2[CH_3^\bullet][C_3H_8] + k_4\left(\frac{k_1}{k_5 + k_6}\right)^{1/2}[C_3H_8]^{1/2} = 0$$

$$\therefore \quad k_2[CH_3^\bullet][C_3H_8] = k_1[C_3H_8] + k_4\left(\frac{k_1}{k_5 + k_6}\right)^{1/2}[C_3H_8]^{1/2}$$

$$\therefore \quad [CH_3^\bullet] = \frac{k_1}{k_2} + \frac{k_4}{k_2}\left(\frac{k_1}{k_5 + k_6}\right)^{1/2}[C_3H_8]^{-1/2}$$

$$-\frac{d[C_3H_8]}{dt} = k_2[CH_3^\bullet][C_3H_8]$$

$$= k_1[C_3H_8] + k_4\left(\frac{k_1}{k_5 + k_6}\right)^{1/2}[C_3H_8]^{1/2}$$

Because of long chains

$$k_1[C_3H_8] \ll k_4\left(\frac{k_1}{k_5 + k_6}\right)^{1/2}[C_3H_8]^{1/2}$$

$$\therefore \quad \frac{d[C_3H_8]}{dt} = k_4\left(\frac{k_1}{k_5 + k_6}\right)^{1/2}[C_3H_8]^{1/2}$$

If the simpler procedure of looking at the rate of production of the major product, $[C_2H_4]$, is used

$$+\frac{d[C_2H_4]}{dt} = k_4[C_3H_7^{\bullet}]$$

and by substituting for $[C_3H_7^{\bullet}]$

$$= k_4\left(\frac{k_1}{k_5 + k_6}\right)^{1/2}[C_3H_8]^{1/2}$$

Note. The overall rate of reaction must not be expressed in terms of production of C_2H_6 as this is only a minor side reaction and not a rate of propagation.

$$k_{obs} = k_4\left(\frac{k_1}{k_5 + k_6}\right)^{1/2}$$

6. Reaction without inhibition:

$$+\frac{d[Br^{\bullet}]}{dt} = 2k_1[Br_2][M] - k_2[Br^{\bullet}][RH] + k_3[R^{\bullet}][Br]_2 - 2k_4[Br^{\bullet}]^2[M]$$

$$= 0 \qquad \text{(I)}$$

$$+\frac{d[R^{\bullet}]}{dt} = k_2[Br^{\bullet}][RH] - k_3[R^{\bullet}][Br_2] = 0 \qquad \text{(II)}$$

Add (I) and (II):

$$2k_1[Br_2][M] = 2k_4[Br^{\bullet}]^2[M]$$

$$[Br^{\bullet}] = \left(\frac{k_1}{k_4}\right)^{1/2}[Br_2]^{1/2} \qquad \text{(III)}$$

$$+\frac{d[HBr]}{dt} = k_2[Br^{\bullet}][RH]$$

$$= k_2\left(\frac{k_1}{k_4}\right)^{1/2}[Br_2]^{1/2}[RH] \qquad \text{(IV)}$$

Note in this mechanism that Equation (II) implies that 'the rates of the two propagation steps are equal'.
Reaction with inhibition:

$$+\frac{d[Br^{\bullet}]}{dt} = 2k_1[Br_2][M] - k_2[Br^{\bullet}][RH] + k_{-2}[R^{\bullet}][HBr]$$

$$+ k_3[R^{\bullet}][Br_2] - 2k_4[Br^{\bullet}]^2[M] = 0 \qquad \text{(V)}$$

$$+\frac{d[R^{\bullet}]}{dt} = k_2[Br^{\bullet}][RH] - k_{-2}[R^{\bullet}][HBr] - k_3[R^{\bullet}][Br_2] = 0 \qquad \text{(VI)}$$

Add (V) and (VI):

$$2k_1[\mathrm{Br_2}][\mathrm{M}] = 2k_4[\mathrm{Br^\bullet}]^2[\mathrm{M}]$$

$$[\mathrm{Br^\bullet}] = \left(\frac{k_1}{k_4}\right)^{1/2}[\mathrm{Br_2}]^{1/2} \qquad \text{(VII)}$$

Note: this equation is identical to the corresponding equation for the mechanism without inhibition. This is as it should be since addition of the steady state equations should generate the algebraic statement that 'the total rate of formation of the chain carriers should be balanced by the total rate of removal of the chain carriers'.

This should hold whether there is inhibition or not.

However, equation (VI) differs from equation (II). Here the rate term for the inhibition step is included, so that the rates of the two propagation steps are no longer equal. The long chains approximation in this form cannot be used for the inhibition mechanism, though it is valid for the reaction without inhibition.

$$+\frac{\mathrm{d[HBr]}}{\mathrm{d}t} = k_2[\mathrm{Br^\bullet}][\mathrm{RH}] - k_{-2}[\mathrm{R^\bullet}][\mathrm{HBr}]$$

$$= k_2\left(\frac{k_1}{k_4}\right)^{1/2}[\mathrm{Br_2}]^{1/2}[\mathrm{RH}] - k_{-2}[\mathrm{R^\bullet}][\mathrm{HBr}]$$

From equation (VI),

$$k_2\left(\frac{k_1}{k_4}\right)^{1/2}[\mathrm{Br_2}]^{1/2}[\mathrm{RH}] = [\mathrm{R^\bullet}]\{k_{-2}[\mathrm{HBr}] + k_3[\mathrm{Br_2}]\}$$

$$[\mathrm{R^\bullet}] = k_2\left(\frac{k_1}{k_4}\right)^{1/2}\left(\frac{[\mathrm{Br_2}]^{1/2}[\mathrm{RH}]}{k_{-2}[\mathrm{HBr}] + k_3[\mathrm{Br_2}]}\right)$$

$$+\frac{\mathrm{d[HBr]}}{\mathrm{d}t} = k_2\left(\frac{k_1}{k_4}\right)^{1/2}[\mathrm{Br_2}]^{1/2}[\mathrm{RH}]$$

$$- k_{-2}[\mathrm{HBr}]k_2\left(\frac{k_1}{k_4}\right)^{1/2}\left(\frac{[\mathrm{Br_2}]^{1/2}[\mathrm{RH}]}{k_{-2}[\mathrm{HBr}] + k_3[\mathrm{Br_2}]}\right)$$

$$= k_2\left(\frac{k_1}{k_4}\right)^{1/2}[\mathrm{Br_2}]^{1/2}[\mathrm{RH}]$$

$$- k_{-2}k_2\left(\frac{k_1}{k_4}\right)^{1/2}\left(\frac{[\mathrm{Br_2}]^{1/2}[\mathrm{RH}][\mathrm{HBr}]}{k_{-2}[\mathrm{HBr}] + k_3[\mathrm{Br_2}]}\right)$$

$$= k_2\left(\frac{k_1}{k_4}\right)^{1/2}[\mathrm{Br_2}]^{1/2}[\mathrm{RH}]\left\{1 - k_{-2}\left[\frac{1}{k_{-2} + k_3\frac{[\mathrm{Br_2}]}{[\mathrm{HBr}]}}\right]\right\}$$

Note: the presence of the inhibition step dramatically complicates the rate expression.

7. Step 1 is the unimolecular decomposition of a diatomic molecule, and step -1 is a recombination of two atoms. This recombination is a highly exothermic reaction, and this energy will reside in the newly formed bond, which is thus highly excited. If this energy is not removed in the time of one vibration, the molecule will split up on vibration and recombination will not occur. If a collision occurs before the first vibration, then energy transfer takes place and the bond will be stabilized. The third body fulfils this function. This is more effective at high pressures and low temperatures.

   The decomposition of the diatomic molecule is the reverse of the recombination step, and so will also require a third body. The physical process involved is transfer of energy by collision to form an activated molecule which has enough energy to react. This is more efficient at high pressures and low temperatures.

   Steps 4 and 5 are recombinations of large radicals. Here there are sufficient normal modes of vibration for the newly formed bond to be stabilized by transfer of the excess energy around the molecule via the normal modes of vibration. The product in step 4 has 36 normal modes, while the product in step 5 has 18. It is only at very low pressures and very high temperatures that 18 normal modes would be insufficient to stabilize the C—C bond that is formed.

$$OH^{\bullet} + H^{\bullet} \rightarrow H_2O$$

   The product has three normal modes only, and a third body is necessary under all normal conditions.

$$CH_3^{\bullet} + CH_3^{\bullet} \rightarrow C_2H_6$$

   The product has 18 normal modes, and for ordinary pressures and temperatures these will be sufficient to stabilize the C—C bond. This may not be the case at very low pressures and very high temperatures; compare with $Cl^{\bullet} + C_2H_4Cl^{\bullet}$.

$$H^{\bullet} + I^{\bullet} \rightarrow HI$$

   The product is a diatomic molecule, and a third body is necessary under all conditions; compare with $Cl^{\bullet} + Cl^{\bullet}$

$$CH_3^{\bullet} + (CH_3)_3C^{\bullet} \rightarrow (CH_3)_3CCH_3$$

   The product has 45 normal modes of vibration, and under all conditions a third body will not be required. It is unlikely that there is a range of pressures and temperatures for which this number of normal modes would be insufficient for efficient energy transfer around the molecule.

8. This is a photochemical reaction, hence the initiation step is likely to be splitting of Cl_2 into two $Cl^{\bullet}$ atoms.

$$Cl_2 + h\nu \rightarrow 2\,Cl^{\bullet}$$

There is no need for a third body to supply activation energy since this is achieved by the radiation. It is also a chain reaction as is shown by the quantum yield of >100. Only one product is observed, and this must be formed in propagation. It has two Cl added and these must be added one at a time to fit a propagation step. This suggests that a $Cl^\bullet$ must add in the first propagation step. There are no H atoms in the reactant so this first propagation step cannot be a H abstraction reaction. Since only one product is formed there can be no major product formed in the first propagation step. All of this suggests

$$Cl^\bullet + C_6F_{10} \rightarrow C_6F_{10}Cl^\bullet$$

as the first propagation step.

The second propagation step must add a further Cl, regenerate $Cl^\bullet$ and form the product:

$$Cl_2 + C_6F_{10}Cl^\bullet \rightarrow C_6F_{10}Cl_2 + Cl^\bullet$$

The termination step must produce no other product, and possible typical steps are radical/radical or atom/atom recombinations.

- $Cl^\bullet + Cl^\bullet \rightarrow Cl_2$, possible since Cl_2 is a reactant and would not be detected as a minor product.
- $C_6F_{10}Cl^\bullet + Cl^\bullet \rightarrow C_6F_{10}Cl_2$. This is the major product produced in propagation, and would not be detected as a minor product.
  It is now necessary to see whether either of these possible mechanisms fit the observed kinetics. This requires a steady state treatment on both mechanisms to find the predicted rate expressions to compare with the observed expression. The observed kinetics are

$$-\frac{d[C_6F_{10}]}{dt} = k_{obs}I_{abs}^{1/2}[Cl_2]$$

- Mechanism 1.

$$Cl_2 + h\nu \rightarrow 2\,Cl^\bullet$$

$$Cl^\bullet + C_6F_{10} \xrightarrow{k_2} C_6F_{10}Cl^\bullet$$

$$Cl_2 + C_6F_{10}Cl^\bullet \xrightarrow{k_3} C_6F_{10}Cl_2 + Cl^\bullet$$

$$Cl^\bullet + Cl^\bullet + M \xrightarrow{k_4} Cl_2 + M$$

- Mechanism 2.

$$Cl_2 + h\nu \rightarrow 2\,Cl^\bullet$$

$$Cl^\bullet + C_6F_{10} \xrightarrow{k_2} C_6F_{10}Cl^\bullet$$

$$Cl_2 + C_6F_{10}Cl^\bullet \xrightarrow{k_3} C_6F_{10}Cl_2 + Cl^\bullet$$

$$C_6F_{10}Cl^\bullet + Cl^\bullet \xrightarrow{k_5} C_6F_{10}Cl_2$$

- Steady state treatment on mechanism 1.

$$\frac{d[Cl^\bullet]}{dt} = 2I_{abs} - k_2[Cl^\bullet][C_6F_{10}] + k_3[C_6F_{10}Cl^\bullet][Cl_2] - 2k_4[Cl^\bullet]^2[M] = 0$$

$$\frac{d[C_6F_{10}Cl^\bullet]}{dt} = k_2[Cl^\bullet][C_6F_{10}] - k_3[C_6F_{10}Cl^\bullet][Cl_2] = 0$$

Adding the steady state equations gives

$$2I_{abs} = 2k_4[Cl^\bullet]^2[M]$$

$$\therefore \quad [Cl^\bullet] = \left(\frac{I_{abs}}{k_4}\right)^{1/2}[M]^{-1/2}$$

$$-\frac{d[C_6F_{10}]}{dt} = k_2[Cl^\bullet][C_6F_{10}] = k_2\left(\frac{I_{abs}}{k_4}\right)^{1/2}[M]^{-1/2}[C_6F_{10}]$$

This does not fit the observed kinetics, and so mechanism 1 is not correct.

- Steady state treatment on mechanism 2.

$$\frac{d[Cl^\bullet]}{dt} = 2I_{abs} - k_2[Cl^\bullet][C_6F_{10}] + k_3[C_6F_{10}Cl^\bullet][Cl_2] - k_5[Cl^\bullet][C_6F_{10}Cl^\bullet] = 0$$

$$\frac{d[C_6F_{10}Cl^\bullet]}{dt} = k_2[Cl^\bullet][C_6F_{10}] - k_3[C_6F_{10}Cl^\bullet][Cl_2] - k_5[Cl^\bullet][C_6F_{10}Cl^\bullet] = 0$$

Adding the steady state equations gives

$$2I_{abs} = 2k_5[Cl^\bullet][C_6F_{10}Cl^\bullet]$$

$$\therefore \quad [Cl^\bullet] = \frac{I_{abs}}{k_5[C_6F_{10}Cl^\bullet]}$$

Assuming long chains, the rates of the two propagation steps are equal:

$$k_2[Cl^\bullet][C_6F_{10}] = k_3[C_6F_{10}Cl^\bullet][Cl_2]$$

$$[Cl^\bullet] = \frac{k_3[C_6F_{10}Cl^\bullet][Cl_2]}{k_2[C_6F_{10}]}$$

Taking the two expressions for $[Cl^\bullet]$ gives

$$[C_6F_{10}Cl^\bullet] = \left(\frac{k_2 I_{abs}}{k_3 k_5}\right)^{1/2}\left(\frac{[C_6F_{10}]}{[Cl_2]}\right)^{1/2}$$

Expressing the rate in a form which involves this radical concentration,

$$-\frac{d[Cl_2]}{dt} = I_{abs} + k_3[C_6F_{10}Cl^\bullet][Cl_2] \approx k_3[C_6F_{10}Cl^\bullet][Cl_2]$$

$$= \left(\frac{k_2 k_3 I_{abs}}{k_5}\right)^{1/2}[C_6F_{10}]^{1/2}[Cl_2]^{1/2}$$

The term in I drops out because of the long chains approximation.
This expression does not fit the observed kinetics, and mechanism 2 is not correct.

Neither of the two possible termination steps fit the observed kinetics. The third possibility would give a product that was not detected:

$$C_6F_{10}Cl^\bullet + C_6F_{10}Cl^\bullet \rightarrow C_{12}F_{20}Cl_2$$

However, although recombination steps are very common termination steps, disproportionation reactions are also found.

$$C_6F_{10}Cl^\bullet + C_6F_{10}Cl^\bullet \xrightarrow{k_6} C_6F_{10} + C_6F_{10}Cl_2$$

This would give as minor products the reactant and the major product with the following mechanism.

- Mechanism 3.

$$Cl_2 + h\nu \rightarrow 2\,Cl^\bullet$$

$$Cl^\bullet + C_6F_{10} \xrightarrow{k_2} C_6F_{10}Cl^\bullet$$

$$Cl_2 + C_6F_{10}Cl^\bullet \xrightarrow{k_3} C_6F_{10}Cl_2 + Cl^\bullet$$

$$C_6F_{10}Cl^\bullet + C_6F_{10}Cl^\bullet \xrightarrow{k_6} C_6F_{10}Cl_2 + C_6F_{10}$$

- Steady state treatment on mechanism 3.

$$\frac{d[Cl^\bullet]}{dt} = 2\,I_{abs} - k_2[Cl^\bullet][C_6F_{10}] + k_3[C_6F_{10}Cl^\bullet][Cl_2] = 0$$

$$\frac{d[C_6F_{10}Cl^\bullet]}{dt} = k_2[Cl^\bullet][C_6F_{10}] - k_3[C_6F_{10}Cl^\bullet][Cl_2] - 2\,k_6[C_6F_{10}Cl^\bullet]^2 = 0$$

Adding the steady state equations gives

$$2\,I_{abs} = 2\,k_6[C_6F_{10}Cl^\bullet]^2$$

$$\therefore \quad [C_6F_{10}Cl^\bullet] = \left(\frac{I_{abs}}{k_6}\right)^{1/2}$$

$$-\frac{d[Cl_2]}{dt} = I_{abs} + k_3[C_6F_{10}Cl^\bullet][Cl_2] \approx k_3[C_6F_{10}Cl^\bullet][Cl_2]$$

$$= k_3\left(\frac{I_{abs}}{k_6}\right)^{1/2}[Cl_2]$$

Long chains allows the term in I_{abs} to drop out.

This expression fits the observed kinetics, and so mechanism 3 involving a disproportionation termination step is a possible mechanism for this reaction.

Chapter 7

1. For the reaction

$$A + B + C \rightarrow \text{activated complex}$$

$$k = k_0 \frac{\gamma_A \gamma_B \gamma_C}{\gamma_{\neq}}$$

$$\log_{10} k = \log_{10} k_0 + \log_{10} \gamma_A + \log_{10} \gamma_B + \log_{10} \gamma_C - \log_{10} \gamma_{\neq}$$

where $\log_{10} \gamma_i = -A\, z_i^2 \sqrt{I}$

For the reaction quoted two mechanisms can be written:

- $HE^+ + H_3O^+ \rightarrow$

  $HE^+ + H_3O^+ + SO_4^{2-} \rightarrow$

  and

- $HE^+ + H_3O^+ \rightarrow$

  $HE^+ + HSO_4^- \rightarrow$

For the first mechanism $HE^+ + H_3O^+ + SO_4^{2-} \rightarrow$ an activated complex which is overall neutral.

$$\begin{aligned}\log_{10} k_{SO_4^{2-}} &= \log_{10} k_0 + \log_{10} \gamma_{HE^+} + \log_{10} \gamma_{H_3O^+} + \log_{10} \gamma_{SO_4^{2-}} - \log_{10} \gamma_{\neq} \\ &= \log_{10} k_0 - A \times 1^2 \times \sqrt{I} - A \times 1^2 \times \sqrt{I} - A \times 2^2 \times \sqrt{I} \\ &\quad + A \times 0^2 \times \sqrt{I} \\ &= \log_{10} k_0 - 6A\sqrt{I}\end{aligned}$$

For the second mechanism $HE^+ + HSO_4^- \rightarrow$ an activated complex which is overall neutral.

$$\begin{aligned}\log_{10} k_{HSO_4^-} &= \log_{10} k_0 + \log_{10} \gamma_{HE^+} + \log_{10} \gamma_{HSO_4^-} - \log_{10} \gamma_{\neq} \\ &= \log_{10} k_0 - A \times 1^2 \times \sqrt{I} - A \times 1^2 \times \sqrt{I} + A \times 0^2 \times \sqrt{I} \\ &= \log_{10} k_0 - 2A\sqrt{I}\end{aligned}$$

The ionic strength dependencies for both are different, but this does not mean that the two mechanisms can be distinguished on this basis. The differences are merely a reflection of the different ways of expressing the overall observed rate constant. These are kinetically equivalent, and so the ionic strength dependencies are merely a reflection of the two ways of expressing the catalytic contribution to the overall rate.

2.

$$k = \frac{\kappa k T}{h} K^{\neq *}$$

$$\log_e k = \log_e \frac{\kappa k T}{h} + \log_e K^{\neq *}$$

$$= \log_e \frac{\kappa k T}{h} - \frac{\Delta G^{\neq *}}{RT}$$

$$\left(\frac{\partial \log_e k}{\partial p}\right)_T = -\frac{1}{RT}\left(\frac{\partial \Delta G^{\neq *}}{\partial p}\right)_T$$

$$= -\frac{\Delta V^{\neq *}}{RT}$$

A plot of $\log_e k$ versus p should have slope $= -\Delta V^{\neq *}/RT$.

$\frac{10^{-5} p}{\text{N m}^{-2}}$	0.98	2000	4000	6000
$\frac{10^5 k}{\text{mol}^{-1}\,\text{dm}^3\,\text{s}^{-1}}$	6.9	10.3	15.4	22.0
$\log_e\left(\frac{k}{\text{mol}^{-1}\,\text{dm}^3\,\text{s}^{-1}}\right)$	−9.58	−9.18	−8.78	−8.42

The graph of $\log_e k$ versus p is linear, and has a slope $= +1.98 \times 10^{-9}\,\text{m}^2\,\text{N}^{-1}$

$$\text{slope} = -\frac{\Delta V^{\neq *}}{RT}$$

$$\therefore \quad \Delta V^{\neq *} = -8.3145\,\text{J mol}^{-1}\,\text{K}^{-1} \times 333.15\,\text{K} \times 1.98 \times 10^{-9}\,\text{m}^2\,\text{N}^{-1}$$

$$= -5.5 \times 10^{-6}\,\text{m}^2\,\text{N}^{-1}\,\text{J mol}^{-1}$$

$$= -5.5 \times 10^{-6}\,\text{m}^2\,\text{N}^{-1}\,\text{N m mol}^{-1}$$

$$= -5.5 \times 10^{-6}\,\text{m}^3\,\text{mol}^{-1}$$

$$= -5.5\,\text{cm}^3\,\text{mol}^{-1}$$

In this reaction, the charge on the activated complex is greater than the charge on either of the reactants. Solvent molecules will become more tightly bound on forming the activated complex, giving an overall decrease in the volume on activation. This is consistent with the experimental observation.

3.
$$k = \frac{\kappa k T}{h} K^{\neq *} = \frac{\kappa k T}{h} \exp\left(\frac{-\Delta H^{\neq *}}{RT}\right) \exp\left(\frac{+\Delta S^{\neq *}}{R}\right)$$

$$\log_e \frac{k}{T} = \log_e \frac{\kappa k}{h} - \frac{\Delta H^{\neq *}}{RT} + \frac{\Delta S^{\neq *}}{R}$$

A plot of $\log_e(k/T)$ versus $1/T$ should be linear with slope $= -\Delta H^{\neq *}/R$. The intercept should be $\log_e(\kappa k/h) + \Delta S^{\neq *}/R$. But because this would involve a ridiculous extrapolation, $\Delta S^{\neq *}$ will have to be calculated.

$\frac{T}{\text{K}}$	313.15	323.15	333.15	343.15
$\frac{10^3 \times 1/T}{\text{K}^{-1}}$	3.193	3.095	3.002	2.914
$\frac{10^6(k/T)}{\text{mol}^{-1}\,\text{dm}^3\,\text{s}^{-1}\,\text{K}^{-1}}$	1.92	3.62	6.45	10.64
$\log_e \frac{(k/T)}{\text{mol}^{-1}\,\text{dm}^3\,\text{s}^{-1}\text{K}^{-1}}$	−13.17	−12.53	−11.95	−11.45

The graph of $\log_e(k/T)$ versus $1/T$ is linear with slope $= -6.23 \times 10^3$ K.

$$\text{slope} = -\frac{\Delta H^{\neq *}}{R}$$

$$\therefore \quad \Delta H^{\neq *} = +8.3145 \times 6.23 \times 10^3\ \text{J mol}^{-1}$$
$$= +51\,800\ \text{J mol}^{-1}$$
$$= +51.8\ \text{kJ mol}^{-1}$$

When $T = 333.15$ K, $k = 2.15 \times 10^{-3}\ \text{mol}^{-1}\,\text{dm}^3\,\text{s}^{-1}$

$$\frac{k}{T} = 6.45 \times 10^{-6}\ \text{mol}^{-1}\,\text{dm}^3\,\text{s}^{-1}\,\text{K}^{-1} \text{ and } \log_e \frac{k}{T} = -11.95$$

$$\frac{\kappa k}{h} = 2.08 \times 10^{10}\ \text{s}^{-1}\,\text{K}^{-1} \text{ and } \log_e\left(\frac{\kappa k}{h}\right) = 23.76$$

$$\log_e\left(\frac{k}{T}\right) = \log_e \frac{\kappa k}{h} - \frac{\Delta H^{\neq *}}{RT} + \frac{\Delta S^{\neq *}}{R}$$

$$\frac{\Delta S^{\neq *}}{R} = \log_e\left(\frac{k}{T}\right) - \log_e \frac{\kappa k}{h} + \frac{\Delta H^{\neq *}}{RT}$$

$$= -11.95 - 23.76 + \frac{51.8 \times 10^3}{8.3145 \times 333.15}$$

$$\frac{\Delta S^{\neq *}}{R} = -11.95 - 23.76 + 18.70$$

$$= -17.01$$

$$\Delta S^{\neq *} = -17.01 \times 8.3145\ \text{J mol}^{-1}\,\text{K}^{-1}$$

$$= -141\ \text{J mol}^{-1}\,\text{K}^{-1}$$

The large and negative value for $\Delta S^{\neq *}$ for this reaction suggests a considerable degree of charge separation in the activated complex.

4. • At $I = 0.100\,\text{mol}^{-1}\,\text{dm}^3$, the reaction of $CH_2BrCOO^-(aq)$ with $S_2O_3^{2-}(aq)$ has a larger rate constant than that for $PtdienX^+(aq) + Y^-(aq)$.

• $CH_2BrCOO^-(aq) + S_2O_3^{2-}(aq) \rightarrow$

$$z_A z_B = (-1) \times (-2) = 2$$

$$\log_{10} k = \log_{10} k_0 + 2 \times 2 \times 0.510 \frac{\sqrt{0.100}}{1 + \sqrt{0.100}}$$

$$\log_{10} k_0 = \log_{10}(1.07 \times 10^{-2}\,\text{mol}^{-1}\,\text{dm}^3\,\text{s}^{-1}) - 0.490 = -2.461$$

$$k_0 = 3.5 \times 10^{-3}\,\text{mol}^{-1}\,\text{dm}^3\,\text{s}^{-1}$$

• $PtdienX^+(aq) + Y^-(aq) \rightarrow$

$$z_A z_B = 1 \times (-1) = -1$$

$$\log_{10} k = \log_{10} k_0 + 2 \times (-1) \times 0.510 \frac{\sqrt{0.100}}{1 + \sqrt{0.100}}$$

$$\log_{10} k_0 = \log_{10}(3.6 \times 10^{-3}\,\text{mol}^{-1}\,\text{dm}^3\,\text{s}^{-1}) + 0.245 = -2.199$$

$$k_0 = 6.3 \times 10^{-3}\,\text{mol}^{-1}\,\text{dm}^3\,\text{s}^{-1}.$$

• When the rate constants at zero ionic strength are compared, the reaction between $PtdienX^+(aq)$ and $Y^-(aq)$ now has the larger rate constant. If conclusions and inferences were made on the basis of non-ideal rate constants (i.e. non-zero ionic strengths) they would be misleading. This problem shows that the relative order of magnitude for these two reactions changes as the ionic strength increases, and emphasizes the need to use ideal rate constants whenever possible. This is particularly so when the products of the charges on the reactants are different. However, other less specific salt effects, such as charge separation in the activated complex, may also be important even for reactions of the same charge type.

5.

$$k = \frac{\kappa k T}{h} K^{\neq *}$$

$$\log_e k = \log_e \frac{\kappa k T}{h} + \log_e K^{\neq *}$$

$$= \log_e \frac{\kappa k T}{h} - \frac{\Delta G^{\neq *}}{RT}$$

$$\left(\frac{\partial \log_e k}{\partial p}\right)_T = -\frac{1}{RT}\left(\frac{\partial \Delta G^{\neq *}}{\partial p}\right)_T$$

$$= -\frac{\Delta V^{\neq *}}{RT}$$

A plot of $\log_e k$ versus p should have slope $= -\Delta V^{\neq *}/RT$.
Likewise, for the equilibrium constant,

$$\left(\frac{\partial \log_e K}{\partial p}\right)_T = -\frac{\Delta V^\theta}{RT}$$

A plot of $\log_e K$ versus p should have slope $= -\Delta V^\theta/RT$.

$\frac{p}{\text{atm}}$	1	800	1500
$\frac{k_1}{\text{mol}^{-1}\,\text{dm}^3\,\text{s}^{-1}}$	10 000	6900	4800
$\log_e k_1$	9.21	8.84	8.48
$\frac{K_1}{\text{mol}^{-1}\,\text{dm}^3}$	1920	940	510
$\log_e K_1$	7.56	6.85	6.23

The plot of $\log_e k_1$ versus p is linear, with

$$\text{slope} = -4.8 \times 10^{-4}\,\text{atm}^{-1}$$

$$\therefore \quad \Delta V^{\neq *} = +4.8 \times 10^{-4}\,\text{atm}^{-1} \times 8.314\,\text{J}\,\text{mol}^{-1}\,\text{K}^{-1} \times 298\,\text{K}$$

$$= +\frac{4.8 \times 10^{-4} \times 8.314\,\text{J}\,\text{mol}^{-1}\,\text{K}^{-1} \times 298\,\text{K}}{1.013\,25 \times 10^5\,\text{N}\,\text{m}^{-2}}$$

$$= +\frac{4.8 \times 10^{-4} \times 8.314\,\text{N}\,\text{m}\,\text{mol}^{-1}\,\text{K}^{-1} \times 298\,\text{K}}{1.013\,25 \times 10^5\,\text{N}\,\text{m}^{-2}}$$

$$= +1.17 \times 10^{-5}\,\text{m}^3\,\text{mol}^{-1}$$

$$= +11.7\,\text{cm}^3\,\text{mol}^{-1}$$

The plot of $\log_e K_1$ versus p is linear, with

$$\text{slope} = -8.94 \times 10^{-4}\,\text{atm}^{-1}$$

$$\therefore \quad \Delta V^\theta = +\left(\frac{8.94 \times 10^{-4}\,\text{N}^{-1}\,\text{m}^2}{1.013\,25 \times 10^5}\right) \times 8.314\,\text{J}\,\text{mol}^{-1}\,\text{K}^{-1} \times 298\,\text{K}$$

$$= +2.19 \times 10^{-5}\,\text{m}^3\,\text{mol}^{-1}$$

$$= +21.9\,\text{cm}^3\,\text{mol}^{-1}$$

Note: take care with the units as this is probably the most difficult part of the question.

ΔV^θ is positive: the product is less highly charged than the reactants, and is therefore less heavily solvated. Solvent is thus released from its ordered and packed structure, therefore the volume increases.

$\Delta V^{\neq *}$ is positive, but less so than ΔV^θ: the activated complex has a lower charge than the reactants and so the solvent is less packed around it. Solvent is thus released from the solvation sheaths, and therefore the volume increases on forming the activated complex.

The magnitude of $\Delta V^{\neq *}$ is less than that for ΔV^{θ}, indicating that the solvation arrangement and charge neutralization is not so fully developed on forming the activated complex as when the final product is formed. The activated complex does resemble the products somewhat, and is probably half-way between reactants and products.

Note how solvation is dominant here: two species $\rightarrow$ one activated complex $\rightarrow$ one product. On this basis a decrease in volume for both processes would be expected. This is not observed.

6. For reactions 1–3, $\Delta V^{\neq *}$ and ΔV^{θ} are very similar with $\Delta V^{\neq *}$ only slightly less negative than ΔV^{θ}. This suggests that the activated complex is very similar to the products, but not quite as compact.

   For reactions 4 and 5, $\Delta V^{\neq *}$ is more negative than ΔV^{θ}, which suggests an activated complex which resembles the product, but is even more compact than the product. Such possibilities complicate possible interpretations of values of $\Delta V^{\neq *}$. For the reverse reactions

   $\Delta V_b^{\neq *}$ values are +6.1, +6.8, +4.3, −5.0 and $-1.5\,\text{cm}^3\,\text{mol}^{-1}$, and
   ΔV_b^{θ} values are +36.9, +37.0, +37.2 + 36.3 and $+35.5\,\text{cm}^3\,\text{mol}^{-1}$.

   None of the $\Delta V_b^{\neq *}$ values is very large and all are very different from the corresponding ΔV_b^{θ} values. The reverse reactions are $1 \rightarrow 2$, and so the $\Delta V_b^{\neq *}$ values would be expected to be positive and fairly large. This is not the case, which suggests that the activated complexes are fairly similar to the reactant C.

   A similar trend in $\Delta S^{\neq *}$ values for both forward and back reactions would be expected, though the $\Delta S^{\neq *}$ values are complicated by the fact that the complexity of the reactants must be considered.

Chapter 8

1. Mechanistic rate $= k_1[\text{A}][\text{NH}_3] + k_2[\text{A}]$

   Note: there is no $[H_2O]$ in the last term since H_2O is the solvent, and is thus in excess.

   Observed rate $= k_{\text{obs}}[\text{A}]$
   where k_{obs} is a pseudo-first order rate constant, see units in table to confirm.

   $$\therefore \quad k_{\text{obs}} = k_1[\text{NH}_3] + k_2$$

   $\therefore$ k_{obs} versus $[NH_3]$ should be linear, with slope $= k_1$ and intercept $= k_2$.

The graph is linear, confirming the analysis, and gives
$k_1 = \text{slope} = 3.55 \times 10^{-4}\ \text{mol}^{-1}\ \text{dm}^3\ \text{s}^{-1}$, and $k_2 = \text{intercept} = 4.2 \times 10^{-5}\ \text{s}^{-1}$.

2. Mechanistic rate $= k_2[\text{E}]_{\text{actual}}[\text{OH}^-]_{\text{actual}} + k_3[\text{HE}^+]_{\text{actual}}[\text{OH}^-]_{\text{actual}} + k_4[\text{E}]_{\text{actual}}[\text{OH}^-]_{\text{actual}}[\text{M}^{2+}]_{\text{actual}}$

The most sensible way in which to follow this reaction would be at constant pH, and so the observed rate can be expressed as

$$\text{observed rate} = k_{\text{obs}}[\text{E}]_{\text{total}}[\text{OH}^-]_{\text{actual}}$$

with k_{obs} being a second order rate constant as required.

$$k_{\text{obs}}[\text{E}]_{\text{total}}[\text{OH}^-]_{\text{actual}} = k_2[\text{E}]_{\text{actual}}[\text{OH}^-]_{\text{actual}} + k_3[\text{HE}^+]_{\text{actual}}[\text{OH}^-]_{\text{actual}} + k_4[\text{E}]_{\text{actual}}[\text{OH}^-]_{\text{actual}}[\text{M}^{2+}]_{\text{actual}}$$

$$k_{\text{obs}} = k_2\frac{[\text{E}]_{\text{actual}}}{[\text{E}]_{\text{total}}} + k_3\frac{[\text{HE}^+]_{\text{actual}}}{[\text{E}]_{\text{total}}} + k_4[\text{M}^{2+}]\frac{[\text{E}]_{\text{actual}}}{[\text{E}]_{\text{total}}}$$

$$[\text{E}]_{\text{total}} = [\text{E}]_{\text{actual}} + [\text{HE}^+]_{\text{actual}}$$

$$\text{E} + \text{H}_2\text{O} \underset{k_{-1}}{\overset{k_1}{\rightleftharpoons}} \text{HE}^+ + \text{OH}^-$$

$$K_1 = \left(\frac{[\text{HE}^+][\text{OH}^-]}{[\text{E}]}\right)_{\text{equil}}$$

$$[\text{HE}^+]_{\text{actual}} = \frac{K_1[\text{E}]_{\text{actual}}}{[\text{OH}^-]_{\text{actual}}}$$

$$\therefore\quad k_{\text{obs}} = k_2\frac{[\text{E}]_{\text{actual}}}{[\text{E}]_{\text{total}}} + k_3 K_1\frac{[\text{E}]_{\text{actual}}}{[\text{E}]_{\text{total}}[\text{OH}^-]_{\text{actual}}} + k_4[\text{M}^{2+}]\frac{[\text{E}]_{\text{actual}}}{[\text{E}]_{\text{total}}}$$

$\text{p}K_1 = 6.3$, and $K_1 = 5.0 \times 10^{-7}\ \text{mol dm}^{-3}$, hence no extensive protonation will occur, and so

$$[\text{E}]_{\text{total}} \approx [\text{E}]_{\text{actual}} \text{ and } [\text{HE}^+]_{\text{actual}} = \frac{K_1[\text{E}]_{\text{actual}}}{[\text{OH}^-]_{\text{actual}}} \approx \frac{K_1[\text{E}]_{\text{total}}}{[\text{OH}^-]_{\text{actual}}}$$

$$\therefore\quad k_{\text{obs}} = k_2\frac{[\text{E}]_{\text{total}}}{[\text{E}]_{\text{total}}} + k_3 K_1\frac{[\text{E}]_{\text{total}}}{[\text{E}]_{\text{total}}[\text{OH}^-]_{\text{actual}}} + k_4[\text{M}^{2+}]\frac{[\text{E}]_{\text{total}}}{[\text{E}]_{\text{total}}}$$

$$\therefore\quad k_{\text{obs}} = k_2 + k_3\frac{K_1}{[\text{OH}^-]_{\text{actual}}} + k_4[\text{M}^{2+}]_{\text{actual}}$$

k_2 and k_3 can be found from values of k_{obs} found at various pH values, i.e. $[\text{OH}^-]$ in the absence of added M^{2+}.

$$\therefore\quad k_{\text{obs}} = k_2 + k_3\frac{K_1}{[\text{OH}^-]_{\text{actual}}}$$

A plot of k_{obs} versus $1/[OH^-]_{actual}$ should be linear with intercept $= k_2$ and slope $= k_3 K_1$. Since K_1 is known, then k_3 can be found.

k_4 can then be found by following the hydrolysis at one pH, i.e. at one value of $[OH^-]$, but in the presence of varying $[M^{2+}]$.

$$\therefore\ k_{obs} - \left(k_2 + k_3 \frac{K_1}{[OH^-]_{actual}}\right) = k_4[M^{2+}]_{actual}$$

All quantities on the left hand side of this equation can either be measured or calculated. A graph of $k_{obs} - k_2 - k_3 K_1/[OH]_{actual}$ versus $[M^{2+}]_{actual}$ should be linear, going through the origin and with slope k_4.

3. A solution containing $Tl^+(aq)$ will also contain the ion pair TlOH(aq).

$$Tl^+(aq) + OH^-(aq) \rightleftharpoons TlOH(aq)$$

$$K = \left(\frac{[TlOH]}{[Tl^+][OH^-]}\right)_{equil}$$

If there is a dependence of the rate of hydrolysis on $[Tl^+]$, it is likely that the ion pair also reacts with the ester.

$$E + OH^- \xrightarrow{k} \text{products}$$

$$E + TlOH \xrightarrow{k_{TlOH}} \text{products}$$

Mechanistic rate:

$$-\frac{d[E]_{total}}{dt} = k[E][OH^-]_{actual} + k_{TlOH}[E][TlOH]_{actual}$$

Observed rate:

$$-\frac{d[E]_{total}}{dt} = k_{obs}[E][OH^-]_{actual}$$

$$\therefore\ k_{obs}[E][OH^-]_{actual} = k[E][OH^-]_{actual} + k_{TlOH}[E][TlOH]_{actual}$$

$$\therefore\ k_{obs} = k + k_{TlOH}\left(\frac{[TlOH]_{actual}}{[OH^-]_{actual}}\right) = k + k_{TlOH}\, K[Tl^+]_{actual}$$

$K = 6\,\text{mol}^{-1}\,\text{dm}^3$. This value for the association constant indicates that the fraction of Tl^+ associated is low when $[OH^-]$ is sufficiently low, and so the relation

$$[Tl^+]_{total} = [Tl^+]_{actual} + [TlOH]_{actual}$$

becomes

$$[\mathrm{Tl}^+]_{\mathrm{actual}} \approx [\mathrm{Tl}^+]_{\mathrm{total}}$$
$$\therefore \quad k_{\mathrm{obs}} = k + k_{\mathrm{TlOH}}\,K[\mathrm{Tl}^+]_{\mathrm{total}}$$

A plot of k_{obs} versus $[\mathrm{Tl}^+]_{\mathrm{total}}$ should be linear with intercept $= k$, and slope $= k_{\mathrm{TlOH}}\,K$.

If the reaction were followed by monitoring the decrease in pH with time, then, since the $[\mathrm{OH}^-]_{\mathrm{actual}}$ will be altering with time, so will the fraction of Tl^+ associated with the ion pair. If the reaction is studied at constant pH, using a pH-stat, then the fraction associated remains constant throughout reaction, thereby simplifying the kinetic analysis.

4. Mechanistic rate $= k_2[\mathrm{NH_3}]_{\mathrm{actual}}[\mathrm{HNCO}]_{\mathrm{actual}}$

Since $[\mathrm{NH_4^+}]_0 = [\mathrm{OCN}^-]_0$, then $[\mathrm{NH_3}]_t = [\mathrm{HNCO}]_t$,

and

$$K_1 = \left(\frac{[\mathrm{NH_3}][\mathrm{HNCO}]}{[\mathrm{NH_4^+}][\mathrm{OCN}^-]}\right)_{\mathrm{equil}} = \left(\frac{[\mathrm{NH_3}]^2}{[\mathrm{NH_4^+}]^2}\right)_{\mathrm{actual}}$$

$$\therefore \quad [\mathrm{NH_3}]_{\mathrm{actual}} = (K_1)^{1/2}[\mathrm{NH_4^+}]_{\mathrm{actual}}, \text{ giving}$$

$$\text{mechanistic rate} = k_2[\mathrm{NH_3}]^2_{\mathrm{actual}} = k_2\,K_1\,[\mathrm{NH_4^+}]^2_{\mathrm{actual}}$$
$$\text{observed rate} = k_{\mathrm{obs}}[\mathrm{NH_4^+}]^2_{\mathrm{total}}$$

Since K_1 is small, very little deprotonation will occur, and $[\mathrm{NH_4^+}]_{\mathrm{actual}} \approx [\mathrm{NH_4^+}]_{\mathrm{total}}$.

$$\text{mechanistic rate} = k_2[\mathrm{NH_3}]^2_{\mathrm{actual}} = k_2\,K_1\,[\mathrm{NH_4^+}]^2_{\mathrm{total}}$$
$$\text{observed rate} = k_{\mathrm{obs}}[\mathrm{NH_4^+}]^2_{\mathrm{total}}$$
$$\therefore \quad k_{\mathrm{obs}}[\mathrm{NH_4^+}]^2_{\mathrm{total}} = k_2\,K_1\,[\mathrm{NH_4^+}]^2_{\mathrm{total}}$$
$$\therefore \quad k_{\mathrm{obs}} = k_2\,K_1$$

Since the mechanistic rate $= k_2[\mathrm{NH_3}]^2_{\mathrm{actual}}$, reaction is second order. The data quoted is concentration/time data, and so the integrated rate expression is appropriate, and a plot of $1/[\mathrm{NH_4^+}]_{\mathrm{total}}$ versus time should be linear with slope $= k_{\mathrm{obs}}$.

$$\frac{1}{[\mathrm{NH_4^+}]_t} = \frac{1}{[\mathrm{NH_4^+}]_0} + k_{\mathrm{obs}}\,t$$

$\frac{10^3[NH_4^+]}{\text{mol dm}^{-3}}$	8.85	8.26	7.75	7.30	6.94	6.62
$\frac{1}{[NH_4^+]\text{mol}^{-1}\text{ dm}^3}$	113	121	129	137	144	151
$\frac{\text{time}}{\text{min}}$	0	170	340	500	650	790

This graph is linear, with slope $= 0.0478\,\text{mol}^{-1}\,\text{dm}^3\,\text{min}^{-1}$

$$\therefore\ k_2 = \frac{0.0478}{1.3\times 10^{-5}}\,\text{mol}^{-1}\,\text{dm}^3\,\text{min}^{-1} = 3.68\times 10^3\,\text{mol}^{-1}\,\text{dm}^3\,\text{min}^{-1}.$$

5. • Step 3 is a fast step, and so the rate of formation of $[L_4Co(NHR_2)OH]^{2+}$ is governed by the production of $[L_4Co(NR_2)]^{2+}$ in the slow step 2.

$$[L_4Co(NHR_2)X^{2+}]_{total} = [L_4Co(NHR_2)X^{2+}]_{actual} + [L_4Co(NR_2)X^{+}]_{actual}$$

$$\text{increase in}[L_4Co(NR_2)X^{+}]_{actual} = \text{decrease in}[L_4Co(NHR_2)X^{2+}]_{actual}$$
$$-\ \text{increase in}[L_4Co(NHR_2)OH^{2+}]_{actual}$$

$$+\frac{d[L_4Co(NR_2)X^{+}]_{actual}}{dt} = -\frac{d[L_4Co(NHR_2)X^{2+}]_{actual}}{dt}$$
$$-\frac{d[L_4Co(NHR_2)OH^{2+}]_{actual}}{dt}$$

$$\therefore\ +\frac{d[L_4Co(NHR_2)OH^{2+}]_{actual}}{dt}$$
$$= -\left(\frac{d[L_4Co(NHR_2)X^{2+}]_{actual}}{dt} + \frac{d[L_4Co(NR_2)X^{+}]_{actual}}{dt}\right)$$

$$\therefore +\frac{d[L_4Co(NHR_2)OH^{2+}]_{actual}}{dt} = -\frac{d[L_4Co(NHR_2)X^{2+}]_{total}}{dt}$$

• Rate of reaction $= k_2[L_4Co(NR_2)X^{+}]_{actual}$

$$[L_4Co(NHR_2)X^{2+}]_{total} = [L_4Co(NHR_2)X^{2+}]_{actual} + [L_4Co(NR_2)X^{+}]_{actual}$$

$$K_1 = \frac{[L_4Co(NR_2)X^{+}]_{actual}}{[L_4Co(NHR_2)X^{2+}]_{actual}[OH^-]_{actual}}$$

$$[L_4Co(NHR_2)X^{2+}]_{total} = \frac{[L_4Co(NR_2)X^{+}]_{actual}}{K_1[OH^-]_{actual}} + [L_4Co(NR_2)X^{+}]_{actual}$$

$$= [L_4Co(NR_2)X^{+}]_{actual}\left\{\frac{1 + K_1[OH^-]_{actual}}{K_1[OH^-]_{actual}}\right\}$$

$$[L_4Co(NR_2)X^{+}]_{actual} = \frac{[L_4Co(NHR_2)X^{2+}]_{total}K_1[OH^-]_{actual}}{1 + K_1[OH^-]_{actual}}$$

$$\text{rate of reaction} = \frac{k_2[L_4Co(NHR_2)X^{2+}]_{total}K_1[OH^-]_{actual}}{1 + K_1[OH^-]_{actual}}$$

Since the units of k_{obs} are time^{-1}, k_{obs} is a pseudo-first order rate constant, which implies that $L_4Co(NHR_2)X^{2+}$, or OH^-, is being held in excess. The fact that $[OH^-]_{total}$, and not $[L_4Co(NHR_2)X^{2+}]_{total}$, appears in the expression for k_{obs} given in the question indicates that the observed rate expression has to be first order in $L_4Co(NHR_2)X^{2+}$ and zero order in OH^-. This can only be achieved by having OH^- held in excess, so that

$$[OH^-]_{actual} = [OH^-]_{total}$$

$$\therefore \text{ rate of reaction} = \frac{k_2[L_4Co(NHR_2)X^{2+}]_{total}K_1[OH^-]_{total}}{1 + K_1[OH^-]_{total}}$$

$$\text{Observed rate of reaction} = k[L_4Co(NHR_2)X^{2+}]_{total}[OH^-]_{total}$$
$$= k_{obs}[L_4Co(NHR_2)X^{2+}]_{total}$$

where k_{obs} is a pseudo-first order rate constant $= k[OH^-]_{total}$

$$\therefore \quad k_{obs}[L_4Co(NHR_2)X^{2+}]_{total} = \frac{k_2[L_4Co(NHR_2)X^{2+}]_{total}K_1[OH^-]_{total}}{1 + K_1[OH^-]_{total}}$$

$$\therefore \quad k_{obs} = \frac{k_2 K_1[OH^-]_{total}}{1 + K_1[OH^-]_{total}}$$

6. If the pre-equilibrium is set up rapidly and maintained throughout the reaction, then step 2 will be the rate-determining step.

$$\text{Rate of reaction} = k_2[ML]_{actual}[X]$$

$$K_1 = \left(\frac{[ML]_{actual}}{[M]_{actual}[L]_{actual}}\right)_{equil}$$

$$[M]_{total} = [M]_{actual} + [ML]_{actual}$$

and

$$[M]_{actual} = \frac{[ML]_{actual}}{K_1[L]_{actual}}$$

Need to get $[ML]_{actual}$ in terms of $[M]_{total}$ and K_1

$$\therefore \quad [M]_{total} = \frac{[ML]_{actual}}{K_1[L]_{actual}} + [ML]_{actual}$$

$$= [ML]_{actual}\left(\frac{1}{K_1[L]_{actual}} + 1\right)$$

$$= [ML]_{actual}\left(\frac{1 + K_1[L]_{actual}}{K_1[L]_{actual}}\right)$$

$$[\mathrm{ML}]_{\mathrm{actual}} = \frac{[\mathrm{M}]_{\mathrm{total}} K_1 [\mathrm{L}]_{\mathrm{actual}}}{1 + K_1 [\mathrm{L}]_{\mathrm{actual}}}$$

$$\therefore \quad \text{rate} = \frac{k_2 [\mathrm{M}]_{\mathrm{total}} K_1 [\mathrm{L}]_{\mathrm{actual}} [\mathrm{X}]}{1 + K_1 [\mathrm{L}]_{\mathrm{actual}}}$$

$$\text{observed rate} = k[\mathrm{M}]_{\mathrm{total}} [\mathrm{L}][\mathrm{X}]$$

This is third order, but the quoted observed rate constants are given in terms of a second order rate constant, as indicated by the units, $\mathrm{mol}^{-1}\,\mathrm{dm}^3\,\mathrm{s}^{-1}$. The standard procedure in reactions of the type being considered is to hold the ligand in excess. That this is the case is confirmed by the question asking what would happen at high and low concentrations of ligand.

$$\therefore \quad \text{observed rate} = k_{\mathrm{obs}} [\mathrm{M}]_{\mathrm{total}} [X]$$

where k_{obs} is a pseudo-second order rate constant $= k[\mathrm{L}]_{\mathrm{total}}$

L is held in excess, $\therefore \quad [\mathrm{L}]_{\mathrm{actual}} = [\mathrm{L}]_{\mathrm{total}}$

$$\therefore \quad k_{\mathrm{obs}} = \frac{k_2 K_1 [\mathrm{L}]_{\mathrm{actual}}}{1 + K_1 [\mathrm{L}]_{\mathrm{actual}}} = \frac{k_2 K_1 [\mathrm{L}]_{\mathrm{total}}}{1 + K_1 [\mathrm{L}]_{\mathrm{total}}}$$

At very high values of the excess L,

$K_1 [\mathrm{L}]_{\mathrm{total}} \gg 1$, and $\therefore\ k_{\mathrm{obs}} = k_2$, and the reaction is zero order in ligand.

At low values of the excess ligand:

$K_1 [\mathrm{L}]_{\mathrm{total}} \ll 1$, and $\therefore\ k_{\mathrm{obs}} = k_2 K_1 [\mathrm{L}]_{\mathrm{total}}$, and the reaction is now first order in ligand.

To find k_2 and K_1,

$$\frac{1}{k_{\mathrm{obs}}} = \frac{1 + K_1 [\mathrm{L}]_{\mathrm{total}}}{k_2 K_1 [\mathrm{L}]_{\mathrm{total}}}$$

$$= \frac{1}{k_2 K_1 [\mathrm{L}]_{\mathrm{total}}} + \frac{1}{k_2}$$

A plot of $1/k_{\mathrm{obs}}$ versus $1/[\mathrm{L}]_{\mathrm{total}}$ should be linear with intercept $= 1/k_2$ and slope $= 1/k_2 K_1$.

$\frac{[\mathrm{L}]}{\mathrm{mol\,dm^{-3}}}$	0.008 56	0.0192	0.0301	0.0423	0.0654	0.0988
$\frac{1}{[\mathrm{L}]/\mathrm{mol\,dm^{-3}}}$	116.8	52.1	33.2	23.6	15.3	10.1
$\frac{10^4 k_{\mathrm{obs}}}{\mathrm{mol^{-1}\,dm^3\,s^{-1}}}$	0.83	1.28	1.51	1.67	1.84	1.96
$\frac{10^{-4}}{k_{\mathrm{obs}}/\mathrm{mol^{-1}\,dm^3\,s^{-1}}}$	1.205	0.781	0.662	0.599	0.543	0.510

The graph is linear confirming the kinetic analysis:

intercept $= 1/k_2 = 0.43 \times 10^4$ mol dm^{-3} s $\therefore\ k_2 = 2.33 \times 10^{-4}$ mol^{-1} dm^3 s^{-1}

$$\text{slope} = 1/k_2 K_1 = 0.66 \times 10^2 \text{ mol}^2 \text{ dm}^{-6} \text{ s} \ \therefore\ K_1 = \frac{1}{k_2 \times 0.66 \times 10^2} = 65 \text{ mol}^{-1} \text{ dm}^3$$

7. (a) rate of reaction $= k_2[\mathrm{ML_5}]_{\mathrm{actual}}[\mathrm{L}^*]_{\mathrm{actual}}$

$$[\mathrm{ML_6}]_{\mathrm{total}} = [\mathrm{ML_6}]_{\mathrm{actual}} + [\mathrm{ML_5}]_{\mathrm{actual}}$$

$$K_1 = \frac{[\mathrm{ML_5}]_{\mathrm{actual}}[\mathrm{L}]_{\mathrm{actual}}}{[\mathrm{ML_6}]_{\mathrm{actual}}} \quad \text{and} \quad [\mathrm{ML_6}]_{\mathrm{actual}} = \frac{[\mathrm{ML_5}]_{\mathrm{actual}}[\mathrm{L}]_{\mathrm{actual}}}{K_1}$$

We need to find $[\mathrm{ML_5}]_{\mathrm{actual}}$ in terms of $[\mathrm{ML_6}]_{\mathrm{total}}$ and K_1.

$$[\mathrm{ML_6}]_{\mathrm{total}} = \frac{[\mathrm{ML_5}]_{\mathrm{actual}}[\mathrm{L}]_{\mathrm{actual}}}{K_1} + [\mathrm{ML_5}]_{\mathrm{actual}}$$

$$= [\mathrm{ML_5}]_{\mathrm{actual}}\left\{\frac{[\mathrm{L}]_{\mathrm{actual}} + K_1}{K_1}\right\}$$

$$[\mathrm{ML_5}]_{\mathrm{actual}} = [\mathrm{ML_6}]_{\mathrm{total}}\left\{\frac{K_1}{[\mathrm{L}]_{\mathrm{actual}} + K_1}\right\}$$

$$\therefore \quad \text{rate of reaction} = k_2[\mathrm{ML_5}]_{\mathrm{actual}}[\mathrm{L}^*]_{\mathrm{actual}}$$

$$= \frac{k_2 K_1 [\mathrm{ML_6}]_{\mathrm{total}}[\mathrm{L}^*]_{\mathrm{actual}}}{[\mathrm{L}]_{\mathrm{actual}} + K_1}$$

Since L^* is held in excess, then the amount removed by reaction is very small compared with $[\mathrm{L}^*]_{\mathrm{total}}$, $\therefore\ [\mathrm{L}^*]_{\mathrm{actual}} \approx [\mathrm{L}^*]_{\mathrm{total}}$

$$\therefore \quad \text{rate} = \frac{k_2 K_1 [\mathrm{ML_6}]_{\mathrm{total}}[\mathrm{L}^*]_{\mathrm{total}}}{[\mathrm{L}]_{\mathrm{actual}} + K_1}$$

$$\text{observed rate} = k_{\mathrm{obs}}[\mathrm{ML_6}]_{\mathrm{total}}[\mathrm{L}^*]_{\mathrm{total}}$$

$$k_{\mathrm{obs}} = \frac{k_2 K_1}{[\mathrm{L}]_{\mathrm{actual}} + K_1}$$

Note: k_{obs} here is a true second order rate constant.
Usually, with an equation having a denominator which is a sum or difference, the reciprocal of both sides is taken:

$$\frac{1}{k_{\mathrm{obs}}} = \frac{K_1 + [\mathrm{L}]_{\mathrm{actual}}}{k_2 K_1} = \frac{1}{k_2} + \frac{[\mathrm{L}]_{\mathrm{actual}}}{k_2 K_1}$$

The standard graph would be a plot of $1/k_{\mathrm{obs}}$ versus $[\mathrm{L}]_{\mathrm{actual}}$. This is not possible since L is the product of the pre-equilibrium, which is not removed by further reaction, and so builds up continuously throughout the reaction.

(b) If L is held in excess by adding a known large excess, then $[\mathrm{L}]_{\mathrm{actual}} = [\mathrm{L}]_{\mathrm{total}}$, and values of k_{obs} could be found for several experiments where a different value of excess L is used for each. A plot of $1/k_{\mathrm{obs}}$ versus $[\mathrm{L}]_{\mathrm{total}}$ would then be linear with intercept $= 1/k_2$ and slope $= 1/k_2 K_1$. From these k_2 and k_1 can be found individually.

8. (a) Since the equilibrium is set up rapidly and is maintained throughout reaction, the slow step is step 2, and

$$\text{rate of reaction} = k_2[\text{adduct}]_{\text{actual}}[\mathrm{Br}_2]_{\text{actual}}$$

$$K_1 = \frac{[\text{adduct}]_{\text{actual}}}{[\text{alkene}]_{\text{actual}}[\mathrm{Br}_2]_{\text{actual}}} \text{ and } [\text{alkene}]_{\text{actual}} = \frac{[\text{adduct}]_{\text{actual}}}{K_1[\mathrm{Br}_2]_{\text{actual}}}$$

$$\begin{aligned}
[\text{alkene}]_{\text{total}} &= [\text{alkene}]_{\text{actual}} + [\text{adduct}]_{\text{actual}} \\
&= \frac{[\text{adduct}]_{\text{actual}}}{K_1[\mathrm{Br}_2]_{\text{actual}}} + [\text{adduct}]_{\text{actual}} \\
&= [\text{adduct}]_{\text{actual}}\left\{\frac{1 + K_1[\mathrm{Br}_2]_{\text{actual}}}{K_1[\mathrm{Br}_2]_{\text{actual}}}\right\} \\
[\text{adduct}]_{\text{actual}} &= [\text{alkene}]_{\text{total}}\left\{\frac{K_1[\mathrm{Br}_2]_{\text{actual}}}{1 + K_1[\mathrm{Br}_2]_{\text{actual}}}\right\} \\
\text{rate} &= \frac{k_2 K_1[\mathrm{Br}_2]^2_{\text{actual}}[\text{alkene}]_{\text{total}}}{1 + K_1[\mathrm{Br}_2]_{\text{actual}}} \\
\text{observed rate} &= k_{\text{obs}}[\text{alkene}]_{\text{total}}[\mathrm{Br}_2]^2_{\text{total}}
\end{aligned}$$

Note: the observed rate is second order in Br_2, since the reaction involves, in effect, one molecule of alkene and two molecules of Br_2.
Since Br_2 is in excess, $[\mathrm{Br}_2]_{\text{actual}} \approx [\mathrm{Br}_2]_{\text{total}}$

$$\text{rate} = \frac{k_2 K_1[\mathrm{Br}_2]^2_{\text{total}}[\text{alkene}]_{\text{total}}}{1 + K_1[\mathrm{Br}_2]_{\text{total}}}$$

$$\therefore \quad k_{\text{obs}} = \frac{k_2 K_1}{1 + K_1[\mathrm{Br}_2]_{\text{total}}}$$

(b) If the Br_2 excess is small, then $K_1[\mathrm{Br}_2]_{\text{total}} \ll 1$, and
rate $\approx k_2 K_1[\text{alkene}]_{\text{total}}[\mathrm{Br}_2]^2_{\text{total}}$, and reaction is second order in Br_2.
If the Br_2 excess is large, then $K_1[\mathrm{Br}_2]_{\text{total}} \gg 1$, and
rate $\approx k_2[\text{alkene}]_{\text{total}}[\mathrm{Br}_2]_{\text{total}}$, and reaction is first order in Br_2.

(c)

$$\begin{aligned}
k_{\text{obs}} &= \frac{k_2 K_1}{1 + K_1[\mathrm{Br}_2]_{\text{total}}} \\
\frac{1}{k_{\text{obs}}} &= \frac{1 + K_1[\mathrm{Br}_2]_{\text{total}}}{k_2 K_1} \\
&= \frac{1}{k_2 K_1} + \frac{[\mathrm{Br}_2]_{\text{total}}}{k_2}
\end{aligned}$$

A plot of $1/k_{obs}$ versus $[Br_2]_{total}$ should be linear with intercept $= 1/(k_2 K_1)$ and slope $= 1/k_2$. Hence k_2 and K_1 can be found.

(d) If K_1 is very small, then $K_1[Br_2]_{total} \ll 1$, and $k_{obs} = k_2 K_1$. Unless K_1 is known independently, k_2 cannot be found.

9.

$$-\frac{d[H_2A]_{total}}{dt} = k_{uncat}[H_2A]_{total} + k_{cat}[ZnA]_{actual}$$

$$-\frac{d[H_2A]_{total}}{dt} = k_{obs}[H_2A]_{total}$$

$$k_{obs} = k_{uncat} + k_{cat}\frac{[ZnA]_{actual}}{[H_2A]_{total}}$$

$$= k_{uncat} + k_{cat} K\left(\frac{[Zn^{2+}]_{actual}[A^{2-}]_{actual}}{[H_2A]_{total}}\right)$$

since

$$K = \frac{[ZnA]_{actual}}{[Zn^{2+}]_{actual}[A^{2-}]_{actual}}$$

The reaction is carried out at constant pH and so the fraction of the acid ionized to the dianion, α, is constant throughout the experiment. Since the Zn^{2+} is in excess then $[Zn^{2+}]_{actual} \approx [Zn^{2+}]_{total}$, giving

$$k_{obs} = k_{uncat} + k_{cat} K \alpha [Zn^{2+}]_{total}$$

where

$$\alpha = \frac{[A^{2-}]_{actual}}{[H_2A]_{total}}$$

If the mechanism is correct and the approximations justified, a plot of k_{obs} versus $[Zn^{2+}]_{total}$ should be linear with slope $= k_{cat} K \alpha$ and intercept $= k_{uncat}$.

The graph is linear with intercept $= k_{uncat} = 0.725 \times 10^{-4}\ s^{-1}$ and slope $= k_{cat} K \alpha = 1.475 \times 10^{-2}\ mol^{-1}\ dm^3\ s^{-1}$.

Since $K = 1580\ mol^{-1}\ dm^3$ and

$$\alpha = [A^{2-}]_{actual}/[H_2A]_{total} = 6 \times 10^{-5}/0.03 = 0.002\,00, \text{then}$$

$$k_{cat} = \frac{1.475 \times 10^{-2}\ mol^{-1}\ dm^3\ s^{-1}}{1580\ mol^{-1}\ dm^3 \times 0.002\,00}$$

$$= 4.67 \times 10^{-3}\ s^{-1}$$

List of Specific Reactions Discussed or Analysed as a Means of Illustrating Various Points

Gas phase reactions

$C_2H_5NH_2$ decomposition – calculation of rate constant from total pressure with time, 23–24

N_2O decomposition – demonstration of first order kinetics from concentration with time data, 64

C_3H_8 decomposition – formulation of steady state equations and prediction of rate expression, 85–86

Cis/trans isomerisation of 1-ethyl-2-methylcyclopropane – example of finding the rate constant for a reversible reaction, 91

Calculation of p factors for a variety of reaction types, 109–110

A unimolecular decomposition – calculation of observed first order plots from partial pressure with time data, and illustration of the falling off in the first order rate constant with decrease in pressure, 146–147

Cyclopropane isomerisation – demonstration of falling off in the first order rate constant with decrease in pressure, 149–150, 150–151

Demonstration of the dependence of k_1 on s for a unimolecular decomposition, 157–158

Photochemical decomposition of CH_3COCH_3 – deduction of a mechanism to fit experimental observation, 189–192

Decomposition of di-2-methyl propan-2-yl peroxide – Formulation of steady state equations and rate expression, 193–195

N_2O_5 decomposition – a steady state analysis, 195–198

Reaction of NO(g) with O_2(g) – analysis and demonstration of kinetically equivalent mechanisms, 198–201

The H_2/I_2 reaction in the gas phase – a mechanistic and kinetic analysis, 206–208

CH_3OCH_3 decomposition – determination of the mechanism, 210–211; the steady state analysis, 219–220

An Introduction to Chemical Kinetics. Margaret Robson Wright
© 2004 John Wiley & Sons, Ltd. ISBNs: 0-470-09058-8 (hbk) 0-470-09059-6 (pbk)

CH_3CHO decomposition – determination of the mechanism, 211–213; the steady state analysis, 233–238; setting up of the steady state expression for the overall activation energy in terms of the activation energies for the individual steps, 238–239

H_2/Br_2 reaction – a steady state analysis on the reaction with inhibition, 213–216; without inhibition 216–217; determination of the individual rate constants, 217–218

Stylised Rice-Herzfeld mechanisms, 221–224, with surface termination, 240–243

RH/Br_2 reaction – a steady state analysis, 225–227

Oxidation of hydrocarbons – mechanism and steady state analysis, 229–232

A stylised branched chain mechanism – a steady state analysis, and kinetic criteria for the explosive region, 246–249

H_2/O_2 reaction – an example of a reaction with explosion limits, 249–252

Cool flames in oxidation of hydrocarbons – an analysis of the chemical behaviour and reactions involved, 254–259

Solution phase reactions

Hydrolysis of an ester – illustration of pseudo order, 75

CCl_3COOH(aq) decarboxylation – demonstration of first order kinetics from volume of CO_2(g) produced with time, 77–79

Conversion of thiourea to ammonium thiocyanate – determination of activation energy from experimental data, 93–95

Reaction of alkyl halides with anions in solution, $RX(aq) + Y^-(aq)$, – a steady state treatment and comparison with a pre-equilibrium treatment, 202–204, 360–363

Calculation of ionic strength 272–273

$PtLX^+(aq) + Y^-(aq)$ – analysis of kinetic data demonstrating the primary salt effect, 277–279

$CH_2BrCOO^-(aq) + S_2O_3^{2-}(aq)$ – kinetic analysis on experimental data demonstrating dependence of rate constant on relative permittivity, 282–283

$(C_2H_5)_3N^+CH_2COOC_2H_5(aq) + OH^-(aq)$ – kinetic analysis on experimental data demonstrating the primary salt effect, 285–289

$V^{2+}(aq) + Co(NH_3)_5SO_4^+(aq)$ – determination of $\Delta H^{\neq *}$ and $\Delta S^{\neq *}$ from rate constant as a function of temperature, 290–291

Comment on magnitude and sign of $\Delta S^{\neq *}$ for a variety of reactions, 300–301

Determination of $\Delta V^{\neq *}$ from dependence of rate constant on pressure, 304–305

Information deducible from the magnitude of $\Delta H^{\neq *}$, $\Delta S^{\neq *}$ and $\Delta V^{\neq *}$ for a variety of reactions 312–313

Comparison of $\Delta V^{\neq *}$ and $\Delta S^{\neq *}$ for a variety of reactions, 314

Base hydrolysis of an ester – deduction of the rate expression from the mechanism, 318–320

$Pt(dien)Cl^+(aq)$ + pyridine(aq) – analysis of experimental data in terms of a given mechanism, 320–321

Base hydrolysis of glycine ethyl ester – mechanistic analysis and use of experimental data to confirm the mechanism, an approximate analysis, 321–328; a full analysis, 336–339

Base hydrolysis of an ester in presence of metal ions, metal ion catalysis – analysis in terms of kinetically equivalent mechanisms, 330–331
Acid hydrolysis of a charged ester in the presence of SO_4^{2-}(aq) anion catalysis – analysis in terms of kinetically equivalent mechanisms, 332–336
Decarboxylation of β-ketomonocarboxylic acids – formulation of the rate expression from the mechanism, 339–341
Decarboxylation of β-ketodicarboxylic acids – approximations to the overall rate expression deduced from the mechanism, 341–342
Decarboxylation of oxaloacetic acid – contributions to the overall rate, 342–343
Glycine ethyl ester, metal ion catalysed hydrolysis, formulation of the rate expression, 344–346
$A \rightleftarrows B \rightarrow C$ – a pre-equilibrium kinetic analysis and predictions, 347–351, 363–365
$A + B \rightleftarrows C \rightarrow D$ – a pre-equilibrium kinetic analysis and predictions, 351–356, 363–365
$Fe^{III}(CN)_6^{3-}$(aq) + $Co^{II}(edta)^{2-}$(aq) analysis of experimental data using the schematic mechanism – $A + B \rightleftarrows C \rightarrow D$, 356–358
Enzyme mechanism analysis, $E + S \rightleftarrows ES \rightleftarrows E$ + product, 365–368

Index

Bold type indicates definitions

A factor **93**, 104, 108–109, 136–137, 138–140, 140, 143–144, 265–267, 289, 293, 294, 295, 297, 298, 299, 300, 339
  determination of 104, 108, 349
Absorbance **8**, 8–13, 20, 29
Absorption coefficient, molar **8**, 11
Absorption of radiation 2, 7, 8–9, 13, 14, 15, 19, 20, 204–206
Absorption of sound 35–38
Accelerated flow 29
Acid catalysed hydrolysis 26, 75, 331–336
Acid–base catalysis 359, 365
Activated complex 4, 100, 125–127, 129, 130, 131, 132, 224
  degrees of freedom 135, 137–139, 143–145
  in unimolecular theory 155, 158–159, 160–161
  thermodynamic aspects 140–142, 143–145
  with respect to PE surfaces 165–168, 169, 171–180
  statistical mechanical aspects 132–139
  reactions in solution 265, 267, 269–272, 281, 284, 289, 292, 293, 294, 296–297, 297–300, 303–304, 305, 306, 308–309, 310–313, 329, 331–332, 335–336
  structure, reactions in solution 331–332, 335–336
Activated intermediate 129–130, 132
Activated molecule 2–3, 122, 129, 147–148, 152, 153, 154, 155, 156, 158, 159, 160–161, 196, 223
Activating collision 3, 153
Activation 2, 3, 4, **87–89**, 122, **135**, 147, 148
  in unimolecular reactions 152–153, 153–154, 155, 157, 158–160
  in complex gas reactions 185, 196–198, 222–223, 224, 227–229, 235
  in reactions in solution 293, 296, 297–300, 302, 303–304, 304, 308–309, 310–313
Activation energy **2**, **92–95**, 103, 104–108, 134, 136, 156–158, 224, 227, 238–239, 255–257, 267–268, 303–304
  determination of 92–95
  calculations using 104–108
  see also critical energy
Activity 270–271, 342
Activity coefficient 270–271, 271–272
Actual and total concentration 89–90, **284–285**, 321–328, 330–331, 332–336, 336–339, 340–341, 341–342, 342–343, 344–346, 346–351, 352–356, 356–358, 360–363, 363–365, 365–368
Actual ionic strength 284–285
Adsorption at surface 240–243
Angle of scattering 110–112, 112–122
Angular velocity 103
Anharmonicity 124–125, 161, 228
Anion catalysis 317, 334, 334–336
"Approach to equilibrium" 66, **90**
Arrhenius 2
Arrhenius equation **92–95**, 100, 103, 104–106, 108–109, 136, 154–159, 161, 279–280, 339
Association 264, 281, 284–285, 289, 313

An Introduction to Chemical Kinetics. Margaret Robson Wright
 ISBNs: 0-470-09058-8 (hbk) 0-470-09059-6 (pbk)

Associative mechanism **292**, 300–301, 301, 310–313
Atom–atom recombination 160, 214, 228–229, 240, 243
Atomic mechanism, H_2/I_2 reaction, 206–208
Atoms
degrees of freedom 137–140
light and heavy 170, 172–177, 178
H_2/I_2 reaction 206–208
Attractive forces 110, 172, 265, 327–328
Attractive interaction 119, 123
Attractive surface **167–168**, **170**–172, 177, 178
Autocatalysis 85, 244
Average rate **45**

Back reaction 87, 91, 127–128, 170, 214, 215, 255–257, 360–363
Backward scattering 111, 112, 114, 174, 184
see also rebound
Barrier 130, 131, 165–168, 169, 170–179
Base hydrolysis 51, 318–320, 321–325, 325–328, 328–331, 331, 336–339, 344–346
Basin **178–180**
Beer's Law, **8–9**, 9–13
Bimolecular **3**, **87–89**, **185**, 206, 213, **222**
Bimolecular activation **3**, 159–160
Bimolecular deactivation **3**
Bimolecular reaction **3**, **87–89**, 102, 104, 152, 185, 206, 213
Br_2-stat 26–27
Branched chain **244**
Branched chain explosions 247–252
Branched chain reaction 244–252, 252–259
Branching coefficient **244**
Brownian motion 265
Build-up
to explosion 243, 244–246, 247–249, 253–254, 257–259
to steady state 14, 82–84, 85, 186, 209–210, 219, 234, 244

Catalysis 18, 329–331, 331, 332–336, 344–346
see also Acid/base catalysis, Metal ion catalysis, Anion catalysis, Enzyme catalysis
Chain carriers 208–209, 210–211, 211–213, 213–218, 219, 221, 223–224, 224–227, 230, 234, 236
Chain length 221,
see also Chain reactions
Chain reactions 47, 87, 183, **186**, 188, 204, 205–206, **208–209**, 209–239
with branching 244–252
with degenerate branching 252–259
with surface termination 240–243
Charge 109–110, 263–268, 270, 271–272, 272–274, 296–297
effect of on thermodynamic quantities, 301–313
in reactions of charged species 271–299, 300–301, 302, 306–308, 310, 310–313, 317–365
Charge separation 264, 265, 296–297, 299–300, 309, 310–313
Charge/solvent interactions 282
Charged activated complex/transition state 264, 265
Charge-distribution 293, 297, 299
Charge-separated activated complexes 265, 293, 303, 309, 310, 311
Charge-separated species 264, 265, 271, 274, 297, 299
Chelation 285
Chemical potential **270**
Chemiluminescence **100**, 165, 170, 177
Christiansen 3
Chromatographic techniques/methods 6, 7, 15, 16, 27, 29, 30
Collision complex 119–121, 178–180, 184
Collision frequency **102**
Collision number **102**
Collision rate 2, 88, 100, 102–108, 109–110, 265–268
Collision theory 2, 4, 99–110, 110–122, 280, 294
comparison with transition state theory 136–137, 142–143
for reactions in solution 265–268, 280, 294
Collision theory – modified 4, 99, 110–122
Collision theory – simple 99–110, 136–137, 142–143
Complex equilibria 35, 36
see also Ultrasonics
Complex kinetics 79–84, 85–86, 87–89, 89–90, 147–152, 195–204, 206–208, 213–224, 225–227, 229–232, 233–238, 238–239, 241–243, 247–249, 317–368
Complex mechanism *see* Complex reactions and Complex kinetics
Complex mixtures, analysis of 6–7

Complex model (unimolecular), 155–159
Complex reactions/mechanisms 3, 6–7, 13–14, 15–16, 85–86, 189–192, 192–195, 195–198, 198–201, 202–204, 206–208, 209, 210, 210–211, 211–213, 213–218, 219–220, 225–227, 229–232, 233–238, 238–239, 241–243, 247–249, 249–250, 254–255, 318–320, 320–321, 321–325, 325–328, 330–331, 332–336, 336–339, 339–342, 342–343, 344–346, 346–358, 360–363, 363–365, 365–368
Complexes, complexing 264, 285, 320–321, 344, 351
Complex kinetic schemes 79–81
Complexity of structure
in unimolecular reactions 159–160, 223
$\Delta H^{\neq *}$ for solution reactions 302
$\Delta V^{\neq *}$ for reactions in solution 310
and p factors and $\Delta S^{\neq *}$ 140, 143, 292–295, 300–301, 312–313
Composite rate – definition **209–210**
Computer analysis 36, 58
Computer simulation 81, 189
Concurrent reactions 317–321, 321–328, 328–331, 332–336, 336–339, 339–343, 344–346
Conductance methods – use in measuring rates 24–25, 29
Configuration – geometrical 3, **122–123**, 123–131, 133–135, 135, 147, 152, 155, 158–159, 160–161, 165, 167, 171, 172, 173, 176, 178, 184, 185, 263, 271
Consecutive reactions 35, **186–187**
Continuous flow 27–28, 29, 33
Continuum, solvent as 281
Conventional methods for rate measurement 5, 20–27
Conventional radiation 13, 204–206
use for H_2/I_2 reaction 206–208
Conventional rates **17–18**, 19
Conventional source of radiation 8, 13, 19
Cool flames 255–259
Critical configuration 3, 4, 125, 129, 131, 133, 135, 147, 152, 158, 160, 165, 171, 172, 173, 174, 178, 185, 271
see also Activated complex and Transition state
Critical energy 2–3, 19, 99–100, 102, 104, 122, 125–131, 133, 151–154, 155–160
see also Activation energy
Cross section 101, 103, 111, 112–113, 117, 119, 120, 172, 175, 177
"Cutting the corner" trajectory 172–174, 175–177
Cycle in chain reactions **186**, **208–209**, 211, 221, 257

Deactivation 2, 3, 4, 87–89, 122, 147–148, 148, 151, 153, 155, 158, 185, 222–223, 227–229
Debye–Hückel constant 270, 272
Debye–Hückel equation (law) 270, 272
Debye–Hückel theory 270–279
Decarboxylation reactions 339–343
Decomposition – unimolecular 145–161, 185, 195, 196–198, 204, 227–229
Decomposition – specific examples of 21, 23–24, 64, 85–86, 189–192, 193–195, 195–198, 210–211, 211–213, 219–220, 222, 233–238, 238–239
Deflection 100, 110–112, 112–122, 177, 179–180
Degenerate branching 244, 252–259
Degree of orientation of solvent 297–300, 303–304, 308–309
Degrees of freedom 123–125, 135, 137–139, 144–145, 224, 228–229, 235
internal for reactions in solution 292, 294, 295, 298, 299, 300–301, 302, 306, 310–312
Degrees of freedom non-linear activated complex 137–139, 144–145
Degrees of freedom, linear activated complex 137–139
Degrees of freedom, linear molecule 135, 137–139, 144–145, 153
Degrees of freedom, non-linear molecule 135, 137–139, 144–145, 153, 229
Dependence of rate constant on pressure, reactions in solution **141–142**, 304–309
Dependence of rate on temperature **92**, 92–95, 250–252
see Arrhenius equation
Derivative **58–59**, 141–142
Detection of species 5, 6–17, 28–29, 30, 188
Determination of concentrations 5, 6–17, 30, 77–78, 188
Diatomic molecule, degrees of freedom 123–125, 137–139, 228–229

Diatomic molecule, Morse curve 123–125
Dielectric 263
Differential rate equation 59, 62, 66, 68, 71, 73–77, 79–81, 84, 89–90, 186, 189, and Tables 3.2, 3.4
see also Steady state equations
Differential rate methods 52–58
Diffusion 8, 240, 241, 243, 251–252, 265
Dilatometry 27
Dipole moment **295**
Dipole/dipolar 263, 264, 282, 285, 295, 296, 297, 298, 299–301, 308, 312–313
Displacement from equilibrium 30, 31, 33–35
see Relaxation methods and "Approach to Equilibrium"
Disproportionation **209**
Dissociation energy 123–125
Dissociation limit 124
Dissociative mechanism 292, 301, 305–306, 310–313
Dissociative step 292, **305–306**, 310–313
Distribution of velocities in products 112–114

Early barrier 167–168, 170–172, 178
Effect of solvent 280–283, 293–296, 302–303, 308
see also Solvation pattern
Electric field 33, 264
Electric field jump 33–35
Electron spin resonance esr 15, 38, 189
Electron withdrawing 339–340, 341, 344
Electronic distribution 264
Electronic properties 264
Electronic states 7, 100, 132, 253, 257
Electrostatic energy of interaction 264, 266–268, 280–281
Electrostatic interactions 264, 266–268, 269, 274, 280–282, 284, 293, 295, 299, 302
Electrostatically modified collision theory 265–268
Electrostatically modified transition state theory 263, 269–272, 279–284, 289–313
Elementary reaction 3, 87–89, 122, 147–148, 168–170, 184–185, 186–188, 201
Elementary step 185, 186–188, 204–208, 222
Emf methods 26–27
Emission of radiation 2, 7, 9, 100, 205
see Chemiluminescence, Fluorescence, Laser induced fluorescence
Emission spectra 7
Endothermic 127–128, 177–178, 255–257
Energised molecules, *see* Activated molecules, Excited molecules
Energy disposal on reaction 172–173, 173–177, 177–178, 179–180
Energy of activation, *see* Activation energy, Critical energy
Energy transfer 2, 4, 35–38, 122, 147–148, 151, 152–153, 155, 158–159, 159–160, 161, 196–197, 222–223, 227–229
Enhancement of rate 172, 174, 175–176
Enthalpy of activation, $\Delta H^{\neq *}$, 141–143, 224, 264, 279, 282, 290–292, 301–304, 311–313
Entrance valley 125, 131, 165–168, 171–172, 174, 178
Entropy of activation, $\Delta S^{\neq *}$, 142–144, 144–145, 224, 264, 279, 282, 289–301, 303, 308–309, 310–311, 312–313
Enzyme kinetics 359, 365–368
Equilibrium constant in transition state theory 132, 134, 141, 269–272
Equilibrium internuclear distance in transition state theory 123, 125, 173, 173–174, 178
Equilibrium position 30, 31, 33–34, 89–90
Equilibrium statistical mechanics 132, 133, 134
Excitation 9, 13, 19–20, 100, 152, 204
see also Activation
Excited molecules 9, 19–20, 100, 151, 191, 204–208, 228–229
see also Activated molecules, Energised molecules
Excited states 7, 9, 19–20, 38, 100, 172, 253, 257
Exit valley 125, 131, 165, 167, 171, 172, 173, 176, 178
Exothermic 100, 118, **127–128**, 170–177, 178, 227–229, 244, 255–257, 257–259
Exothermicity 228–229
Explosion 210, 243–259
Explosion limits 249–252, 258–259

Fast reactions 4, 5, 14, 17–18, 25, 27–38, 82–84, 210
Field strength *see* Electric field

Fine structure 7
First order integrated rate equation **62–66**, **Tables 3.3 and 3.4**
First order rate constant, meaning of **48–49**
First order, meaning of **48**, 48, 49
Flash photolysis 7, 13, 14, 31–32, 33
Flow methods 13, 27–30, 33
see Accelerated flow, Continuous flow, Stopped flow
Flow of energy around molecule 158, 159, 161, 227–229
Flow rate 28, 29
Fluorescence **9**, 13, 14–15, 83, 100, 205
Forces 110, 117, 119, 172, 264, 265, 327
Forward reaction 127–129, 170, 255–257
Forward scattering 117–118, 172, 185
see Stripping
Fractional life-time 64–66
see also Half-life, Table 3.3
Free energy 269–270
Free energy of activation, $\Delta G^{\neq *}$, 141–142, 280–281, 282, 290, 292, 293, 295, 302, 304, 307–308
Free radical reactions, *see* radical reactions
Free radical, *see* Radicals
Free translation (internal) **130**, 134–135, 137–139, 142, 144–145, 269–271
Frequency 7, 9, 13, 14, 15, 35–38, 132, 138
see also Vibration

Gas phase termination (distinct from surface termination) 240–243, 247–249, 249, 251–252, 255, 257
see also Termination and Straight chains
General acid catalysis 334
General base catalysis 327, 334
Geometrical configuration, *see* Configuration
Grazing 117–118
see Stripping

H abstraction 191, 212, 237, 238, 239, 255
Half-life 17–18, **18**, 31, 32, 43, **59–62**, 63–64, 64–66, 67– 68, 70, **Tables 3.3 and 3.4**
Harmonic oscillator/vibration 103, 123–124, 135, 161
Heavy atom 170, 172, 175–177
Herzfeld *see* Rice–Herzfeld
High pressures (unimolecular reactions) 2, 3, 87–89, 145–147, 148–149, 149–151, 154–155, 155–157, 159–160, 161, 222–223, 227–229
High temperature explosion limit 259
Hinshelwood Theory 155–158
Hydrogen bonding 264
Hydrolysis 75, 285–289, 318–320, 321–328, 328–331, 331–336, 336–339, 344–346

Ideality 268, 270
Identification 6–17, 36, 204
Impact parameter **101–102**, 111–112, 117–118, 118–119, 172, 175, 177
Incident intensity 8
Individual rate constants in complex kinetics, determination of 217–218, 218–220, 233, 237–238
see also Reactions in solutions 317–368
Induction period **210**, 253
Infra red 8, 13, 14, 16
Inhibition 18, 191, 209, 213–214, 214–216, 221
Initial rate **46–47**, 58, 188, 215, 216, 319, 365, 367–368
Initiation methods of 19–20, 27
see also Photochemical initiation and Thermal initiation
Initiation step *see* Chain reactions
Initiator 229
Inorganic chain mechanisms 213–218
Instantaneous rate **45–46**
Integrated rate equation/expression 35, 43, **58–73**, 79–84, 89–90
Integrated rate methods 58–73
Intensity of radiation 7–9, 13–15, 20, 31–32, 205, 208, 214, 216, 218
see also Flash photolysis, Lasers
Interaction 117, 119, 123, 125, 177, 179
see Electrostatic interaction, Interionic interaction, Intermolecular interaction
Interchange mechanism 301, 310, 311, 312–313
Interchange step 306
Interionic interactions 264, 268, 270, 277, 281, 284
see Electrostatic interaction
Intermediates 6, 7, 13–14, 15, 31, 81–87, 186, 188, 192–201, 205, 208–209, 211, 222, 229–232, 264, 346–365, 365–368

see Steady state and Chain carriers
Intermolecular interactions 100, 110, 111
Internal degrees of freedom *see* Degrees of freedom
Internal free translation *see* Free translation
Internal modes for reactions in solution 310, 311
see also Degrees of freedom
Internal motions 137, 138, 292, 299, 302, 306
see also Free translation
Internal states 7, 9, 100
Internal structure 110, 137, 294, 296, 298, 299
Inverse first order 50, 202–204
Ion pair 264, 284–285, 285–289, 328–331
Ion–dipole reactions 295
Ionic strength 19, **270**, 271–279, 281, 284–289, 293, 329, 332, 342
Ion–ion reactions 7, 265–268, 269–272, 273–279, 279–289, 293–294, 296–298, 300–301, 302–304, 306–307, 308–309, 310–313
Ion–molecule reactions 7, 185, 271, 274, 280, 295, 296, 298–299, 302–303, 308, 309, 310–313

Kassel Theory 158–160, 161
Kinetic analysis 5, 31, **43–95**, 346–358
Kinetic energy 102–103, 104, 123
see also Rotation, Translation, Vibration
Kinetic theory 102, 131, 133
Kinetically equivalent mechanisms 183, 189, 198–201, 206–208, 317, 328–331, 332–336, 363–365

Large perturbation 31–33
Laser photolysis 14, 19, 31, 32, 204
Laser-induced fluorescence 9, 100
Lasers 4, 9, 13, 13–14, 15, 19, 20, 31, 32, 33, 83, 100, 204
Late barrier 165, 167, 167–169, 172–177, 178
Life time 186
Light atoms 170, 172–175, 176
Like charge/sign 109–110, 268, 281, 293–294, 297–298, 300, 302, 304–307, 310, 311, 312
Lindemann Mechanism 3, 153–155
Line broadening 38
Line width 7, 38
Linear activated complex – degrees of freedom *see* Degrees of Freedom, linear activated complex
Linear approach/linear recession 123
Linear configuration 123
Linear molecule – degrees of freedom *see* Degrees of freedom, linear molecule
Long chains approximation 215, 217, 220, **221**, 229–232, 234, 236
Low pressure – unimolecular reactions 2–3, 88, 145–147, 148, 150, 159–160, 196–198, 222–223, 227–228, 229, 235
Low pressure surface termination 240, 251–252, 254
Lower explosion limit 251–252
Luminescence 252, 257, 258

Major intermediate 188
Major products 188, 189, 211, 211–212
Mass 102, 130, 132, 138, 170
Mass/charge ratio 6
Mass spectrometry MS 6–7, 17, 27, 29, 30, 189
Mass type 170, 172–177
see also Heavy atom, Light atom
Master mechanism 3, 87–89, 147–148, 151–160, 185, 222–223
Mathematically complex schemes 81
Maximum rate **47**
Maximum scattering **113–114**
Maxwell–Boltzmann distribution 2, 102–103, 125, 154, 155–156, 157–158
Mechanism 2, 3, 5, 6
see Complex reactions
Medium 264, 282
Metal ion catalysis 317, 329, 331, 334, 344–346
Metal ion complexes 344, 345
Michaelis constant 368
Michaelis–Menten equation 368
Microscopic 4, 122, 281, 283, 289
Microscopically inhomogeneous 283
Microwave 7, 8, 13, 15
Mild explosions 252, 253, 254
see also Cool flames
Minimum energy path 4, 165, 172–173, 176
see Reaction coordinate
Minor product 6, 7, 188, 189, 191, 209, 211, 212, 221, 222

Modified solvent–solvent interaction 268
Molar absorption coefficient 8, 11
Molar units 103, 139–140, 157, 289
Molecular beam contour diagrams 112–122, 180
Molecular beams 4, 13, 100, 110, 110–122, 151, 170, 172, 174, 176–177, 179–180, 206
Molecular dynamics 4
Molecular units 103, 139–140, 157, 289
Molecularity **3**, 292, 301, 305–306
Molecule–molecule reactions in solution 296, 299–300, 310–312
Moment of inertia 138
Morse curve 123–125
Most probable route **125**, 172–173, 176
  see Minimum energy path, Reaction coordinate
Most probable velocity in products 113–116, 117, 119
Multi-step relaxation equilibria 35

Negative activation energy 240
Negative temperature coefficient/dependence 253, 254, 259
Non-chain complex reactions in gas phase 189–192, 192–195, 195–198, 198–201, 206–208
Non-equilibrium position 30
  see Relaxation methods
Non-explosive region in branched chain reactions 247–249, 250–252, 258–259
Non-ideality 263, 268, 269–280, 280, 284, 334, 342
Non-linear activated complex – degrees of freedom, *see* Degrees of freedom non-linear activated complex
Non-linear molecule – degrees of freedom, *see* Degrees of freedom non-linear molecule
Non-radical intermediates in steady state treatments 202–204, 359–363
Non-reactive collision 100
Non-steady state 14, 188, 209–210
  see Build up, Explosions
Normal mode
  in transition state theory 123, 132, 135, 144
  in unimolecular theory 153, 155, 157, 158–159, 159–160, 160, 161
  with reference to third bodies 224, 227–229, 233, 235
Normal mode of vibration 123, 132, 135, 144, 153, 155, 157, 158–159, 159–160, 160–161, 224, 227–229, 233, 235
Nuclear magnetic resonance, nmr 15, 38
Nuclear spin states 7
Numerical integration 81, 189

Observed activation energy in terms of E_A's of individual steps 238–239
Oppositely charged, *see* Unlike charge
Order (entropy) 298
Order meaning of **48–50**
Order, determination of
  Differential methods 52–58
  Integrated methods 58–84
Orientation of solvent 297–300, 303–304, 308–309
Oscillate/oscillatory reaction 85, 255–258
Oscillator 123–124, 135, 161
  see also Vibrational modes
Overall order **48**
Oxidation of hydrocarbons 229–232
  see also Cool flames

p factor **108–109**, 109–110, 136, 140, 142–143, 145, 295, 296, 298
  p factors, *A*, $\Delta S^{\neq *}$ for gases 109–110, 136, 140, 142–143, 145
  p factors, *A*, $\Delta S^{\neq *}$ for solution reactions 265, 294, 295, 296, 298
Parallel reactions 186–188
Partial derivative 141–142
Partial pressure in stoichiometry 23–24
Partition function 99, 132–133, 134–135, 135–136, 137–140, 142–143, 265, 269, 271
Partition function per unit volume 133, 135
Path-length 8
PE contour diagram *see* Potential energy contour diagram, Entrance valley, Exit valley
PE maximum 4, 125
PE profile 126–128, 129–130, 133, 165–180
Period of sound wave 35
Periodic (displacements) 30, 35–38

Permittivity, *see* Relative permittivity
Perturbation 30, 31, 33–35, 66, 90
pH 26, 26–27, 51, 319–320, 323, 324, 325–328, 331, 337–339, 341, 342, 342–343, 345
Photochemical equivalence, law of 205
Photochemical initiation 19, 214, 216
Photochemical rate 216, 218
see H_2/I_2 reaction 206–208
Photochemistry (photochemical) 2, 19, 19–20, 30, 31–32, 33, 188, 189–192, 204–206, 206–208, 210, 214, 216, 218, 237, 238
Photoelectron spectroscopy 15–16
Photolysis 7, 13, 14, 19, 31–32, 33, 204, 218
pH-stat 26–27, 319, 331, 345
Physical mechanism 87–89, 147–148, 185, 222
see Master mechanism
Physical methods 21–27, 27–38, 79
Polarisability 263–264, 313
Polyatomic molecule or ion, degrees of freedom 137, 153, 228–229, 294, 295
see also Degrees of freedom, linear molecule and non-linear molecule
Polyatomic molecule PE surface 122
Position of equilibrium 33, 35, 255–257, 322
Potential energy 3, 4, 103, 122–131, 133, 134, 135, 154, 165–180, 264, 271, 285
Potential energy barrier 130, 165–168
Potential energy contour diagram 126, 129, 131, 165, 166, 167, 169, 171, 173, 174, 175, 176
Potential energy surface 3, 122–131, 133, 134, 135, 142, 165–180, 184, 271
Potential energy well 178–179
Pre-equilibrium **92**, 201, 202–204, 346–358, 359–365
Pre-exponential factor, *see* A factor
Pressure jump 33–35
Pressure limits in explosions 250–252, 254
Primary process, photochemical 204–206
Product valley, *see* Exit valley
Propagation 185, **186**, 205–206
Straight chains 208–239
Branched chains, 244–259
Pseudo order 66, **74–77**
Pseudo rate constant **74–77**, 245, 318, 320, 320–321, 351, 354, 355, 358
Pulsed laser 14, 29, 32
Pulsed radiolysis 32

Quadrupole 282
Quantum (of radiation) in photochemistry 204–206
Quantum mechanical calculations 2, 123, 165
Quantum mechanical tunnelling 129–130
Quantum unimolecular theory 159, 161
Quantum yield 189–192, 204–206, 210
Quenching 20

Radiation 2, 7–15, 15–16, 19, 19–20, 29, 31–33, 100, 204–206, 208
Radicals
detection and identification of 6, 7, 13–17, 19, 31–33, 204
generation of 19–20, 31–33, 204
in solution 264, 359
see also Chain reactions, Chain carriers
Raman spectra 7, 8
Rate constant
meaning of **48–50**
determination, differential method 52–58
determination, integral method 58–84
Rate, meaning of **44–47**
determination of 17–38
Rate equation
Differential, *see* Differential rate equation
Integrated, *see* Integrated rate equation
Rate expression, meaning of **48–50**
Rate of activation in unimolecular reactions 147–148, 151–152, 153, 155, 157–158, 158, 159–160
Rate of branching 245–246, 257, 258
Rate of change of configuration 130, 131
Rate of collision, *see* Collision rate
Rate of deactivation in unimolecular reactions 147–148, 151–152, 153–154, 155, 158, 159–160
Rate of initiation, *see* Steady state, Explosions
Rate of propagation, *see* Propagation, Steady state
Rate of reaction step in unimolecular reactions 147–148, 154, 155, 158, 159–160

Rate of termination, *see* Steady state, Explosions
Rate, dependence on temperature *see* Dependence of rate on temperature, Arrhenius equation
Rate-determining step **82–83**, 87–89, 92, 148, 154, 159–160, 222, 227–228, 235, 240, 243, 346–347, 348, 352, 357
Reactant valley, *see* Entrance valley
Reaction coordinate 4, 125, 126, 127, 129, 130, 133, 134, 135, 165, 172, 176, 178
Reaction unit/Reaction entity 122, 123, 125, 126, 130, 131, 133, 165, 171, 173, 174, 176, 178, 263
Reactions in solution 19, 20, 21, 38, 141–142, 263–368
Reactive collision 103, 111–112
Rebound mechanism 118–119, 174, 184
Recycle, *see* Regeneration
Redistribution of energy 100, 155
Reduced mass **102**, 105–106, 133–135
Regeneration of chain carriers 186, 188, 192, 193, 208, 209, 211, 212
Regeneration of catalyst 331, 332, 365
Relative permittivity 19, 263, 264, 265, 266, 267, 268, 269, 280–283, 285, 293, 295, 296, 299, 302, 303, 306 307, 308
Relative translational motion 102, 104, 152
Relative velocity 100–101
Relaxation methods 30–32, 32–38
Relaxation time 35–36, **64–66**, **Table 3.3**
Repulsion 110, 119, 171, 178, 265
Repulsive barrier *see* Late barrier
Reversible reactions 81, 89–91, 92, 198–201, 202–204, 255–257, 257–258, 346–358, 359–368
Rice–Herzfeld Mechanisms 221–243
Rotation 7, 100, 103, 110, 119, 135, 137, 138–139, 140, 144–145, 153, 178–179, 224, 292, 294, 306
Rotational degrees of freedom 135, 137, 138–139, 144–145, 153, 224, 292, 294, 306
Rotational mode 103, 135, 144–145, 153
Rotational partition function 132, 135, 138–139, 140
Rotational state 7, 100
Rotational symmetry number 132
Rotational energy 153
Rotational structure 100, 294

Saturation 354
see Swamping
Scattered products 111, 111–122, 172, 174–175, 177, 178–180, 184
Scattering 100, 110–122, 172, 174–175, 177, 178–180, 184
Scattering diagram (long lived complex) 119–120, 178–180, 184
Second explosion limit 252
Second order meaning of **47–49**
Second order integrated rate equation **66–68**, 68, **Table 3.4**
Second order rate constant meaning of **47–49**
Secondary process, photochemical 204–205
Self-heating 244, 252, 257–258
Shock tube 7, 13, 31, 33, 34
Simple model (unimolecular) 153–155, 157
Sine waves 161
Single impulse 30, 33–35, 90
Slater Theory 160–161
Small perturbation 30, 33–35, 35–38, 66
Solute/solute interactions 268
see also Ionic interactions
Solute/solvent interactions 268
Solvation 263, 267, 268, 269, 281, 282, 283, 285, 296–301, 303–304, 308–309, 310, 311, 312, 312–313
Solvation pattern 263, 269, 281, 296–301, 303–304, 308–309, 310, 311, 312, 312–313
Solvent/solvent interactions 263, 268
Sound wave 35–38
Specific base catalysis 329
Spectrophotometry 8–13, 14, 27, 29
Spectroscopic methods of analysis/ determination/estimation 6, 7–16, 19, 27, 29, 30, 31–32, 38, 132, 170, 204–206
Spectroscopic quantities used in statistical mechanics 132, 135, 271
Spin resonance 15, 38
Squared terms 102–103, 154, 155–156, 157–158
Standard state 142, 307
Steady state conditions/region **84–89**, 246, 247–249, 257

Steady state method/treatment/analysis
84–89, 89, 147–149, 186, 192–195, 196–198, 198–201, 202–204, 207–208, 213, 214–218, 218–221, 222, 225–227, 229–232, 233–238, 238–239, 241–243, 246, 247–249, 257, 359–360, 360–362, 362–363, 363–365, 365–368
Stoichiometric arguments 23–24, 77–79
Stoichiometric concentration **284–285**
Stoichiometric ionic strength **284–285**, 289
Stopped flow 28–29
Straight chain 208–239, 240–243
comparisons with Branched chain 244, 246, 247–248, 249–250, 254, 254–255, 255–257, 258–259
Stripping 117–118, 119, 120, 170–172, 177
Structureless continuum 281
Substrate in enzyme kinetics **365**
Successful collisions 103, 111–112
Surface reactions 19, 210, 240–243, 247–249, 249, 251–252, 255, 257
Surface termination 210, 240–243, 247–249, 249, 251–252, 255, 257
Swamping 354
see Saturation
Symmetrical PE barrier 131, 165–167
Symmetrical PE contour diagram 131, 166
Symmetrical PE profile 131, 166–167
Symmetrical scattering 119–122
Symmetry number 132

Temperature effect of 1, 92–95, 100, 104–108, 154–155, 157, 159, 160, 161, 250–252, 254, 255–257, 257–259, 268, 339
see also Arrhenius equation
Temperature, effect on relative permittivity 293
Temperature jump 33–35
Temperature limit, cool flames 254, 258–259
Termination 208–239, 239–243, 243–259
Termolecular reaction **3**, **87**, 185, 199, 201
Thermal explosions 244
Thermal initiation 19–20, 216
Thermal rate, *see* H_2/I_2 reaction 206–208
Thermodynamic formulation of transition state theory 140–145, 265, 269, 269–272, 279–313
Thermodynamic functions/quantities **140–142**
Thermoneutral 127–128
Third bodies 183, 227–229, 241, 243
Third explosion limit 252
Third order meaning of **48**, **Table 3.4**
Three-halves (3/2) order **48**, **and Table 3.2**
Time lag 3, 147, 154, 158, 159
Total energy in transition state theory 123, 125, 129
Total molecular partition function 134
Trace 6, 188, 211, 212, 253
Trajectory 172–173, 176
Transition state 4, **125**
see Activated complex
Transition state theory 4, 99, 122–145, 224, 237, 263, 265, 269–272, 279–284, 289–313
Transition state theory expression **132**, **136**, **269**
Translational degrees of freedom 135, 137, 138–139, 144–145, 224
Translational energy 118, 119, 120, 172, 174, 175–177, 178, 179
Translational partition function 132, 134–135, 137, 138–139
Transmission coefficient 130, 133–134, 269
Tunable laser 13
Tunnelling 129–130

Ultrasonic relaxation techniques 35–38, 66
Ultraviolet 7, 8, 13, 14, 16
Undeflected molecules 111, 172
Unimolecular 2, **3**, **87**, 185, 222
Unimolecular theories 2, 99, **145**, 145–161, 195
see Lindemann mechanism, Hinshelwood theory, Kassel theory, Slater theory
Unimolecular reaction 2, 3, 19, 87, 99, 145–161, 185, 195–198, 222–223, 227–229, 233, 235, 240, 243, 292
Units **47**, 49–50, 53, 55, 75–77, 95, 103, 107–108, 109, 139–140, 157–158, 289, 305, 318, 332–334, 354, 355, 358
Units of rate **47**
Units of rate constants **49–50**, 53
of pseudo rate constants 75–77, 318, 332–334, 354, 355, 358

Unlike charge 265, 273–274, 281, 284, 293–294, 297–298, 302, 303, 307, 310, 311, 312, 327
Unscattered molecules 111, 172
Upper explosion limit 252

Velocity of sound 35–38
Vibration 103, 110, 118, 119, 120, 123–124, 125, 132, 135, 137, 138–140, 144–145, 151, 152–153, 153–161, 172, 174, 175–177, 178, 178–179, 224, 227–229, 235, 292, 302, 306
Vibrational degrees of freedom 103, 123–124, 133, 135–137, 138–140, 144–145, 153, 153–161, 224, 227–229, 235, 292, 302, 306
Vibrational energy 103, 118, 119, 120, 123–124, 125, 151, 152–153, 153–161, 172, 174, 175–177, 178, 178–179
Vibrational frequency 132, 138
Vibrational modes 103, 123–124, 132, 135, 137, 138–140, 144–145, 153, 153–161, 224, 227–229, 235, 292, 302, 306
Vibrational states 7, 100, 151, 153–154, 155, 158–159
Vibrational structures 100
Viscosity 19, 263, 264
Visible radiation 7, 8, 13, 14, 31

Water rate 318–320
Wavelength 7, 8, 11, 13
Well 129–130, 131, 178–180

X-ray photoelectron spectroscopy 15–16

Zero ionic strength 274, 279, 280, 281, 282, 285, 293
Zero order meaning of **48–49**
Zero order integrated rate equation **68–70**, **Table 3.4**
Zero order rate constant meaning of **48–49**